AF478444

Current Topics in Behavioral Neurosciences

Volume 29

Series editors

Mark A. Geyer, La Jolla, CA, USA
Bart A. Ellenbroek, Wellington, New Zealand
Charles A. Marsden, Nottingham, UK
Thomas R.E. Barnes, London, UK

About this Series

Current Topics in Behavioral Neurosciences provides critical and comprehensive discussions of the most significant areas of behavioral neuroscience research, written by leading international authorities. Each volume offers an informative and contemporary account of its subject, making it an unrivalled reference source. Titles in this series are available in both print and electronic formats.

With the development of new methodologies for brain imaging, genetic and genomic analyses, molecular engineering of mutant animals, novel routes for drug delivery, and sophisticated cross-species behavioral assessments, it is now possible to study behavior relevant to psychiatric and neurological diseases and disorders on the physiological level. The *Behavioral Neurosciences* series focuses on "translational medicine" and cutting-edge technologies. Preclinical and clinical trials for the development of new diagnostics and therapeutics as well as prevention efforts are covered whenever possible.

More information about this series at http://www.springer.com/series/7854

Richard M. Kostrzewa · Trevor Archer
Editors

Neurotoxin Modeling of Brain Disorders— Life-long Outcomes in Behavioral Teratology

 Springer

Editors
Richard M. Kostrzewa
Department of Biomedical Sciences
East Tennessee State University
Johnson City, TN
USA

Trevor Archer
Department of Psychology
University of Gothenburg
Gothenburg
Sweden

ISSN 1866-3370   ISSN 1866-3389   (electronic)
Current Topics in Behavioral Neurosciences
ISBN 978-3-319-34134-7   ISBN 978-3-319-34136-1   (eBook)
DOI 10.1007/978-3-319-34136-1

Library of Congress Control Number: 2016942033

Printed on acid-free paper

This Springer imprint is published by Springer Nature
The registered company is Springer International Publishing AG Switzerland

Preface

The perinatal period of neuro-ontogeny is a critical stage in organization and modeling of brain and peripheral nervous system. During this developmental stage nerve growth to target sites is under the influence of neurotrophins; neurotransmitter receptors on nerves, activated or not, govern in-part their ultimate viability/ survivability, as redundancy leads to nerve cell death; and the residual neural organization of the neural network becomes the life-long template of neural organization. It is in this melding period wherein the effects of neurotoxins and receptor agonists/antagonists intervene to produce permanent effects in development. This book on *Neurotoxin Modeling of Brain Disorders—Life-long Outcomes in Behavioral Teratology* is intended to present an overview on some of the most common agents that are known to effect the neural organization of brain. Abnormalities in some instances are obvious from histopathologic evidence, also from neural dynamics or receptor deficiencies; and these altered neural patterns are manifest in inherent behavioral expressions or in altered behavioral responses to challenge by associated receptor agonists or antagonists. Unfortunately, brief exposure to multiple substances during neural ontogeny can often result in life-long patterns of expression. This important topic, mainly studied in animals, is reflective of the kinds of neural changes that can occur in human fetuses when a mother is exposed during pregnancy to a variety of substances including drugs of abuse. The chapters in this book highlight the types of life-long changes in brain that can occur when substances act on the brain during ontogeny. The book is thus a brief compendium on neuroteratologic agents; the book is instructive to those engaged in studies in neuro-ontogeny; the book is also important to clinicians involved in the care of pregnant women or children.

Contents

Neuroteratology and Animal Modeling of Brain Disorders 1
Trevor Archer and Richard M. Kostrzewa

Part I Neuroteratogenic Agents

**Perinatal Lesioning and Lifelong Effects of the Noradrenergic
Neurotoxin 6-Hydroxydopa** 43
Richard M. Kostrzewa

**Selective Lifelong Destruction of Brain Monoaminergic Nerves
Through Perinatal DSP-4 Treatment** 51
Przemysław Nowak

**Noradrenergic–Dopaminergic Interactions Due to DSP-4–MPTP
Neurotoxin Treatments: Iron Connection** 73
Trevor Archer

Perinatal Domoic Acid as a Neuroteratogen 87
Tracy A. Doucette and R. Andrew Tasker

Perinatal 192 IgG-Saporin as Neuroteratogen 111
Laura Petrosini, Paola De Bartolo, Debora Cutuli and Francesca Gelfo

**NGF in Early Embryogenesis, Differentiation, and Pathology
in the Nervous and Immune Systems** 125
Luisa Bracci-Laudiero and Maria Egle De Stefano

Part II Neuroteratogens to Model Human Disorders

**Disrupted Circadian Rhythm as a Common Player
in Developmental Models of Neuropsychiatric Disorders** 155
Eva M. Marco, Elena Velarde, Ricardo Llorente and Giovanni Laviola

Neurobehavioral Effects from Developmental Methamphetamine Exposure 183
Sarah A. Jablonski, Michael T. Williams and Charles V. Vorhees

Early-Life Toxic Insults and Onset of Sporadic Neurodegenerative Diseases—an Overview of Experimental Studies 231
Anna Maria Tartaglione, Aldina Venerosi and Gemma Calamandrei

The Use of Perinatal 6-Hydroxydopamine to Produce a Rodent Model of Lesch–Nyhan Disease 265
Darin J. Knapp and George R. Breese

Perinatal 6-Hydroxydopamine Modeling of ADHD 279
John P. Kostrzewa, Rose Anna Kostrzewa, Richard M. Kostrzewa, Ryszard Brus and Przemysław Nowak

Attention-Deficit/Hyperactivity Disorder: Focus upon Aberrant N-Methyl-D-Aspartate Receptors Systems 295
Trevor Archer and Danilo Garcia

Perinatal 6-Hydroxydopamine to Produce a Lifelong Model of Severe Parkinson's Disease 313
John P. Kostrzewa, Rose Anna Kostrzewa, Richard M. Kostrzewa, Ryszard Brus and Przemysław Nowak

Exercise and Nutritional Benefits in PD: Rodent Models and Clinical Settings 333
Trevor Archer and Richard M. Kostrzewa

Lifelong Rodent Model of Tardive Dyskinesia—Persistence After Antipsychotic Drug Withdrawal 353
Richard M. Kostrzewa and Ryszard Brus

Perinatal Influences of Valproate on Brain and Behaviour: An Animal Model for Autism 363
Peter Ranger and Bart A. Ellenbroek

Applications of the Neonatal Quinpirole Model to Psychosis and Convergence upon the Dopamine D_2 Receptor 387
Russell W. Brown and Daniel J. Peterson

Postnatal Phencyclidine (PCP) as a Neurodevelopmental Animal Model of Schizophrenia Pathophysiology and Symptomatology: A Review 403
B. Grayson, S.A. Barnes, A. Markou, C. Piercy, G. Podda and J.C. Neill

Pathological Implications of Oxidative Stress in Patients and Animal Models with Schizophrenia: The Role of Epidermal Growth Factor Receptor Signaling 429
Tadasato Nagano, Makoto Mizuno, Keisuke Morita and Hiroyuki Nawa

Neuroteratology and Animal Modeling of Brain Disorders

Trevor Archer and Richard M. Kostrzewa

Abstract Over the past 60 years, a large number of selective neurotoxins were discovered and developed, making it possible to animal-model a broad range of human neuropsychiatric and neurodevelopmental disorders. In this paper, we highlight those neurotoxins that are most commonly used as neuroteratologic agents, to either produce lifelong destruction of neurons of a particular phenotype, or a group of neurons linked by a specific class of transporter proteins (i.e., dopamine transporter) or body of receptors for a specific neurotransmitter (i.e., NMDA class of glutamate receptors). Actions of a range of neurotoxins are described: 6-hydroxydopamine (6-OHDA), 6-hydroxydopa, DSP-4, MPTP, methamphetamine, IgG-saporin, domoate, NMDA receptor antagonists, and valproate. Their neuroteratologic features are outlined, as well as those of nerve growth factor, epidermal growth factor, and that of stress. The value of each of these neurotoxins in animal modeling of human neurologic, neurodegenerative, and neuropsychiatric disorders is discussed in terms of the respective value as well as limitations of the derived animal model. Neuroteratologic agents have proven to be of immense importance for understanding how associated neural systems in human neural disorders may be better targeted by new therapeutic agents.

Keywords Neuro-ontogeny · Neurotoxins · Neurotoxicity · Neuroteratology · Epidermal growth factor · Nerve growth factor · 6-Hydroxydopa · 6-Hydroxydopamine · DSP-4 · MPTP · 5,7-DHT · Methamphetamine · NMDA · IgG-saporin · Domoic acid · Quinpirole · Valproate · Parkinson's disease · Schizophrenia · Autism · Lesch–Nyhan disease · Tardive dyskinesia

T. Archer (✉)
Department of Psychology, University of Gothenburg, Box 500,
430 50 Gothenburg, Sweden
e-mail: trevor.archer@psy.gu.se

R.M. Kostrzewa
Department of Biomedical Sciences, Quillen College of Medicine,
East Tennessee State University, PO Box 70577, Johnson City, TN 37614, USA
e-mail: kostrzew@etsu.edu

© Springer International Publishing Switzerland 2015
Curr Topics Behav Neurosci (2016) 29: 1–40
DOI 10.1007/7854_2015_434
Published Online: 09 February 2016

Contents

1　Introduction 2
　1.1　Developmental Inflammatory Processes 4
　1.2　Perinatal Insult and Neurologic Neurodegenerative Disorders 5
　1.3　Neurotrophins and Neuronal Development 6
2　Actions and Mechanisms of Selective Neurotoxins 7
　2.1　6-Hydroxydopamine 7
　2.2　6-Hydroxydopa 8
　2.3　DSP-4 8
　2.4　Co-administration of DSP-4 and MPTP 9
　2.5　Methamphetamine 9
　2.6　Domoic Acid 10
　2.7　192 IgG-Saporin 10
　2.8　Quinpirole 11
3　Animal Modeling with Neuroteratologic Agents 11
　3.1　Rodent Model of PD Produced by Perinatal 6-OHDA Treatment 11
　3.2　Exercise Effectiveness in Improving Behavioral Deficits in a Rodent Model
　　　of PD 12
　3.3　Rodent Model of ADHD Produced by Perinatal 6-OHDA with Adulthood
　　　5,7-DHT Lesions 13
　3.4　ADHD and NMDA-R Systems 14
　3.5　Rodent Model of Lesch–Nyhan Disease Produced by Perinatal
　　　6-OHDA Treatment 15
　3.6　Permanent Animal Model of Tardive Dyskinesia 15
　3.7　Valproate Modeling of Autism Spectrum Disorder 16
4　Perinatal Insults that Model Psychosis Schizophrenia 18
　4.1　Epidermal Growth Factor and Schizophrenia Modeling 18
　4.2　Phencyclidine and Schizophrenia Modeling 18
　4.3　Methamphetamine and Schizophrenia Modeling 19
　4.4　Quinpirole and Schizophrenia Modeling 20
　4.5　Stress and Neuropsychiatric Disorders 21
　4.6　Genetic Model of Alzheimer's Disease 21
5　Conclusion 21
References 22

1　Introduction

The concept posed in this paper is that perinatal insult can influence later-life neural survival or later-life susceptibility to toxic challenge. Also, perinatal insult can alone result in lifelong neural and behavioral abnormalities in humans, able to be modeled by appropriate treatments of animals. The series of topics highlighted here provide support for what was once a concept, but which is now recognized fact, supported by experimental and observational data in animal subjects and humans.

There are an encyclopedic number of studies demonstrating that perinatal exposure of noxious or seemingly innocuous agents alter the pattern of neural ontogenetic development and produce permanent neuroanatomical, neurochemical, and/or behavioral abnormalities. Examples of this are provided for what might be termed selective neurotoxins.

One general and surviving notion has been that the neurodegenerative and, for that matter, even the neurodevelopmental psychiatric disorders are induced by specific agents that degenerate one or more clearly defined population(s) of neurons, circuits, and/or regions, e.g., dopaminergic neurons in parkinsonism. However, the much more prevalent issue of senescence is a wider phenomenon affecting cells throughout the body, including spontaneous dying of neurons, as per pars compacta substantia nigra (SNpc) dopaminergic neurons and onset of Parkinson's disease (PD) (Rodriguez et al. 2015). Neuropsychiatric disorders, such as schizophrenia, attention deficit/hyperactivity disorder (ADHD), autism and depression, and Lesch–Nyhan disease (LND), stem from abnormalities and disruptions, both genetic and environmental, of the normal courses of the developmental cycles (Grados et al. 2014; Groves et al. 2014). For example, exposure to femtomolar concentrations (fM, 10^{-15} mol/dm^3 or 10^{-12} mol/m^3) of fragrances (generally sweet or pleasant smell) results in morphological changes at the light microscopic level in fetal neuroblastoma cell lines pertaining to reduced oxytocin-positive and arginine vasopressin-positive neurons in male but not female neuroblastoma cell lines (Sealey et al. 2015). Stressor exposure during early life has the potential to increase an individual's susceptibility to a number of neuropsychiatric conditions such as mood and anxiety disorders and schizophrenia in adulthood. Epigenetic processes exert cellular/tissue-specific changes in regulating expression of genes, providing potential biomarkers for examining the developmental trajectory of early stress-induced susceptibility to adult neuropsychiatric and/or neurologic disorders (Ibi and González-Maeso 2015; Marco et al. 2016). With the proliferation of gene-based models and etiology-based models for studying brain disorders (Bezard et al. 2013), the predictability of human and animal in vivo outcomes for neurotoxicity and retardation of developmental trajectories proceeds apace.

Under conditions of chronic inflammation, "mediator" molecules like cytokines may be disadvantageous to organism development over prolonged or exaggerated periods. Neuroprotective or neurotoxic outcomes evolving from interactions between cytokines and/or metabolites of tryptophan catabolism, the neuroactive kynurenines, partly influenced by corticosteroid action, contribute to the fate of several signaling pathways, e.g., serotonergic, dopaminergic, and glutamatergic transmissions, and receptor functions such as N-methyl-D-aspartate receptor (NMDA-R) or α_7-nicotinic acetylcholine receptor (Myint 2013). For instance, altered kynurenine metabolism is implicated in the pathogenesis of Alzheimer's disease (AD), PD, and Huntington's disease (HD), whereas the metabolites and key enzymes, analogs of the metabolites, and small-molecule enzyme inhibitors, preventing the formation of neurotoxic compounds, confer both neuroprotective and therapeutic properties (Tan et al. 2012). Inflammatory mediators activate the kynurenine metabolic pathway and immobilize the production of neuroactive metabolites, thereby initiating a pathogenic cascade with neuropsychiatric consequences (Allison and Ditor 2015; Brundin et al. 2015; Meier et al. 2015). According to the (genetic) Vulnerability-stress-inflammation developmental notion of schizophrenia, stress generates immune alterations (proinflammatory cytokines) that influence dopaminergic, serotonergic, noradrenergic, and glutamatergic neurotransmission through the activation of the enzyme indoleamine 2,3-dioxygenase (IDO) of tryptophan/kynurenine metabolites, leading to kynurenic acid, with the concomitant activation of microglia, a veritable cascade of neuroinflammatory events (Müller et al. 2015).

1.1 Developmental Inflammatory Processes

Immune activation through prenatal or early postnatal exposure to viruses or bacterial products (e.g., lipopolysaccharide (LPS)) consistently impairs brain development and influences behavioral, emotional, and cognitive functional domains (Kirsten et al. 2013; Xia et al. 2014; Zhu et al. 2014a, b) with consequences for the pathogenesis of neuropsychiatric conditions (Delany et al. 2015; Kelly et al. 2015; Mossakowski et al. 2015; Pariante 2015). Maternal inflammation is similarly reflected in neuroinflammatory events resulting in structural and functional disturbances to the developing offspring brain.

Prenatal LPS-induced reorganization of the dendritic architecture was found in both L2PC-A and L2PC-B types, predominantly in the L2PC-A type in mouse offspring (Gao et al. 2015); there was also a differential alteration of intrinsic electrophysiological properties of the two L2PC types. As the resting membrane potential of L2PC-A neurons became hyperpolarized, these neurons were less excitable, whereas the resting membrane potential of L2PC-B neurons was partially depolarized and more excitable. Thus, morphological and electrophysiological abnormalities were linked to pyramidal neuron dysfunction stemming from inflammatory events during pregnancy. Parental microglia-induced neuroinflammation, triggered by bacterial or viral infections, may induce features of neuropsychiatric/neurologic disorders, such as ADHD, schizophrenia, and autism in offspring (Byrnes et al. 2009). In mice exposed prenatally to LPS at gestational days 15 and 17, there was downregulation of peripheral benzodiazepine receptors (PBRs), mediated by the activation of mGluR5 in astrocytes (Arsenault et al. 2015). In addition, the mGluR5–PBR interaction in a mouse model of schizophrenia (Basta-Kaim et al. 2015; Wischhof et al. 2015) was applicable to brain disorder pathophysiology. Thus, LPS-driven ontogenetic effects at mGluR5 have implications in later-life onset of neuropsychiatric disorders.

Inflammatory cytokines are able to affect neuronal ontogeny indirectly by acting at glia and subverting their imbued neuroprotective action to one that is adverse for neurons. By this means, gestational inflammation can indirectly affect neural function—and thereby pose a risk for later age development of neurological or psychiatric disorders (Fukushima et al. 2015; Jo et al. 2015; Steardo et al. 2015). A number of mechanisms may come into play: (i) stimulation of the phagocyte NADPH oxidase (PHOX) to produce superoxide and derivative oxidants, (ii) expression of inducible nitric oxide synthase (iNOS) that produces NO and derivative oxidants, (iii) release of glutamate and glutaminase, (iv) release of tumor necrosis factor alpha (TNF-α), (v) release of cathepsin B, (vi) phagocytosis of stressed neurons, and finally, (vii) decreased release of nutritive brain-derived neurotrophic factor (BDNF) and insulin-like growth factor-1 (IGF-1) (Brown and Vilalta 2015).

Despite all the evidence that neuroinflammation and reactive gliosis feature prominently in most brain and CNS disorders, the notion of glial cells as passive responders to neuronal damage rather than drivers of synaptic dysfunction is changing. Glia have active signaling activity with neurons and influence synaptic development, transmission, and plasticity by mobilizing a plethora of secreted and contact-dependent signals (Chung et al. 2015). Reactive astrogliosis, a feature of AD, presents a continuum of neuropathological processes with accompanying morphological, functional, genetic, and epigenetic events (Jain et al. 2015; Pekny et al. 2014; Steardo et al. 2015; Verkhratsky et al. 2015). Calcium, proteoglycan, TGF-β, NFκB, and complement mediate the neuron–glia interactions under physiological and neurodegenerative states (Lian and Zheng 2015). Although the influences of astrocytes on the aging process are more suspected than implicated, they appear to adopt different functions dependent on disease progression and the extent of accompanying parenchymal inflammation. Astrocytes enable clearance of Aβ and restrict the spread of inflammation in brain, yet astrocytes promote neurodegeneration in AD by releasing neurotoxins and negating crucial metabolic roles (Birch 2014). Using an experimental model of small subcortical infarcts in mice for studying pathophysiological changes in the corticospinal tract and assessing long-term neurologic outcomes and behavioral performance, Uchida et al. (2015) administered the vasoconstrictor peptide, endothlin-1 (ET-1), and the NOS inhibitor N(G)-nitro-L-arginine methyl ester (L-NAME), into the internal capsule of mice. At two months, they observed a loss of axons and myelin surrounded by reactive gliosis in the region of the injection and severe neurological deficits.

1.2 Perinatal Insult and Neurologic Neurodegenerative Disorders

The Latent Early-life Associated Regulation (LEARn) model poses environmental exposures as "hits," which, when sufficient in strength and/or number as a fetal insult, leads to altered neural development and later-life disorder or susceptibility to disorder (Lahiri et al. 2009)—giving support to the "developmental origins of health and disease" (DOHaD) hypothesis (Barker 2007). This topic has been recently reviewed, in reference to perinatal insult and the ultimate development of neurodegenerative disorders (Tartaglione et al. 2016).

For example, perinatal exposure of mice or monkeys to lead (Pb) results in later-life cognitive deficits accompanied by the upregulation of amyloid precursor protein (APP), Aβ deposits, and phosphorylated tau in brain—features of **Alzheimer's disease (AD)** (Bihaqi and Zawia 2013; Bihaqi et al. 2014). Similarly, perinatal exposure of rats to lead (Pb) leads to a similar pattern of deficits along with an increase in the brain level of 8-hydroxy-2'-deoxyguanosine (oxo8dG), a major DNA oxidation metabolite reflecting oxidative stress (Bolin et al. 2006). Other heavy metals (arsenic, cadmium) and pesticide exposure during perinatal development produce similar dysfunctions in animals (Baldi et al. 2011; Ashok et al. 2015). This series of examples supports the contention that early-life insults can have permanent effects in brain and behavior.

In an analogous manner, perinatal exposure or treatment with iron leads to behavioral indices of **PD** in mice (Fredriksson et al. 1999, 2000) and rats (Dal-Pizzol et al. 2001), with effects thought to be associated with observed oxidative stress in brain. Manganese (Mn) had a direct effect but also increased the susceptibility of brain to later-life toxic insult (Cordova et al. 2012).

These examples give credence to the likelihood that there are multiple kinds of perinatal insults that produce lifelong neural dysfunctions, some of which lead to a greater incidence of neurological, neurodegenerative, and psychiatric disorders in humans.

1.3   Neurotrophins and Neuronal Development

In a long series of studies beginning in the first half of the twentieth century, R. Levi-Montalcini discovered that there were proteins termed neurotrophins that were essential for the development of the nervous system. One of these neurotrophins, nerve growth factor (NGF), was shown to promote the growth and development of the sympathetic nervous system during ontogeny, now known to act by regulating the expression of genes associated with axonal growth and synaptogenesis (Miller and Kaplan 2001). NGF likewise has a prominent effect on the maintenance and development of cholinergic nerves in basal forebrain (Niewiadomska et al. 2009). Impaired cleavage of proNGF to NGF has been suggested as one of the possible causes of degeneration of basal forebrain cholinergic nuclei in AD (Tuszynski and Blesch 2004). Reduced neuronal responding to NGF is another of many other possibilities related to the loss of cholinergic nerves in AD (Cooper et al. 1994). The multifactorial effect of NGF on the nervous system and on the immune system development has been recently reviewed (Bracci-Laudiero and De Stefano 2016).

Synthesis of NGF in brain cells and in the peripheral nervous system is upregulated by the catecholamines (Barra et al. 2014; Hasan and Smith 2014; Sygnecka et al. 2015), which is in keeping with the physiological relation between the level of NGF mRNA and the density of innervation in the peripheral sympathetic nervous systems (Furukawa 2015). NGF is essential for the survival and functional maintenance of forebrain cholinergic neurons projecting mainly to the cortex and hippocampus (Hohsfield et al. 2014; Iulita and Cuella 2014; Perez et al. 2015), with particular importance for the relative levels of pro-NGF and mature

NGF. Thus, for example, diabetic encephalography has been characterized by deteriorations in the maturation of NGF (Soligo et al. 2015). NGF increases low-density lipoprotein receptor levels in PC6.3 cells and in cultured septal neurons from embryonic rat brain (Do et al. 2015), indicating that NGF and simvastatin, which is used to decrease unhealthy lipid levels, stimulates lipoprotein uptake by neurons with a positive effect on neurite outgrowth. Increases in low-density lipoprotein receptors and lipoprotein particles in neurons may exert a functional role during the brain development, as well as in neuroregenerative processes and following traumatic brain injuries. Although aging is a normal physiological process accompanied, more often than not, by deteriorations in certain cognitive domains, alterations in the levels of neurotrophic factors NGF, BDNF, and GDNF (glia-derived neurotrophic factor) are implicated in this decline, which implicates lowered neurotrophic levels in the pathogenesis of AD and other age-related disorders (Budni et al. 2015).

2 Actions and Mechanisms of Selective Neurotoxins

2.1 6-Hydroxydopamine

6-Hydroxydopamine (6-OHDA), the first selective neurotoxin to come into common use, was discovered in the late 1960s by H Thoenen and JP Tranzer during their search for norepinephrine (NE) analogs that might provide dark osmophilic "staining" of noradrenergic nerves during the electron microscopic observation (Thoenen and Tranzer 1968a, b). 5-Hydroxydopamine (5-OHDA) fulfilled that criterion, but 6-OHDA to their surprise produced overt destruction of noradrenergic nerves and its action was selective, leaving surrounding tissues and other nerves intact. 6-OHDA was eventually found to produce its neurotoxicity by generating intraneuronal oxidative stress and by an action on mitochondrial cytochromes, thereby blocking ATP formation and energy depletion of neurons (Cohen and Heikkila 1974). Later, 6-OHDA neurotoxicity was extended to dopaminergic nerves, as well (Ungerstedt 1968, 1971).

6-OHDA has found extensive use in neuroscience research, being cited (as "6-hydroxydopamine OR 6-OHDA") in ~12,000 papers in PubMed. 6-OHDA is a useful agent for uncovering effects of noradrenergic and dopaminergic nerves and for studying neurotoxic processes and mechanisms and reactive neuroprotective strategic mechanisms of these nerves. As a neurotoxin, 6-OHDA destruction of pars compacta SNpc in adult species (rodents, non-human primates) is of value for producing animal modeling of PD. As a neuroteratogen—6-OHDA administration during ontogeny—6-OHDA has effectively modeled several neural disorders including PD, ADHD, and LND, each of which is described subsequently.

2.2 6-Hydroxydopa

Following the discovery of 6-OHDA as a neurotoxin, 6-hydroxydopa (6-OHDOPA) was developed with the rationale that (1) 6-OHDOPA would be able to cross the blood–brain barrier (6-OHDA does not), (2) to be decarboxylated to 6-OHDA in brain; thus, 6-OHDOPA would actually be a protoxin, and (3) 6-OHDOPA-derived 6-OHDA would then destroy noradrenergic and/or dopaminergic nerves deep in brain, (4) while obviating unintentional damage to other nerves which would otherwise occur during injection of 6-OHDA per se into brain (Ong et al. 1969; Berkowitz et al. 1970). Subsequently, 6-OHDOPA was confirmed as a neurotoxin, able to produce destruction to noradrenergic sympathetic nerves (Kostrzewa and Jacobowitz 1972; Sachs and Jonsson 1972a, b) and noradrenergic nerves in brain (Jacobowitz and Kostrzewa 1971; Kostrzewa and Jacobowitz 1973; Zieher and Jaim-Etcheverry 1973, 1975a, b). 6-OHDOPA also proved to be a unique neuroteratologic agent, able to destroy noradrenergic nerves in brain (Kostrzewa and Harper 1974)—with preference for locus coeruleus nuclei (Kostrzewa and Harper 1974; Tohyama et al. 1974a, b; Clark et al. 1979) and the dorsal bundle providing noradrenergic innervation to dorsal brain (Kostrzewa and Garey 1976, 1977)—while leaving dopaminergic innervation to rodent striatum virtually intact (Kostrzewa et al. 1988). This specificity of 6-OHDOPA for noradrenergic nerves provided a unique advantage in mapping noradrenergic nerves in brain in the 1970s (Jacobowitz and Kostrzewa 1971; Sachs et al. 1973).

6-OHDOPA, however, had specifically low potency and also lethality at high dose (Kostrzewa and Garey 1976). Part of the lethal effect may reside in additional agonist action of 6-OHDOPA at non-NMDA glutamatergic receptors (Rosenberg et al. 1991). At the time of its discovery 35 years ago, 6-OHDOPA was useful as a selective noradrenergic neurotoxin. Important discoveries were made by the use of this neurotoxin on noradrenergic systems in brain, including mapping of the dorsal noradrenergic bundle to forebrain, cerebellum, and spinal cord. 6-OHDOPA likewise was useful in uncovering the labile nature of locus coeruleus neurons. At this time, the inherent limitations of 6-OHDOPA relegate it to secondary status; tagged antibodies for marker enzymes (e.g., immunotoxin for dopamine-β-hydroxylase) also provide a more advantageous means to assess noradrenergic nerves. 6-OHDOPA mechanisms and actions were recently reviewed (Kostrzewa 2016).

2.3 DSP-4

DSP-4 [N-(2-chloroethyl)-N-ethyl-2-bromobenzylamine] is another neurotoxin discovered by S Ross and colleagues in the early 1970s during their search for bretylium-related compounds (Ross et al. 1973; Ross and Renyi 1976). DSP-4 was initially found to cross the blood–brain barrier and cyclize to a reactive aziridinium targeted to the NE transporter (NET) and taken up primarily by locus coeruleus noradrenergic nerves, leading to NE depletion (Jonsson et al. 1981, 1982) and overt destruction (Lyons et al. 1989). DSP-4 was recently reviewed (Bortel 2014; Nowak 2016; Ross and Stenfors 2015).

As a neuroteratogen, DSP-4 has relatively selective action on locus coeruleus projections to neocortex, hippocampus, cerebellum, and spinal cord, while leaving peripheral sympathetic nerves relatively unaffected (Zieher and Jaim-Etcheverry 1975a). Typically, reactive sprouting of noradrenergic innervation to hindbrain and cerebellum occurs consequent to relative inactivation or destruction of locus coeruleus-derived innervation to neocortex, hippocampus and spinal cord (Jonsson et al. 1981, 1982; Dabrowska et al. 2007; Bortel et al. 2008; Sanders et al. 2011). Effects are lifelong. DSP-4 has been used to study the neurotoxic and neuroprotective mechanisms of noradrenergic neurons and to determine the association between early loss of noradrenergic innervation and brain and behavioral outcomes.

2.4   Co-administration of DSP-4 and MPTP

When noradrenergic nerves are lesioned with DSP-4 prior to MPTP (1-methyl-4-phenyl-1,2,3,6-tetrahydropyridine) treatment of C57/Bl6 mice, dopaminergic lesioning is enhanced (Fornai et al. 1997). Prenatal iron (Fe^{2+}, 7.5 mg/kg, on postnatal days 19–12) further exacerbates the effects on dopaminergic neurons and extent of movement disordering produced by the combination of DSP-4 and MPTP (Archer and Fredriksson 2006). Noradrenergic neuronal dysfunction is considered to add to the motor dysfunction in PD, as indicated by enhanced dysfunction of the subthalamic nucleus following the combination of DSP4 and 6-OHDA (Wang et al. 2014). This topic was recently reviewed (Archer 2016b).

2.5   Methamphetamine

The AMPH analog methamphetamine (METH) replicates many of the effects of AMPH. METH (like AMPH), with high affinity for the NET, DAT (dopamine transporter), and the serotonin (5-HT) transporter SERT, is accumulated by these nerves to evoke non-exocytotic release of NE, DA, and 5-HT (Sitte and Freissmuth 2010). Acute effects reflect sympathomimetic and serotoninergic actions at their respective receptor sites (de la Torre et al. 2000). Chronic METH is associated with neurotoxicity (Seiden et al. 1976), being related to acute METH-induced hyperthermia, promotion of reactive oxygen species (ROS), and excitotoxicity (Krasnova and Cadet 2009). Neuroteratologic effects of METH are expressed in large part by a spectrum of behavioral alterations, as outlined later in this paper.

2.6 Domoic Acid

Domoic acid, an agonist at AMPA/kainate-R, is an excitotoxin in high dose (Verdoorn et al. 1994; Tasker et al. 1996), producing neuronal loss and astrocytosis in hippocampus and amygdala, as well as prefrontal cortex and thalamus (Teitelbaum 1990). In both humans and laboratory rodents, neuropathological disturbances are characterized also by reactive gliosis and loss of neurons detected from 24 h onward and most severely after a week or two (Ananth et al. 2001, 2003). Immunohistochemical and histopathologic evidence of chronic inflammation from rats treated with domoic acid indicated severe neuronal degeneration, astrocytosis, microgliosis, universal NOS expression, and dystrophic calcification from 5 days to 54 days after administration (Vieira et al. 2015).

When administered to perinatal rats, domoic acid produces head tremor with vacuous chewing, "wet-dog shakes," circling, forelimb tremor, hindlimb hyperextension, and hind-paw biting. At very high dose myoclonic jerks and clonic-tonic convulsions (Xi et al. 1997; Doucette et al. 2000).

When administered in low dose during the second week of postnatal ontogeny, these rats in adulthood displayed prominent cell loss in hippocampal CA1 and CA3 regions (Doucette et al. 2004; Bernard et al. 2007), a reduction in GABA neurons (Gill et al. 2010) with prominent mossy fiber sprouting (Holmes et al. 1999), and behaviorally, these rats, as adults, had stage 2/3 seizure (Racine 1972) induced by a novel/stressful environment (Doucette et al. 2004). Other behavioral deficits have been noted (Pérez-Gómez and Tasker 2014). The neurotoxic and behavioral outcome of perinatal domoic acid was recently reviewed (Pérez-Gómez and Tasker 2014; Doucette and Tasker 2016).

2.7 192 IgG-Saporin

The immunotoxin 192 IgG-saporin consists of the monoclonal antibody 192 IgG conjugated to the ribosome-inactivating protein (RIP) saporin. In perinatal rats, 192 IgG targets the low-affinity rat NGF receptor ($p75^{NGF}$), which is expressed solely on cholinergic neurons in the nucleus basalis magnocellularis (NBM) and diagonal band of Broca (DBBh) in rat basal forebrain. Saporin, being then internalized by receptor-mediated endocytosis, travels by retrograde axonal transport to the neuronal perikaryon and inactivates ribosomes to inhibit protein synthesis, leading to neuronal cell death (Wenk et al. 1994; Leanza et al. 1995; Pappas et al. 1996; Robertson et al. 1998). Perinatal IgG-saporin selectively destroys 70–75 % of cholinergic in NBM/DBBh (Leanza et al. 1996), leading to ~70 % cholinergic denervation of hippocampus. In contrast to 192 IgG-saporin treatment of adult rats, which also destroys cerebellar Purkinje cells (Leanza et al. 1995; Waite et al. 1995; De Bartolo et al. 2009, 2010), perinatal 192 IgG-saporin spares Purkinje cells which have a lower expression of $p75^{NGF}$.

192 IgG-saporin reduces ultrasonic vocalization (Kehoe et al. 2001; Ricceri et al. 2007) and impairs passive avoidance learning in rat pups (Ricceri et al. 2002). In adulthood, 192 IgG-saporin-lesioned rats spent less time exploring a novel environment (Ricceri et al. 1997; Scattoni et al. 2003), but otherwise there was limited impairment in learning and memory (Leanza et al. 1996; Pappas et al. 1996) except at 22 months (Pappas et al. 2005). The perinatal effects of IgG-saporin were recently reviewed (Petrosini et al. 2014, 2016).

2.8 Quinpirole

Acute administration of the DA D_2-R agonist quinpirole produces several short-lived behavioral effects, the most prominent being yawning with penile erection in male rats (Kostrzewa and Brus 1991). However, when administered to rats once a day for several days during postnatal ontogeny, quinpirole produces DA D_2-R supersensitization that is manifested as enhanced yawning, locomotor activity, altered pain threshold, and vertical jumping with paw treading (Kostrzewa et al. 1991, 1993a, b; Kostrzewa 1995; Kostrzewa and Kostrzewa 2012). DA D_2-R supersensitivity persists for the duration of the life span (Oswiecimska et al. 2000) and is associated with enhanced AMPH-induced release of DA in rat striatum (Nowak et al. 2001; Cope et al. 2010).

Rats that had been quinpirole-primed during the first week or more of postnatal life have cognitive impairment (Brown et al. 2002) and deficits in prepulse inhibition (Maple et al. 2015). These behavioral effects are accompanied by a reduced brain level of BDNF (Brown et al. 2008; Thacker et al. 2006) and reduced expression level of the regulator of G-protein-signaling (RGS) RGS 9 gene which functions to terminate D_2-R agonist action. Because these effects are largely attenuated by olanzapine (Thacker et al. 2006), rats with permanent D_2-R supersensitivity have been posited as an animal of schizophrenia (Brown et al. 2012; Brown and Peterson 2016; Kostrzewa et al. 2016c; Maple et al. 2015).

3 Animal Modeling with Neuroteratologic Agents

3.1 Rodent Model of PD Produced by Perinatal 6-OHDA Treatment

Perinatal intracerebral (icv) treatment of rats with 6-OHDA (134 μg, half on each side) produces near-total lesioning of SNpc and lifelong near-total dopaminergic denervation of striatum. Acutely there is no discernible behavioral effect, and rats develop into adulthood with no motor deficit. Permanent serotoninergic hyperinnervation of striatum occurs as rats develop into adulthood (Berger et al. 1985; Snyder et al. 1986). Repeated DA D_1 agonist treatments in adulthood prime DA D_1-R, which remain supersensitized for the remainder of life, while DA D_2-R are less affected (Breese et al. 1984a, 1985a, b, 1987; Criswell et al. 1989; Hamdi and Kostrzewa 1991; Kostrzewa and Gong 1991; Gong et al. 1992, 1993a, 1994; see Kostrzewa 1995; Kostrzewa et al. 1998). The perinatal 6-OHDA-lesioned rats represents a suitable animal model of severe PD (Kostrzewa et al. 2006; Kostrzewa et al. 2016b).

An alternative animal model of PD is produced by administering 6-OHDA unilaterally in adult rats to lesion the SNpc; effectiveness of known and putative anti-parkinsonian agents can be assessed by counting the numbers of rotations produced by those treatments (Ungerstedt 1971). Bilateral 6-OHDA adulthood treatment would produce aphasia, adipsia, lack of grooming, and immobility—with consequent death in a matter of day, except with prolonged special care. Still, these rats remain fragile.

In contrast to adulthood 6-OHDA-lesioned rats, the perinatally 6-OHDA-lesioned rats provide immense advantages for assessing anti-parkinsonian agents. Because perinatally 6-OHDA-lesioned rats are able to eat, drink, groom, and ambulate—as per intact controls, even in the relative absence of SNpc dopaminergic neurons, this neurochemical/neuroanatomical model of PD is behaviorally robust and demonstrates ambulatory enhancement when treated with anti-parkinsonian agents; motor dyskinesia produced by high-dose L-DOPA is able to be discerned (Kostrzewa et al. 2006; Kostrzewa et al. 2016b). These rats have been used to assess the elevation of tissue levels of striatal DA after acute L-DOPA treatment (Kostrzewa et al. 2002, 2005); also the effect of acute AMPH on striatal exocytosis (Nowak et al. 2005); and the effect of acute L-DOPA on striatal levels of ROS (Kostrzewa et al. 2000, 2002; Nowak et al. 2010). The perinatal 6-OHDA-lesioned rat, as a modeling of PD, is described in detail in a recent paper (Kostrzewa et al. 2016b).

3.2 Exercise Effectiveness in Improving Behavioral Deficits in a Rodent Model of PD

Physical exercise has proven to be effective and is a recommended alternative for ameliorating, even reversing motor and behavioral dysfunctions in neurodegenerative disorders (Archer 2011, 2012, 2014; Archer et al. 2014a, b; Archer and Garcia 2015; Archer and Kostrzewa 2012; Archer et al. 2011a, b, 2014a, b). In a rodent model of PD, exercise produced profound ameliorative effects (Archer and Fredriksson 2010, 2012, 2013; Archer et al. 2011a, b, 2014a, b; Fredriksson et al. 2011). Exercise is a particularly useful intervention in PD patients in sedentary occupations. The several links between exercise and quality of life, disorder progression and staging, risk factors, and symptom biomarkers in PD all endow a promise for improved prognosis. Nutrition provides a strong determinant for disorder vulnerability and prognosis, with fish oils and vegetables with a Mediterranean diet offering both protection and resistance, whereas exercise increases synaptic strength and influences neurotransmission. Nevertheless, the heterogeneity of exercise/activity programs, including stretching, muscle strengthening, balance, postural exercises, occupational therapy, cueing, and/or treadmill training, remains an issue and consensus concerning the optimal approach (Abbruzzese et al. 2015; but see also Uhrbrand et al. 2015). Three factors determining the effects of exercise on disorder severity of patients may be presented: (i) exercise effects on motor impairment, gait, posture, and balance; (ii) exercise reduction of oxidative stress, stimulation of mitochondrial biogenesis, and upregulation of autophagy; and (iii) exercise stimulation of dopaminergic neurochemistry and trophic factors.

Running-wheel performance, as measured by distance run by control and parkinsonian-modeled mice from different treatment groups, was related to dopaminergic system integrity, indexed by striatal DA levels (Archer and Kostrzewa 2016). Support for these notions (regarding the almost finite advantages to be gleaned from exercise) continues to emerge. Exercise triggers plasticity-related events in the human PD brain, such as corticomotor excitation, increases in gray matter volume, and an elevation in BDNF levels (Hirsch et al. 2015). Finally, both nutrition and exercise may facilitate positive epigenetic outcomes, such as lowering the dosage of L-DOPA required for a therapeutic effect. Exercise, as a potent epigenetic regulator, implies a potential to counteract pathophysiological processes and alterations, notwithstanding a paucity of understanding in the underlying molecular mechanisms and dose–response relationships (Archer 2015).

3.3 Rodent Model of ADHD Produced by Perinatal 6-OHDA with Adulthood 5,7-DHT Lesions

In the 1970s, B Shaywitz and colleagues produced an animal of "minimal brain dysfunction," akin to today's nomenclature for ADHD, by 6-OHDA lesioning of perinatal rats. These rats demonstrate attentional deficits with spontaneous hyper-locomotor activity, each of which is attenuated by acute AMPH treatment (Shaywitz et al. 1976a, b). Over the past 40 years, this has remained the gold standard for rodent modeling of ADHD.

A variation of this model consists of perinatal 6-OHDA lesioning (134 µg, half on each side), followed by adulthood (10 weeks of age) lesioning with 5,7-dihydroxytryptamine (5,7-DHT, 75 µg icv). Treatment of 6-OHDA-lesioned rats with 5,7-DHT had the effect of reducing striatal serotoninergic hyperinnervation by 30 % and suppressing D_1-R supersensitivity while enhancing 5-HT_{2C}-R sensitivity. Behaviorally, these rats displayed enhanced hyperlocomotor activity (vs rats lesioned solely with 6-OHDA), and this activity was attenuated by AMPH (Kostrzewa et al. 1994). Moreover, this animal model of ADHD was able to discern the ability of m-chlorophenylpiperazine (mCPP), a 5-HT agonist, to suppress the hyperlocomotor activity and thereby indicate a new approach toward ADHD treatment (Brus et al. 2004). In vivo microdialysis study indicates that the activity-suppressant effects of AMPH and mCPP are unrelated to exocytosis of striatal DA and 5-HT (Nowak et al. 2007). The higher level of hyperlocomotor activity in rats with the dual 6-OHDA + 5,7-DHT lesions represents a more robust model of ADHD in testing agents with the potential for ADHD treatment (Paterak and Stefański 2014; Kostrzewa et al. 2008). This animal model for ADHD was recently reviewed (Kostrzewa et al. 2016a). A non-pharmacological approach toward abating features of ADHD has been demonstrated (Archer and Kostrzewa 2012).

3.4 ADHD and NMDA-R Systems

An imbalance between central inhibitory/excitatory neurotransmitters and relative activity/connectivity between brain regions, with concomitant disturbances of higher cognitive function, is considered to reflect the pathogenesis of ADHD (He et al. 2015; Mohl et al. 2015; Monden et al. 2015; Roman-Urrestarazu et al. 2015).

Dysfunction of the default-mode network in ADHD patients is considered together with some of the animal models used to examine the neurobiological aspects of ADHD. Much evidence indicates that compounds/interventions that antagonize/block glutamate receptors and/or block glutamate signaling during the "brain growth spurt" (or in the adult animal model) may induce functional and biomarker deficits. Mice treated with glutamate receptor antagonist (MK-801, dizocilpine; ketamine) during the "brain growth spurt" fail to display exploratory activity when placed in a novel environment (the test cages) and later fail to adapt to the environment with locomotor suppression, implying a cognitive dysfunction. A disturbance of glutamate signaling during a critical stage of neural ontogeny may contribute to the ADHD pathophysiology. In a functional magnetic resonance imaging (fMRI) study of executive functioning in ADHD adults and matched controls, it was observed that in people with ADHD, there was a failure of deactivation of the medial prefrontal cortex (Salavert et al. 2015). In another study of ADHD adults, using a rest-to-take switching task, there was a disturbed reinitiation of a rest state.

"Hot" and "cool" cognitive functions present a dichotomy within executive function whereby the former refers to affective domains and the latter to cognitive domains (Doebel and Zelazo 2013; Hongwanishkul et al. 2006; Zelazo et al. 2003, 2004). Top-down processes that operate in more affectively neutral contexts have been termed "cool" executive functioning, whereas those operating in motivationally and emotionally significant situations are referred to as "hot" (Zelazo and Carlson 2012). ADHD children exhibited "cool" executive function deficits which appeared to be unrelated to comorbid oppositional defiant disorder (Antonini et al. 2015). Finally, Babenko et al. (2015) have highlighted the intricate interplay between prenatal stress exposure, associated changes in miRNA expression, and DNA methylation in placenta and brain with possible links to greater risks for incidence of ADHD later in life. The association of studies with NMDA-R antagonists and ADHD has been reviewed recently (Archer 2016a; Archer and Garcia 2016).

3.5 Rodent Model of Lesch–Nyhan Disease Produced by Perinatal 6-OHDA Treatment

Lesch–Nyhan disease (LND), a relatively rare neuroteratologic disorder attributable to a mutation in the HPRT 1 gene, is characterized by deficiency in hypoxanthine–guanine phosphoribosyltransferase (HGPRT). Abnormality in purine recycling leads to high serum levels of uric acid, the end product of purine metabolism, and gout—deposition of uric acid crystals in joints and soft tissue. Neurological symptoms represent a range of stages from mild to severe, but often being associated with self-biting and self-mutilation (Abel et al. 2014; Fu et al. 2015; Schroeder et al. 2001). In five different strains of mice with an HPRT gene knockout—characterized by one of two different HPRT gene mutations (Jinnah et al. 1999)—the nigrostriatal dopaminergic tract was found to be incompletely developed and the striatum had both reduced DA content and increased oxidative stress (Visser et al. 2001). While the HPRT-deficient mouse represents a viable model for the enzymatic deficiency in LND, the behavioral counterpart representing self-mutilation, however, is better modeled in rats that were perinatally lesioned with 6-OHDA (Breese et al. 1984b, 1986, 1989, 1990a, b; 1994; 2005). In these rats, DA D_1-R are overtly supersensitive (for some behaviors) (Kostrzewa and Gong 1991; Kostrzewa et al. 1992; Gong et al. 1993a, b; 1994) and are further able to be supersensitized by repeated treatments with L-DOPA or a D_1-R agonist—a priming process (Breese et al. 1984a, 1985a, b, 1987). When perinatal 6-OHDA-lesioned rats are acutely treated as adults with L-DOPA or with a DA D_1-R agonist, there is prominent self-biting and self-mutilation that can be counteracted with a DA D_1-R antagonist (see Wong et al. 1996; Papadeas and Breese 2014). Curiously, LND individuals have a DA deficiency in basal ganglia (as per 6-OHDA rats), and this apparently accrues from inadequate development of dopaminergic innervation (Göttle et al. 2014). The perinatal 6-OHDA-lesioned rat as a model of LND has recently been reviewed (Knapp and Breese 2016).

3.6 Permanent Animal Model of Tardive Dyskinesia

Tardive dyskinesia (TD) is a movement disorder produced in primates and other mammalian species by repeated treatments, over a period of months, with a DA D_2-

R antagonist. In humans, the D_2-R antagonist is a common feature of antipsychotic agents used to treat schizophrenia. TD presents as involuntary repetitive purposeless movements, most often of the lower face—resembling someone chewing gum and sometimes also with tongue thrusting (Casey 1987; Jeste and Caligiuri 1993). In rats, TD is most reasonably produced by including haloperidol or other D_2-R antagonist in the drinking water (Waddington et al. 1983; Waddington 1990). After a period of ~ 3 months, these rats, behaviorally, display spontaneous purposeless (vacuous) chewing movements (VCMs) which persist for as long as the D_2-R antagonist is present in the drinking water. After withdrawal of the D_2-R antagonist from drinking water, VCMs gradually disappear over a period of 4 to 6 weeks. This latter feature in rats—relating to the regression of TD upon D_2-R antagonist withdrawal—contrasts with human TD, in which the TD persists and is often permanent even after the D_2-R antagonist withdrawal.

In an attempt to produce a permanent model of TD, rats were first lesioned as perinates with 6-OHDA (134 µg, half on each side). When these rats (and controls) reached adulthood, haloperidol was added to the drinking water for a period of nearly one year. While intact control rats developed TD (i.e., increased number of VCMs) after ~ 3 months, 6-OHDA-lesioned rats developed TD only after 2 months. Moreover, the number of VCMs in haloperidol/6-OHDA rats was 2-fold greater than the number of VCMs in haloperidol/intact control rats. Significantly, after the removal of haloperidol from drinking water (i.e., haloperidol withdrawn stage), VCMs gradually disappeared in haloperidol/intact rat over a period of ~ 2 months, while VCMs persisted in 6-OHDA-lesioned rats, at the same elevated level and until the experiment ended 8 months later. At that time, it was determined that the D_2-R number (i.e., V_{max}) had been increased during the haloperidol phase and that D_2-R number had reverted to normal by 8 months—signifying that numbers of VCMs were unrelated to numbers of strital D_2-R (Huang et al. 1997).

The advantage of persistent VCMs in the withdrawal phase is that it becomes possible to test agents that might have the ability to suppress VCMs. To this end, it was found that agonists and antagonist at both the D_2-R and D_1-R had no effect, nor did agonists or antagonists at a number of other types of receptors. Only antagonists at 5-HT-R attenuated VCMs in rats in the withdrawal phase, and the common feature of each of these antagonists was that they have affinity for the 5-HT_{2C}-R, a likely site that can be targeted to reduce TD in humans during the antipsychotic withdrawal phase (Kostrzewa et al. 2007). This animal model of TD is described in detail in a recent paper (Kostrzewa and Brus 2016).

3.7 Valproate Modeling of Autism Spectrum Disorder

Prenatal/postnatal/perinatal etiologies, ranging from exposures involving drugs to infections, as well as genetic factors, are complicit in autism spectrum disorder

(ASD) that affects roughly 1–2 % of all children, according to the current analyses (Pelly et al. 2015). Several maternal diseases during pregnancy are linked to ASD, pregestationally and/or gestationally, including diabetes mellitus, maternal infections (i.e., rubella, cytomegalovirus), prolonged fever, and maternal inflammation, inducing changes in a variety of inflammatory cytokines (Ornoy et al. 2015); among external agents affecting ASD outcome are drugs such as valproic acid (VPA), the anticonvulsant agent and mood stabilizer, and antiepileptic compounds (Kulaga et al. 2011; Jacobsen et al. 2014). VPA is associated with poorer longer-term child developmental outcomes (Galbally et al. 2010).

Several aspects of animal models, generally and specifically pertaining to ASD, are scrutinized and surveyed, including construct validity, face validity, ASD-like behavioral and neurochemical alterations, histone deacetylase inhibition which elevates ROS, oxidative stress, and the status of experimental models and mitigating factors. These above processes relate to an altered epigenetic landscape in ASDs via altered methylation/hydroxymethylation patterns, local histone modification patterns, and chromatin remodeling (Banerjee et al. 2014; Grayson and Guidotti 2015; Siniscalco 2015).

ASD is characterized by deficits in social interaction and restricted or repetitive behaviors, but often accompanied by other behavioral (e.g., aggression), intellectual (e.g., lower IQ), neurological (e.g., epilepsy), or psychiatric (e.g., anxiety, depression) symptoms (Levy et al. 2009). The antiepileptic drug valproate (VPA), when used clinically to treat epilepsy and bipolar disorder in pregnant women (Lloyd 2013), is associated with a 4 % risk for offspring to develop ASD (Christianson et al. 1994; Christensen et al. 2013), with the incidence being 4–5 times greater in males (Wingate et al. 2014). Several types of animal models of ASD have been produced, but the most common model is produced by VPA treatment of perinatal rats (Rodier et al. 1997; Ranger and Ellenbroek 2016).

When pregnant rats are treated with VPA on gestation day 12, the time of fetal neural tube closure (Kim et al. 2011), the brain of offspring has notable abnormalities, including increased neocortical thickness with a higher number of cortical neurons (Sabers et al. 2015), reduced spine density in the hippocampus (Takuma et al. 2014), hyperserotonemia (Narita et al. 2002), and other defects. Behaviorally in rats and mice, there is hyperactivity, repetitive behaviors, and social deficits (Kim et al. 2014), resembling the behavioral spectrum in humans with ASD.

VPA is thought to act by inhibiting histone deacetylase (Phiel et al. 2001), resulting in hyperacetylated histones and associated increased transcriptional activity of multiple genes (Lloyd 2013), which is thought to account for the neuroteratologic effects. Secondarily, VPA increases the production of ROS in brain (Winn and Wells 1999), which may be detrimental to DNA integrity.

Animal modeling of ASD by VPA has been reviewed recently (Roullet et al. 2013; Ranger and Ellenbroek 2016).

4   Perinatal Insults that Model Psychosis Schizophrenia

There are a plethora of agents that, when administered to animals during ontogenetic development, model features of schizophrenia in the adulthood stage. Some of the more common agents having such an effect include epidermal growth factor (EGF) and its homologue neuregulin (NRG-1), METH, phencyclidine (PCP), and quinpirole. Details regarding these substances and their respective roles in animal modeling of psychosis and schizophrenia are described in the following section.

4.1   Epidermal Growth Factor and Schizophrenia Modeling

When administered to perinatal rats and mice, both EGF and NRG-1 produce adulthood effects that mirror some of the features common in schizophrenia: PPI deficit, altered sensorimotor gating and social interaction, exploratory suppression, cognitive deficit, sensitization to psychostimulants (METH; MK-801, dizocilpine), and other behavioral effects (Sotoyama et al. 2011, 2013; Sakai et al. 2014). Most deficits are reversed by atypical antipsychotics such as clozapine and risperidone but not by typical antipsychotics such as haloperidol (Sotoyama et al. 2013). Yet, in the EGF and NRG-1 models, learning is not compromised, as demonstrated by testing for context fear learning and passive avoidance learning (Futamura et al. 2003; Tohmi et al. 2005).

EGF is thought to exert its major effect on dopaminergic neurons in the SN, increasing dopaminergic activity in the globus pallidum (Sotoyama et al. 2011), while NRG-1 is more selective for dopaminergic neurons in the VTA (Abe et al. 2009; Iwakura et al. 2011a, b), producing enhanced dopaminergic activity in the prefrontal cortex (Kato et al. 2011). ROS formation is considered as a primary process in mediating these effects, as antioxidants suppress some of the adulthood behavioral deficits (Mizuno et al. 2008, 2010). This topic has been reviewed recently (Nagano et al. 2016).

4.2   Phencyclidine and Schizophrenia Modeling

In rodents, prolonged non-competitive NMDA-R antagonism by ketamine or PCP evokes a change in biomarkers in brain accompanied by a spectrum of behavioral activities that model schizophrenia—with non-classical antipsychotics acutely reversing many of the deficits (Barnes et al. 2015; Pyndt Jørgensen et al. 2015). Acute and subchronic treatments with PCP affect differentially the neuronal activity of different brain regions: basal DA, but not serotonin. Output in the medial prefrontal cortex is markedly reduced, and tyrosine hydroxylase expression in the ventral tegmental area is decreased, thereby accounting in part for concomitant behavioral alterations expressed through locomotor sensitization and cognitive deficits (Castañé et al. 2015).

Perinatal administration of the NMDA-R antagonist PCP to rodents produces a spectrum of neuropathological and behavioral effects that model some of the features of schizophrenia. Disruption of glutamate signaling during ontogeny by PCP is thought to impede development of the GABAergic system in brain (Ben-Ari et al. 1997; Le Magueresse and Monyer 2013), resulting in an overall imbalance in neuronal excitation and inhibition in brain in adulthood (Hoftman and Lewis 2011). In perinatal PCP-treated rats and mice, there is an adulthood reduction in fast-spiking GABAergic interneurons in medial prefrontal cortex, nucleus accumbens, and hippocampus (Nakatani-Pawlak et al. 2009; Kaalund et al. 2013; Radonjic et al. 2013; Kjaerby et al. 2014), mimicking reduced GABAergic interneuronal activity in the brain of schizophrenic patients (Reynolds et al. 2004). In the nucleus accumbens, there is also a prominent reduction in dendritic spine density of spiny neurons (Nakatani-Pawlak et al. 2009). Anatomic and neurochemical changes in PCP rodents include a decrease in the number of parvalbumin-positive cells and spine density in the frontal cortex, nucleus accumbens, and hippocampus (Nakatani-Pawlak et al. 2009). Also in brain, glutathione and antioxidant defenses are reduced (Radonjic et al. 2010; Stojkovic et al. 2012).

Behaviorally, there are cognitive deficits in the adulthood rodents that were treated perinatally with PCP, as demonstrated by impaired working memory (Morris water maze testing (Sircar 2003) and rate of learning (delayed spontaneous alternation task) (Wang et al. 2001), sensorimotor dysfunction (deficit in prepulse inhibition) (Anastasio and Johnson 2008; Broberg et al. 2010, 2013; Chen et al. 2011; Kjaerby et al. 2013), social withdrawal (White et al. 2009), reduced attention in a social novelty discrimination paradigm (Terranova et al. 2005), and executive function (attentional set-shifting task for executive function) (Broberg et al. 2008). Many of the behavioral deficits are reversed by atypical antipsychotics. This topic was recently reviewed (Neill et al. 2014; Grayson et al. 2016).

4.3 *Methamphetamine and Schizophrenia Modeling*

METH, used and abused illicitly as an aphrodisiac and euphoriant, produces elevated mood, increased alertness and concentration, "energy" in fatigued individuals, and reduced appetite and promotes (initial) weight loss at lower doses, whereas at higher doses the drug induces psychosis, affective disorders, and rhabdomyolysis (Ago et al. 2006; De Carolis et al. 2015; Harro 2015; Mouton et al. 2015). METH use by pregnant women is associated with cognitive, attentional, and mood dysfunctions in offspring (Hrebíčková et al. 2014; McDonell-Dowling and Kelly 2015; Smith et al. 2015).

Ontogenetic effects of METH are diverse and heavily reliant on gestational age in terms of long-lived alterations in behavior, epigenetic expression, neuronal organization, and overall neurotransmission and receptor parameters (Roos et al. 2015; Vrajová et al. 2014). The prenatal effects on cognitive and emotional behavior provide evidence of drastic disruptions of normal behavioral patterns (Fialová et al. 2015; Malinová-Ševčíková et al. 2014; Šlamberová et al. 2014, 2015). Long-term behavioral alterations induced by chronic METH use imply alterations in gene and protein expression within specific brain subregions involved in the reward circuitry and accompanied by major epigenetic modifications—histone acetylation and methylation (Desplats et al. 2014; Godino et al. 2015). Although epigenetic changes have not as yet been detected following prenatal METH exposures, these findings are awaited (Cadet 2014; Cadet and Jayanthi 2013).

Perinatal METH treatment has a range of effects on adulthood behaviors in rodents, depending upon whether METH is pre- and/or postnatal (Graham et al. 2013; Jablonski et al. 2016). Postnatal METH treatment in the range of birth through the postweaning period has the most pronounced effects, generally suppressing adulthood spontaneous locomotor activity and increasing acoustic startle reactivity (Vorhees et al. 2009). Given at the critical postnatal period, METH produces learning impairment and spatial memory impairment (Vorhees et al. 1994a, b, 2009).

Perinatal METH produces a persistent reduction in brain levels of DA and 5-HT, inhibiting tyrosine hydroxylase activity (Ricaurte et al. 1982; Bowyer et al. 1998), also 5-HT transporters (Kokoshka et al. 1998), and also other neurotransmitter systems. This topic was recently reviewed (Bisagno and Cadet 2014; Jablonski et al. 2016).

4.4 Quinpirole and Schizophrenia Modeling

Repeated daily postnatal quinpirole treatments of rats produce permanent DA D_2-R supersensitivity (Kostrzewa 1995; Kostrzewa et al. 2003, 2004, 2008, 2011, 2016c). In adulthood, these rats display enhanced D_2-R agonist-evoked behaviors and a spectrum of behavioral alterations. Rats exhibit improved active avoidance responding (Brus et al. 1998b), learning and memory deficits (Brus et al. 1998a) in the Morris water maze task, on place, and on match-to-place versions of this task (Brown et al. 2002, 2004a, 2005), and a deficit in prepulse inhibition (PPI) to acute startle (Maple et al. 2007). In the hippocampus on these rats, BDNF and NGF were reduced (Thacker et al. 2006; Maple et al. 2007), while in the striatum, nucleus accumbens, and frontal cortex expression of RGS9, a transcript regulating G-protein coupling to the D_2-R was reduced (Maple et al. 2007). Long-term olanzapine treatment reversed the cognitive deficits, reversed the PPI deficit, and normalized otherwise reduced BDNF and NGF levels in hippocampus (Thacker et al. 2006; Maple et al. 2007) and RGS9 expression (Maple et al. 2007). Because nicotine likewise reverses D_2-R supersensitization, drugs acting on α_7 nAChRs (e.g., nicotine) have been suggested for the treatment of schizophrenia (Tizabi et al. 1999;

Brown et al. 2004b, 2006; Perna and Brown 2013). Quinpirole modeling of schizophrenia was recently reviewed (Kostrzewa et al. 2016a, b, 2016c; Brown and Peterson 2016).

4.5 Stress and Neuropsychiatric Disorders

Prenatal restraint stress (PRS) during the last week of gestation is associated with postweaned offspring displaying attentional deficits, increased anxiety, impaired spatial learning (Lemaire et al. 2000) and deficit in working memory (Maccari et al. 2003), reduction in social play behavior, increased latency in approaching a novel object (Laviola et al. 2004), and a syndrome complex resembling features of ASD (see Weinstock 2008). Clearly, glucocorticoids are implicated in these outcomes. Disruption in the circadian rhythm also has analogous effects to PRS, as each is posed as a means to model psychiatric disorders (Marco et al. 2016).

4.6 Genetic Model of Alzheimer's Disease

In the laboratory mouse model for AD, APPswe/PS1dE9, with mutant transgenes of APP and presenilin-1 (PS1), chronic inflammation provokes amyloid plaque formation as early as 4 months of age, with numbers of plaques increasing with aging (Ruan et al. 2009). CD11b-positive microglia clusters appeared in hippocampus and neocortex at the same period of development and these also proliferated with age. Clustered glial fibrillary acidic protein (GFAP)-positive astrocytes were observed in hippocampus and cortex after six months of age and became more numerous with aging. Astrocytes appear to be central to AD pathophysiology since the β-amyloid peptide Aβ suppresses cholinergic innervation and synaptic function, subsequent to astrocytic glutamate gliotransmission. Further, Aβ causes neuronal hyperexcitability (Hertz et al. 2015). Other developmental animal models of AD are expected to be introduced and to become more commonplace.

5 Conclusion

Neurotoxins have become paramount in exploring neuronal function in relation to neuroscience research and, in particular, in animal modeling of neurological, psychiatric, and behavioral dysfunctional states. This concise review highlights the mechanisms and action of the most commonly used neurotoxins and reviews the use of individual neurotoxins in animal modeling of PD, ADHD, LND, autism, TD, and psychotic and schizophrenic states. The influence of neurotrophins, EGF in particular, on ontogenetic is outlined, and the influence of perinatal stress as well as disrupted circadian cycling on neuronal ontogeny is described. Animal modeling of human disorders is likely to be used to an ever greater extent and through use of neurotoxins yet to be discovered.

References

Abbruzzese G, Marchese R, Avanzino L, Pelosin E (2015) Rehabilitation for Parkinson's disease: Current outlook and future challenges. Parkinsonism Relat Disord pii:S1353-8020(15)00380-6

Abe Y, Namba H, Zheng Y, Nawa H (2009) In situ hybridization reveals developmental regulation of ErbB1-4 mRNA expression in mouse midbrain: implication of ErbB receptors for dopaminergic neurons. Neuroscience 161(1):95–110

Abel TJ, Dalm BD, Grossbach AJ, Jackson AW, Thomsen T, Greenlee JD (2014) Lateralized effect of pallidal stimulation on self-mutilation in Lesch-Nyhan disease. J Neurosurg Pediatr 14(6):594–597

Ago M, Ago K, Hara K, Kashimura S, Ogata M (2006) Toxicological and histopathological analysis of a patient who died nine days after a single intravenous dose of methamphetamine: a case report. Leg Med (Tokyo) 8(4):235–239

Allison DJ, Ditor DS (2015) Targeting inflammation to influence mood following spinal cord injury: a randomized clinical trial. J Neuroinflammation 12(1):204

Ananth C, Thameem Dheen S, Gopalakrishnakone P, Kaur C (2001) Domoic acid-induced neuronal damage in the rat hippocampus: changes in apoptosis related genes (bcl-2, bax, caspase-3) and microglial response. J Neurosci Res 66(2):177–190

Ananth C, Gopalakrishnakone P, Kaur C (2003) Induction of inducible nitric oxide synthase expression in activated microglia following domoic acid (DA)-induced neurotoxicity in the rat hippocampus. Neurosci Lett 338(1):49–52

Anastasio NC, Johnson KM (2008) Atypical anti-schizophrenic drugs prevent changes in cortical N-methyl-D-aspartate receptors and behavior following sub-chronic phencyclidine administration in developing rat pups. Pharmacol Biochem Behav 90(4):569–577

Antonini TN, Becker SP, Tamm L, Epstein JN (2015) Hot and cool executive functions in children with attention-deficit/hyperactivity disorder and comorbid oppositional defiant disorder. J Int Neuropsychol Soc 21(8):584–595

Archer T (2011) Physical exercise alleviates debilities of normal aging and Alzheimer's disease. Acta Neurol Scand 123:221–238

Archer T (2012) Influence of physical exercise on traumatic brain injury deficits: scaffolding effect. Neurotox Res 21(4):418–434

Archer T (2014) Health benefits of physical exercise for children and adolescents. J Novel PysioTher 4:203

Archer T (2015) Physical exercise as an epigenetic factor determining behavioral outcomes. Clin Exp Psychol 1:1

Archer T (2016a) NMDA-R blockers and ADHD modeling, In: Kostrzewa RM, Archer T (eds.) Neurotoxin modeling of brain disorders-lifelong outcomes in behavioral teratology. Springer, New York

Archer T (2016b) Noradrenergic-dopaminergic interactions due to DSP4—MPTP neurotoxin treatments: iron connection. In: Kostrzewa RM, Archer T (eds) Neurotoxin modeling of brain disorders-lifelong outcomes in behavioral teratology. Springer, New York

Archer T, Fredriksson A (2006) Influence of noradrenaline denervation on MPTP-induced deficits in mice. J Neural Transm (Vienna) 113(9):1119–1129

Archer T, Fredriksson A (2010) Physical exercise attenuates MPTP-induced deficits in mice. Neurotox Res 18(3–4):313–327

Archer T, Fredriksson A (2012) Delayed exercise-induced functional and neurochemical partial restoration following MPTP. Neurotox Res 21(2):210–221

Archer T, Fredriksson A (2013) The yeast product Milmed enhances the effect of physical exercise on motor performance and dopamine neurochemistry recovery in MPTP-lesioned mice. Neurotox Res 24(3):393–406

Archer T, Garcia D (2015) Exercise and dietary restriction for promotion of neurohealth benefits. Health 7:136–152

Archer T, Garcia D (2016) Attention-deficit/hyperactive disorder: focus upon aberrant N-methyl-D-aspartate receptor systems. In: Kostrzewa RM, Archer T (eds) Neurotoxin modeling of brain disorders-lifelong outcomes in behavioral teratology. Springer, New York

Archer T, Kostrzewa RM (2012) Physical exercise alleviates ADHD symptoms: regional deficits and development trajectory. Neurotox Res 21(2):195–209

Archer T, Kostrzewa RM (2016) Exercise and nutritional benefits in PD: rodent models and clinical settings. In: Kostrzewa RM, Archer T (eds) Neurotoxin Modeling of Brain Disorders-Lifelong Outcomes in Behavioral Teratology. Springer, New York

Archer T, Fredriksson A, Johansson B (2011a) Exercise alleviates Parkinsonism: clinical and laboratory evidence. Acta Neurol Scand 123(2):73–84

Archer T, Fredriksson A, Schütz E, Kostrzewa RM (2011b) Influence of physical exercise on neuroimmunological functioning and health: aging and stress. Neurotox Res 20(1):69–83

Archer T, Garcia D, Fredriksson A (2014a) Restoration of MPTP-induced deficits by exercise and Milmed($^{®}$) co-treatment. PeerJ 2:e531

Archer T, Josefsson T, Lindwall M (2014b) Effects of physical exercise on depressive symptoms and biomarkers in depression. CNS Neurol Disord: Drug Targets 13(10):1640–1653

Arsenault D, Coulombe K, Zhu A, Gong C, Kil KE, Choi JK, Poutiainen P, Brownell AL (2015) Loss of metabotropic glutamate receptor 5 function on peripheral benzodiazepine receptor in mice prenatally exposed to LPS. PLoS One 10(11):e0142093

Ashok A, Rai NK, Tripathi S, Bandyopadhyay S (2015) Exposure to As-, Cd-, and Pb-mixture induces Aβ, amyloidogenic APP processing and cognitive impairments via oxidative stress-dependent neuroinflammation in young rats. Toxicol Sci 143(1):64–80

Babenko O, Kovalchuk I, Metz GA (2015) Stress-induced perinatal and transgenerational epigenetic programming of brain development and mental health. Neurosci Biobehav Rev 48:70–91

Baldi I, Gruber A, Rondeau V, Lebailly P, Brochard P, Fabrigoule C (2011) Neurobehavioral effects of long-term exposure to pesticides: results from the 4-year follow-up of the PHYTONER study. Occup Environ Med 68(2):108–115

Banerjee S, Riordan M, Bhat MA (2014) Genetic aspects of autism spectrum disorders: insights from animal models. Front Cell Neurosci 8:58

Barker DJ (2007) The origins of the developmental origins theory. J Intern Med 261(5):412–417. Review

Barnes SA, Sawiak SJ, Caprioli D, Jupp B, Buonincontri G, Mar AC, Harte MK, Fletcher PC, Robbins TW, Neill JC, Dalley JW (2015) Impaired limbic cortico-striatal structure and sustained visual attention in a rodent model of schizophrenia. Int J Neuropsychopharmacol 18(2). pii: pyu010. doi: 10.1093/ijnp/pyu010

Barra R, Cruz G, Mayerhofer A, Paredes A, Lara HE (2014) Maternal sympathetic stress impairs follicular development and puberty of the offspring. Reproduction 148(2):137–145

Basta-Kaim A, Fijał K, Ślusarczyk J, Trojan E, Głombik K, Budziszewska B, Leśkiewicz M, Regulska M, Kubera M, Lasoń W, Wędzony K (2015) Prenatal administration of liposaccharide induces sex-dependent changes in glutamic acid decarboxylase and parvalbumin in the adult rat brain. Neuroscience 287:78–92

Ben-Ari Y, Khazipov R, Leinekugel X, Caillard O, Gaiarsa JL (1997) GABA$_A$, NMDA and AMPA receptors: a developmentally regulated 'ménage à trois'. Trends Neurosci 20(11):523–529. Review

Berger TW, Kaul S, Stricker EM, Zigmond MJ (1985) Hyperinnervation of the striatum by dorsal raphe afferents after dopamine-depleting brain lesions in neonatal rats. Brain Res 336(2):354–358

Berkowitz BA, Spector S, Brossi A, Focella A, Teitel S (1970) Preparation and biological properties of (-) and (+)-6-hydroxydopa. Experientia 26(9):982–983

Bernard PB, MacDonald DS, Gill DA, Ryan CL, Tasker RA (2007) Hippocampal mossy fiber sprouting and elevated trkB receptor expression following systemic administration of low dose domoic acid during neonatal development. Hippocampus 17:1121–1133

Bezard E, Yue Z, Kirik D, Spillantini MG (2013) Animal models of Parkinson's disease: limits and relevance to neuroprotection studies. 28(1):61–70

Bihaqi SW, Zawia NH (2013) Enhanced taupathy and AD-like pathology in aged primate brains decades after infantile exposure to lead (Pb). Neurotoxicology 39:95–101

Bihaqi SW, Bahmani A, Adem A, Zawia NH (2014) Infantile postnatal exposure to lead (Pb) enhances tau expression in the cerebral cortex of aged mice: relevance to AD. Neurotoxicology 44:114–120

Birch AM (2014) The contribution of astrocytes to Alzheimer's disease. Biochem Soc Trans 42(5):1316–1320

Bisagno V, Cadet JL (2014) Methamphetamine and MDMA neurotoxicity: biochemical and molecular mechanisms. In: Kostrzewa RM (ed) Handbook of neurotoxicity. Springer New York, pp 219–236. ISBN 978-1-4614-5835-7 (print); ISBN 978-1-4614-5836-4 (eBook); ISBN 978-1-4614-7458-6 (print and electronic bundle). doi:10.1007/978-1-4614-5836-4_75

Bolin CM, Basha R, Cox D, Zawia NH, Maloney B, Lahiri DK, Cardozo-Pelaez F (2006) Exposure to lead and the developmental origin of oxidative DNA damage in the aging brain. FASEB J 20(6):788–790

Bortel A (2014) Nature of DSP-4 neurotoxicity. In: Kostrzewa RM (ed) Handbook of neurotoxicity. Springer, New York, pp 347–363. ISBN 978-1-4614-5835-7 (print); ISBN 978-1-4614-5836-4 (eBook); ISBN 978-1-4614-7458-6 (print and electronic bundle). doi:10.1007/978-1-4614-5836-4_80

Bortel A, Słomian L, Nitka D, Swierszcz M, Jaksz M, Adamus-Sitkiewicz B, Nowak P, Jośko J, Kostrzewa RM, Brus R (2008) Neonatal N-(-2-chloroethyl)-N-ethyl-2-bromobenzylamine (DSP-4) treatment modifies the vulnerability to phenobarbital- and ethanol-evoked sedative-hypnotic effects in adult rats. Pharmacol Rep 60:331–338

Bowyer JF, Frame LT, Clausing P, Nagamoto-Combs K, Osterhout CA, Sterling CR, Tank AW (1998) Long-term effects of amphetamine neurotoxicity on tyrosine hydroxylase mRNA and protein in aged rats. J Pharmacol Exp Ther 286(2):1074–1085

Bracci-Laudiero L, De Stefano ME (2016) NGF in early embryogenesis, differentiation and pathology in the nervous and immune systems. In: Kostrzewa RM, Archer T (eds) Neurotoxin modeling of brain disorders-lifelong outcomes in behavioral teratology. Springer, New York

Breese GR, Baumeister AA, McCown TJ, Emerick SG, Frye GD, Mueller RA (1984a) Neonatal-6-hydroxydopamine treatment: model of susceptibility for self-mutilation in the Lesch-Nyhan syndrome. Pharmacol Biochem Behav 21(3):459–461

Breese GR, Baumeister AA, McCown TJ, Emerick SG, Frye GD, Crotty K, Mueller RA (1984b) Behavioral differences between neonatal and adult 6-hydroxydopamine-treated rats to dopamine agonists: relevance to neurological symptoms in clinical syndromes with reduced brain dopamine. J Pharmacol Exp Ther 231(2):343–354

Breese GR, Baumeister A, Napier TC, Frye GD, Mueller RA (1985a) Evidence that D-1 dopamine receptors contribute to the supersensitive behavioral responses induced by L-dihydroxyphenylalanine in rats treated neonatally with 6-hydroxydopamine. J Pharmacol Exp Ther 235(2):287–295

Breese GR, Napier TC, Mueller RA (1985b) Dopamine agonist-induced locomotor activity in rats treated with 6-hydroxydopamine at differing ages: functional supersensitivity of D-1 dopamine receptors in neonatally lesioned rats. J Pharmacol Exp Ther 234(2):447–455

Breese GR, Mueller RA, Napier TC, Duncan GE (1986) Neurobiology of D_1 dopamine receptors after neonatal-6-OHDA treatment: relevance to Lesch-Nyhan disease. Adv Exp Med Biol 204:197–215

Breese GR, Duncan GE, Napier TC, Bondy SC, Iorio LC, Mueller RA (1987) 6-hydroxydopamine treatments enhance behavioral responses to intracerebral microinjection of D_1- and D_2-dopamine agonists into nucleus accumbens and striatum without changing dopamine antagonist binding. J Pharmacol Exp Ther 240(1):167–176

Breese GR, Criswell HE, Duncan GE, Mueller RA (1989) Dopamine deficiency in self-injurious behavior. Psychopharmacol Bull 25(3):353–357. Review. Erratum in: Psychopharmacol Bull 1990; 26(3):296

Breese GR, Criswell HE, Duncan GE, Mueller RA (1990a) A dopamine deficiency model of Lesch-Nyhan disease—the neonatal-6-OHDA-lesioned rat. Brain Res Bull 25(3):477–484. Review

Breese GR, Criswell HE, Mueller RA (1990b) Evidence that lack of brain dopamine during development can increase the susceptibility for aggression and self-injurious behavior by influencing D_1-dopamine receptor function. Prog Neuropsychopharmacol Biol Psychiatry 14 (Suppl):S65–S80

Breese GR, Criswell HE, Johnson KB, O'Callaghan JP, Duncan GE, Jensen KF, Simson PE, Mueller RA (1994) Neonatal destruction of dopaminergic neurons. Neurotoxicology 15(1):149–159

Breese GR, Knapp DJ, Criswell HE, Moy SS, Papadeas ST, Blake BL (2005) The neonate-6-hydroxydopamine-lesioned rat: a model for clinical neuroscience and neurobiological principles. Brain Res Brain Res Rev 48(1):57–73

Broberg BV, Dias R, Glenthøj BY, Olsen CK (2008) Evaluation of a neurodevelopmental model of schizophrenia—early postnatal PCP treatment in attentional set-shifting. Behav Brain Res 190(1):160–163

Broberg BV, Oranje B, Glenthøj BY, Fejgin K, Plath N, Bastlund JF (2010) Assessment of auditory sensory processing in a neurodevelopmental animal model of schizophrenia—gating of auditory-evoked potentials and prepulse inhibition. Behav Brain Res 213(2):142–147

Broberg BV, Madsen KH, Plath N, Olsen CK, Glenthøj BY, Paulson OB, Bjelke B, Søgaard LV (2013) A schizophrenia rat model induced by early postnatal phencyclidine treatment and characterized by magnetic resonance imaging. Behav Brain Res 250:1–8

Brown RW, Peterson DJ (2016) Applications of the neonatal quinpirole model to psychosis and convergence upon the dopamine D_2 receptor. In: Kostrzewa RM, Archer T (eds) Neurotoxin modeling of brain disorders-lifelong outcomes in behavioral teratology. Springer, New York

Brown GC, Vilalta A (2015) How microglia kill neurons. Brain Res 1628(Pt B):288–297

Brown RW, Gass JT, Kostrzewa RM (2002) Ontogenetic quinpirole treatments produce spatial memory deficits and enhance skilled reaching in adult rats. Pharmacol Biochem Behav 72(3):591–600

Brown RW, Flanigan TJ, Thompson KN, Thacker SK, Schaefer TL, Williams MT (2004a) Neonatal quinpirole treatment impairs Morris water task performance in early postweanling rats: relationship to increases in corticosterone and decreases in neurotrophic factors. Biol Psychiatry 56(3):161–168

Brown RW, Thompson KD, Thompson KN, Ward JJ, Thacker SK, Williams MT, Kostrzewa RM (2004b) Adulthood nicotine treatment alleviates behavioural impairments in rats neonatally treated with quinpirole: possible roles of acetylcholine function and neurotrophic factor expression. Eur J Neurosci 19(6):1634–1642

Brown RW, Thompson KN, Click IA, Best RA, Thacker SK, Perna MK (2005) The effects of eticlopride on Morris water task performance in male and female rats neonatally treated with quinpirole. Psychopharmacology 180(2):234–240

Brown RW, Perna MK, Schaefer TL, Williams MT (2006) The effects of adulthood nicotine treatment on D_2-mediated behavior and neurotrophins of rats neonatally treated with quinpirole. Synapse 59(5):253–259

Brown RW, Perna MK, Maple AM, Wilson TD, Miller BE (2008) Adulthood olanzapine treatment fails to alleviate decreases of ChAT and BDNF RNA expression in rats quinpirole-primed as neonates. Brain Res 1200:66–77

Brown RW, Maple AM, Perna MK, Sheppard AB, Cope ZA, Kostrzewa RM (2012) Schizophrenia and substance abuse comorbidity: nicotine addiction and theneonatal quinpirole model. Dev Neurosci 34(2–3):140–151

Brundin L, Erhardt S, Bryleva EY, Achtyes ED, Postolache TT (2015) The role of inflammation in suicidal behaviour. Acta Psychiatr Scand 132(3):192–203

Brus R, Szkilnik R, Nowak P, Kasperska A, Oświęcimska J, Kostrzewa RM, Shani J (1998a) Locomotor sensitization of dopamine receptors by their agonists quinpirole and SKF-38393, during maturation and aging in rats. Pharmacol Rev Commun 10:25–30

Brus R, Szkilnik R, Nowak P, Kostrzewa RM, Shani J (1998b) Sensitivity of central dopamine receptors in rats to quinpirole and SKF-38393, administered at their early stages of ontogenicity, evaluated by learning and memorizing a conditioned avoidance reflex. Pharmacol Rev Commun 10:31–36

Brus R, Nowak P, Szkilnik R, Mikolajun U, Kostrzewa RM (2004) Serotoninergics attenuate hyperlocomotor activity in rats. Potential new therapeutic strategy for hyperactivity. Neurotox Res 6(4):317–325

Budni J, Bellettini-Santos T, Mina F, Garcez ML, Zugno AI (2015) The involvement of BDNF, NGF and GDNF in aging and Alzheimer's disease. Aging Dis 6(5):331–341

Byrnes KR, Stoica B, Loane DJ, Riccio A, Davis MI, Faden A (2009) Metabotropic glutamate receptor 5 activation inhibits microglial associated inflammation and neurotoxicity. Glia 57(5):550–560

Cadet JL (2014) Epigenetics of stress, addiction, and resilience: Therapeutic implications. Mol Neurobiol. 53(1):545–560

Cadet JL, Jayanthi S (2013) Epigenetics of methamphetamine-induced changes in glutamate function. Neuropsychopharmacology 38(1):248–249

Casey DE (1987) Tardive dyskinesia. In: Meltzer HY (ed) Psychopharmacology: the third generation of progress. Raven Press, New York, pp 1411–1419

Castañé A, Santana N, Artigas F (2015) PCP-based mice models of schizophrenia: differential behavioral, neurochemical and cellular effects of acute and subchronic treatments. Psychopharmacology 232(21–22):4085–4097

Chen J, Wang Z, Li M (2011) Multiple 'hits' during postnatal and early adulthood periods disrupt the normal development of sensorimotor gating ability in rats. J Psychopharmacol 25(3):379–392

Christensen J, Grønborg TK, Sørensen MJ, Schendel D, Parner ET, Pedersen LH, Vestergaard M (2013) Prenatal valproate exposure and risk of autism spectrum disorders and childhood autism. JAMA 309(16):1696–1703

Christianson AL, Chester N, Kromberg JG (1994) Fetal valproate syndrome: clinical and neuro-developmental features in two sibling pairs. Dev Med Child Neurol 36:361–369

Chung WS, Welsh CA, Barres BA, Stevens B (2015) Do glia drive synaptic and cognitive impairment in disease? Nat Neurosci 18(11):1539–1545

Clark MB, King JC, Kostrzewa RM (1979) Loss of nerve cell bodies in caudal locus coeruleus following treatment of neonates with 6-hydroxydopa. Neurosci Lett 13(3):331–336

Cohen G, Heikkila RE (1974) The generation of hydrogen peroxide, superoxide radical, and hydroxyl radical by 6-hydroxydopamine, dialuric acid, and related cytotoxic agents. J Biol Chem 249(8):2447–2452

Cooper JD, Lindholm D, Sofroniew MV (1994) Reduced transport of ^{125}I-NGF by cholinergic neurons and downregulated TrkA expression in the medial septum of aged rats. Neuroscience 62:625–629

Cope ZA, Huggins KN, Sheppard AB, Noel DM, Roane DS, Brown RW (2010) Neonatal quinpirole treatment enhances locomotor activation and dopamine release in the nucleus accumbens core in response to amphetamine treatment in adulthood. Synapse 64(4):289–300

Cordova FM, Aguiar AS Jr, Peres TV, Lopes MW, Gonçalves FM, Remor AP, Lopes SC, Pilati C, Latini AS, Prediger RD, Erikson KM, Aschner M (2012) Leal RB (2012) In vivo manganese exposure modulates Erk, Akt and Darpp-32 in the striatum of developing rats, and impairs their motor function. PLoS One 7(3):e33057

Criswell H, Mueller RA, Breese GR (1989) Priming of D_1-dopamine receptor responses: long-lasting behavioral supersensitivity to a D_1-dopamine agonist following repeated administration to neonatal 6-OHDA-lesioned rats. J Neurosci 9(1):125–133

Dabrowska J, Nowak P, Brus R (2007) Desensitization of 5-HT(1A) autoreceptors induced by neonatal DSP-4 treatment. Eur Neuropsychopharmacol 17:129–137

Dal-Pizzol F, Klamt F, Frota ML Jr, Andrades ME, Caregnato FF, Vianna MM, Schröder N, Quevedo J, Izquierdo I, Archer T, Moreira JC (2001) Neonatal iron exposure induces oxidative stress in adult Wistar rat. Brain Res Dev Brain Res 130(1):109–114

De Bartolo P, Gelfo F, Mandolesi L, Foti F, Cutuli D, Petrosini L (2009) Effects of chronic donepezil treatment and cholinergic deafferentation on parietal pyramidal neuron morphology. J Alzheimers Dis 17(1):177–191

De Bartolo P, Cutuli D, Ricceri L, Gelfo F, Foti F, Laricchiuta D, Scattoni ML, Calamandrei G, Petrosini L (2010) Does age matter? Behavioral and neuro-anatomical effects of neonatal and adult basal forebrain cholinergic lesions. J Alzheimers Dis 20(1):207–227

de la Torre R, Farré M, Ortuño J, Mas M, Brenneisen R, Roset PN, Segura J, Camí J (2000) Non-linear pharmacokinetics of MDMA ('ecstasy') in humans. Br J Clin Pharmacol 49(2):104–109

De-Carolis C, Boyd GA, Mancinelli L, Pagano S, Eramo S (2015) Methamphetamine abuse and "meth mouth" in Europe. Med Oral Patol Oral Cir Bucal 20(2):e205–e210

Delany FM, Byrne ML, Whittle S, Simmons JG, Olsson C, Mundy LK, Patton GC, Allen NB (2015) Depression, immune function, and early adrenarche in children. Psychoneuroendocrinology 63:228–234

Desplats P, Dumaop W, Cronin P, Gianella S, Woods S, Letendre S, Smith D, Masliah E, Grant I (2014) Epigenetic alterations in the brain associated with HIV-1 infection andmethamphetamine dependence. PLoS One 9(7):e102555

Do HT, Bruelle C, Pham DD, Jauhiainen M, Eriksson O, Korhonen LT, Lindholm D (2015) Nerve growth factor (NGF) and pro-NGF increase low-density lipoprotein (LDL) receptors in neuronal cells partly by different mechanisms: role of LDL in neurite outgrowth. J Neurochem 2015 Oct 20. doi:10.1111/jnc.13397. [Epub ahead of print]

Doebel S, Zelazo PD (2013) Bottom-up and top-down dynamics in young children's executive function: Labels aid 3-year-olds' performance on the Dimensional Change Card Sort. Cogn Dev 28(3):222–232

Doucette TA, Tasker RA (2016) Perinatal domoic acid as a neuroteratogen. In: Kostrzewa RM, Archer T (eds) Neurotoxin modeling of brain disorders-lifelong outcomes in behavioral teratology. Springer, New York

Doucette TA, Strain SM, Allen GV, Ryan CL, Tasker RAR (2000) Comparative behavioural toxicity of domoic acid and kainic acid in neonatal rats. Neurotoxicol Teratol 22:863–869

Doucette TA, Bernard PB, Husum H, Perry MA, Ryan CL, Tasker RA (2004) Low doses of domoic acid during postnatal development produce permanent changes in rat behaviour and hippocampal morphology. Neurotox Res 6(7,8):555–563

Fialová M, Šírová J, Bubeníková-Valešová V, Šlamberová R (2015) The effect of prenatal methamphetamine exposure on recognition memory in adult rats. Prague Med Rep 116(1):31–39

Fornai F, Bassi L, Bonaccorsi I, Giorgi F, Corsini GU (1997) Noradrenaline loss selectivity exacerbates nigrostriatal toxicity in different species of rodents. Funct Neurol 12(3–4):193–198

Fredriksson A, Schröder N, Eriksson P, Izquierdo I, Archer T (1999) Neonatal iron exposure induces neurobehavioural dysfunctions in adult mice. Toxicol Appl Pharmacol 159(1):25–30

Fredriksson A, Schröder N, Eriksson P, Izquierdo I, Archer T (2000) Maze learning and motor activity deficits in adult mice induced by iron exposure during a critical postnatal period. Brain Res Dev Brain Res 119(1):65–74

Fredriksson A, Stigsdotter IM, Hurtig A, Ewalds-Kvist B, Archer T (2011) Running wheel activity restores MPTP-induced functional deficits. J Neural Transm 118(3):407–420

Fu R, Sutcliffe D, Zhao H, Huang X, Schretlen DJ, Benkovic S, Jinnah HA (2015) Clinical severity in Lesch-Nyhan disease: the role of residual enzyme and compensatory pathways. Mol Genet Metab 114(1):55–61

Fukushima S, Furube E, Itoh M, Nakashima T, Miyata S (2015) Robust increase in microglia proliferation in the fornix of hippocampal axonal pathway after single LPS stimulation. J Neuroimmunol 285:31–40

Furukawa S (2015) Basic research on neurotrophic factors and its application to medical uses. Yakugaku Zasshi 135(11):1213–1226

Futamura T, Kakita A, Tohmi M, Sotoyama H, Takahashi H, Nawa H (2003) Neonatal perturbation of neurotrophic signaling results in abnormal sensorimotor gating and social interaction in adults: implication for epidermal growth factor in cognitive development. Mol Psychiatry 8(1):19–29

Galbally M, Roberts M, Buist A; Perinatal Psychotropic Review Group (2010) Mood stabilizers in pregnancy: a systematic review. Aust NZJ Psychiatry 44(11):967–977

Gao Y, Liu L, Li Q, Wang Y (2015) Differential alterations in the morphology and electrophysiology of layer II pyramidal cells in the primary visual cortex of a mouse model prenatally exposed to LPS. Neurosci Lett 591:138–143

Gill DA, Ramsay R, Tasker RA (2010) Selective reductions in subpopulations of GABAergic neurons in a developmental rat model of epilepsy. Brain Res 1331:114–123

Godino A, Jayanthi S, Cadet JL (2015) Epigenetic landscape of amphetamine and methamphetamine addiction in rodents. Epigenetics 10(7):574–580

Gong L, Kostrzewa RM, Fuller RW, Perry KW (1992) Supersensitization of the oral response to SKF 38393 in neonatal 6-OHDA-lesioned rats is mediated through a serotonin system. J Pharmacol Exp Ther 261:1000–1007

Gong L, Kostrzewa RM, Brus R, Fuller RW, Perry KW (1993a) Ontogenetic SKF 38393 treatments sensitize dopamine D_1 receptors in neonatal 6-OHDA-lesioned rats. Brain Res Dev Brain Res 76(1):59–65

Gong L, Kostrzewa RM, Perry KW, Fuller RW (1993b) Dose-related effects of a neonatal 6-OHDA lesion on SKF 38393- and m-chlorophenylpiperazine-induced oral activity responses of rats. Brain Res Dev Brain Res 76(2):233–238

Gong L, Kostrzewa RM, Li C (1994) Neonatal 6-OHDA and adult SKF 38393 treatments alter dopamine D_1 receptor mRNA levels: absence of other neurochemical associations with the enhanced behavioral responses of lesioned rats. J Neurochem 63:1282–1290

Göttle M, Prudente CN, Fu R, Sutcliffe D, Pang H, Cooper D, Veledar E, Glass JD (2014) Loss of dopamine phenotype among midbrain neurons in Lesch-Nyhan disease. Ann Neurol 76(1):95–107

Grados M, Sung HM, Kim S, Srivastava S (2014) Genetic findings in obsessive-compulsive disorder connect to brain-derived neutrophic factor and mammalian target of rapamycin pathways: implications for drug development. Drug Dev Res 75(6):372–383

Graham DL, Amos-Kroohs RM, Braun AA, Grace CE, Schaefer TL, Skelton MR, Williams MT, Vorhees CV (2013) Neonatal +-methamphetamine exposure in rats alters adult locomotor responses to dopamine D_1 and D_2 agonists and to a glutamate NMDA receptor antagonist, but not to serotonin agonists. Int J Neuropsychopharmacol 16(2):377–391

Grayson DR, Guidotti A (2015) Merging data from genetic and epigenetic approaches to better understand autistic spectrum disorder. Epigenomics. 2015 Nov 9. [Epub ahead of print] PMID: 26551091

Grayson B, Barnes SA, Markou A, Piercy C, Podda G, Neill JC (2016) Postnatal phencyclidine (PCP) as a neurodevelopmental model of schizophrenia pathophysiology and symptpmatology: a review. In: Kostrzewa RM, Archer T (eds) Neurotoxin modeling of brain disorders-lifelong outcomes in behavioral teratology. Springer, New York

Groves NJ, McGrath JJ, Burne TH (2014) Vitamin D as a neurosteroid affecting the developing and adult brain. Annu Rev Nutr 34:117–141

Hamdi A, Kostrzewa RM (1991) Ontogenic homologous supersensitization of dopamine D_1 receptors. Eur J Pharmacol 203:115–120

Harro J (2015) Neuropsychiatric adverse effects of amphetamine and methamphetamine. Int Rev Neurobiol 120:179–204

Hasan W, Smith PG (2014) Decreased adrenoceptor stimulation in heart failure rats reduces NGF expression by cardiac parasympathetic neurons. Auton Neurosci 181:13–20

He N, Li F, Li Y, Guo L, Chen L, Huang X, Lui S, Gong Q (2015) Neuroanatomical deficits correlate with executive dysfunction in boys with attention deficit hyperactivity disorder. Neurosci Lett 600:45–49

Hertz L, Chen Y, Waagepetersen HS (2015) Effects of ketone bodies in Alzheimer's disease in relation to neural hypometabolism, β-amyloid toxicity and astrocyte function. J Neurochem 134(1):7–20

Hirsch MA, Iyer SS, Sanjak M (2015) Exercise-induced neuroplasticity in human Parkinson's disease: what is the evidence telling us? Parkinsonism Relat Disord 22(Suppl 1):S78–S81

Hoftman GD, Lewis DA (2011) Postnatal developmental trajectories of neural circuits in the primate prefrontal cortex: identifying sensitive periods for vulnerability to schizophrenia. Schizophr Bull 37(3):493–503

Hohsfield LA, Ehrlich D, Humpel C (2014) Intravenous infusion of nerve growth factor-secreting monocytes supports the survival of cholinergic neurons in the nucleus basalis of Meynert in hypercholesterolemia Brown-Norway rats. J Neurosci Res 92(3):298–306

Holmes GL, Sarkisian M, Ben-Ari Y, Chevassus-Au-Louis N (1999) Mossy fiber sprouting after recurrent seizures during early development in rats. J Comp Neurol 22:537–553

Hongwanishkul D, Happaney KR, Lee WS, Zelazo PD (2006) Assessment of hot and cool executive function in young children: age-related changes and individual differences. Dev Neuropsychol 28(2):617–644

Hrebíčková I, Malinová-Ševčíková M, Macúchová E, Nohejlová K, Šlamberová R (2014) Exposure to methamphetamine during first and second half of prenatal period and its consequences on cognition after long-term application in adulthood. Physiol Res 63(Suppl 4): S535–S545

Huang N-Y, Kostrzewa RM, Li C, Perry KW, Fuller RW (1997) Increased spontaneous oral dyskinesias persist in haloperidol-withdrawn rats neonatally lesioned with 6-hydroxydopamine: absence of an association with the B_{max} for [^{3}H]raclopride binding to neostriatal homogenates. J Pharmacol Exp Ther 280:268–276

Ibi D, González-Maeso J (2015) Epigenetic signaling in schizophrenia. Cell Signal 27(10):2131–2136

Iulita MF, Cuello AC (2014) Nerve growth factor metabolic dysfunction in Alzheimer's disease and Down syndrome. Trends Pharmacol Sci 35(7):338–348

Iwakura Y, Zheng Y, Sibilia M, Abe Y, Piao YS, Yokomaku D, Wang R, Ishizuka Y, Takei N, Nawa H (2011a) Qualitative and quantitative re-evaluation of epidermal growth factor-ErbB1 action on developing midbrain dopaminergic neurons in vivo and in vitro: target-derived neurotrophic signaling (Part 1). J Neurochem 118(1):45–56

Iwakura Y, Wang R, Abe Y, Piao YS, Shishido Y, Higashiyama S, Takei N, Nawa H (2011b) Dopamine-dependent ectodomain shedding and release of epidermal growth factor in developing striatum: target-derived neurotrophic signaling (Part 2). J Neurochem 118(1):57–68

Jablonski SA, Williams MT, Vorhees CV (2016) Neurobehavioral effects from developmental methamphetamine exposure. In: Kostrzewa RM, Archer T (eds) Neurotoxin modeling of brain disorders-lifelong outcomes in behavioral teratology. Springer, New York

Jacobowitz D, Kostrzewa R (1971) Selective action of 6-hydroxydopa on noradrenergic terminals: mapping of preterminal axons of the brain. Life Sci I 10(23):1329–1342

Jacobsen PE, Henriksen TB, Haubek D, Østergaard JR (2014) Prenatal exposure to antiepileptic drugs and dental agenesis. PLoS One 9(1):e84420

Jain P, Wadhwa PK, Jadhav HR (2015) Reactive astrogliosis: role in Alzheimer's disease. CNS Neurol Disord Drug Targets 14(7):872–879

Jeste DV, Caligiuri MP (1993) Tardive dyskinesia. Schizophr Bull 19:303–315

Jinnah HA, Jones MD, Wojcik BE, Rothstein JD, Hess EJ, Friedmann T, Breese GR (1999) Influence of age and strain on striatal dopamine loss in a genetic mouse model of Lesch-Nyhan disease. J Neurochem 72(1):225–229

Jo WK, Zhang Y, Emrich HM, Dietrich DE (2015) Glia in the cytokine-mediated onset of depression: fine tuning the immune response. Front Cell Neurosci 9:268

Jonsson G, Hallman H, Ponzio F, Ross S (1981) DSP4 (N-(2-chloroethyl)-N-ethyl-2-bromobenzylamine)—a useful denervation tool for central and peripheral noradrenaline neurons. Eur J Pharmacol 72:173–188

Jonsson G, Hallman H, Sundström E (1982) Effects of the noradrenaline neurotoxin DSP4 on the postnatal development of central noradrenaline neurons in the rat. Neuroscience 7(11):2895–2907

Kaalund SS, Riise J, Broberg BV, Fabricius K, Karlsen AS, Secher T, Plath N, Pakkenberg B (2013) Differential expression of parvalbumin in neonatal phencyclidine-treated rats and socially isolated rats. J Neurochem 124(4):548–557

Kato T, Abe Y, Sotoyama H, Kakita A, Kominami R, Hirokawa S, Ozaki M, Takahashi H, Nawa H (2011) Transient exposure of neonatal mice to neuregulin-1 results in hyperdopaminergic states in adulthood: implication in neurodevelopmental hypothesis for schizophrenia. Mol Psychiatry 16(3):307–320

Kehoe P, Callahan M, Daigle A, Mallinson K, Brudzynski S (2001) The effect of cholinergic stimulation on rat pup ultrasonic vocalizations. Dev Psychobiol 38(2):92–100

Kelly JR, Kennedy PJ, Cryan JF, Dinan TG, Clarke G, Hyland NP (2015) Breaking down the barriers: the gut microbiome, intestinal permeability and stress-related psychiatric disorders. Front Cell Neurosci 9:392

Kim KC, Kim P, Go HS, Choi CS, Yang SI, Cheong JH, Shin CY, Ko KH (2011) The critical period of valproate exposure to induce autistic symptoms in Sprague-Dawley rats. Toxicol Lett 201(2):137–142

Kim JW, Seung H, Kwon KJ, Ko MJ, Lee EJ, Oh HA, Choi CS, Kim KC, Gonzales EL, You JS, Choi DH, Lee J, Han SH, Yang SM, Cheong JH, Shin CY, Bahn GH (2014) Subchronic treatment of donepezil rescues impaired social, hyperactive, and stereotypic behaviour in valproic acid-induced animal model of autism. PLoS One 9(8):e104927

Kirsten TB, Lippi LL, Bevilacqua E, Bernardi MM (2013) LPS exposure increases maternal corticosterone levels, causes placental injury and increases Il-1B levels in adult rat offspring: relevance to autism. PLoS One 8(12):e82244

Kjaerby C, Bundgaard C, Fejgin K, Kristiansen U, Dalby NO (2013) Repeated potentiation of the metabotropic glutamate receptor 5 and the alpha 7 nicotinic acetylcholine receptor modulates behavioural and GABAergic deficits induced by early postnatal phencyclidine (PCP) treatment. Neuropharmacology 72:157–168

Kjaerby C, Broberg BV, Kristiansen U, Dalby NO (2014) Impaired GABAergic inhibition in the prefrontal cortex of early postnatalphencyclidine (PCP)-treated rats. Cereb Cortex 24(9):2522–2532

Knapp DJ, Breese GR (2016) Perinatal 6-hydroxydopamine to produce a rodent model of Lesch-Nyhan disease. In: Kostrzewa RM, Archer T (eds) Neurotoxin modeling of brain disorders-lifelong outcomes in behavioral teratology. Springer, New York

Kokoshka JM, Metzger RR, Wilkins DG, Gibb JW, Hanson GR, Fleckenstein AE (1998) Methamphetamine treatment rapidly inhibits serotonin, but not glutamate, transporters in rat brain. Brain Res 799(1):78–83

Kostrzewa RM (1988) Reorganization of noradrenergic neuronal systems following neonatal chemical and surgical injury. Prog Brain Res 73:405–423. Review

Kostrzewa RM (1995) Dopamine receptor supersensitivity. Neurosci Biobehav Rev 19:1–17

Kostrzewa RM (2016) Perinatal effects of 6-hydroxydopa, a noradrenergic neurotoxin and AMPA receptor excitotoxin. In: Kostrzewa RM, Archer T (eds) Neurotoxin modeling of brain disorders-lifelong outcomes in behavioral teratology. Springer, New York

Kostrzewa RM, Brus R (1991) Ontogenic homologous supersensitization of quinpirole-induced yawning in rats. Pharmacol Biochem Behav 39:517–519

Kostrzewa RM, Brus R (2016) Lifelong rodent model of tardive dyskinesia-persistence after antipsychotic drug withdrawal. In: Kostrzewa RM, Archer T (eds) Neurotoxin modeling of brain disorders-lifelong outcomes in behavioral teratology. Springer, New York

Kostrzewa RM, Garey RE (1976) Effects of 6-hydroxydopa on noradrenergic neurons in developing rat brain. J Pharmacol Exp Ther 197:105–118

Kostrzewa RM, Garey RE (1977) Sprouting of noradrenergic terminals in rat cerebellum following neonatal treatment with 6-hydroxydopa. Brain Res 124:385–391

Kostrzewa RM, Gong L (1991) Supersensitized D_1 receptors mediate enhanced oral activity after neonatal 6-OHDA. Pharmacol Biochem Behav 39(3):677–682

Kostrzewa RM, Harper JW (1974) Effect of 6-hydroxydopa on catecholamine-containing neurons in brains of newborn rats. Brain Res 69(1):174–181

Kostrzewa R, Jacobowitz D (1972) The effect of 6-hydroxydopa on peripheral adrenergic neurons. J Pharmacol Exp Ther 183(2):284–297

Kostrzewa R, Jacobwitz D (1973) Acute effects of 6-hydroxydopa on central monoaminergic neurons. Eur J Pharmacol 21(1):70–80

Kostrzewa RM, Kostrzewa FP (2012) Neonatal 6-hydroxydopamine lesioning enhances quinpirole-induced vertical jumping in rats that were quinpirole primed during postnatal ontogeny. Neurotox Res 21(2):231–235

Kostrzewa RM, Brus R, Kalbfleisch J (1991) Ontogenetic homologous sensitization to the antinociceptive action of quinpirole in rats. Eur J Pharmacol 209:157–161

Kostrzewa RM, Gong L, Brus R (1992) Serotonin (5-HT) systems mediate dopamine (DA) receptor supersensitivity. Acta Neurobiol Exp 53:31–41

Kostrzewa RM, Brus R, Rykaczewska M, Plech A (1993a) Low dose quinpirole ontogenically sensitizes to quinpirole-induced yawning in rats. Pharmacol Biochem Behav 44:487–489

Kostrzewa RM, Guo J, Kostrzewa FP (1993b) Ontogenetic quinpirole treatments induce vertical jumping activity in rats. Eur J Pharmacol 239:183–187

Kostrzewa RM, Brus R, Kalbflesich JH, Perry KW, Fuller RW (1994) Proposed animal model of attention deficit hyperactivity disorder. Brain Res Bull 34:161–167

Kostrzewa RM, Reader TA, Descarries L (1998) Serotonin neural adaptations to ontogenetic loss of dopamine neurons in rat brain. J Neurochem 70:889–898

Kostrzewa RM, Kostrzewa JP, Brus R (2000) Dopaminergic denervation enhances susceptibility to hydroxyl radicals in rat neostriatum. Amino Acids 19:183–199

Kostrzewa RM, Kostrzewa JP, Brus R (2002) Neuroprotective and neurotoxic roles of levodopa (L-DOPA) in neurodegenerative disorders relating to Parkinson's disease. Amino Acids 23:57–63

Kostrzewa RM, Kostrzewa JP, Brus R (2003) Dopamine receptor supersensitivity: an outcome and index of neurotoxicity. Neurotox Res 5:111–118

Kostrzewa RM, Nowak P, Kostrzewa JP, Kostrzewa RA, Brus R (2004) Dopamine D_2 agonist priming in intact and dopamine-lesioned rats. Neurotox Res 6:457–462

Kostrzewa RM, Nowak P, Kostrzewa JP, Kostrzewa RA, Brus R (2005) Peculiarities of L-DOPA treatment of Parkinson's disease. Amino Acids 28:157–164

Kostrzewa RM, Kostrzewa JP, Brus R, Kostrzewa RA, Nowak P (2006) Proposed animal model of severe Parkinson's disease: neonatal 6-hydroxydopamine lesion of dopaminergic innervation of striatum. J Neural Transm Suppl 70:277–279

Kostrzewa RM, Huang N-Y, Kostrzewa JP, Nowak P, Brus R (2007) Modeling tardive dyskinesia: predictive 5-HT_{2C} receptor antagonist treatment. Neurotox Res 11(1):41–50

Kostrzewa RM, Kostrzewa JP, Kostrzewa RA, Nowak P, Brus R (2008) Pharmacological models of ADHD. J Neural Transm 115(2):287–298

Kostrzewa RM, Kostrzewa JP, Kostrzewa RA, Kostrzewa FP, Brus R, Nowak P (2011) Stereotypic progressions in psychotic behavior. Neurotox Res 19:243–252

Kostrzewa JP, Kostrzewa RA, Kostrzewa RM, Brus R, Nowak P (2016a) Perinatal 6-hydroxydopamine modeling of ADHD. In: Kostrzewa RM, Archer T (eds) Neurotoxin modeling of brain disorders-lifelong outcomes in behavioral teratology. Springer, New York

Kostrzewa JP, Kostrzewa RA, Kostrzewa RM, Brus R, Nowak P (2016b) Perinatal 6-hydroxydopamine to produce a lifelong model of severe Parkinson's disease. In: Kostrzewa RM, Archer T (eds) Neurotoxin modeling of brain disorders-lifelong outcomes in behavioral teratology. Springer, New York

Kostrzewa RM, Nowak P, Brus R, Brown RW (2016c) Perinatal treatments with the dopamine D_2-receptor agonist quinpirole produces permanent D_2-receptor supersensitization: a model of schizophrenia. Neurochem Res 2015 Nov 7. [Epub ahead of print] PMID: 26547196 In press

Krasnova IN, Cadet JL (2009) Methamphetamine toxicity and messengers of death. Brain Res Rev 60:379–407

Kulaga S, Sheehy O, Zargarzadeh AH, Moussally K, Bérard A (2011) Antiepileptic drug use during pregnancy: perinatal outcomes. Seizure 20(9):667–672

Lahiri DK, Maloney B, Zawia NH (2009) The LEARn model: an epigenetic explanation for idiopathic neurobiological diseases. Mol Psychiatry 14(11):992–1003

Laviola G, Adriani W, Rea M, Aloe L, Alleva E (2004) Social withdrawal, neophobia, and stereotyped behavior in developing rats exposed to neonatal asphyxia. Psychopharmacology 175:196–205

Le Magueresse C, Monyer H (2013) GABAergic interneurons shape the functional maturation of the cortex. Neuron 77(3):388–405

Leanza G, Nilsson OG, Wiley RG, Björklund A (1995) Selective lesioning of the basal forebrain cholinergic system by intraventricular 192 IgG-saporin: behavioural, biochemical and stereological studies in the rat. Eur J Neurosci 7(2):329–343

Leanza G, Nilsson OG, Nikkhah G, Wiley RG, Björklund A (1996) Effects of neonatal lesions of the basal forebrain cholinergic system by 192 immunoglobulin G-saporin: biochemical, behavioural and morphological characterization. Neuroscience 74(1):119–141

Lemaire V, Koehl M, Le Moal M, Abrous DN (2000) Prenatal stress produces learning deficits associated with an inhibition of neurogenesis in the hippocampus. Proc Natl Acad Sci USA 97(20):11032–11037

Levy SE, Mandell DS, Schultz RT (2009) Autism. Lancet 374(9701):1627–1638

Lian H, Zheng H (2015) Signaling pathways regulating neuron glia interaction and their implications in Alzheimer's disease. J Neurochem. 6 Nov 2015. doi:10.1111/jnc.13424. [Epub ahead of print] Review

Lloyd KA (2013) A scientific review: mechanisms of valproate-mediated teratogenesis. Bioscience Horizons 6:hzt003

Lyons WE, Fritschy JM, Grzanna R (1989) The noradrenergic neurotoxin DSP-4 eliminates the coeruleospinal projection but spares projections of the A5 and A7 groups to the ventral horn of the rat spinal cord. J Neurosci 9:1481–1489

Maccari S, Darnaudery M, Morley-Fletcher S, Zuena AR, Cinque C, Van Reeth O (2003) Prenatal stress and long-term consequences: implications of glucocorticoid hormones. Neurosci Biobehav Rev 27:119–127

Malinová-Ševčíková M, Hrebíčková I, Macúchová E, Nová E, Pometlová M, Šlamberová R (2014) Differences in maternal behavior and development of their pups depend on the time of methamphetamine exposure during gestation period. Physiol Res 63(Suppl 4):S559–S572

Maple AM, Perna MK, Parlaman JP, Stanwood GD, Brown RW (2007) Ontogenetic quinpirole treatment produces long-lasting decreases in the expression of Rgs9, but increases Rgs17 in the striatum, nucleus accumbens and frontal cortex. Eur J Neurosci 26(9):2532–2538

Maple AM, Smith KJ, Perna MK, Brown RW (2015) Neonatal quinpirole treatment produces prepulse inhibition deficits in adult male and female rats. Pharmacol Biochem Behav 137:93–100

Marco EM, Velarde E, Llorente R, Laviola G (2016) Disrupted circadian rhythm as a common player in developmental models of neuropsychiatric disorders. In: Kostrzewa RM, Archer T (eds) Neurotoxin modeling of brain disorders-lifelong outcomes in behavioral teratology. Springer, New York

McDonnell-Dowling K, Kelly JP (2015) The consequences of prenatal and/or postnatal methamphetamine exposure on neonatal development and behaviour in rat offspring. Int J Dev Neurosci 47(Pt B):147–156

Meier TB, Drevets WC, Wurfel BE, Ford BN, Morris HM, Victor TA, Bodurka J, Kent Teague T, Dantzer R, Savitz J (2015) Relationship between neurotoxic kynurenine metabolites and reductions in right medial prefrontal cortical thickness in major depressive disorder. Brain Behav Immun pii: S0889-1591(15) 30048-30049

Miller FD, Kaplan DR (2001) Neurotrophin signalling pathways regulating neuronal apoptosis. Cell Mol Life Sci 58:1045–1053

Mizuno M, Kawamura H, Takei N, Nawa H (2008) The anthraquinone derivative emodin ameliorates neurobehavioral deficits of a rodent model for schizophrenia. J Neural Transm 115(3):521–530

Mizuno M, Kawamura H, Ishizuka Y, Sotoyama H, Nawa H (2010) The anthraquinone derivative emodin attenuates methamphetamine-induced hyperlocomotion and startle response in rats. Pharmacol Biochem Behav 97(2):392–398

Mohl B, Ofen N, Jones LL, Robin AL, Rosenberg DR, Diwadkar VA, Casey JE, Stanley JA (2015) Neural dysfunction in ADHD with reading disability during a word rhyming continuous performance task. Brain Cogn 99:1–7

Monden Y, Dan I, Nagashima M, Dan H, Uga M, Ikeda T, Tsuzuki D, Kyutoku Y, Gunji Y, Hirano D, Taniguchi T, Shimoizumi H, Watanabe E, Yamagata T (2015) Individual classification of ADHD children by right prefrontal hemodynamic responses during a go/no-go task as assessed by fNIRS. Neuroimage Clin 9:1–12

Mossakowski AA, Pohlan J, Bremer D, Lindquist R, Millward JM, Bock M, Pollok K, Mothes R, Viohl L, Radbruch M, Gerhard J, Bellmann-Strobl J, Behrens J, Infante-Duarte C, Mähler A, Boschmann M, Rinnenthal JL, Füchtemeier M, Herz J, Pache FC, Bardua M, Priller J, Hauser AE, Paul F, Niesner R, Radbruch H (2015) Tracking CNS and systemic sources of oxidative stress during the course of chronic neuroinflammation. Acta Neuropathol 130(6):799–814

Mouton M, Harvey BH, Cockeran M, Brink CB (2015) The long-term effects of methamphetamine exposure during pre-adolescence on depressive-like behaviour in a genetic animal model of depression. Metab Brain Dis 18 Nov 2015 [Epub ahead of print]

Müller N, Weidinger E, Leitner B, Schwarz MJ (2015) The role of inflammation in schizophrenia. Front Neurosci 9:372

Myint AM (2013) Inflammation, neurotoxins and psychiatric disorders. Mod Trends Pharmacopsychiatri 28:61–74

Nagano T, Mizuno M, Morita K, Nawa H (2016) Pathological implications of oxidative stress in patients and animal models with schizophrenia: the role of epidermal growth factor receptor signaling. In: Kostrzewa RM, Archer T (eds) Neurotoxin modeling of brain disorders-lifelong outcomes in behavioral teratology. Springer, New York

Nakatani-Pawlak A, Yamaguchi K, Tatsumi Y, Mizoguchi H, Yoneda Y (2009) Neonatal phencyclidine treatment in mice induces behavioral, histological and neurochemical abnormalities in adulthood. Biol Pharm Bull 32(9):1576–1583

Narita N, Kato M, Tazoe M, Miyazaki K, Narita M, Okado N (2002) Increased monoamine concentration in the brain and blood of fetal thalidomide- and valproic acid-exposed rat: putative animal models for autism. Pediatr Res 52(4):576–579

Neill JC, Harte MK, Haddad PM, Lydall ES, Dwyer DM (2014) Acute and chronic effects of NMDA receptor antagonists in rodents, relevance to negative symptoms of schizophrenia: a translational link to humans. Eur Neuropsychopharmacol 24(5):822–835

Niewiadomska G, Baksalerska-Pazera M, Riedel G (2009) The septo-hippocampal system, learning and recovery of function. Prog Neuropsychopharmacol Biol Psychiatry 33:791–805

Nowak P (2016) Selective lifelong destruction of brain monoaminergic nerves through perinatal DSP-4 treatment. In: Kostrzewa RM, Archer T (eds) Neurotoxin modeling of brain disorders-lifelong outcomes in behavioral teratology. Springer, New York

Nowak P, Brus R, Kostrzewa RM (2001) Amphetamine-induced enhancement of neostriatal in vivo microdialysate dopamine content in rats, quinpirole-primed as neonates. Pol J Pharmacol 53:319–329

Nowak P, Kostrzewa RM, Kwięcinski A, Bortel A, Łabus L, Brus R (2005) Neurotoxic action of 6-hydroxydopamine on the nigrostriatal dopaminergic pathway in rats sensitized with D-amphetamine. J Physiol Pharmacol 56(2):325–333

Nowak P, Bortel A, Dąbrowska J, Oświęcimska J, Drosik M, Kwieciński A, Opara J, Kostrzewa RM, Brus R (2007) Amphetamine and mCPP effects on dopamine and serotonin striatal in vivo microdialysates in an animal model of hyperactivity. Neurotox Res 11(2):131–144

Nowak P, Kostrzewa RA, Skaba D, Kostrzewa RM (2010) Acute L-DOPA effect on hydroxyl radical- and DOPAC-levels in striatal microdialysates of Parkinsonian rats. Neurotox Res 17 (3):299–304

Ong HH, Creveling CR, Daly JW (1969) The synthesis of 2,4,5-trihydroxyphenylalanine (6-hydroxydopa). A centrally active norepinephrine-depleting agent. J Med Chem 12(3):458–462

Ornoy A, Weinstein-Fudim L, Ergaz Z (2015) Prenatal factors associated with autism spectrum disorder (ASD). Reprod Toxicol 56:155–169

Oswiecimska J, Brus R, Szkilnik R, Nowak P, Kostrzewa RM (2000) 7-OH-DPAT, unlike quinpirole, does not prime a yawning response in rats. Pharmacol Biochem Behav 67(1):11–15

Papadeas ST, Breese GR (2014) 6-Hydroxydopamine lesioning of dopamine neurons in neonatal and adult rats induces age-dependent consequences. In: Kostrzewa RM (ed) Handbook of neurotoxicity. Springer, New York, pp 133–198. ISBN 978-1-4614-5835-7 (print); ISBN 978-1-4614-5836-4 (eBook); ISBN 978-1-4614-7458-6 (print and electronic bundle). doi:10.1007/978-1-4614-5836-4_59

Pappas BA, Davidson CM, Fortin T, Nallathamby S, Park GA, Mohr E, Wiley RG (1996) 192 IgG-saporin lesion of basal forebrain cholinergic neurons in neonatal rats. Brain Res Dev Brain Res 96(1–2):52–61

Pappas BA, Payne KB, Fortin T, Sherren N (2005) Neonatal lesion of forebrain cholinergic neurons: further characterization of behavioral effects and permanency. Neuroscience 133(2):485–492

Pariante CM (2015) Neuroscience, mental health and the immune system: overcoming the brain-mind-body trichotomy. Epidemiol Psychiatr Sci 27:1–5

Paterak J, Stefański R (2014) 5,6- and 5,7-dihydroxytryptamines as serotoninergic neurotoxins. In: Kostrzewa RM (ed) Handbook of neurotoxicity. Springer, New York, pp 299–325. ISBN 978-1-4614-5835-7 (print); ISBN 978-1-4614-5836-4 (eBook); ISBN 978-1-4614-7458-6 (print and electronic bundle). doi:10.1007/978-1-4614-5836-4_76

Pekny M, Wilhelmsson U, Pekna M (2014) The dual role of astrocyte activation and reactive gliosis. Neurosci Lett 565:30–38

Pelly L, Vardy C, Fernandez B, Newhook LA, Chafe R (2015) Incidence and cohort prevalence for autism spectrum disorders in the Avalon Peninsula, Newfoundland and Labrador. CMAJ Open 3(3):E276–E280. doi:10.9778/cmajo.20140056

Perez SE, He B, Nadeem M, Wuu J, Scheff SW, Abrahamson EE, Ikonomovic MD, Mufson EJ (2015) Resilience of precuneus neurotrophic signaling pathways despite amyloid pathology in prodromal Alzheimer's disease. Biol Psychiatry 77(8):693–703

Pérez-Gómez A, Tasker A (2014) Domoic acid as a neurotoxin. In: Kostrzewa RM (ed) Handbook of neurotoxicity. Springer, New York, pp 399–419. ISBN 978-1-4614-5835-7 (print); ISBN 978-1-4614-5836-4 (eBook); ISBN 978-1-4614-7458-6 (print and electronic bundle). doi:10.1007/978-1-4614-5836-4_87

Perna MK, Brown RW (2013) Adolescent nicotine sensitization and effects of nicotine on accumbal dopamine release in a rodent model of increased dopamine D_2 receptor sensitivity. Behav Brain Res 242:102–109

Petrosini L, De Bartolo P, Tutuli D (2014) Neurotoxic effects, mechanisms and outcome of 192-IgG saporin. In: Kostrzewa RM (ed) Handbook of neurotoxicity. Springer, New York, pp 591–609. ISBN 978-1-4614-5835-7 (print); ISBN 978-1-4614-5836-4 (eBook); ISBN 978-1-4614-7458-6 (print and electronic bundle). doi:10.1007/978-1-4614-5836-4_79

Petrosini L, De Bartolo P, Cutuli D, Gelfo F (2016) Perinatal IgG saporin as neuroteratogen. In: Kostrzewa RM, Archer T (eds) Neurotoxin modeling of brain disorders-lifelong outcomes in behavioral teratology. Springer, New York

Phiel CJ, Zhang F, Huang EY, Guenther MG, Mitchell AL, Klein PS (2001) Histone deacetylase is a direct target of valproic acid, a potent anticonvulsant, mood stabilizer, and teratogen. J Biol Chem 276:36734–36741

Pyndt Jørgensen B, Krych L, Pedersen TB, Plath N, Redrobe JP, Hansen AK, Nielsen DS, Pedersen CS, Larsen C, Sørensen DB (2015) Investigating the long-term effect of subchronic phencyclidine-treatment on novel object recognition and the association between the gut microbiota and behavior in the animal model of schizophrenia. Physiol Behav 141:32–39

Racine RJ (1972) Modification of seizure activity by electrical stimulation. II. Motor seizure. Electroencephalogr Clin Neurophysiol 32:281–294

Radonjić NV, Knezević ID, Vilimanovich U, Kravić-Stevović T, Marina LV, Nikolić T, Todorović V, Bumbasirević V, Petronijević ND (2010) Decreased glutathione levels and altered antioxidant defense in an animal model of schizophrenia: long-term effects of perinatal phencyclidine administration. Neuropharmacology 58(4–5):739–745

Radonjić NV, Jakovcevski I, Bumbaširević V, Petronijević ND (2013) Perinatal phencyclidine administration decreases the density of cortical interneurons and increases the expression of neuregulin-1. Psychopharmacology 227(4):673–683

Ranger P, Ellenbroek BA (2016) Perinatal influences of valproate on brain and behavior: an animal model for autism. In: Kostrzewa RM, Archer T (eds) Neurotoxin modeling of brain disorders-lifelong outcomes in behavioral teratology. Springer, New York

Reynolds GP, Abdul-Monim Z, Neill JC, Zhang ZJ (2004) Calcium binding protein markers of GABA deficits in schizophrenia—postmortem studies and animal models. Neurotox Res 6 (1):57–61

Ricaurte GA, Guillery RW, Seiden LS, Schuster CR, Moore RY (1982) Dopamine nerve terminal degeneration produced by high doses of methylamphetamine in the rat brain. Brain Res 235(1):93–103

Ricceri L, Calamandrei G, Berger-Sweeney J (1997) Different effects of postnatal day 1 versus 7 192 immunoglobulin G-saporin lesions on learning, exploratory behaviors, and neurochemistry in juvenile rats. Behav Neurosci 111(6):1292–1302

Ricceri L, Hohmann C, Berger-Sweeney J (2002) Early neonatal 192 IgG saporin induces learning impairments and disrupts cortical morphogenesis in rats. Brain Res 954(2):160–172

Ricceri L, Cutuli D, Venerosi A, Scattoni ML, Calamandrei G (2007) Neonatal basal forebrain cholinergic hypofunction affects ultrasonic vocalizations and fear conditioning responses in preweaning rats. Behav Brain Res 183(1):111–117

Robertson RT, Gallardo KA, Claytor KJ, Ha DH, Ku KH, Yu BP, Lauterborn JC, Wiley RG, Yu J, Gall CM, Leslie FM (1998) Neonatal treatment with 192 IgG-saporin produces long-term forebrain cholinergic deficits and reduces dendritic branching and spine density of neocortical pyramidal neurons. Cereb Cortex 8(2):142–155

Rodier PM, Ingram JL, Tisdale B, Croog VJ (1997) Linking etiologies in humans and animal models: studies of autism. Reprod Toxicol 11(2–3):417-422. Review

Rodriguez M, Rodriguez-Sabate C, Morales I, Sanchez A, Sabate M (2015) Parkinson's disease as a result of aging. Aging Cell 14(3):293–308

Roman-Urrestarazu A, Lindholm P, Moilanen I, Kiviniemi V, Miettunen J, Jääskeläinen E, Mäki P, Hurtig T, Ebeling H, Barnett JH, Nikkinen J, Suckling J, Jones PB, Veijola J, Murray GK (2015) Brain structural deficits and working memory fMRI dysfunction in young adults who were diagnosed with ADHD in adolescence. Eur Child Adolesc Psychiatry 26 Aug 2015. [Epub ahead of print]

Roos A, Kwiatkowski MA, Fouche JP, Narr KL, Thomas KG, Stein DJ, Donald KA (2015) White matter integrity and cognitive performance in children with prenatalmethamphetamine exposure. Behav Brain Res 279:62–67

Rosenberg PA, Loring R, Xie Y, Zaleskas V, Aizenman E (1991) 2,4,5-trihydroxyphenylalanine in solution forms a non-N-methyl-D-aspartate glutamatergic agonist and neurotoxin. Proc Natl Acad Sci USA 88(11):4865–4869

Ross SB, Renyi AL (1976) On the long-lasting inhibitory effect of N-(2-chloroethyl)-N-ethyl-2-bromobenzylamine (DSP 4) on the active uptake of noradrenaline. J Pharm Pharmacol 28:458–459

Ross SB, Stenfors C (1915) DSP4, a selective neurotoxin for the locus coeruleus noradrenergic system. A review of its mode of action. Neurotox Res 27(1):15–30. Review

Ross SB, Johansson JG, Lindborg B, Dahlbom R (1973) Cyclizing compounds. I. Tertiary N-(2-bromobenzyl)-N-haloalkylamines with adrenergic blocking action. Acta Pharm Sueccica 10:29–42

Roullet FI, Lai JK, Foster JA (2013) In utero exposure to valproic acid and autism–a current review of clinical and animal studies. Neurotoxicol Teratol 36:47–56. Review

Ruan L, Kang Z, Pei G, Le Y (2009) Amyloid deposition and inflammation in APPswe/PSIde9 mouse model of Alzheimer's disease. Curr Alz Res 6:531–540

Sabers A, Bertelsen FC, Scheel-Krüger J, Nyengaard JR, Møller A (2015) Corrigendum to "Long-term valproic acid exposure increases the number of neocortical neurons in the developing rat brain" [Neurosci.Lett. 580 (2014) 12–16] A possible new animal model of autism. Neurosci Lett 588:203–207

Sachs C, Jonsson G (1972a) Degeneration of central and peripheral noradrenaline neurons produced by 6-hydroxy-DOPA. J Neurochem 19(6):1561–1575

Sachs C, Jonsson G (1972b) Selective 6-hydroxy-DOPA induced degeneration of central and peripheral noradrenaline neurons. Brain Res 40(2):563–568

Sachs C, Jonsson G, Fuxe K (1973) Mapping of central noradrenaline pathways with 6-hydroxy-DOPA. Brain Res 63:249–261

Sakai M, Kashiwahara M, Kakita A, Nawa H (2014) An attempt of non-human primate modeling of schizophrenia with neonatal challenges of epidermal growth factor. J Addict Res Ther 5:1

Salavert J, Ramos-Quiroga JA, Moreno-Alcázar A, Caseras X, Palomar G, Radua J, Bosch R, Salvador R, McKenna PJ, Casas M, Pomarol-Clotet E (2015) Functional imaging changes in the medial prefrontal cortex in adult ADHD. J Atten Disord 2015 Oct 29. pii: 1087054715611492. [Epub ahead of print]

Sanders JD, Happe HK, Bylund DB, Murrin LC (2011) Changes in postnatal norepinephrine alter alpha-2 adrenergic receptor development. Neuroscience 192:761–772

Scattoni ML, Calamandrei G, Ricceri L (2003) Neonatal cholinergic lesions and development of exploration upon administration of the GABAa receptor agonist muscimol in preweaning rats. Pharmacol Biochem Behav 76(2):213–221

Schroeder SR, Oster-Granite ML, Berkson G, Bodfish JW, Breese GR, Cataldo MF, Cook EH, Crnic LS, DeLeon I, Fisher W, Harris JC, Horner RH, Iwata B, Jinnah HA, King BH, Lauder JM, Lewis MH, Newell K, Nyhan WL, Rojahn J, Sackett GP, Sandman C, Symons F, Tessel RE, Thompson T, Wong DF (2001) Self-injurious behavior: gene-brain-behavior relationships. Ment Retard Dev Disabil Res Rev 7(1):3–12

Sealey LA, Hughes BW, Steinemann A, Pestaner JP, Gene Pace D, Bagasra O (2015) Environmental factors may contribute to autism development and male bias: Effects of fragrances on developing neurons. Environ Res 142:731–738

Seiden LS, Fischman MW, Schuster CR (1976) Long-term methamphetamine induced changes in brain catecholamines in tolerant rhesus monkeys. Drug Alcohol Depend 1:215–219

Shaywitz BA, Klopper JH, Yager RD, Gordon JW (1976a) Paradoxical response to amphetamine in developing rats treated with 6-hydroxydopamine. Nature 261(5556):153–155

Shaywitz BA, Yager RD, Klopper JH (1976b) Selective brain dopamine depletion in developing rats: an experimental model of minimal brain dysfunction. Science 191(4224):305–308

Siniscalco D (2015) Commentary: the impact of neuroimmune alterations in autism spectrum disorder. Front Psychiatry 6:145

Sircar R (2003) Postnatal phencyclidine-induced deficit in adult water maze performance is associated with N-methyl-D-aspartate receptor upregulation. Int J Dev Neurosci 21(3):159–167

Sitte HH, Freissmuth M (2010) The reverse operation of Na$^+$/Cl$^-$-coupled neurotransmitter transporters—why amphetamines take two to tango. J Neurochem 112:340–355

Šlamberová R, Vrajová M, Schutová B, Mertlová M, Macúchová E, Nohejlová K, Hrubá L, Puskarčíková J, Bubeníková-Valešová V, Yamamotová A (2014) Prenatal methamphetamine exposure induces long-lasting alterations in memory and development of NMDA receptors in the hippocampus. Physiol Res 63(Suppl 4):S547–S558

Šlamberová R, Pometlová M, Macúchová E, Nohejlová K, Stuchlík A, Valeš K (2015) Do the effects of prenatal exposure and acute treatment of methamphetamineon anxiety vary depending on the animal model used? Behav Brain Res 292:361–369

Smith LM, Diaz S, LaGasse LL, Wouldes T, Derauf C, Newman E, Arria A, Huestis MA, Haning W, Strauss A, Della Grotta S, Dansereau LM, Neal C, Lester BM (2015) Developmental and behavioral consequences of prenatal methamphetamineexposure: a review of the Infant Development, Environment, and Lifestyle (IDEAL) study. Neurotoxicol Teratol 51:35–44

Snyder AM, Zigmond MJ, Lund RD (1986) Sprouting of serotoninergic afferents into striatum after dopamine-depleting lesions in infant rats: a retrograde transport and immunocytochemical study. J Comp Neurol 245(2):274–281

Soligo M, Protto V, Florenzano F, Bracci-Laudiero L, De Benedetti F, Chiaretti A, Manni L (2015) The mature/pro nerve growth factor ratio is decreased in the brain of diabetic rats: analysis by ELISA methods. Brain Res 1624:455–468

Sotoyama H, Zheng Y, Iwakura Y, Mizuno M, Aizawa M, Shcherbakova K, Wang R, Namba H, Nawa H (2011) Pallidal hyperdopaminergic innervation underlying D_2 receptor-dependent behavioral deficits in the schizophrenia animal model established by EGF. PLoS One 6(10): e25831

Sotoyama H, Namba H, Chiken S, Nambu A, Nawa H (2013) Exposure to the cytokine EGF leads to abnormal hyperactivity of pallidal GABA neurons: implications for schizophrenia and its modeling. J Neurochem 126(4):518–528

Steardo L Jr, Bronzuoli MR, Iacomino A, Esposito G, Steardo L, Scuderi C (2015) Does neuroinflammation turn on the flame in Alzheimer's disease? Focus on astrocytes. Front Neurosci 9:259

Stojković T, Radonjić NV, Velimirović M, Jevtić G, Popović V, Doknić M, Petronijević ND (2012) Risperidone reverses phencyclidine induced decrease in glutathione levels and alterations of antioxidant defense in rat brain. Prog Neuropsychopharmacol Biol Psychiatry 39(1):192–199

Sygnecka K, Heine C, Scherf N, Fasold M, Binder H, Scheller C, Franke H (2015) Nimodipine enhances neurite outgrowth in dopaminergic brain slice co-cultures. Int J Dev Neurosci 40:1–11

Takuma K, Hara Y, Kataoka S, Kawanai T, Maeda Y, Watanabe R, Takano E, Hayata-Takano A, Hashimoto H, Ago Y, Matsuda T (2014) Chronic treatment with valproic acid or sodium butyrate attenuates novel object recognition deficits and hippocampal dendritic spine loss in a mouse model of autism. Pharmacol Biochem Behav 126:43–49

Tan L, Yu JT, Tan L (2012) The kynurenine pathway in neurodegenerative diseases: mechanistic and therapeutic considerations. J Neurol Sci 323(1–2):1–8

Tartaglione AM, Venerosi A, Calamandrei G (2016) Anna Maria Tartaglione1,2, Aldina Venerosi 1, Gemma Calamandrei1Early life toxic insults and onset of sporadic neurodegenerative diseases. an overview of experimental studies. In: Kostrzewa RM, Archer T (eds) Neurotoxin modeling of brain disorders-lifelong outcomes in behavioral teratology. Springer, New York

Tasker RA, Strain SM, Drejer J (1996) Selective reduction in domoic acid toxicity in vivo by a novel non-N-methyl-D-aspartate receptor antagonist. Can J Physiol Pharmacol 74:1047–1054

Teitelbaum J (1990) Acute manifestations of domoic acid poisoning: case presentations. Can Dis Wkly Rep 16:5–6

Terranova JP, Chabot C, Barnouin MC, Perrault G, Depoortere R, Griebel G, Scatton B (2005) SSR181507, a dopamine D(2) receptor antagonist and 5-HT(1A) receptor agonist, alleviates disturbances of novelty discrimination in a social context in rats, a putative model of selective attention deficit. Psychopharmacology 181(1):134–144

Thacker SK, Perna MK, Ward JJ, Schaefer TL, Williams MT, Kostrzewa RM, Brown RW (2006) The effects of adulthood olanzapine treatment on cognitive performance and neurotrophic factor content in male and female rats neonatally treated with quinpirole. Eur J Neurosci 24(7):2075–2083

Thoenen H, Tranzer JP (1968a) Chemical sympathectomy by selective destruction of adrenergic nerve endings with 6-hydroxydopamine. Naunyn Schmiedebergs Arch Exp Pathol Pharmakol 261(3):271–288

Thoenen H, Tranzer JP (1968b) [On the possibility of chemical sympathectomy by selective destruction of adrenergic nerve endings with 6-hydroxydopamine (6-OH-DA)]. Naunyn Schmiedebergs Arch Exp Pathol Pharmakol 260(2):212–213. German

Tizabi Y, Copeland RL Jr, Brus R, Kostrzewa RM (1999) Nicotine blocks quinpirole-induced behavior in rats: psychiatric implications. Psychopharmacol 145:433–441

Tohmi M, Tsuda N, Mizuno M, Takei N, Frankland PW, Nawa H (2005) Distinct influences of neonatal epidermal growth factor challenge on adult neurobehavioral traits in four mouse strains. Behav Genet 35(5):615–629

Tohyama M, Maeda T, Kashiba A, Shimizu N (1974a) Fluorescence and electron microscopic analysis of axonal change of coerulo-cortical noradrenaline neuron system following destruction of locus coeruleus and administration of 6-hydroxydopa in the rat brain. Med J Osaka Univ 24(4):205–221

Tohyama M, Maeda T, Shimizu N (1974b) Detailed noradrenaline pathways of locus coeruleus neuron to the cerebral cortex with use of 6-hydroxydopa. Brain Res 79(1):139–144

Tuszynski MH, Blesch A (2004) Nerve growth factor: from animal models of cholinergic neuronal degeneration to gene therapy in Alzheimer's disease. Prog Brain Res 146:441–449

Uchida H, Sakata H, Fujimura M, Niizuma K, Kushida Y, Dezawa M, Tominaga T (2015) Experimental model of small subcortical infarcts in mice with long-lasting functional disabilities. Brain Res 1629:318–328

Uhrbrand A, Stenager E, Pedersen MS, Dalgas U (2015) Parkinson's disease and intensive exercise therapy—a systematic review and meta-analysis of randomized controlled trials. J Neurol Sci 353(1–2):9–19

Ungerstedt U (1968) 6-Hydroxy-dopamine induced degeneration of central monoamine neurons. Eur J Pharmacol 5(1):107–110

Ungerstedt U (1971) Postsynaptic supersensitivity after 6-hydroxy-dopamine induced degeneration of the nigro-striatal dopamine system. Acta Physiol Scand Suppl 367:69–93

Verdoorn TA, Johansen TH, Drejer J, Nielsen EO (1994) Selective block of recombinant GluR6 receptors by NS-102, a novel non-NMDA receptor antagonist. Eur J Pharmacol 269:43–49

Verkhratsky A, Parpura V, Pekna M, Pekny M, Sofroniew M (2015) Glia in the pathogenesis of neurodegerative disease. Biochem Soc Trans 42(5):1291–1301

Vieira AC, Aleman N, Cifuentes JM, Bermudez R, Lopez Pena M, Botana LM (2015) Brain pathology in adult rats treated with domoic acid. Veterinary Pathol 52:1077–1086

Visser JE, Smith DW, Moy SS, Breese GR, Friedmann T, Rothstein JD, Jinnah HA (2001) Oxidative stress and dopamine deficiency in a genetic mouse model of Lesch-Nyhan disease. Brain Res Dev Brain Res 133(2):127–139

Vorhees CV, Ahrens KG, Acuff-Smith KD, Schilling MA, Fisher JE (1994a) Methamphetamine exposure during early postnatal development in rats: I. Acoustic startle augmentation and spatial learning deficits. Psychopharmacology 114(3):392–401

Vorhees CV, Ahrens KG, Acuff-Smith KD, Schilling MA, Fisher JE (1994b) Methamphetamine exposure during early postnatal development in rats: II. Hypoactivity and altered responses to pharmacological challenge. Psychopharmacology 114(3):402–408

Vorhees CV, Skelton MR, Grace CE, Schaefer TL, Graham DL, Braun AA, Williams MT (2009) Effects of (+)-methamphetamine on path integration and spatial learning, but not locomotor activity or acoustic startle, align with the stress hyporesponsive period in rats. Int J Dev Neurosci 27(3):289–298

Vrajová M, Schutová B, Klaschka J, Stěpánková H, Rípová D, Šlamberová R (2014) Age-related differences in NMDA receptor subunits of prenatallymethamphetamine-exposed male rats. Neurochem Res 39(11):2040–2046

Waddington JL (1990) Spontaneous orofacial movements induced in rodents by very long-term neuroleptic drug administration: phenomenology, pathophysiology and putative relationship to tardive dyskinesia. Psychopharmacology 101:431–447

Waddington JL, Cross AJ, Gamble SJ, Bourne RC (1983) Spontaneous orofacial dyskinesia and dopaminergic function in rats after 6 months of neuroleptic treatment. Science 220:530–532

Waite JJ, Chen AD, Wardlow ML, Wiley RG, Lappi DA, Thal LJ (1995) 192 immunoglobulin G-saporin produces graded behavioral and biochemical changes accompanying the loss of cholinergic neurons of the basal forebrain and cerebellar Purkinje cells. Neuroscience 65(2):463–476

Wang C, McInnis J, Ross-Sanchez M, Shinnick-Gallagher P, Wiley JL, Johnson KM (2001) Long-term behavioral and neurodegenerative effects of perinatal phencyclidine administration: implications for schizophrenia. Neuroscience 107(4):535–550

Wang Y, Musich PR, Serrano MA, Zou Y, Zhang J, Zhu MY (2014) Effects of DSP4 on the noradrenergic phenotypes and its potential molecular mechanisms in SH-SY5Y cells. Neurotox Res 25(2):193–207

Weinstock M (2008) The long-term behavioural consequences of prenatal stress. Neurosci Biobehav Rev 32:1073–1086

Wenk GL, Stoehr JD, Quintana G, Mobley S, Wiley RG (1994) Behavioral, biochemical, histological, and electrophysiological effects of 192 IgG-saporin injections into the basal forebrain of rats. J Neurosci 14(10):5986–5995

White IM, Minamoto T, Odell JR, Mayhorn J, White W (2009) Brief exposure to methamphetamine (METH) and phencyclidine (PCP) during late development leads to long-term learning deficits in rats. Brain Res 1266:72–86

Wingate M, Kirby RS, Pettygrove S, Cunniff C, Schulz E, Ghosh T, Robinson C, Lee LC, Landa R, Constantino J, Fitzgerald R, Zahorodny W, Daniels J, Nicholas J, Charles J, McMahon W, Bilder D, Durkin M, Baio J, Christensen D, Braun KV, Clayton H, Goodman A, Doernberg N, Yeargin-Allsopp M, Lott E, Mancilla KC, Hudson A, Kast K, Jolly K, Chang A, Harrington R, Fitzgerald R, Shenouda J, Bell P, Kingsbury C, Bakian A, Henderson A, Arneson C, Washington A, Frenkel G, Wright V (2014) Prevalence of autism spectrum disorder among children aged 8 years—autism and developmental disabilities monitoring network, 11 sites, United States, 2010. MMWR Surveill Summ. 63(2):1–21

Winn LM, Wells PG (1999) Maternal administration of superoxide dismutase and catalase in phenytoin teratogenicity. Free Radic Biol Med 26(3–4):266–274

Wischhof L, Irrsack E, Osorio C, Koch M (2015) Prenatal LPS-exposure—a neurodevelopmental rat model of schizophrenia—differentially affects cognitive functions, myelination, and parvalbumin expression in male and female offspring. Prog Neuropsychopharmacol Biol Psychiatry 57:17–30

Wong DF, Harris JC, Naidu S, Yokoi F, Marenco S, Dannals RF, Ravert HT, Yaster M, Evans A, Rousset O, Bryan RN, Gjedde A, Kuhar MJ, Breese GR (1996) Dopamine transporters are markedly reduced in Lesch-Nyhan disease in vivo. Proc Natl Acad Sci USA 93(11):5539–5543

Xi D, Peng YG, Ramsdell JS (1997) Domoic acid is a potent neurotoxin to neonatal rats. Nat Toxins 5:74–79

Xia Y, Qi F, Zou J, Yang J, Yao Z (2014) Influenza vaccination during early pregnancy contributes to neurogenesis and behavioral function in offspring. Brain Behav Immun 42:212–221

Zelazo PD, Carlson SM (2012) Hot and cool executive function in childhood and adolescence: development and plasticity. Child Devel Perspec. doi:10.1111/j.1750-8606.2012.00246.x

Zelazo PD, Müller U, Frye D, Marcovitch S, Argitis G, Boseovski J, Chiang JK, Hongwanishkul D, Schuster BV, Sutherland A (2003) The development of executive function in early childhood. Monogr Soc Res Child Dev 68(3):vii–137

Zelazo PD, Craik FI, Booth L (2004) Executive function across the life span. Acta Psychol 115 (2–3):167–183

Zhu F, Zheng Y, Ding YQ, Liu Y, Zhang X, Wu R, Guo X, Zhao J (2014a) Minocycline and risperidone prevent microglia activation and rescue behavioral deficits induced by neonatal intrahippocampal injection of liposaccharide in rats. PLoS One 9(4):e93966

Zhu F, Zhang L, Ding YQ, Zhao J, Zheng Y (2014b) Neonatal intrahippocampal injection of liposaccharide induces deficits in social behavior and prepulse inhibition and microgial activation in rats: implication for a new schizophrenia animal model. Brain Behav Immun 38:166–174

Zieher LM, Jaim-Etcheverry G (1973) Regional differences in the long-term effect of neonatal 6-hydroxydopa treatment on rat brain noradrenaline. Brain Res 60(1):199–207

Zieher LM, Jaim-Etcheverry G (1975a) 6-hydroxydopa during development of central adrenergic neurons produces different long-term changes in rat brain noradrenaline. Brain Res 86(2):271–281

Zieher LM, Jaim-Etcheverry G (1975b) Different alterations in the development of the noradrenergic innervation of the cerebellum and the brain stem produced by neonatal 6-hydroxydopa. Life Sci 17(6):987–991

Part I
Neuroteratogenic Agents

Perinatal Lesioning and Lifelong Effects of the Noradrenergic Neurotoxin 6-Hydroxydopa

Richard M. Kostrzewa

Abstract 6-hydroxydopa (6-OHDOPA) was synthesized with the expectation that it would be able to cross the blood–brain barrier to be enzymatically decarboxylated to 6-hydroxydopamine (6-OHDA), the newly discovered neurotoxin for noradrenergic and dopaminergic neurons. In part, 6-OHDOPA fulfilled these criteria. When administered experimentally to rodents, 6-OHDOPA destroyed peripheral sympathetic noradrenergic nerves and did exert neurotoxicity to noradrenergic nerves in brain—in large part, from its conversion to 6-OHDA. However, the efficacy of 6-OHDOPA was less than that of 6-OHDA; also, 6-OHDOPA was relatively selective for noradrenergic neurons; near-lethal doses of 6-OHDOPA were required to damage dopaminergic nerves; and ultimately, 6-OHDOPA was found to be an agonist at AMPA receptors, thus accounting for more non-specificity. Nevertheless, 6-OHDOPA was found to be a particularly valuable tool in uncovering processes and mechanisms associated with noradrenergic nerve regeneration and sprouting, particularly when administered to perinatal rodents. Also, 6-OHDOPA was a good tool for selective mapping of noradrenergic nerve tracts in brain, since dopaminergic tracts were unaffected and did not interfere with the histofluorescent methodology used for this purpose in the early 1970s. As an experimental research tool, 6-OHDOPA was valuable in a short time-window, but its utility is largely limited because of newer research technologies that provide better means today for nerve tract mapping, and for experimental approaches engaged toward study of processes and mechanisms attending nerve regeneration. AMPA actions of 6-OHDOPA have not been extensively studied, so this avenue may enliven use of 6-OHDOPA in the future.

Keywords 6-hydroxydopa · 6-hydroxydopamine · Nerve regeneration · Nerve sprouting · Noradrenergic nerves · Neuroteratogen · AMPA

R.M. Kostrzewa (✉)
Department of Biomedical Sciences, Quillen College of Medicine, East Tennessee State University, PO Box 70577, Johnson City, TN 37614, USA
e-mail: kostrzew@etsu.edu

© Springer International Publishing Switzerland 2015
Curr Topics Behav Neurosci (2016) 29: 43–50
DOI 10.1007/7854_2015_414
Published Online: 12 December 2015

Contents

1   Introduction 44
2   Mechanism of Action of 6-OHDOPA 44
3   6-OHDOPA as a Neuroteratogen 45
4   6-OHDOPA Agonist Action at AMPA Receptors 46
5   Non-specific Effects of Perinatal 6-OHDOPA 46
6   Summary 47
References 47

1   Introduction

The norepinephrine (NE) isomer 6-hydroxydopamine (6-OHDA) was initially found to produce long-lasting depletion of norepinephrine in heart (Porter et al. 1963, 1965). Subsequently, 6-OHDA was shown by electron microscopy to overtly destroy noradrenergic nerves (Thoenen and Tranzer 1968a, b) and dopaminergic nerves (Ungerstedt 1968). One limitation of 6-OHDA, however, was its inability to cross the blood–brain barrier (BBB) (Kostrzewa and Jacobowitz 1974).

6-hydroxydopa (6-OHDOPA) was synthesized as an expected pro-toxin and, like levodopa, able to cross the BBB prior to its decarboxylation to 6-OHDA (Ong et al. 1969; Berkowitz et al. 1970; Evans and Cohen 1989, 1993). As expected, 6-OHDOPA produced norepinephrine (NE) depletion in brain and peripheral tissues (Jonsson and Sachs 1973; Kostrzewa and Jacobowitz 1972, 1973; Richardson and Jacobowitz 1973) and subsequently was shown by loss of tyrosine hydroxylase activity and by histochemical, electronmicroscopic, and silver degeneration staining to overtly destroy noradrenergic nerves (Jacobowitz and Kostrzewa 1971; Kostrzewa and Harper 1974, 1975; Sachs et al. 1973; Kostrzewa et al. 1978; Tohyama et al. 1974; Toyama et al. 1974), with preference for noradrenergic perikarya in caudal locus coeruleus (LC) (Clark et al. 1979). Peripheral noradrenergic (i.e., sympathetic) nerves, in contrast to central noradrenergic nerves, appear to fully recover from 6-OHDOPA damage (Kostrzewa and Jacobowitz 1972; Sachs and Jonsson 1972a, b) and can be protected from damage by administering a peripherally acting dopa decarboxylase inhibitor, namely carbidopa (Kostrzewa et al. 2000).

2   Mechanism of Action of 6-OHDOPA

The destructive effects of 6-OHDOPA on noradrenergic and dopaminergic neurons are attributable to its conversion to 6-OHDA, which is known to auto-oxidize to ortho- and para-quinones, aminochromes, and hydroxyindoles (Adams et al. 1972;

Blank et al. 1972; Saner and Thoenen 1971; Senoh and Witkop 1959; Wehrli et al. 1972)—reactive species leading to formation of intraneuronal peroxide (Heikkila and Cohen 1971, 1972a, b), superoxide, and hydroxyl radical (Cohen and Heikkila 1974; Heikkila and Cohen 1973).

3 6-OHDOPA as a Neuroteratogen

When administered to perinatal rats, 6-OHDOPA (60 µg/g at P0 + P2 + P4) produced lifelong alterations in noradrenergic innervation of brain (Kostrzewa 1975; Kostrzewa and Garey 1976). The nucleus LC providing the major portion of noradrenergic innervation of dorsal brain was directly damaged, with there being loss of one-third of the approximately 1500 perikarya, and with half the numbers of cells in the caudal portion of the LC undergoing degeneration (Clark et al. 1979). Hippocampal noradrenergic innervation was reduced by >95 %, and neocortex, by ∼70 % (Kostrzewa and Harper 1974, 1975; Kostrzewa 1975; Kostrzewa and Garey 1976, 1977). As a consequence of damage to the dorsal bundle, the major ascending noradrenergic tract to forebrain, noradrenergic fibers projecting to regions near the LC per se sprouted and hyperinnervated midbrain, pons, medulla, and cerebellum (Jaim-Etcheverry and Zieher 1977; Jaim-Etcheverry et al. 1975; Kostrzewa and Harper 1974, 1975; Kostrzewa 1975; Kostrzewa and Garey 1976, 1977; Kostrzewa et al. 1978; 1982; Zieher and Jaim-Etcheverry 1979). This reactive sprouting resulted in as much as a twofold increase in numbers of fibers' innervation of caudal brain regions. In contrast, innervation to hypothalamus was slightly altered. The pairing of noradrenergic hypoinnervation of forebrain with noradrenergic hyperinnervation of hindbrain is replicated by knife cuts of the dorsal bundle shortly after birth, suggesting that hindbrain hyperinnervation is an outcome of forebrain noradrenergic hypoinnervation (Klisans-Fuenmayor et al. 1986; Kostrzewa et al. 1988).

This spectrum of effects is replicated (1) by single 6-OHDOPA treatment of rats at birth, also (2) by prenatal 6-OHDOPA, administered to pregnant rats at G14 or later, and (3) by prenatal 6-OHDOPA to pregnant mice at G13 or later (Jaim-Etcheverry et al. 1975; Kostrzewa 2007; Kostrzewa et al. 1978; Zieher and Jaim-Etcheverry 1973, 1975a, b). When 6-OHDOPA is administered solely at P3, there is an absence of noradrenergic sprouting to cerebellum. Described perinatal 6-OHDOPA effects on noradrenergic neurons persist throughout life (Jaim-Etcheverry et al. 1975; McLean et al. 1976, 1980; Zieher and Jaim-Etcheverry 1973, 1975a, b). A single 6-OHDOPA treatment at P5 fails to produce noradrenergic hyperinnervation of midbrain, while single 6-OHDOPA treatment as late as P14 still produces hyperinnervation of pons–medulla (Kostrzewa and Garey 1977).

Agonists at µ-opioid receptors (i.e., morphine, met-/leu-enkephalin, beta-endorphin, and d-ala-enkephalinamide) appeared to enhance perinatal 6-OHDOPA neurotoxicity and thereby enhance noradrenergic hyperinnervation of hindbrain and cerebellum—the effect being attenuated by the opioid receptor antagonist naloxone (Harston et al. 1980, 1981; Kostrzewa and Klisans-Fuenmayor 1984).

Despite the consequences of 6-OHDOPA on noradrenergic innervation, dopaminergic innervation to neostriatum is unaltered through the duration of postnatal ontogeny and for the life span by perinatal low-dose 6-OHDOPA (Kostrzewa and Garey 1976); at high-dose perinatal 6-OHDOPA, the tuberoinfundibular dopaminergic tract (Lin et al. 1993) is more susceptible to damage than the nigrostriatal dopaminergic tract (Nomura and Segawa 1979). Also, serotoninergic nerves are resistant to 6-OHDOPA neurotoxicity (Richardson et al. 1974).

Perinatal 6-OHDOPA treatment initially damaged sympathetic noradrenergic nerves innervating peripheral organs (i.e., heart, salivary glands), but by maturity all organs were fully innervated.

6-OHDOPA actions on noradrenergic neurons have been reviewed elsewhere (Kostrzewa 1988, 1988, 2014).

4   6-OHDOPA Agonist Action at AMPA Receptors

Although 6-OHDOPA was shown, in 1976, to produce more of an excitatory action than glutamate on frog spinal neurons (Biscoe et al. 1976), not until 1990 was it discovered that 6-OHDOPA exerts agonist action at alpha-amino-3-hydroxy-5-methylisoxazole-4-propionic acid receptors (AMPA-R) (Cha et al. 1991; Kunig et al. 1994a, b; Olney et al. 1990). Actually, 6-OHDOPA-quinone is the suspected agonist (Aizenman et al. 1990, 1992; Rosenberg et al. 1991). AMPA-R activity would represent a confound to the actions of 6-OHDOPA per se on noradrenergic neurons.

5   Non-specific Effects of Perinatal 6-OHDOPA

In rats treated with perinatal 6-OHDOPA, cholineacetyltransferase activity is reduced in brainstem (Jaim-Etcheverry et al. 1975), and atropine-induced locomotor activity at P20 and P50 is enhanced, while pilocarpine catalepsy is abated (Nomura and Segawa 1979; Nomura et al. 1979). This cholinergic subsensitivity is reflected in the B_{max} for [^{3}H]QNB (quinuclidinyl benzilate) binding at muscarinic receptor sites in mesolimbic and striatal brain regions and also in heart (Nomura et al. 1979).

High-dose 6-OHDOPA treatment is associated with the production of methemoglobinemia (Corrodi et al. 1971), which of itself promotes in vivo formation of 6-OHDOPA from tyrosine, a process enhanced by hydrogen peroxide formation (Agrup et al. 1983) and known to increase after 6-OHDOPA treatment.

6 Summary

6-OHDOPA was envisioned as an experimental tool, able to cross the BBB to be decarboxylated to 6-OHDA, and thus exert effects on noradrenergic and/or dopaminergic nerves in brain. By this means, 6-OHDA effects on brain could be realized without the necessity to otherwise inject 6-OHDA directly into brain— since 6-OHDA does not cross the BBB. However, the neurotoxicity action (i.e., efficacy) of 6-OHDOPA is far less than that of 6-OHDA. Moreover, as a means of minimizing global effects, experimental 6-OHDA is generally applied to specific brain nuclei or specified tracts, not intraventricularly or intracisternally. Consequently, the overall utility of 6-OHDOPA is greatly restricted in biomedical research.

The non-specific effects of 6-OHDOPA, namely AMPA-R agonist activity by non-enzymatically formed 6-OHDOPA-quinone, and the methemoglobinemia arising from 6-OHDOPA and its quinone, further restrict the usefulness of 6-OHDOPA as an experimental tool.

Nevertheless, the early work with 6-OHDOPA did validate its role as a relatively selective noradrenergic neurotoxin, and the actions of 6-OHDOPA in perinates led to discovery of processes and mechanisms associated with nerve sprouting and nerve regeneration. Conceivably, action at the AMPA-R could still be advantageous in 6-OHDOPA use as a research tool.

Acknowledgments Studies for this paper were provided by the National Institutes of Mental Health; NINDS; March of Dimes; and Scottish Rite Foundation.

References

Adams RN, Murrill E, McCreery R, Blank L, Karolczak M (1972) 6-Hydroxydopamine, a new oxidation mechanism. Eur J Pharmacol 17(2):287–292

Agrup G, Hansson C, Rorsman H, Rosengren E, Tegner E (1983) Methaemoglobin-catalysed formation of dopa and 6-OH-dopa from tyrosine. Acta Derm Venereol 63(2):152–155

Aizenman E, White WF, Loring RH, Rosenberg PA (1990) A 3,4-dihydroxyphenylalanine oxidation product is a non-N-methyl-D-aspartate glutamatergic agonist in rat cortical neurons. Neurosci Lett 116(1–2):168–171

Aizenman E, Boeckman FA, Rosenberg PA (1992) Glutathione prevents 2,4,5-trihydroxyphenylalanine excitotoxicity by maintaining it in a reduced, non-active form. Neurosci Lett 144(1–2):233–236

Berkowitz BA, Spector S, Brossi A, Focella A, Teitel S (1970) Preparation and biological properties of (−)- and (+)-6-hydroxydopa. Experientia 26(9):982–983

Biscoe TJ, Evans RH, Headley PM, Martin MR, Watkins JC (1976) Structure-activity relations of excitatory amino acids on frog and rat spinal neurones. Br J Pharmacol 58(3):373–382

Blank CL, Kissinger PT, Adams RN (1972) 5,6-Dihydroxyindole formation from oxidized 6-hydroxydopamine. Eur J Pharmacol 19(3):391–394

Cha JH, Dure LS IV, Sakurai SY, Penney JB, Young AB (1991) 2,4,5-Trihydroxyphenylalanine (6-hydroxy-dopa) displaces [^{3}H]AMPA binding in rat striatum. Neurosci Lett 132(1):55–58

Clark MB, King JC, Kostrzewa RM (1979) Loss of nerve cell bodies in caudal locus coeruleus following treatment of neonates with 6-hydroxydopa. Neurosci Lett 13(3):331–336

Cohen G, Heikkila RE (1974) The generation of hydrogen peroxide, superoxide radical, and hydroxyl radical by 6-hydroxydopamine, dialuric acid, and related cytotoxic agents. J Biol Chem 249(8):2447–2452

Corrodi H, Clark WG, Masuoka DI (1971) The synthesis and effects of DL-6-hydroxydopa. In: Malmfors T, Thoenen H (eds) 6-Hydroxydopamine and catecholamine neurons. North Holland, Amsterdam, pp 187–192

Evans JM, Cohen G (1989) Studies on the formation of 6-hydroxydopamine in mouse brain after administration of 2,4,5-trihydroxyphenylalanine (6-hydroxyDOPA). J Neurochem 52(5):1461–1467

Evans J, Cohen G (1993) Catecholamine uptake inhibitors elevate 6-hydroxydopamine in brain after administration of 6-hydroxydopa. Eur J Pharmacol 232(2–3):241–245

Harston CT, Morrow A, Kostrzewa RM (1980) Enhancement of sprouting and putative regeneration of central noradrenergic fibers by morphine. Brain Res Bull 5(4):421–424

Harston CT, Clark MB, Hardin JC, Kostrzewa RM (1981) Opiate-enhanced toxicity and noradrenergic sprouting in rats treated with 6-hydroxydopa. Eur J Pharmacol 71(4):365–373

Heikkila R, Cohen G (1971) Inhibition of biogenic amine uptake by hydrogen peroxide: a mechanism for toxic effects of 6-hydroxydopamine. Science 172(3989):1257–1258

Heikkila R, Cohen G (1972a) Further studies on the generation of hydrogen peroxide by 6-hydroxydopamine. Potentiation by ascorbic acid. Mol Pharmacol 8(2):241–248

Heikkila RE, Cohen G (1972b) In vivo generation of hydrogen peroxide from 6-hydroxydopamine. Experientia 28(10):1197–1198

Heikkila RE, Cohen G (1973) 6-Hydroxydopamine: evidence for superoxide radical as an oxidative intermediate. Science 181(4098):456–457

Jacobowitz D, Kostrzewa R (1971) Selective action of 6-hydroxydopa on noradrenergic terminals: mapping of preterminal axons of the brain. Life Sci I 10(23):1329–1342

Jaim-Etcheverry G, Zieher LM (1977) Differential effect of various 6-hydroxydopa treatments on the development of central and peripheral noradrenergic neurons. Eur J Pharmacol 45(2):105–116

Jaim-Etcheverry G, Teitelman G, Zieher LM (1975) Choline acetyltransferase activity increases in the brain stem of rats treated at birth with 6-hydroxydopa. Brain Res 100(3):699–704

Jonsson G, Sachs C (1973) Pharmacological modifications of the 6-hydroxy-dopa induced degeneration of central noradrenaline neurons. Biochem Pharmacol 22(14):1709–1716

Klisans-Fuenmayor D, Harston CT, Kostrzewa RM (1986) Alterations in noradrenergic innervation of the brain following dorsal bundle lesions in neonatal rats. Brain Res Bull 16(1):47–54

Kostrezewa RM, Klara JW, Robertson J, Walker LC (1978) Studies on the mechanism of sprouting of noradrenergic terminals in rat and mouse cerebellum after neonatal 6-hydroxydopa. Brain Res Bull 3(5):525–531

Kostrzewa RM (1975) Effects of neonatal 6 hydroxydopa treatment on monamine content of rat brain and peripheral tissues. Res Commun Chem Pathol Pharmacol 11(4):567–579

Kostrzewa RM (1988) Reorganization of noradrenergic neuronal systems following neonatal chemical and surgical injury. Prog Brain Res 73:405–423. Review. PMID: 3138742

Kostrzewa RM (1998) 6-Hydroxydopa, a catecholamine neurotoxin and endogenous excitotoxin at non-NMDA receptors. In: Kostrzewa RM (ed) Highly selective neurotoxins: basic and clinical applications. Humana Press, Totowa NJ, pp 109–129

Kostrzewa RM (2007) The blood-brain barrier for catecholamines—revisited. Neurotoxicity Res 11:261–271

Kostrzewa RM (2014) Survey of selective neurotoxins, in Section on Selective Neurotoxins, In: Kostrzewa RM (ed) Handbook of neurotoxicity, Springer New York, Heidelberg, Dordrecht, London, pp 3–67. ISBN 978-1-4614-5835-7 (print); ISBN 978-1-4614-5836-4 (eBook); ISBN 978-1-4614-7458-6 (print and electronic bundle). doi:10.1007/978-1-4614-5836-4_53

Kostrzewa RM, Garey RE (1976) Effects of 6-hydroxydopa on noradrenergic neurons in developing rat brain. J Pharmacol Exp Ther 197:105–118

Kostrzewa RM, Garey RE (1977) Sprouting of noradrenergic terminals in rat cerebellum following neonatal treatment with 6-hydroxydopa. Brain Res 124:385–391

Kostrzewa RM, Harper JW (1974) Effect of 6-hydroxydopa on catecholamine-containing neurons in brains of newborn rats. Brain Res 69(1):174–181

Kostrzewa RM, Harper JW (1975) Comparison of the neonatal effects of 6-hydroxydopa and 6-hydroxydopamine on growth and development of noradrenergic neurons in the central nervous system, In: Jonsson G, Malmfors T, Sachs C (eds) Chemical tools in catecholamine research, vol I. North Holland Publ Co, Amsterdam, The Netherlands, pp 181–188

Kostrzewa R, Jacobowitz D (1972) The effect of 6-hydroxydopa on peripheral adrenergic neurons. J Pharmacol Exp Ther 183(2):284–297

Kostrzewa R, Jacobwitz D (1973) Acute effects of 6-hydroxydopa on central monoaminergic neurons. Eur J Pharmacol 21(1):70–80

Kostrzewa RM, Jacobowitz DM (1974) Pharmacological actions of 6-hydroxydopamine. Pharmacol Rev 26(3):199–288. Review. PMID: 4376244

Kostrzewa RM, Klisans-Fuenmayor D (1984) Development of an opioid-specific action of morphine in modifying recovery of neonatally-damaged noradrenergic fibers in rat brain. Res Commun Chem Pathol Pharmacol 46(1):3–11

Kostrzewa RM, Harston CT, Fukushima H, Brus R (1982) Noradrenergic fiber sprouting in the cerebellum. Brain Res Bull 9(1-6):509–517. PMID: 7172038

Kostrzewa RM, Hardin JC, Jacobowitz DM (1988) Destruction of cells in the midportion of the locus coeruleus by a dorsal bundle lesion in neonatal rats. Brain Res 442(2):321–328

Kostrzewa RM, Kostrzewa JP, Brus R (2000) Dopaminergic denervation enhances susceptibility to hydroxyl radicals in rat neostriatum. Amino Acids 19(1):183–199

Künig G, Hartmann J, Niedermeyer B, Deckert J, Ransmayr G, Heinsen H, Beckmann H, Riederer P (1994a) Excitotoxins L-beta-oxalyl-amino-alanine (L-BOAA) and 3,4,6-trihydroxyphenylalanine (6-OH-DOPA) inhibit [^{3}H] alpha-amino-3-hydroxy-5-methyl-4-isoxazole-propionic acid (AMPA) binding in human hippocampus. Neurosci Lett 169(1–2):219–222

Künig G, Niedermeyer B, Krause F, Hartmann J, Deckert J, Ransmayr G, Heinsen H, Beckmann H, Riederer P (1994b) Interactions of neurotoxins with non-NMDA glutamate receptors: an autoradiographic study. J Neural Transm Suppl 43:59–62

Lin JY, Mai LM, Pan JT (1993) Effects of systemic administration of 6-hydroxydopamine, 6-hydroxydopa and 1-methyl-4-phenyl-1,2,3,6-tetrahydroxypyridine (MPTP) on tuberoin-fundibular dopaminergic neurons in the rat. Brain Res 624(1–2):126–130

McLean JH, Kostrzewa RM, May JG (1976) Behavioral and biochemical effects of neonatal treatment of rats with 6-hydroxydopa. Pharmacol Biochem Behav 4(5):601–607

McLean JH, Glasser RS, Kostrzewa RM, May JG (1980) Effects of neonatal 6-hydroxydopa on behavior in female rats. Pharmacol Biochem Behav 13(6):863–868

Morgan DN, McLean JH, Kostrzewa RM (1979) Effects of 6-hydroxydopamine and 6-hydroxydopa on development of behavior. Pharmacol Biochem Behav 11(3):309–312

Nomura Y, Segawa T (1979) Striatal dopamine content reduced in developing rats treated with 6-hydroxydopa. Jpn J Pharmacol 29(2):306–309

Nomura Y, Kajiyama H, Segawa T (1979) Decrease in muscarinic cholinergic response of the rat heart following treatment with 6-hydroxydopa. Eur J Pharmacol 60(4):323–327

Olney JW, Zorumski CF, Stewart GR, Price MT, Wang GJ, Labruyere J (1990) Excitotoxicity of L-dopa and 6-OH-dopa: implications for Parkinson's and Huntington's diseases. Exp Neurol 108(3):269–272

Ong HH, Creveling CR, Daly JW (1969) The synthesis of 2,4,5-trihydroxyphenylalanine (6-hydroxydopa). A centrally active norepinephrine-depleting agent. J Med Chem 12(3): 458–462

Porter CC, Totaro JA, Stone CA (1963) Effect of 6-hydroxydopamine and some other compounds on the concentration of norepinephrine in the hearts of mice. J Pharmacol Exp Ther 140: 308–316

Porter CC, Totaro JA, Burcin A (1965) The relationship between radioactivity and norepinephrine concentrations in the brains and hearts of mice following administration of labeled methyldopa or 6-hydroxydopamine. J Pharmacol Exp Ther 150(1):17–22

Richardson JS, Jacobowitz DM (1973) Depletion of brain norepinephrine by intraventricular injection of 6-hydroxydopa: a biochemical, histochemical and behavioral study in rats. Brain Res 58(1):117–133

Richardson JS, Cowan N, Hartman R, Jacobowitz DM (1974) On the behavioral and neurochemical actions of 6-hydroxydopa and 5,6-dihydroxytryptamine in rats. Res Commun Chem Pathol Pharmacol 8(1):29–44

Rosenberg PA, Loring R, Xie Y, Zaleskas V, Aizenman E (1991) 2,4,5-trihydroxyphenylalanine in solution forms a non-N-methyl-D-aspartate glutamatergic agonist and neurotoxin. Proc Natl Acad Sci USA 88(11):4865–4869

Sachs C, Jonsson G (1972a) Degeneration of central and peripheral noradrenaline neurons produced by 6-hydroxy-DOPA. J Neurochem 19(6):1561–1575

Sachs C, Jonsson G (1972b) Selective 6-hydroxy-DOPA induced degeneration of central and peripheral noradrenaline neurons. Brain Res 40(2):563–568

Sachs C, Jonsson G, Fuxe K (1973) Mapping of central noradrenaline pathways with 6-hydroxy-DOPA. Brain Res 63:249–261

Saner A, Thoenen H (1971) Model experiments on the molecular mechanism of action of 6-hydroxydopamine. Mol Pharmacol 7(2):147–154

Senoh S, Witkop B (1959) Formation and rearrangements of aminochromes from a new metabolite of dopamine and some of its derivatives. J Am Chem Soc 81:6231–6235

Thoenen H, Tranzer JP (1968a) Chemical sympathectomy by selective destruction of adrenergic nerve endings with 6-Hydroxydopamine. Naunyn Schmiedebergs Arch Exp Pathol Pharmakol 261(3):271–288

Thoenen H, Tranzer JP (1968b) On the possibility of chemical sympathectomy by selective destruction of adrenergic nerve endings with 6-hydroxydopamine (6-OH-DA). Naunyn Schmiedebergs Arch Exp Pathol Pharmakol 260(2):212–213. German. PMID: 4239240

Tohyama M, Maeda T, Kashiba A, Shimizu N (1974) Fluorescence and electron microscopic analysis of axonal change of coerulo-cortical noradrenaline neuron system following destruction of locus coeruleus and administration of 6-hydroxydopa in the rat brain. Med J Osaka Univ 24(4):205–221

Toyama M, Maeda T, Shimizu N (1974) Detailed noradrenaline pathways of locus coeruleus neuron to the cerebral cortex with use of 6-hydroxydopa. Brain Res 79(1):139–144

Ungerstedt U (1968) 6-Hydroxy-dopamine induced degeneration of central monoamine neurons. Eur J Pharmacol 5(1):107–110

Wehrli PA, Pigott F, Fischer U, Kaiser A (1972) Oxidation products of 6-hydroxy-dopamine. Helv Chim Acta 55(8):3057–3061 (Article in German)

Zieher LM, Jaim-Etcheverry G (1973) Regional differences in the long-term effect of neonatal 6-hydroxydopa treatment on rat brain noradrenaline. Brain Res 60(1):199–207

Zieher LM, Jaim-Etcheverry G (1975a) 6-hydroxydopa during development of central adrenergic neurons produces different long-term changes in rat brain noradrenaline. Brain Res 86(2): 271–281

Zieher LM, Jaim-Etcheverry G (1975b) Different alterations in the development of the noradrenergic innervation of the cerebellum and the brain stem produced by neonatal 6-hydroxydopa. Life Sci 17(6):987–991

Zieher LM, Jaim-Etcheverry G (1979) 6-Hydroxydopamine during development: relation between opposite regional changes in brain noradrenaline. Eur J Pharmacol 58(3):217–223. PMID: 510355

Selective Lifelong Destruction of Brain Monoaminergic Nerves Through Perinatal DSP-4 Treatment

Przemysław Nowak

Abstract *N*-(2-chloroethyl)-*N*-ethyl-2-bromobenzylamine (DSP-4) is a highly selective neurotoxin for noradrenergic projections originating from the locus coeruleus (LC). The outcome of the systemic DSP-4 treatment of newborn rats is an alteration in postnatal development of the noradrenergic system, involving the permanent denervation of distal noradrenergic projection areas (neocortex, hippocampus, spinal cord), accompanied by noradrenergic hyperinnervation in regions proximal to the LC cell bodies (cerebellum, pons–medulla). DSP-4 is well tolerated by developing rats and does not increase the mortality rate. Permanent noradrenergic denervation in the cerebral cortex and spinal cord is present at all developmental stages, although this effect is more pronounced in rats treated with DSP-4 at an early age, i.e., up to postnatal day 5 (PND 5). Notably, regional hyperinnervation is a hallmark of neonatal DSP-4 treatment, which is not observed after either prenatal or adult DSP-4 application. In contrast to robust biochemical changes in the brain, DSP-4 treatment of newborn rats has a marginal effect on arousal and cognition functions assessed in adulthood, and these processes are critically influenced by the action of the noradrenergic neurotransmitter, norepinephrine (NE). Conversely, neonatal DSP-4 does not significantly affect 5-hydroxytryptamine (serotonin; 5-HT), dopamine (DA), gamma-aminobutyric acid (GABA), and histamine levels in brain. However, as a consequence of altering the functional efficacy of 5-HT$_{1A}$, 5-HT$_{1B}$, DA, and GABA receptors, these neurotransmitter systems are profoundly affected in adulthood. Thus, the noradrenergic lesion obtained with neonatal DSP-4 treatment represents a unique neurobiological technique for exploring the interplay between various neuronal phenotypes and examining the pathomechanism of neurodevelopmental disorders.

Keywords DSP-4 · Neurotoxin · Degeneration · Depletion · Locus coeruleus · Noradrenergic neurons · Perinatal period

P. Nowak (✉)
Department of Toxicology and Addiction, Department of Toxicology and Health Protection, School of Public Health in Bytom, Medical University of Silesia, Medyków 18 Street, 40-752, Katowice, Poland
e-mail: pnowak@sum.edu.pl

© Springer International Publishing Switzerland 2015
Curr Topics Behav Neurosci (2016) 29: 51–72
DOI 10.1007/7854_2015_398
Published Online: 02 October 2015

Contents

1 Introduction 52
2 DSP-4 Treatment of Newborn Rats 53
3 Noradrenergic System 53
    3.1 Biochemical Alterations 53
    3.2 Behavioral and Biological Rhythm Alterations 55
4 Serotoninergic System 56
    4.1 Biochemical Alterations 56
    4.2 Behavioral Alterations 58
5 Dopaminergic System 59
    5.1 Biochemical Alterations 59
    5.2 Behavioral Alterations 59
6 GABAergic System 60
    6.1 Biochemical Alterations 60
    6.2 Behavioral Alterations 60
7 Other Neuronal Phenotypic Systems 61
8 Prenatal DSP-4 Treatment 62
9 Summary 62
    9.1 Noradrenergic System 62
    9.2 Serotoninergic System 63
    9.3 Dopaminergic System 64
    9.4 GABAergic System 64
    9.5 Other Neuronal Phenotypic Systems 64
10 Conclusions 65
References 66

1 Introduction

During the search of tertiary haloalkylamines related to bretylium and capable, once in the brain, of cyclizing by an intramolecular reaction to quaternary ammonium derivative with distinctive adrenergic neuron blocking activity, a compound was identified with a long-lasting inhibitory effect on [^{3}H]-labeled noradrenaline (NE) uptake by brain slices (Ross et al. 1973; Ross and Renyl 1976). This compound, N-(2-chloroethyl)-N-ethyl-2-bromobenzylamine (DSP-4), interacts with the NE transporter (NET), whereby DSP-4 is then accumulated intraneuronally to trigger the degeneration of noradrenergic terminals (Lyons et al. 1989). On the molecular level, a cellular energy collapse, bringing about lack of ATP, may serve as an explanation of DSP-4 neurotoxicity (Wenge and Bönisch 2009). Another hypothesis assumes that DSP-4 down-regulates the noradrenergic phenotypes by its actions on DNA replication, leading to replication stress and cell cycle arrest (Wang et al. 2014). However, the precise mechanism of DSP-4s' action still remains unclear. Markers of terminal loss are observed as a reduction in NE brain tissue content (Jonsson et al. 1982; Dabrowska et al. 2007; Nowak et al. 2008), decrease in NET density (Jonsson et al. 1981; Sanders et al. 2011), and temporary increase in α_2-adrenergic autoreceptor (Wolfman et al. 1994; Sanders et al. 2011). The changes in these noradrenergic markers in specific brain regions suggest a selective loss of afferents from locus coeruleus (LC) noradrenergic neurons (Jonsson et al. 1981). The animal age at the time of DSP-4 administration is crucial for its biological response and long-lasting changes in noradrenergic system and other neuronal phenotypes (e.g., serotoninergic and dopaminergic) (Jonsson et al. 1981, 1982; Nowak et al. 2006; Dabrowska et al. 2008).

2　DSP-4 Treatment of Newborn Rats

DSP-4 is a highly selective neurotoxin for noradrenergic projections originating from the LC. DSP-4 is typically injected intraperitoneally (i.p.), occasionally subcutaneously (s.c.), or intravenously (i.v.) in adult rats or mice. Considering technical and methodological limitations, newborn rats, for systemic administration, are routinely administered s.c. injections (Donnerer et al. 1990; Jonsson et al. 1981). The systemic administration of this compound has a rapid dose-related effect on brain NE levels (Jaim-Etcheverry and Zieher 1980), sustained throughout the remaining lifetime (Jonsson et al. 1981). In contrast to the central nervous system (CNS), the absence of alterations in sympathetic neurons in the periphery (superior cervical ganglia) has been reported (Jaim-Etcheverry and Zieher 1980). The lowest tested DSP-4 dose (10 mg/kg s.c.) produces a NE reduction of approximately 50 % in the cerebral cortex and spinal cord, with negligible changes in the cerebellum and pons–medulla. Conversely, a large DSP-4 dose (50 mg/kg s.c.) incurs near complete NE depletion in the cerebral cortex and spinal cord, with concomitant increases in NE levels, both in the pons–medulla and cerebellum, although the effects are more pronounced in the latter region (Jonsson et al. 1982). A dose of 50 mg/kg of DSP-4 is well tolerated by developing rats, does not increase the mortality rate, and is the most efficient in producing a long-lasting reduction of NE reuptake in the brain, both in male and female rats (Jonsson et al. 1981). Similarly, DSP-4 administration schedules also differ between studies. A single injection scheme at postnatal days (PND) 1–3 is the most effective and frequently employed technique for the selective generation of NE phenotypes. However, DSP-4 can be administered to newborn rats once on postnatal day 1 (PND 1), twice on PND 1 and 3, and three times on PND 1, 3, and 5, but this compound loses selectivity with increased frequency dosing (Brus et al. 2004).

3　Noradrenergic System

3.1　Biochemical Alterations

The outcome of systemic DSP-4 treatment of newborn rats is an alteration in the postnatal development of the noradrenergic system involving the permanent denervation of distal NE projection areas, accompanied by NE hyperinnervation in regions proximal to the LC cell bodies, likely 1980reflecting the collateral sprouting of noradrenergic neurons. Notably, the total number of NE nerve terminals remains unaltered (Jaim-Etcheverry and Zieher ; Jonsson and Sachs 1982).

Neonatal DSP-4 treatment leads to persistent changes of endogenous NE levels in several tested brain regions. In adulthood, NE is depleted approximately 90–98 % in the cerebral cortex (frontal, cingulate, parietal, occipital), hippocampus, spinal cord (lumbar) (Jonsson et al. 1982; Nowak et al. 2008, 2009), and olfactory bulb (Cornwell et al. 1996), while only 35 % depletion is observed in the striatum, following neonatal treatment (Dabrowska et al. 2007). Negligible effects are observed in the hypothalamus (Jonsson et al. 1982; Dabrowska et al. 2007). The hypothalamus receives the main noradrenergic input from the lateral tegmentum (LT) (Cryan et al. 2002), supporting previous findings that systemic pretreatment with DSP-4 selectively destroys noradrenergic terminals emanating from the LC. No significant changes in NE concentration are noted in the thalamus and medulla oblongata (Jonsson et al. 1982; Korossy-Mruk et al. 2013), while a pronounced increase in NE levels is observed in the cerebellum, brainstem, mesencephalon, and pons–medulla after neonatal DSP-4 treatment (Jonsson et al. 1982; Bortel et al. 2008c). Basically, following neonatal DSP-4 treatment, the level of the major NE metabolite, 3-methoxyl-4-hydroxyphenylethylene glycol (MHPG), is not affected in frontal cortex, hippocampus, hypothalamus, and striatum; consequently, MOPEG/NE metabolic ratios are significantly increased in frontal cortex, hippocampus, and striatum (Dabrowska et al. 2007) or unchanged in the brain stem and cerebellum (Bortel et al. 2008c).

Treating neonatal animals (on PND 3) with DSP-4 and harvesting brains on PND 32 produce a near complete elimination of NE in the cerebral cortex and hippocampus with a parallel reduction in the noradrenergic transporter (NET), as determined through autoradiographic analysis with [^{3}H]nisoxetine, a highly specific ligand for the NET (Sanders et al. 2011). When administered at birth, DSP-4 also reduces [^{3}H]NE reuptake in the spinal cord (93 %) but increases [^{3}H]NE reuptake in the cerebellum (94 %), with no significant alterations in the striatum. However, when measuring NE uptake in the entire brain, no significant changes are observed, suggesting that similar to neonatal 6-hydroxydopamine 6-(OHDA) treatment, the altered development of NE neurons induced through DSP-4 primarily reflects a redistribution of the NE nerve terminal projections of the LC without appreciably affecting the total number of NE terminals (Jonsson et al. 1982; Jonsson and Sachs 1982). Notably, pretreatment with desmethylimipramine (DMI; the NE uptake blocker) or pargyline (non-selective monoamine oxidase inhibitor) effectively counteracts the DSP-4-induced alterations of NE levels in the frontal and occipital cortex, stratum, pons–medulla, cerebellum, and spinal cord, with the exception of the olfactory bulb and hippocampus (only partial protection). Consistent with the results of the NE assay, DMI pretreatment completely abolishes the DSP-4-induced reduction of NE uptake in the cerebral cortex and increase in NE uptake in the cerebellum (Jonsson et al. 1982).

Permanent NE denervation in the cerebral cortex and spinal cord is present at all developmental stages, although this effect is more pronounced in rats treated with

DSP-4 at an early age (up to PND 5). NE reduction is gradually less marked following DSP-4 treatments as late as PND 14, after which the extent of NE depletion is constant into adulthood. Furthermore, NE hyperinnervation in the cerebellum and pons–medulla is only observed when DSP-4 is administered to one- and three-day-old rats, whereas when injected at eight days or older, permanent NE depletion in these structures is observed. Acute NE depletion (one day after DSP-4 treatment) occurs in animals of all ages, and the neurotoxic effect of this compound is rapid, being maximal at 2 h after DSP-4 injection for NE reuptake in vitro and 4–6 h for endogenous NE concentration (Jonsson et al. 1982).

Monoamine neurons respond to lesions with a wide range of compensatory adaptations aimed at preserving functional integrity, ranging from sensitization of the receptor (frequently characterized as the expression of more cell surface receptors or receptor up-regulation) to down-regulation or adaptation of second messengers (Giorgi et al. 2008; Kostrzewa et al. 2008). Basically, this is true when NE denervation is performed in adult rats (Heal et al. 1993; Wolfman et al. 1994; Kreiner et al. 2011) or when other phenotypes, such as dopamine (DA) or 5-hydroxytryptamine (serotonin; 5-HT), are observed (Pranzatelli and Gregory 1993; Sawynok and Reid 1994; Kostrzewa et al. 2011). DSP-4 lesions on PND 3 produce only transient α_2-adrenergic receptor reduction in brain at PND 5. The α_2-adrenergic receptor "recovers" to control levels at PND 15 and PND 25, and no further change in the total receptor density is detected. In addition, on PND 25, there is no alteration in the association of α_2-adrenergic receptors with G proteins (Sanders et al. 2011).

Interestingly, severe damage does not necessarily impair the activity of the NE system in terms of postsynaptic efficacy. There are a number of compensatory mechanisms (increased presynaptic turnover, leading to increased NE synthesis and release per surface unit), which maintain the level of extracellular NE at a sufficient rate to provide appropriate receptor stimulation, at least under basal conditions. This phenomenon is observed either when DSP-4 is administered to neonates (Nowak et al. 2004) or adult rats (Kask et al. 1997), as a consequence of volume transmission and/or reduced clearance due to reductions in NET number, an aftereffect of noradrenergic fiber degeneration (Hughes and Stanford 1998).

3.2 Behavioral and Biological Rhythm Alterations

Learning and memory processes are critically influenced through noradrenergic activity, and numerous studies in adult animals support this hypothesis (Khakpour-Taleghani et al. 2009; Reid and Harley 2010; Gazarini et al. 2013). In neonatal NE-depleted rats, some abnormalities can also be detected. Cornwell-Jones et al. (1990) reported attenuated olfactory learning in adult rats treated at the day of birth (PND 0) with DSP-4 (50 mg/kg), consistent with additional data suggesting that NE depletion impairs the adoption of a new olfactory environment after weaning (Cornwell-Jones et al. 1982). In addition, DSP-4 pretreatment (PND 0) disrupts the acquisition of odor, specifically odor association in the sensory preconditioning paradigm, in 16-day-old rats, but no effects of DSP-4 are observed on first-order conditioning (Chen et al. 1993). In contrast, DSP-4 at PND 0 does not impair the memory consolidation of emotionally arousing tasks, such as inhibitory avoidance learning, when assessed at 28 days later (Cornwell-Jones et al. 1989). Similar studies yield variable and inconsistent results (Archer 1982; Bennett et al. 1990; Hauser et al. 2012). Interestingly, the slight effects of neonatal DSP-4 treatment on learning and memory processes in rats show close parallels and inconsiderable alterations in the biochemical mechanisms underlying cognitive function, e.g., immediate early gene expression in the brain for the regulation of neuroplasticity. Precisely, neonatal DSP-4 administration leads to only moderate increases or no changes in the levels of Arc, c-fos, and zif268 at PNDs 13, 25, and 60, in marked contrast to the effects of similar lesions in the adult brain (Sanders et al. 2008).

LC, the major noradrenergic nucleus in brain also mediates spontaneous and evoked arousal. The injection of DSP-4 into rats at PND 7 does not significantly affect sleep bout duration, although the wake bout duration and percentage of time awake are significantly increased when tested at PND 21 (Gall et al. 2009).

While basic mechanisms controlling sexual development are complex and controversial, the CNS, including the NE system, is an essential participant in this process (Martins-Afférri et al. 2003; Izvolskaia et al. 2009). A single exposure of five-day-old female rat pups to DSP-4 delays sexual maturation, measured as vaginal opening (VO) (i.e., 34.2 days with saline versus 36.1 days with DSP-4). However, DSP-4 rats showed a concomitant delay in weight gain, and both control and treated animals showed similar VO when adjusted to the same body weight (Jacobson et al. 1988). In conclusion, neonatal DSP-4 treatment marginally influences behavioral and biological rhythms in adulthood.

4   Serotoninergic System

4.1   Biochemical Alterations

Despite being regarded as a highly selective neurotoxin for the noradrenergic system, DSP-4, when administered in the absence of a 5-HT reuptake inhibitor (e.g., zimelidine), induces dose-dependent changes in endogenous 5-HT levels in brain. Neonatal DSP-4 treatment (50 mg/kg) significantly reduces endogenous 5-HT concentrations in the cerebral cortex ($\sim 55\ \%$), hippocampus ($\sim 60\ \%$), spinal cord ($\sim 40\ \%$), and cerebellum ($\sim 30\ \%$), when analyzed at the adult stage. In parallel with these observations, 5-HT reuptake in the occipital cortex, hippocampus, and cerebellum is reduced. No significant effects are observed in other brain regions (Jonsson et al. 1981, 1982). Interestingly, neonatal DSP-4 treatment does not produce any significant change in the endogenous 5-HT concentration and 5-HT reuptake up to one month after drug administration, when whole brain is analyzed (Jonsson et al. 1982).

The SERT blocker, zimelidine (10 mg/kg), abolishes the effects of DSP-4 on 5-HT and its metabolite 5-hydroxyindoleacetic acid (5-HIAA) in frontal cortex, hippocampus, striatum, and hypothalamus without interfering with the action of this neurotoxin on noradrenergic nerve terminals (Jonsson et al. 1981, 1982; Dabrowska et al. 2007). In addition, no significant differences in the 5-HT synthesis rate in the frontal cortex, hippocampus, striatum, hypothalamus (Dabrowska et al. 2007), and cerebellum (Roczniak et al. 2015) between control and DSP-4-pretreated animals are observed. Under these conditions, despite the absence of changes in the content and synthesis rate of 5-HT in the brain, other adaptive alterations in the 5-HT system are detected. These alterations are not noticeable until biochemical responses are "enforced" through the stimulation or blocking of 5-HT receptors. Such alterations are obvious due to a mutual interaction between NE and 5-HT; e.g., numerous brain regions are innervated with both noradrenergic (origin, LC) and serotoninergic neurons originating from the dorsal raphe nuclei (DRN) and median raphe nuclei (MRN). Furthermore, LC, the major NE brain stem nucleus, sends projections to the DRN, while the DRN projects to the LC (Sim and Joseph 1993; Peyron et al. 1996) creating ample opportunity for cross-modulation (Millan et al. 2000a, b; Weikop et al. 2004). The scarcity of studies on this subject in neonatal DSP-4-lesioned rats makes it difficult to predict the net balance between NE vs. 5-HT. Moreover, endogenous NE exerts direct tonic stimulatory control on the release of 5-HT through α_1-adrenoceptors and an indirect tonic inhibitory influence through α_2-adrenoceptors located on noradrenergic nerve terminals within the raphe nuclei, although inhibitory α_2-heteroreceptors are also localized on terminals of serotoninergic neurons in corticolimbic structures (Haddjeri et al. 1995).

The stimulation of the somatodendritic 5-HT_{1A} autoreceptors decreases the cell firing rate, synthesis, turnover, and release of 5-HT within raphe nuclei and subsequently within serotonergic projection areas (Blier et al. 1998). Because NE might have an inhibitory effect on cortical 5-HT release (Haddjeri et al. 1995), noradrenergic denervation results in a moderate increase in basal 5-HT microdialysate content in the medial prefrontal cortex of neonatal DSP-4-treated rats. Also, systemic administration of the 5-HT_{1A} receptor agonist (R-(+)-8-OH-DPAT) induces a long-lasting reduction of extracellular 5-HT content in the medial prefrontal cortex of intact and DSP-4-pretreated rats, but this effect is significantly reduced in noradrenergic lesioned animals, likely reflecting the desensitization of 5-HT_{1A} autoreceptors in the raphe nuclei of such rats (Dabrowska et al. 2008). Conversely, in the same experimental model, the stimulation of 5-HT_{1B} receptors through the systemic administration of a specific agonist (CP 94253) does not evoke significant changes in 5-HT release in the medial prefrontal cortex between control and DSP-4-treated rats (Ferdyn-Drosik et al. 2010).

In the chromatographic assay of the 5-HT synthesis rate, the systemic administration of R-(+)-8-OH-DPAT to control rats inhibits 5-HT synthesis in the prefrontal cortex, hypothalamus, and striatum 42, 20, and 46 %, respectively, and this effect is antagonized through pretreatment with the 5-HT_{1A} receptor antagonist. Interestingly, R-(+)-8-OH-DPAT does not significantly inhibit the 5-HT synthesis rate in all examined brain structures of neonatal DSP-4-treated rats. These results, together with microdialysis study, suggest the desensitization of 5-HT_{1A} autoreceptors in the raphe nuclei as a result of noradrenergic lesions (Dabrowska et al. 2008). Similar results concerning 5-HT_{1B} receptor stimulation have been reported, i.e., 5-HT_{1B} receptor agonist (CP 94253) reduces the 5-HT synthesis rate in medial prefrontal cortex of control rats (~ 33 %), but with no effect on neonatal DSP-treated animals (Ferdyn-Drosik et al. 2010).

4.2 Behavioral Alterations

5-HT_{1A} receptors are located presynaptically on the soma and dendrites of 5-HT neurons of the DRN and MRN as inhibitory autoreceptors, and postsynaptically in the forebrain areas, including the hippocampus, lateral septum, and cortex (Hamon 2000). The stimulation of presynaptic 5-HT_{1A} receptors evokes behavioral responses, such as hyperphagia in satiated rats (Ebenezer 1992) and pigs (Ebenezer et al. 1999) and anxiolytic-like effects in rats (Jolas et al. 1995) and gerbils (File et al. 1996).

Dabrowska et al. (2008) observed that the 5-HT_{1A} receptor agonist (R-(+)-8-OH-DPAT) produces a significant increase in food intake in control rats, but fails to elicit a response in neonatal DSP-4-treated animals. In the anxiolytic-like activity assessment (plus maze test, social interaction test), R-(+)-8-OH-DPAT induces anxiolytic-like activity in control rats; however, this compound does not evoke an effect on DSP-4-lesioned rats. These findings lend support to the hypothesis of 5-HT_{1A} autoreceptor desensitization in neonatal NE-lesioned animals.

Conversely, behavioral responses reflecting postsynaptic 5-HT_{1A} receptor activation include stereotypic behavior in rats (O'Connell and Curzon 1996), an antidepressive effect in mice and rats (Luscombe et al. 1993; Schreiber and De Vry 1993), and hypothermia (Millan et al. 1993). In the forced swimming test and the learned helplessness test, R-(+)-8-OH-DPAT displays antidepressant-like activity to a similar extent in both control and DSP-4-lesioned rats, and this effect is antagonized through the 5-HT_{1A} antagonist, WAY 100635 (Dabrowska et al. 2008). Further studies reveals that R-(+)-8-OH-DPAT induces hypothermia and "5-HT_{1A} syndrome" to a similar extent in both control and NE denervated rats; these effects are completely antagonized by WAY 100635 (Dabrowska et al. 2007). Thus, these data suggest that the reactivity of postsynaptically located 5-HT_{1A} receptors remains unchanged after neonatal DSP-4 lesioning.

Systemic DSP-4 treatment of newborn rats also modulates behavioral responses mediated by the 5-HT_{1B} receptor. Activation of this receptor through a specific agonist (CP 94253) elicits anxiogenic-like effects in intact rats, as demonstrated in the elevated plus maze test (Lin and Parsons 2002). Notably, Ferdyn-Drosik et al. (2008, 2010) also observed that CP 94253 induced anxiogenic-like activity in control rats and depression-like behavior in the Porsolt test. However, no effects are detected in the DSP-4 group, suggesting the existence of regulatory mechanisms in DSP-4-lesioned animals altering the functional efficacy of 5-HT_{1B} receptors and, accordingly, affecting serotoninergic activity, as 5-HT_{1B} receptors are desensitized in these rats.

5 Dopaminergic System

5.1 Biochemical Alterations

Neonatal DSP-4 treatment (50 mg/kg at PND 1 and PND 3) does not acutely (i.e., directly) affect DA and DOPAC levels in the cortex, hippocampus, striatum, mesencephalon, pons–medulla, cerebellum, and spinal cord of adult rats (Jonsson et al. 1981; Nowak et al. 2006). In addition, rats treated with DSP-4 (50 mg/kg) at PND 7 do not exhibit significant changes in the DA or 5-HT concentration in cortical or non-cortical tissues (including medulla, midbrain, and diencephalon, minus the cerebellum) versus saline controls at PND 21 (Gall et al. 2009). In parallel, the DA synthesis rate, assessed as L-DOPA accumulation after aromatic amino acid decarboxylase inhibitor administration (hydroxybenzylhydrazine; NDS-1015), remains unchanged in several tested brain regions (hippocampus, striatum, and cerebellum) (Roczniak et al. 2015). Other studies report an adaptive decrease in the DA content in the occipital cortex (38 %), hippocampus (49 %), and mesencephalon (13 %) (Jonsson et al. 1982) or thalamus (50 %) (Donnerer et al. 1992).

5.2 Behavioral Alterations

The stimulation of specific DA receptors elicits distinct behavioral responses, e.g., acute DA D_1 agonist treatment of rats produces vacuous chewing movements, sniffing, and grooming behavior (Arnt et al. 1987; Hamdi and Kostrzewa 1991; Kostrzewa and Gong 1991). Acute "classical" DA D_2 agonist (e.g., apomorphine, piribedil) administration exerts biphasic effects on behavior, i.e., yawning and hypomotility at low doses and stereotypy and hypermobility at high doses (Butterworth et al. 1993; Eilam and Szechtman 1989; Mogilnicka and Klimek 1997). However, the yawning behavior elicited at low doses of mixed DA D_2/D_3 agonists (e.g., quinpirole) has been associated with DA D_3 receptor stimulation (Kostrzewa and Brus 1991a).

Neonatal DSP-4 treatment does not affect DA D_1 receptor agonist (SKF 38393)-evoked vacuous chewing movements and DA D_2 receptor agonist (quinpirole—high dose)-induced locomotor and exploratory activities in adulthood. Conversely, quinpirole (low dose)-induced yawning behavior is more prominent in the group lesioned with DSP-4 (Nowak et al. 2009).

Notably, repeated treatment with dopaminergic agonists (e.g., amphetamine, apomorphine, and quinpirole) produces an increased response, or sensitization (priming), to the drug-induced effects, referred to as DA receptor supersensitivity (Mattingly and Gotsick 1989; Nowak et al. 2005; Brus et al. 2003). Male rats treated for the first 28 days after birth with the dopamine D_2/D_3 agonist quinpirole develop lifelong sensitization of the central DA D_2/D_3 receptors that manifests as enhanced quinpirole-induced yawning behavior and motor behaviors in adulthood

(Kostrzewa and Brus 1991b). When DSP-4 is administered to rats (PND 1 and PND 3) to largely destroy noradrenergic innervation of the brain, the magnitude of quinpirole-induced action is significantly reduced. This observation suggests that intact central noradrenergic innervation is important for behavioral responses mediated through a variety of DA receptors and the expression of priming phenomenon (Nowak et al. 2006).

6 GABAergic System

6.1 Biochemical Alterations

Neonatal DSP-4 treatment of rats does not alter GABA tissue levels in prefrontal cortex, hippocampus, brain stem, and cerebellum and does not modify phenobarbital- or ethanol-evoked GABA reduction in the aforementioned structures (Bortel et al. 2008c). Similarly, under steady-state conditions, the microdialysate content of GABA in the prefrontal cortex in DSP-4 neonatally lesioned rats does not differ when compared to control rat. However, a twofold greater increase in the extracellular GABA level after vigabatrin (GABA transaminase inhibitor) injection is observed in DSP-4-lesioned rats (Bortel et al. 2008b).

The NE depletion of cerebral cortex and hippocampus produced by DSP-4 treatment is accompanied by a reduction (approximately 20 %) in the number of GABA-A receptors localized on the presynaptic axons and nerve terminals of NE neurons, and a decrease in the Bmax for the low-affinity GABA-B receptor site in cerebral cortex and hippocampus (25 and 28 %, respectively) (Suzdak and Gianutsos 1985). Neonatal DSP-4 treatment also reduces the number of benzodiazepine (BDZ) receptors in the cerebral cortex, while a significant increase in the number of BDZ receptors in the cerebellum occurs (Medina and Novas 1983). These results support the existence of functional coupling between the noradrenergic and GABAergic systems, i.e., decrease or increase in the density of BDZ receptors following the denervation (e.g., cortex) or hyperinnervation of some brain structures (e.g., cerebellum).

6.2 Behavioral Alterations

In the elevated plus maze test, a paradigm based on the conflict between the innate tendencies of rodents to explore novel environments versus an innate tendency to avoid open areas, systemic DSP-4 treatment of newborn rats does not affect spontaneous anxiety-like behavior in adulthood, but diminishes anxiolysis elicited through diazepam administration (Bortel et al. 2007). Neonatal DSP-4 treatment also leads to adulthood changes in the activity of sedative–hypnotic drugs, i.e., extends the time to the loss of righting reflex after phenobarbital and reduces sleep time after ethanol administration (Bortel et al. 2008c). Moreover, reduced vulnerability of the pro-convulsant activity of bicuculline (GABA-A receptor antagonist) is observed in DSP-4-pretreated rats (Bortel et al. 2008a).

7 Other Neuronal Phenotypic Systems

Alterations of NE exocytosis in the thalamus, brain stem, and other nuclei alter the output of nociceptive information in higher brain centers from projection neurons (Post et al. 1985; Pagano et al. 2012; Suehiro et al. 2013). LC stimulation, which increases NE release in the spinal cord, inhibits nociceptive transmission in the dorsal horn via α_2-adrenergic receptors (Margalit and Segal 1979; Delaney et al. 2007; Jiang et al. 2010). In addition, LC neurons possess a high density of post-synaptic mu-opioid receptors (Van Bockstaele and Commons 2001), and cannabinoids modulate noradrenergic neuronal activity. Scavone et al. (2010) provided evidence for the heterogeneous distribution of CB_1 receptors in the LC and demonstrated that this receptor and mu-opioid receptors coexist in cellular profiles in this region, creating ample opportunity for interactions between cannabinoid and noradrenergic systems. Indeed, the destruction of noradrenergic neurons through neonatal DSP-4 treatment significantly decreases the antinociceptive effects of methanandamide (CB_1 receptor agonist) in the tail-immersion test, hot-plate test, and writhing test, with ambiguous results in paw pressure and formalin hind paw tests. Simultaneously, marked changes in the antinociceptive effects of methanandamide in DSP-4-treated rats are not accompanied by increases in CB_1 receptor density in the rat brain (Korossy-Mruk et al. 2013).

Neonatal DSP-4 treatment (PND 1 and PND 3) produces nearly imperceptible effects on the central histaminergic system in rats. Histamine content remains unchanged in the frontal cortex, hypothalamus, cerebellum, and medulla oblongata, and only a modest but significant reduction (31 %) in the hippocampus is observed in DSP-4 rats. In addition, exploratory activity, irritability, and nociceptive activity are unaltered after histamine receptor antagonist treatment, although locomotor activity and vacuous chewing movements are increased compared with control when histamine H_2 and H_3 antagonists are applied (Nowak et al. 2008).

DSP-4 administered to rat pups (PND 4) reduces [^{3}H]NE uptake in the cortex and hypothalamus approximately 90 and 37 %, respectively, at PND 7. Simultaneously, nicotine-stimulated [^{3}H]NE release from the neonatal hypothalamus is almost completely eliminated after prior lesioning of the LC. Hence, systems controlling critical homeostatic functions (stress, feeding, etc.) under nicotinic acetylcholine receptor (nAChR) influence are disrupted through NE denervation (O'Leary and Leslie 2006).

DSP-4 treatment of rats at PND 2 produces adaptive changes in the neuropeptide innervation in adulthood. Vasoactive intestinal polypeptide (VIP) is markedly elevated in the cortex, hippocampus, striatum, and medulla, whereas substance P (SP), neurokinin-like immunoreactivity (NK-Li), and calcitonin gene-related peptide (CGRP) remain unchanged (Donnerer et al. 1992). Perhaps, the increased availability of growth factors in the CNS induces denser innervation spared neuron systems (e.g., VIP) after the elimination of noradrenergic innervation, as discussed for peripheral neurons after chemical sympathectomy (Aberdeen et al. 1990). Conversely, the dorsal and spinal cord neuropeptide Y (NPY) concentration does not differ following neonatal DSP-4 treatment (Donnerer et al. 1990).

8   Prenatal DSP-4 Treatment

DSP-4 crosses the blood–placenta barrier and has a potent neurotoxic effect in the prenatal stage. When the pregnant dam is injected with DSP-4 (20 mg/kg) on gestation days 18 and 19, the NE concentration in frontal and occipital cortex of one-day-old pups is reduced, with no change in NE in cerebellum. The NE level in the frontal cortex remains reduced until adulthood (Jaim-Etcheverry and Zieher 1980; König et al. 1985). The absence of changes in the NE concentration in the cerebellum and pons–medulla of treated prenates (observed as adults) contrasts with the effects observed after DSP-4 treatment of newborn animals (Jonsson et al. 1981).

Neonatal DSP-4 administration does not lead to remarkable changes in spontaneous locomotor and exploratory activity. The results obtained from rotating-rod tests (examining motor coordination) are controversial, and subtle differences between male and female rats can be detected. Thus, the absence of distinct changes in locomotor and exploratory activity suggests that the role of NE in these behavioral patterns might be limited (König et al. 1985).

9   Summary

9.1   Noradrenergic System

Neonatal DSP-4 (50 mg/kg s.c.) incurs near complete NE depletion (90–98 %) in the cerebral cortex, hippocampus, and spinal cord. Minor effects are observed in the striatum and hypothalamus, while no significant changes in NE concentration are noted in the thalamus and medulla oblongata. In contrast, a pronounced increase in NE levels is observed in the cerebellum, brain stem, mesencephalon, and pons–medulla (Jonsson et al. 1981, 1982; Dabrowska et al. 2007; Bortel et al. 2008c; Korossy-Mruk et al. 2013). The "picture" of endogenous NE brain alterations suggests that DSP-4 has a preferential, selective neurotoxic effect on nerve terminal projections emanating from noradrenergic perikarya of LC and does not affect NE axons in brain receiving a dense innervation from non-cerulean noradrenergic cells (Jonsson et al. 1981). A near complete elimination of NE in the cerebral cortex and hippocampus by neonatal DSP-4 treatment is accompanied by a parallel reduction in NET (Sanders et al. 2011) and in vitro [^{3}H]NE reuptake in cerebral cortex and spinal cord with concomitant [^{3}H]NE reuptake increases in the cerebellum and no significant alterations in the striatum (Jonsson et al. 1982; Jonsson and Sachs 1982). Permanent noradrenergic denervation in the cerebral cortex and spinal cord is present at all developmental stages, although this effect is more pronounced in rats treated with DSP-4 at an early age (up to PND 5). NE reduction is gradually less marked following DSP-4 treatments as late as PND 14, after which the extent of NE depletion is constant into adulthood (Jonsson et al. 1982). DSP-4 lesions on PND 3 produce only transient α_2-adrenergic receptor reduction in brain at PND 5. The α_2-adrenergic receptor "recovers" to control levels at PND 15 and PND 25, and no further change in the total receptor density is detected (Sanders et al. 2011). Interestingly, severe damage does not impair the activity of the NE system in terms of postsynaptic efficacy. The compensatory mechanisms maintain the level of extracellular NE at a sufficient rate to provide appropriate receptor stimulation, at least under basal conditions (Nowak et al. 2004). Neonatal DSP-4 treatment marginally influences behavior and biological rhythms in adulthood; some abnormalities in learning and memory processes (Cornwell-Jones et al. 1982, 1989; Chen et al. 1993), spontaneous and evoked arousal (Gall et al. 2009), and sexual development (Jacobson et al. 1988) can be detected.

9.2 Serotoninergic System

DSP-4, when administered in the absence of a 5-HT reuptake inhibitor, reduces (up to 60 %) endogenous 5-HT concentrations in the cerebral cortex, hippocampus, spinal cord, and cerebellum, when analyzed at the adult stage. In parallel with these observations, 5-HT reuptake in the occipital cortex, hippocampus, and cerebellum is reduced (Jonsson et al. 1981, 1982). The SERT blocker, zimelidine, abolishes the effects of DSP-4 on 5-HT and its metabolite 5-HIAA in frontal cortex, hippocampus, striatum, and hypothalamus without interfering with the action of this neurotoxin on noradrenergic nerve terminals (Jonsson et al. 1981, 1982; Dabrowska et al. 2007). Under these conditions, despite the absence of changes in the content and synthesis rate of 5-HT in the brain (Dabrowska et al. 2007; Roczniak et al. 2015), other adaptive alterations in the 5-HT system are detected. These alterations are not noticeable until behavioral or biochemical responses are "enforced" through the stimulation or blocking of 5-HT receptors. Notably, experimental data suggest the existence of regulatory mechanisms in neonatally DSP-4-lesioned animals altering the functional efficacy of 5-HT$_1$ receptors and, accordingly, affecting serotoninergic activity, as 5-HT$_{1A}$ autoreceptors (Dabrowska et al. 2007, 2008) and 5-HT$_{1B}$ receptors (Ferdyn-Drosik et al. 2008, 2010) are desensitized in these rats.

9.3 Dopaminergic System

Neonatal DSP-4 treatment does not affect DA and DOPAC levels in the cortex, hippocampus, striatum, mesencephalon, pons–medulla, cerebellum, and spinal cord of adult rats (Jonsson et al. 1981; Nowak et al. 2006). In parallel, the DA synthesis rate remains unchanged in several tested brain regions (hippocampus, striatum, and cerebellum) (Roczniak et al. 2015). Neonatal DSP-4 treatment does not affect DA D_1 and D_2 receptor agonist-evoked behavioral responses (vacuous chewing movements, locomotor, and exploratory activities), whereas DA D_3 quinpirole-induced yawning behavior as well as lifelong sensitization of the central DA D_2/D_3 receptors are modified. These observations suggest that intact central noradrenergic innervation is important for behavioral responses mediated through a variety of DA receptors and the expression of priming phenomenon (Nowak et al. 2006, 2009).

9.4 GABAergic System

Neonatal DSP-4 treatment of rats does not induce any change in GABA tissue levels (prefrontal cortex, hippocampus, brainstem, cerebellum) and does not modify phenobarbital or ethanol-evoked GABA reduction in the brain (Bortel et al. 2008c). However, DSP-4 treatment increases GABAergic neurotransmission in prefrontal cortex of rats in adulthood (Bortel et al. 2008b). Also, there is a reduction in the number of GABA-A receptors in cerebral cortex and hippocampus (Suzdak and Gianutsos 1985) and BDZ receptors in the cerebral cortex, while a significant increase in the number of BDZ receptors in the cerebellum is noted (Medina and Novas 1983). These results support the existence of functional coupling between the noradrenergic and GABAergic systems, i.e., decrease or increase in the density of BDZ receptors following GABAergic denervation (e.g., cortex) or hyperinnervation of some brain structures (e.g., cerebellum). Moreover, neonatal DSP-4 treatment diminishes anxiolysis elicited through diazepam administration (Bortel et al. 2007), reduces vulnerability of the pro-convulsant activity of bicuculline (Bortel et al. 2008a), and leads to adulthood changes in the activity of sedative–hypnotic drugs (Bortel et al. 2008c).

9.5 Other Neuronal Phenotypic Systems

The destruction of noradrenergic neurons through neonatal DSP-4 treatment significantly decreases the antinociceptive effects of methanandamide (CB_1 receptor agonist) without noticeable changes in CB_1 receptor density in the rat brain (Korossy-Mruk et al. 2013). Conversely, nearly imperceptible effects on the central histaminergic system in rats after DSP-4 treatment are observed. Histamine content remains unchanged in several tested brain structures, and only a modest reduction in the hippocampus is observed (Nowak et al. 2008). It is noteworthy that neonatal DSP-4 treatment produces some adaptive changes in the neuropeptide innervation in adulthood; VIP is markedly elevated in the cortex, hippocampus, striatum, and medulla, whereas SP, NK-Li, CGRP, and NPY remain unchanged (Donnerer et al. 1990, 1992).

10 Conclusions

Studies by Ross and colleagues in 1973 were a stimulus for investigations of DSP-4 activity, over the next 40 years. DSP-4 has attained pharmacological legitimacy and a "strong" position among the other neurotoxins utilized in the neuroscience field, e.g., 6-OHDA, 1-methyl-4-phenyl-1,2,3,6-tetrahydropyridine (MPTP), and rotenone. This is attributable to the following: (1) DSP-4 appears to be a useful experimental tool in studies of the mechanisms of LC neuron degeneration and recovery. This is based on the findings that DSP-4 interacts with NET, whereby it is then accumulated intraneuronally to trigger the degeneration of noradrenergic terminals in the brain (Lyons et al. 1989). Related DSP-4 effects are observed in the peripheral sympathetic neurons as compared with central noradrenergic phenotypes. However, a complete recovery of peripheral noradrenergic nerves and NE tissue level and reuptake is observed within several weeks (Jaim-Etcheverry et al. 1980; Jonsson et al. 1981); (2) DSP-4 may serve as a useful tool to discriminate between LC- and non-LC-dependent behavioral and biochemical responses. This results from the fact that DSP-4 has been considered a LC-selective noradrenergic neurotoxin based on documented alterations in terminal noradrenergic fibers in brain regions innervated chiefly by the LC without affecting non-cerulean noradrenergic axons (Jaim-Etcheverry and Zieher 1980; Jonsson and Sachs 1982; Cryan et al. 2002); (3) DSP-4 in laboratory animals may serve as a convenient model to study the mechanisms and functional consequences of the compensatory but selective noradrenergic hyperinnervation. Typically, DSP-4 is applied to adult animals but may also be administered during perinatal period. Thus, DSP-4 treatment of newborn rats leads to pronounced denervations of distant nerve terminal projections, while innervation areas close to the cell bodies become hyperinnervated, likely reflecting the collateral sprouting of noradrenergic neurons. Notably, regional hyperinnervation is a hallmark of neonatal DSP-4 treatment, which is not observed after either prenatal or adult DSP-4 application (Jonsson et al. 1982; Bortel et al. 2008c); (4) DSP-4 may be used to investigate synaptic biochemistry including mechanisms of noradrenergic receptor up- and down-regulation, in particular the phenomenon of temporary and reversible receptor changes. Notably, neonatal DSP-4 lesions produce only transient adrenergic receptor reduction in brain with short "recovery" phase resulting finally in no change in the total receptor density (Sanders et al. 2011). This strikingly contrasts with dopaminergic or serotoninergic neonatal lesions (Kostrzewa et al. 2008); and (5) DSP-4 may be helpful in clarifying the pharmacological bases for the complex interactions between noradrenergic and other neuronal phenotypes. In fact, neonatal DSP-4 treatment produces desensitization of 5-HT$_{1A}$ and 5-HT$_{1B}$ autoreceptors (Dabrowska et al. 2007, 2008; Ferdyn-Drosik et al. 2010), diminishes GABA-A receptor-mediated behavioral and biochemical responses (Bortel et al. 2007, 2008a, b, c) as well as modifies behavioral responses mediated through a variety of DA receptors and the expression of DA receptor priming phenomenon (Nowak et al. 2006, 2009).

In conclusion, a better understanding of the role of the noradrenergic system in the therapeutic effects of several psychotropic medications as well as introducing new therapies for behavioral and/or psychiatric disorders would not have been possible without apparent (vivid) DSP-4 contribution. Concurrently, it is believed that DSP-4 still possesses the immense pharmacological potential that can be utilized through the next decades in the neuroscience area.

References

Aberdeen J, Corr L, Milner P, Lincoln J, Burnstock G (1990) Marked increases in calcitonin gene-related peptide-containing nerves in the developing rat following long-term sympathectomy with guanethidine. Neuroscience 35:175–184

Archer T (1982) DSP4 (N-2-chloroethyl-N-ethyl-2-bromobenzylamine), a new noradrenaline neurotoxin, and stimulus conditions affecting acquisition of two-way active avoidance. J Comp Physiol Psychol 96:476–490

Arnt J, Hyttel J, Perregaard J (1987) Dopamine D-1 receptor agonists combined with the selective D-2 agonist quinpirole facilitate the expression of oral stereotyped behaviour in rats. Eur J Pharmacol 133:137–145

Bennett MC, Kaleta-Michaels S, Arnold M, McGaugh JL (1990) Impairment of active avoidance by the noradrenergic neurotoxin, DSP4: attenuation by post-training epinephrine. Psychopharmacology 101:505–510

Blier P, Piñeyro G, el Mansari M, Bergeron R, de Montigny C (1998) Role of somatodendritic 5-HT autoreceptors in modulating 5-HT neurotransmission. Ann N Y Acad Sci 861:204–216

Bortel A, Świerszcz M, Jaksz M, Nitka D, Słomian L, Nowak P, Brus R (2007) Anxiety-like behaviour in neonatally DSP-4 treated rats. Behavioural and biochemical studies. Ann Acad Med Siles 61:485–491

Bortel A, Nitka D, Słomian L, Nowak P, Korossy E, Drab J, Brus R, Kostrzewa RM (2008a) Neonatal noradrenergic lesion with DSP-4 modifying the convulsant effect of bicuculline and pentylenetetrazole in adult rats. Bahavioural and biochemical studies. Ann Acad Med Siles 62:46–52

Bortel A, Nowak P, Brus R (2008b) Neonatal DSP-4 treatment modifies GABAergic neurotransmission in the prefrontal cortex of adult rats. Neurotox Res 13:247–252

Bortel A, Słomian L, Nitka D, Swierszcz M, Jaksz M, Adamus-Sitkiewicz B, Nowak P, Jośko J, Kostrzewa RM, Brus R (2008c) Neonatal N-(-2-chloroethyl)-N-ethyl-2-bromobenzylamine (DSP-4) treatment modifies the vulnerability to phenobarbital- and ethanol-evoked sedative-hypnotic effects in adult rats. Pharmacol Rep 60:331–338

Brus R, Kostrzewa RM, Nowak P, Perry KW, Kostrzewa JP (2003) Ontogenetic quinpirole treatments fail to prime for D2 agonist-enhancement of locomotor activity in 6-hydroxydopamine-lesioned rats. Neurotox Res 5:329–338

Brus R, Nowak P, Labus L, Bortel A, Dabrowska J, Kostzrewa RM (2004) Neonatal lesion of noradrenergic neurons in rat brain: interaction with the dopaminergic system. Pol J Pharmacol 56(Suppl):232

Butterworth RF, Poignant JC (1975) Barbeau A (1993) Apomorphine and piribedil in rats: biochemical and pharmacologic studies. Adv Neurol 9:307–326

Chen WJ, Spear LP, Spear NE (1993) Disruptive influence of norepinephrine depletion on sensory preconditioning, but not first-order conditioning, in preweanling rats. Behav Neural Biol 60:110–117

Cornwell CA, Chang JW, Cole B, Fukada Y, Gianulli T, Rathbone EA, McFarlane H, McGaugh JL (1996) DSP-4 treatment influences olfactory preferences of developing rats. Brain Res 711:26–33

Cornwell-Jones CA, Stephens SE, Dunston GA (1982) Early odor preference of rats are preserved by neonatal 6-hydroxydopamine. Behav Neural Biol 35:217–230

Cornwell-Jones CA, Decker MW, Chang JW, Cole B, Goltz KM, Tran T, McGaugh JL (1989) Neonatal 6-hydroxydopa, but not DSP-4, elevates brainstem monoamines and impairs inhibitory avoidance learning in developing rats. Brain Res 493:258–268

Cornwell-Jones C, Palfai T, Young T, Desai J, Krasenbaum D, Morrison J (1990) Impaired hoarding and olfactory learning in DSP-4-treated rats and control cagemates. Pharmacol Biochem Behav 36:707–711

Cryan JF, Page ME, Lucki I (2002) Noradrenergic lesions differentially alter the antidepressant-like effects of reboxetine in a modified forced swim test. Eur J Pharmacol 436:197–205

Dabrowska J, Nowak P, Brus R (2007) Desensitization of 5-HT(1A) autoreceptors induced by neonatal DSP-4 treatment. Eur Neuropsychopharmacol 17:129–137

Dabrowska J, Nowak P, Brus R (2008) Reactivity of 5-HT1A receptor in adult rats after neonatal noradrenergic neurons' lesion—implications for antidepressant-like action. Brain Res 1239:66–76

Delaney AJ, Crane JW, Sah P (2007) Noradrenaline modulates transmission at a central synapse by a presynaptic mechanism. Neuron 56:880–892

Donnerer J, Amann R, Skofitsch G, Lembeck F (1990) Region-specific noradrenaline depletion by neonatal DSP-4: functional consequences and effect on a coexisting neurotransmitter. Eur J Pharmacol 181:147–150

Donnerer J, Humpel C, Saria A (1992) Effect of noradrenergic denervation by neonatal DSP-4 on peptide neurotransmitter systems in the rat brain. Neuropeptides 22:103–106

Ebenezer IS (1992) Effects of the 5-HT1A agonist 8-OH-DPAT on food intake in food-deprived rats. NeuroReport 3:1019–1022

Ebenezer IS, Parrott RF, Vellucci SV (1999) Effects of the 5-HT1A receptor agonist 8-OH-DPAT on operant food intake in food-deprived pigs. Physiol Behav 67:213–217

Eilam D, Szechtman H (1989) Biphasic effect of D-2 agonist quinpirole on locomotion and movements. Eur J Pharmacol 161:151–157

Ferdyn-Drosik M, Jelito K, Korossy E, Bojanek K, Jaksz M, Świerszcz M, Nowak P, Brus R (2008) Impairment of the central serotoninergic system in rats lesioned with DSP-4 as neonates. Adv Clin Exp Med 6:607–614

Ferdyn-Drosik M, Nowak P, Bojanek K, Bałasz M, Kasperski J, Skaba D, Muchacki R, Kostrzewa RM (2010) Neonatal DSP-4 treatment impairs 5-HT(1B) receptor reactivity in adult rats. Behavioral and biochemical studies. Pharmacol Rep 62:608–620

File SE, Gonzalez LE, Andrews N (1996) Comparative study of pre- and postsynaptic 5-HT1A receptor modulation of anxiety in two ethological animal tests. J Neurosci 16:4810–4815

Gall AJ, Joshi B, Best J, Florang VR, Doorn JA, Blumberg MS (2009) Developmental emergence of power-law wake behavior depends upon the functional integrity of the locus coeruleus. Sleep 32:920–926

Gazarini L, Stern CA, Carobrez AP, Bertoglio LJ (2013) Enhanced noradrenergic activity potentiates fear memory consolidation and reconsolidation by differentially recruiting α1- and β-adrenergic receptors. Learn Mem 20:210–219

Giorgi M, D'Angelo V, Esposito Z, Nuccetelli V, Sorge R, Martorana A, Stefani A, Bernardi G, Sancesario G (2008) Lowered cAMP and cGMP signalling in the brain during levodopa-induced dyskinesias in hemiparkinsonian rats: new aspects in the pathogenetic mechanisms. Eur J Neurosci 28:941–950

Haddjeri N, Blier P, de Montigny C (1995) Noradrenergic modulation of central serotonergic neurotransmission: acute and long-term actions of mirtazapine. Int Clin Psychopharmacol 10 (Suppl 4):11–17

Hamdi A, Kostrzewa RM (1991) Ontogenic homologous supersensitization of dopamine D1 receptors. Eur J Pharmacol 203:115–120

Hamon M (2000) The main features of central 5-HT1A receptors. In: Baumgarten HG, Gothert M (eds) Serotonergic neurons and 5-HT receptors in the CNS. Springer, Berlin, pp 239–268

Hauser J, Sontag TA, Tucha O, Lange KW (2012) The effects of the neurotoxin DSP4 on spatial learning and memory in Wistar rats. Atten Defic Hyperact Disord 4:93–99

Heal DJ, Butler SA, Prow MR, Buckett WR (1993) Quantification of presynaptic alpha 2-adrenoceptors in rat brain after short-term DSP-4 lesioning. Eur J Pharmacol 249:37–41

Hughes ZA, Stanford SC (1998) A partial noradrenergic lesion induced by DSP-4 increases extracellular noradrenaline concentration in rat frontal cortex: a microdialysis study in vivo. Psychopharmacology 136:299–303

Izvolskaia M, Duittoz AH, Tillet Y, Ugrumov MV (2009) The influence of catecholamine on the migration of gonadotropin-releasing hormone-producing neurons in the rat foetuses. Brain Struct Funct 213:289–300

Jacobson W, Wilkinson M, Gibson CJ (1988) Effects of the noradrenergic neurotoxin DSP4 on growth and the gonadal steroid- and PMSG-induced release of LH in the prepubertal female rat. Brain Res Bull 20:643–649

Jaim-Etcheverry G, Zieher LM (1980) DSP-4: a novel compound with neurotoxic effects on noradrenergic neurons of adult and developing rats. Brain Res 188:513–523

Jiang LY, Li SR, Zhao FY, Spanswick D, Lin MT (2010) Norepinephrine can act via alpha(2)-adrenoceptors to reduce the hyper-excitability of spinal dorsal horn neurons following chronic nerve injury. J Formos Med Assoc 109:438–445

Jolas T, Schreiber R, Laporte AM, Chastanet M, De Vry J, Glaser T, Adrien J, Hamon M (1995) Are postsynaptic 5-HT1A receptors involved in the anxiolytic effects of 5-HT1A receptor agonists and in their inhibitory effects on the firing of serotonergic neurons in the rat? J Pharmacol Exp Ther 272:920–929

Jonsson G, Sachs C (1982) Changes in the development of central noradrenaline neurons after neonatal axon lesions. Brain Res Bull 9:641–650

Jonsson G, Hallman H, Ponzio F, Ross S (1981) DSP4 (N-(2-chloroethyl)-N-ethyl-2-bromobenzylamine)—a useful denervation tool for central and peripheral noradrenaline neurons. Eur J Pharmacol 72:173–188

Jonsson G, Hallman H, Sundström E (1982) Effects of the noradrenaline neurotoxin DSP4 on the postnatal development of central noradrenaline neurons in the rat. Neuroscience 7:2895–2907

Kask A, Harro J, Tuomaine P, Rägo L, Männistö PT (1997) Overflow of noradrenaline and dopamine in frontal cortex after [N-(2-chloroethyl)-N-ethyl-2-bromobenzylamine] (DSP-4) treatment: in vivo microdialysis study in anaesthetized rats. Naunyn Schmiedebergs Arch Pharmacol 355:267–272

Khakpour-Taleghani B, Lashgari R, Motamedi F, Naghdi N (2009) Effect of reversible inactivation of locus ceruleus on spatial reference and working memory. Neuroscience 158:1284–1291

König N, Serrano JJ, Jonsson G, Malayal F, Szafarczyk A (1985) Prenatal treatment with 6-hydroxydopa and DSP 4: biochemical, endocrinological and behavioural effects. Int J Dev Neurosci 3:501–509

Korossy-Mruk E, Kuter K, Nowak P, Szkilnik R, Rykaczewska-Czerwinska M, Kostrzewa RM, Brus R (2013) Neonatal DSP-4 treatment modifies antinociceptive effects of the CB1 receptor agonist methanandamide in adult rats. Neurotox Res 23:39–48

Kostrzewa RM, Brus R (1991a) Is dopamine-agonist induced yawning behavior a D3 mediated event? Life Sci 48:PL129

Kostrzewa RM, Brus R (1991b) Ontogenic homologous supersensitization of quinpirole-induced yawning in rats. Pharmacol Biochem Behav 39:517–519

Kostrzewa RM, Gong L (1991) Supersensitized D1 receptors mediate enhanced oral activity after neonatal 6-OHDA. Pharmacol Biochem Behav 39:677–682

Kostrzewa RM, Kostrzewa JP, Brown RW, Nowak P, Brus R (2008) Dopamine receptor supersensitivity: development, mechanisms, presentation, and clinical applicability. Neurotox Res 14:121–128

Kostrzewa RM, Kostrzewa JP, Kostrzewa RA, Kostrzewa FP, Brus R, Nowak P (2011) Stereotypic progressions in psychotic behavior. Neurotox Res 19:243–252

Kreiner G, Zelek-Molik A, Kowalska M, Bielawski A, Antkiewicz-Michaluk L, Nalepa I (2011) Effects of the noradrenergic neurotoxin DSP-4 on the expression of α1-adrenoceptor subtypes after antidepressant treatment. Pharmacol Rep 63:1349–1358

Lin D, Parsons LH (2002) Anxiogenic-like effect of serotonin(1B) receptor stimulation in the rat elevated plus-maze. Pharmacol Biochem Behav 71:581–587

Luscombe GP, Martin KF, Hutchins LJ, Gosden J, Heal DJ (1993) Mediation of the antidepressant-like effect of 8-OH-DPAT in mice by postsynaptic 5-HT1A receptors. Br J Pharmacol 108:669–677

Lyons WE, Fritschy JM, Grzanna R (1989) The noradrenergic neurotoxin DSP-4 eliminates the coeruleospinal projection but spares projections of the A5 and A7 groups to the ventral horn of the rat spinal cord. J Neurosci 9:1481–1489

Margalit D, Segal M (1979) A pharmacologic study of analgesia produced by stimulation of the nucleus locus coeruleus. Psychopharmacology 62:169–173

Martins-Afférri MP, Ferreira-Silva IA, Franci CR, Anselmo-Franci JA (2003) LHRH release depends on Locus Coeruleus noradrenergic inputs to the medial preoptic area and median eminence. Brain Res Bull 61:521–527

Mattingly BA, Gotsick JE (1989) Conditioning and experiential factors affecting the development of sensitization to apomorphine. Behav Neurosci 103:1311–1317

Medina JH, Novas ML (1983) Parallel changes in brain flunitrazepam binding and density of noradrenergic innervation. Eur J Pharmacol 88:377–382

Millan MJ, Rivet JM, Canton H, Le Marouille-Girardon S, Gobert A (1993) Induction of hypothermia as a model of 5-hydroxytryptamine1A receptor-mediated activity in the rat: a pharmacological characterization of the actions of novel agonists and antagonists. J Pharmacol Exp Ther 264:1364–1376

Millan MJ, Lejeune F, Gobert A (2000a) Reciprocal autoreceptor and heteroreceptor control of serotonergic, dopaminergic and noradrenergic transmission in the frontal cortex: relevance to the actions of antidepressant agents. J Psychopharmacol 14:114–138

Millan MJ, Lejeune F, Gobert A, Brocco M, Auclair A, Bosc C, Rivet JM, Lacoste JM, Cordi A, Dekeyne A (2000b) S18616, a highly potent spiroimidazoline agonist at alpha(2)-adrenoceptors: II. Influence on monoaminergic transmission, motor function, and anxiety in comparison with dexmedetomidine and clonidine. J Pharmacol Exp Ther 295:1206–1222

Mogilnicka E, Klimek V (1977) Drugs affecting dopamine neurons and yawning behavior. Pharmacol Biochem Behav 7:303–305

Nowak P, Bortel A, Dąbrowska J, Labus Ł, Brus R, Kostrzewa RM (2004) Effect of neonatal DSP-4 administration on biogenic amine levels and amphetamine-induced release into microdialysates of adult rat frontal cortex. Pol J Pharmacol 56(Suppl):248

Nowak P, Kostrzewa RM, Kwieciński A, Bortel A, Labus Ł, Brus R (2005) Neurotoxic action of 6-hydroxydopamine on the nigrostriatal dopaminergic pathway in rats sensitized with D-amphetamine. J Physiol Pharmacol 56:325–333

Nowak P, Labus Ł, Kostrzewa RM, Brus R (2006) DSP-4 prevents dopamine receptor priming by quinpirole. Pharmacol Biochem Behav 84:3–7

Nowak P, Jochem J, Zwirska-Korczala K, Josko J, Noras L, Kostrzewa RM, Brus R (2008) Ontogenetic noradrenergic lesion alters histaminergic activity in adult rats. Neurotox Res 13:79–83

Nowak P, Nitka D, Kwieciński A, Jośko J, Drab J, Pojda-Wilczek D, Kasperski J, Kostrzewa RM, Brus R (2009) Neonatal co-lesion by DSP-4 and 5,7-DHT produces adulthood behavioral sensitization to dopamine D(2) receptor agonists. Pharmacol Rep 61:311–318

O'Connell MT, Curzon G (1996) A comparison of the effects of 8-OH-DPAT pretreatment of different behavioural responses to 8-OH-DPAT. Eur J Pharmacol 312:137–143

O'Leary KT, Leslie FM (2006) Enhanced nicotinic acetylcholine receptor-mediated [3H] norepinephrine release from neonatal rat hypothalamus. Neuropharmacology 50:81–88

Pagano RL, Fonoff ET, Dale CS, Ballester G, Teixeira MJ, Britto LR (2012) Motor cortex stimulation inhibits thalamic sensory neurons and enhances activity of PAG neurons: possible pathways for antinociception. Pain 153:2359–2369

Peyron C, Luppi PH, Fort P, Rampon C, Jouvet M (1996) Lower brainstem catecholamine afferents to the rat dorsal raphe nucleus. J Comp Neurol 364:402–413

Post C, Arweström E, Minor BG, Wikberg JE, Jonsson G, Archer T (1985) Noradrenaline depletion increases noradrenaline-induced antinociception in mice. Neurosci Lett 59:105–109

Pranzatelli MR, Gregory CM (1993) High and low affinity 5-HT2 and 5-HT1C binding sites: responses to neonatal 5,7-DHT lesions in rat brain. Cytobios 75:197–209

Pudovkina OL, Cremers TI, Westerink BH (2003) Regulation of the release of serotonin in the dorsal raphe nucleus by α1 and α2 adrenoceptors. Synapse 50:77–82

Reid AT, Harley CW (2010) An associativity requirement for locus coeruleus-induced long-term potentiation in the dentate gyrus of the urethane-anesthetized rat. Exp Brain Res 200:151–159

Roczniak W, Babuśka-Roczniak M, Kwapuliński J, Brodziak-Dopierała B, Widuchowski W, Cipora E, Nowak P, Oświęcimska JM (2015) The effect of central noradrenergic system lesion on dopamine (DA) and serotonin (5-HT) synthesis rate following administration of 5-HT3 receptor ligands in chosen parts of the rat brain. Pharmacol Rep 67:146–151

Ross SB, Renyl AL (1976) On the long-lasting inhibitory effect of N-(2-chloroethyl)-N-ethyl-2-bromobenzylamine (DSP 4) on the active uptake of noradrenaline. J Pharm Pharmacol 28:458–459

Ross SB, Johansson JG, Lindborg B, Dahlbom R (1973) Cyclizing compounds. I. Tertiary N-(2-bromobenzyl)-N-haloalkylamines with adrenergic blocking action. Acta Pharm Sueccica 10:29–42

Sanders JD, Happe HK, Bylund DB, Murrin LC (2008) Differential effects of neonatal norepinephrine lesions on immediate early gene expression in developing and adult rat brain. Neuroscience 157:821–832

Sanders JD, Happe HK, Bylund DB, Murrin LC (2011) Changes in postnatal norepinephrine alter alpha-2 adrenergic receptor development. Neuroscience 192:761–772

Sawynok J, Reid A (1994) Spinal supersensitivity to 5-HT1, 5-HT2 and 5-HT3 receptor agonists following 5,7-dihydroxytryptamine. Eur J Pharmacol 264:249–257

Scavone JL, Mackie K, Van Bockstaele EJ (2010) Characterization of cannabinoid-1 receptors in the locus coeruleus: relationship with mu-opioid receptors. Brain Res 1312:18–31

Schreiber R, De Vry J (1993) Neuroanatomical basis for the antidepressant-like effects of the 5-HT1A receptor agonists 8-OH-DPAT and ipsapirone in the rat forced swimming test. Behav Pharmacol 4:625–636

Sim LJ, Joseph SA (1993) Dorsal raphe nucleus efferents: termination in peptidergic fields. Peptides 14:75–83

Suehiro K, Funao T, Fujimoto Y, Yamada T, Mori T, Nishikawa K (2013) Relationship between noradrenaline release in the locus coeruleus and antiallodynic efficacy of analgesics in rats with painful diabetic neuropathy. Life Sci 92:1138–1144

Suzdak PD, Gianutsos G (1985) GABA-noradrenergic interaction: evidence for differential sites of action for GABA-A and GABA-B receptors. J Neural Transm 64:163–172

Van Bockstaele EJ, Commons KG (2001) Internalization of mu-opioid receptors produced by etorphine in the rat locus coeruleus. Neuroscience 108:467–477

Wang Y, Musich PR, Serrano MA, Zou Y, Zhang J, Zhu MY (2014) Effects of DSP4 on the noradrenergic phenotypes and its potential molecular mechanisms in SH-SY5Y cells. Neurotox Res 25:193–207

Weikop P, Kehr J, Scheel-Krüger J (2004) The role of alpha1- and alpha2-adrenoreceptors on venlafaxine-induced elevation of extracellular serotonin, noradrenaline and dopamine levels in the rat prefrontal cortex and hippocampus. J Psychopharmacol 18:395–403

Wenge B, Bönisch H (2009) Interference of the noradrenergic neurotoxin DSP4 with neuronal and nonneuronal monoamine transporters. Naunyn Schmiedebergs Arch Pharmacol 380:523–529

Wolfman C, Abó V, Calvo D, Medina J, Dajas F, Silveira R (1994) Recovery of central noradrenergic neurons one year after the administration of the neurotoxin DSP4. Neurochem Int 25:395–400

Noradrenergic–Dopaminergic Interactions Due to DSP-4–MPTP Neurotoxin Treatments: Iron Connection

Trevor Archer

Abstract The investigations of noradrenergic lesions and dopaminergic lesions have established particular profiles of functional deficits and accompanying alterations of biomarkers in brain regions and circuits. In the present account, the focus of these lesions is directed toward the effects upon dopaminergic neurotransmission and expression that are associated with the movement disorders and psychosis-like behavior. In this context, it was established that noradrenergic denervation, through administration of the selective noradrenaline (NA) neurotoxin, DSP-4, should be performed prior to the depletion of dopamine (DA) with the selective neurotoxin, MPTP. Employing this regime, it was shown that (i) following DSP-4 (50 mg/kg) pretreatment of C57/Bl6 mice, both the functional and neurochemical (DA loss) effects of MPTP (2 × 20 and 2 × 40 mg/kg) were markedly exacerbated, and (ii) following postnatal iron (Fe^{2+}, 7.5 mg/kg, on postnatal days 19–12), pretreatment with DSP-4 followed by the lower 2 × 20 mg/kg MPTP dose induced even greater losses of motor behavior and striatal DA. As yet, the combination of NA-DA depletions, and even more so Fe^{2+}–NA-DA depletion, has been considered to present a movement disorder aspect although studies exploring cognitive domains are lacking. With intrusion of iron overload into this formula, the likelihood of neuropsychiatric disorder, as well, unfolds.

Keywords NA-DA denervations · Mice · DSP-4 · MPTP · Function · Biomarkers · Deficits · Iron overload

Contents

1 DSP-4 Lesioning of NA 74
2 MPTP-Lesioning of DA 75
3 NA-DA Lesions and Iron Overload 77
References 80

T. Archer (✉)
Department of Psychology, University of Gothenburg, Gothenburg, Sweden
e-mail: trevor.archer@psy.gu.se

© Springer International Publishing Switzerland 2015
Curr Topics Behav Neurosci (2016) 29: 73–86
DOI 10.1007/7854_2015_411
Published Online: 31 December 2015

On the basis of neuropathological evidence, central noradrenaline (NA) system impairments have been linked to various neurodegenerative disorders including parkinsonism and other movement disorders (Alvord and Forno 1992). For example, in Parkinson's disease (PD) patients' postmortem neurochemical analyses have confirmed huge losses of NA together with the dopamine (DA) depletions (Hornykiewizc and Kish 1987). In analyses of thalamic neuronal discharge patterns, it was indicated that in PD patients NA was severely reduced in the motor thalamus and also in other regions of the thalamus (Pifl et al. 2012). Furthermore, Pifl et al. (2013) analyzed NA, DA, and serotonin (5-HT) in the motor (ventrolateral and ventroanterior) and non-motor (mediodorsal, centromedian, ventroposterior lateral and reticular) thalamic nuclei of MPTP-treated monkeys that had been found to be (i) constantly asymptomatic; (ii) who had recovered from mild parkinsonism; and (iii) animals presenting either with stable, moderate, or severe parkinsonism. They observed that only the symptomatic parkinsonian animals showed significant losses of NA specifically in the motor thalamus, with the ventroanterior motor nucleus being affected only in the severe parkinsonian animals. Contrastingly, the striatal DA loss was identical in both the mild and severe symptom groups. Certainly, all three neurotransmitters, NA, DA, and 5-HT, were reduced significantly in the rostro-caudal regions of the hypothalamus of patients with idiopathic PD (Shannak et al. 1994). The putatively protective role of noradrenergic innervation for DA neurons was implied from a postmortem study of 20 brain areas, dopamine loss in PD was negatively, but strongly, correlated with healthy NA levels, with regions rich in NA (e.g., the noradrenaline-rich portion of the nucleus accumbens) spared from dopamine loss (Tong et al. 2007).

1 DSP-4 Lesioning of NA

The selective NA neuron denervating effects of DSP4 (N-[2-chloroethyl]-N-ethyl-2-bromobenzylamine) within several brain regions, including the forebrain, midbrain, cerebellum, brain stem, and spinal, accompanied alterations in biomarkers and functional parameters have been documented prodigiously (Archer and Fredriksson 2000, 2001; Archer et al. 1982, 1983, 1984a, b, 1986a; Dooley et al. 1983a; Fredriksson and Archer 2000; Jonsson and Hallman 1982; Jonsson et al. 1981, 1982; Hallman and Jonsson 1984; Hallman et al. 1984; Ponzio et al. 1981; Ross 1976; Ross and Renyi 1976; Sundström et al. 1987). Consistently, brain tissue levels of NA are decreased to between 10 and 30 % of control values (Ross and Stenfors 2015). Systemic administration of DSP4 (50 mg/kg and upwards, i. p. or s.c. injections, generally two weeks before testing) induced marked and long-lasting reductions of dopamine β-hydroxylase activity (Archer et al. 1984a, b, 1985; Ross 1976; Ross and Renyi 1976). Profound and permanent depletions of endogenous NA have been obtained repeatedly in several brain regions, such as the cerebral and cerebellar cortex, hippocampus and spinal cord, generally leaving dopamine (DA) and 5-hydroxytryptamine (5-HT) neurons largely intact;

nevertheless, pretreatment with the 5-HT reuptake inhibitor, zimelidine, unfailingly leaves 5-HT neurons intact (Archer 1982; Archer and Fredriksson 2001; Archer et al. 1984a, b, 1986b; Liu et al. 2015a, b; Rényi et al. 1986). Any transient peripheral tissue depletions that may have appeared following the 50 mg/kg dose of DSP-4 (e.g., Archer et al. 1982; Liang et al. 1995) disappeared within 10 days.

Systemic administration of DSP-4 has been shown to cause minor losses of cortical 5-HT, although not always so, in rats and mice (e.g., Archer et al. 1984a, b, 1985; Roczniak et al. 2013, 2015). Invariably, it was demonstrated that pretreatment with 5-HT reuptake inhibitors, whether zimelidine (20 mg/kg) or fluoxetine (20 mg/kg), up to 30 min before DSP-4 administration, prevented any loss of 5-HT, without affecting the extent of NA denervation (Archer et al. 1986b; Bello et al. 2014; Dabrowska et al. 2008; Fowler et al. 1988; Heal et al. 1993; Post et al. 1987). Functional analyses applying receptor ligand agonists of the DA and 5-HT systems have demonstrated that these neurotransmitter pathways were not changed in DSP-4 rats (Dooley et al. 1983b); neither did the systemic DSP-4 administration produce any notable effects upon the normal functioning of the hypothalamic–pituitary–adrenal axis (Bugajski et al. 1995). Nevertheless, in a study of stress-sensitive and genetic factors underlying dependence behavior in Sprague–Dawley and Wistar–Kyoto rats, Fox et al. (2015) using DSP-4 to deplete NA selectively in these strains showed that DSP-4-treated Sprague–Dawley rats demonstrated a dependence-like phenotype, whereas the DSP-4-treated Wistar–Kyoto rats were unchanged. A comprehensive analysis of the neurotoxic actions of DSP-4 that evaluated different patterns of monoamine-producing depletions over various brain regions in different strains of rats and mice from 3 to 14 days after neurotoxin injections has been described (Fornai et al. 1996, 1997; Hurko et al. 2010; Kostrzewa 2009; Kostrzewa et al. 2011; Nowak et al. 2009). With regard to the possible, transient yet small depletions of 5-HT, following DSP-4, both strain and species differences have been described that pertain to various due to use of the same dose of neurotoxin in different strains/species (Aulakh et al. 1992; Fornai et al. 1996, 2001). Long-term pretreatment of rats with DSP-4 induced marked supersensitivity to noradrenergic (Archer and Fredriksson 2000) and cross-sensitivity to opiate (Archer and Fredriksson 2001).

2 MPTP-Lesioning of DA

The selective DA neurotoxin, 1-methyl-4-phenyl-1,2,3,6-tetrahydropyridine (MPTP), has been shown regularly to induce parkinsonism in human and non-human primates (Langston 1985), that resulted in the loss of substantia nigra cells in the pars compacta of adult humans/adults (Bubak et al. 2015; Mathai et al. 2015; Uehara et al. 2015). Systemic administration of MPTP to C57/Bl6 mice, generally at a dose of 2 × 40 mg/kg, s.c., invariably induced hypokinesia, reflected by loss of locomotion, rearing and total activity in standardized activity test chambers, that could be reversed with a 20 mg/kg dose of l-dopa (Fredriksson et al.

1990; Sundström et al. 1990), accompanied by an 85–90, or more, % loss of DA (Bhargava and Perlow 1988; Schultz et al. 1985; Zuddas et al. 1992). Less rigorous dose regimes, e.g., 2 × 20, or 25, or 30 mg/kg of the neurotoxin, have been less effective in inducing hypoactivity although DA concentrations may indicate up to 50–80 % reductions (Giovanni et al. 1994a, b; Heikkila et al. 1989; Sonsalla and Heikkila 1986). Reductions in 5-HT levels are observed in midbrain and hippocampus regions only in acute and subacute MPTP-treated-mice (Pain et al. 2013). The parameters of MPTP neurotoxicity are virtually permanent (up to and beyond 52 weeks after treatment) with marked correlations between the functional deficits, particularly hypoactivity (locomotion), the neurochemical (DA) levels, severe depletions of DA and metabolites (but see Date et al. 1990), and a dose- and time-dependent recovery of several parameters of motor behavior following the administration of l-dopa, the DA precursor (Archer and Fredriksson 2003; Fredriksson and Archer 1994; Fredriksson et al. 1999; Goshima et al. 1991; Jonsson et al. 1985; Youdim and Arraf 2004). Postnatal iron administration (Fe^{2+}, on postnatal day, at doses of 7.5 or 15 mg/kg) exacerbated the functional, hypokinesia, and neurochemical, DA, deficits that were caused by both the lower (2 × 20 mg/kg) and the higher (2 × 40 mg/kg) dose regimes of MPTP, as well as applying other iron–MPTP combinations (Ayton et al. 2014; Fredriksson and Archer 2003; Fredriksson et al. 2001; Hare et al. 2013; You et al. 2015).

In an MPTP-lesioned primate model, apathy scores were elevated following MPTP, and these scores correlated with PET-scan measures of dopaminergic terminals in the dorsal lateral prefrontal cortex, ventromedial prefrontal cortex, and insular cortex of the treated animals (Tian et al. 2015), implicating dopaminergic dysfunction within the ventral tegmental area-insular cortex pathway plays a role in the manifestation of apathetic behaviors, possibly affecting motor activity. Using a type of 'staging' (Archer et al. 2011) to illustrate parkinsonian state (naive, mild, moderate, and severe) and recording the spontaneous local field potentials were recorded throughout the sensorimotor globus pallidus. Connolly et al. (2015) have described a novel mechanistic framework to understand how progressive loss of dopamine translates into abnormal information processing in the pallidum through alterations in oscillatory activity. Liu et al. (2015a, b) have presented a highly progressive model of parkinsonism in the cynomolgus monkey, a primate, through giving intravenous injections of MPTP (0.2 mg/kg) over 15 days; this procedure induced a stable parkinsonism development over 90 days, at which point the symptoms were stable. Noninvasive positron emission tomographic neuroimaging of vesicular monoamine transporter 2 with 9-[(18)F] fluoropropyl-(+)-dihydrotetrabenazine ([(18)F]AV-133) displayed evidence of the progressive loss of striatal uptake of [(18)F]AV-133. The dopaminergic denervation severity showed a significant linear correlation with the clinical rating scores and the bradykinesia subscores. The progressive feature of MPTP-induced parkinsonism is an important current aspect of MPTP-induced deficits (Archer and Fredriksson 2013; Blesa et al. 2010; Nagai et al. 2012).

3 NA-DA Lesions and Iron Overload

Following the early demonstration of NA-DA neuron interactions (Persson and Waldeck 1970), the requirement for intact noradrenergic pathways for dopaminergic functioning has been studied (Andén and Grabowska 1976; Dolphin et al. 1976a, b; Kostowski 1979; O'Donohue et al. 1979; Tassin et al. 1992). In a primate study, it was shown that denervation of NA pathways following insult to the locus coeruleus exacerbated the neurotoxic damage leading to the experimental parkinsonism following administration of MPTP (Bing et al. 1994). Using subtoxic doses of MPTP into unilateral locus coeruleus 6-hydroxydopamine (6-OHDA) lesions in mice, it was observed there was a significant loss of dopaminergic cells only found in the substantia nigra on the side of the locus coeruleus lesions (Bing et al. 1994). Following DSP-4-induced losses of NA exceeding 75 % in brain regions, the concentrations of endogenous DA in microdialysates from the caudate nucleus and nucleus accumbens were reduced by 52 and 28 %, respectively (Lategan et al. 1992). Similarly, the pretreatment of mice with DSP-4 (40 mg/kg) exacerbated striatal DA loss following MPTP (4 × 10 mg/kg) by a factor of 60 %, whereas by itself DSP-4 did not affect striatal DA (Marien et al. 1993). It should be noted that these doses of the neurotoxins were essentially lower that those applied in the studies above (e.g., Archer et al. 1982; Fredriksson and Archer 1994) and therefore rather impressive. Fornai et al. (1997) measured the acute effects of MPTP on the nigrostriatal DA pathway in DSP-4 lesioned C57/6N mice that were compared to non-NA-lesioned mice that received only MPTP. They obtained a more marked acute DA depletion persisting at 12 h in DSP-4 + MPTP mice compared to MPTP only mice. It appears that in the absence of locus coeruleus axons a more pronounced sensitivity, or 'supersensitivity,' possibly to the loss of a 'recovery-capacity' after noradrenergic denervation (Goldstein 2006), in the striatal DA neurons is produced. Nevertheless, despite successful lesioning of both noradrenergic and dopaminergic neurons in rat brains, no changes in catechol-O-methyltransferase (COMT) protein expression or activity were obtained implying that COMT is not present in DA and NA neurons (Schendzielorz et al. 2013).

Several of the studies described above (e.g., Fornai et al. 1997) applied a neurotoxin dose regime that involved acute administrations. The procedure employed by (Nishii et al. 1991) may be termed 'semi-acute.' They treated 7-week-old C57 black mice as follows, in four groups: Group (i) MPTP (30 mg/kg) was administered each day over 10 days; Group (ii) MPTP treatment as for Group (i), + DSP-4 (50 mg/kg) administered on the final (10th) day of MPTP treatment; Group (iii) vehicle administered over 10 days and DSP-4 (50 mg/kg) on the 10th day (as for Group (ii); and Group (iv) vehicle administered over consecutive days 10 days. They then measured spontaneous and l-dopa-induced motor activity over the next 7–10 days. They obtained a marked reduction in spontaneous motor activity during the initial period (1–2 days) following cessation of neurotoxin treatments which had disappeared 7–10 days later. However, the administration of l-dopa (200 mg/kg) combined with DCI (25 mg/kg) induced a marked rise in the motor activity of the

Group (i) mice, treated with MPTP over 10 days; this peak of motor activity was severely attenuated in the Group (ii) mice, treated with MPTP (10 days) then DSP-4 (Day 10), as expressed by a lower peak effect (35 % less) and lesser peak duration (60 %). Striatal loss of DA was quite comparable in the (i) MPTP, and (ii) MPTP + DSP-4 groups, 82 and 85 %, respectively, which is hardly an exacerbating effect of the NA neurotoxin. In the case of the Marien et al. (1993) study, DA loss after MPTP (4 × 10 mg/kg) was 40 % 7 days postneurotoxin and after DSP-4 (40 mg/kg) prior to MPTP was 60 %. As indicated Fornai et al. (1997), the most severe DA losses occurred when DSP-4 was administered before MPTP.

In order to ensure complete recovery from DSP-4, C57/Bl6 mice were not administered MPTP until three weeks later, and the testing of motor activity was not initiated until a further three weeks had passed (Archer and Fredriksson 2006). Thus, three groups of mice were injected DSP-4 (50 mg/kg, 30 min after zimelidine, 20 mg/kg to protect 5-HT neurons) and three groups injected saline. Three weeks later, one DSP-4 group and one saline group were injected a high dose of DSP4 (2 × 40 mg/kg), one DSP-4 group and one saline group were injected a low dose of DSP4 (2 × 20 mg/kg), and one DSP-4 group and one saline group were injected saline. During the spontaneous motor activity tests three weeks, the following was observed: The spontaneous motor activity of the saline–MPTP-treated mice was reduced in a dose-dependent manner, whereas in the DSP-4–MPTP-treated mice was virtually abolished during the 1st 20-min test period and very nearly so during the 2nd 20-min period of testing. Following a suprathreshold 20 mg/kg dose of l-dopa, motor activity was reinstated completely in the saline–MPTP-treated mice but remained markedly reduced in the DSP-4–MPTP-treated mice. Table 1 presents the locomotor and rearing performance of the saline–MPTP-low, saline–MPTP-high, DSP-4–saline, DSP-4–MPTP-low, and DSP-4–MPTP-high expressed as a percentage of the saline–saline group during the 1st 20-min period of testing as well as the percent striatal DA (of saline–saline controls)

Table 1 Locomotion, rearing and striatal DA, as well as l-dopa-induced activity expressed as a percentage of the saline–saline control group by the groups of mice administered either DSP-4 or saline followed three weeks later by either MPTP, low or high dose, or saline (cf. Archer and Frerdiksson 2006)

Spon. activity	Saline– MPTPL (%)	Saline– MPTPH (%)	DSP-4– Saline (%)	[a]DSP-4– MPTPL (%)	[a]DSP-4– MPTPH (%)
Locomotion	27	10	99	4	0.02
Rearing	44	25	98	1	>0.02
Striatal DA	51	13	101	28	5
Saline– MPTPL (%)	Saline– MPTPH (%)	DSP-4– Saline (%)	[a]DSP-4– MPTPL (%)	[a]DSP-4– MPTPH (%)	Saline– MPTPL (%)
Locomotion	89	79	101	43	20
Rearing	89	87	101	84	37

[a]MPTPL = low-dose (20 mg/kg) MPTP; [a]MPTPH = high-dose (40 mg/kg) MPTP

of the same groups. Locomotor and rearing performance, over 360 min, by the same groups following l-dopa (20 mg/kg) is shown also.

These results indicate a severe exacerbation of the MPTP-induced functional and neurochemical deficits following prior administration of DSP-4, i.e., the noradrenergic system denervation. It is evident also that the restorative effects of suprathreshold l-dopa are compromised. In a further pursuance of these DSP-4 pretreatment–MPTP-induced 'ultra-deficits' in DA functional and neurochemical parameters, postnatal iron administration (Fe^{2+}, 7.5 mg/kg, on postnatal days 10–12) was used to induce further the vulnerability of DA neurons. In this study, only the low (20 mg/kg) dose of MPTP was applied (Fredriksson and Archer 2007). Postnatal iron administration further exacerbated all the deficits induced by DSP-4 and MPTP condition. For example, in mice administered MPTP(Low) locomotion was reduced from 25 % of saline–saline control group in the non-iron condition [compare with 27 % in the Archer and Fredriksson study] to 10 % in the iron condition, whereas in the mice administered DSP-4 + MPTP (low) locomotion was reduced from 10 % in the non-iron condition [compare with 5 % in the Archer and Fredriksson study] to 0 % in the iron condition. Similarly, striatal DA concentrations were compromised drastically through the postnatal iron treatment: in the MPTP (low) group, DA was reduced from 52 % in the non-iron condition [compare with 51 % in the Archer and Fredriksson study] to 39 % in the iron condition, whereas in the DSP-4 + MPTP (low) groups DA was reduced from 25 % in the non-iron condition [compare with 28 % in the Archer and Fredriksson study] to 11 % in the iron condition. Postnatal iron administration induced enduring high levels of total iron content of the basal ganglia in all the groups administered postnatal iron following sacrifice and analysis at 100 days-of-age. Nevertheless, the concentration was highest, significantly, in the DSP-4 + MPTP group implying that the combination of these neurotoxins affected brain iron retention. Finally, the co-administration of clonidine (1 mg/kg) with subthreshold l-dopa (5 mg/kg) alleviated the motor deficits of MPTP-treated mice, but pretreatment with DSP-4 reduced markedly the ameliorative effects of clonidine, the α-adrenoceptor agonist (Archer and Fredriksson 2007; Fredriksson and Archer 2007). It has been shown comprehensively that iron overload contributes to the development of neurodegenerative progressions and the acceleration of normal rates of apoptosis primarily due to its participation in the Fenton reaction and production of reactive oxygen species, as well as functional measures in both motor and cognition domains (Fagherazzi et al. 2012; Lavich et al. 2015; Silva et al. 2012). Postnatal iron administration, by itself, induces long-lasting changes in brain function, e.g., transient hypoactivity in activity cages followed hyperactivity and cognitive deficits (de Lima et al. 2007; Fredriksson and Archer 2006). The hyperactivity was exacerbated by apomorphine and abolished by haloperidol (Fredriksson and Archer 2006). The full implications of the balancing role of iron in conditions of loss of DA-NA integrity have yet to be explored and described; nevertheless, the implications of the observations so far offer a neurodegenerative scenario that may be of utility in understanding developmental trajectories of subnormal dimensions.

Loss of integrity, or a 'masked' vulnerability, by DA neurons, and possibly also NA neurons, through postnatal iron overload may well impart an adverse epigenetic condition to the detriment of normal development. Nevertheless, NA neurons in the locus coeruleus appear to be less susceptible to the effects of iron overload than DA neurons in the substantia nigra (Zecca et al. 2004). Dornelles et al. (2010) have shown that transferrin receptor, H-ferritin, and IRP2 mRNA expressions were affected differentially through the aging process and by postnatal iron supplementation in the cortex, hippocampus, and striatum of rats. The epigenetic reprograming of cortical neurons through alterations of dopaminergic circuits (Brami-Cherrier et al. 2014), such as that produced by neurotoxin insults or iron overload, may exert both neurologic and psychiatric expressions of abnormal function as a basis of brain disorder. Finally, Shin et al. (2014), in a rat model of PD, found that NA depletion did not enhance the extent of DA depletion or the loss of tyrosine hydroxylase-positive innervation in the striatum but rather that damage to brainstem NA innervation accelerated development of motor impairments and the onset of l-dopa-induced dyskinesias in 6-OHDA-lesioned (DA-depleted) rats. These observations are reinforced by the finding that additional noradrenergic depletion (in addition to DA depletion) aggravated forelimb akinesia and abnormal subthalamic nucleus activity in the DSP4-6-OHDA rat model of PD (Wang et al. 2014; see also Lindgren et al. 2014).

References

Alvord EC Jr, Forno LS (1992) Pathology. In: Koller WC (ed) Handbook of Parkinson' disease. Marcel Dekker, New York, pp 255–284

Andén N, Grabowska M (1976) Pharmacological evidence for a stimulation of dopamine neurons by noradrenaline neurons in the brain. Eur J Pharmacol 39(2):275–282

Archer T (1982) Serotonin and fear retention in the rat. J Comp Physiol Psychol 96:491–516

Archer T, Fredriksson A (2000) Effects of clonidine and adrenoceptor antagonists on motor activity in DSP-4-treated mice. I. Dose, time-and parameter-dependency. Neurotox Res 1:235–247

Archer T, Fredriksson A (2001) Effects of alpha-adrenoceptor agonists in chronic morphine administered DSP4-treated rats: evidence for functional cross-sensitization. Neurotox Res 3 (5):411–432

Archer T, Fredriksson A (2003) An antihypokinesic action of alpha2-adrenoceptors upon MPTP-induced behaviour deficits in mice. J Neural Transm (Vienna) 110(2):183–200

Archer T, Fredriksson A (2006) Influence of noradrenaline denervation on MPTP-induced deficits in mice. J Neural Transm (Vienna) (9):1119–1129

Archer T, Fredriksson A (2007) Functional consequences of iron overload in catecholaminergic interactions: the Youdim factor. Neurochem Res 32(10):1625–1639

Archer T, Fredriksson A (2013) The yeast product Milmed enhances the effect of physical exercise on motor performance and dopamine neurochemistry recovery in MPTP-lesioned mice. Neurotox Res 24(3):393–406. doi:10.1007/s12640-013-9405-4

Archer T, Ogren SO, Johansson G, Ross SB (1982) DSP4-induced two-way active avoidance impairment: involvement of central and not peripheral noradrenaline depletion. Psychopharmacology 76:303–309

Archer T, Mohammed AK, Ross SB, Söderberg U (1983) T-maze learning, spontaneous activity and food intake recovery following systemic administration of the noradrenaline neurotoxin, DSP4. Pharmacol Biochem Behav 19:121–130

Archer T, Jonsson G, Ross SB (1984a) A parametric study of the effects of the noradrenaline neurotoxin DSP4 on avoidance acquisition and noradrenaline neurons in the CNS of the rat. Br J Pharmacol 82:249–257

Archer T, Ogren SO, Ross SB, Magnusson O (1984b) Retention deficits induced by acute p-chloroamphetamine following fear conditioning in the rat. Psychopharmacology 82(1–2):14–19

Archer T, Jonsson G, Ross SB (1985) Active and passive avoidance following the administration of systemic DSP4, xylamine, or p-chloroamphetamine. Behav Neural Biol 43:238–249

Archer T, Fredriksson A, Jonsson G, Lewander T, Mohammed AK, Ross SB, Söderberg U (1986a) Central noradrenaline depletion antagonizes aspects of d-amphetamine-induced hyperactivity in the rat. Psychopharmacology 88:141–146

Archer T, Jonsson G, Minor BG, Post C (1986b) Noradrenergic-serotonergic interactions and nociception in the rat. Eur J Pharmacol 120:295–307

Archer T, Kostrzewa RM, Beninger RJ, Palomo T (2011) Staging neurodegenerative disorders: structural, regional, biomarker, and functional progressions. Neurotox Res 19(2):211–234. doi:10.1007/s12640-010-9190-2

Aulakh CS, Hill JL, Lesch KP, Murphy DL (1992) Functional subsensitivity of 5-hydroxytryptamine1C or alpha 2 adrenergic heteroreceptors mediating clonidine-induced growth hormone release in the Fawn-Hooded rat strain relative to the Wistar rat strain. J Pharmacol Exp Ther 262(3):1038–1043

Ayton S, Lei P, Adlard PA, Volitakis I, Cherny RA, Bush AI, Finkelstein DI (2014) Iron accumulation confers neurotoxicity to a vulnerable population of nigral neurons: implications for Parkinson's disease. Mol Neurodegener 9:27. doi:10.1186/1750-1326-9-27

Bello NT, Yeh CY, Verpeut JL, Walters AL (2014) Binge-like eating attenuates nisoxetine feeding suppression, stress activation, and brain norepinephrine activity. PLoS ONE 9(4):e93610. doi:10.1371/journal.pone.0093610

Bhargava HN, Perlow MJ (1988) Effect of 1-methyl-4-phenyl-1,2,3,6-tetrahydropyridine on striatal dopamine receptors in C57BL/6 mice. Toxicol Lett 40(3):219–224

Bing G, Zhang Y, Watanabe Y, McEwen BS, Stone EA (1994) Locus coeruleus lesions potentiate neurotoxic effects of MPTP in dopaminergic neurons of the substantia nigra. Brain Res 668(1–2):261–265

Blesa J, Juri C, Collantes M, Peñuelas I, Prieto E, Iglesias E, Martí-Climent J, Arbizu J, Zubieta JL, Rodríguez-Oroz MC, García-García D, Richter JA, Cavada C, Obeso JA (2010) Progression of dopaminergic depletion in a model of MPTP-induced Parkinsonism in non-human primates. An (18)F-DOPA and (11)C-DTBZ PET study. Neurobiol Dis 38(3):456–463. doi:10.1016/j.nbd.2010.03.006

Brami-Cherrier K, Anzalone A, Ramos M, Forne I, Macciardi F, Imhof A, Borrelli E (2014) Epigenetic reprogramming of cortical neurons through alteration of dopaminergic circuits. Mol Psychiatry 19(11):1193–1200. doi:10.1038/mp.2014.67

Bubak AN, Redmond DE Jr, Elsworth JD, Roth RH, Collier TJ, Bjugstad KB, Blanchard BC, Sladek JR Jr (2015) A potential compensatory role for endogenous striatal tyrosine hydroxylase-positive neurons in a nonhuman primate model of Parkinson's disease. Cell Transpl 24(4):673–680. doi:10.3727/096368915X687741

Bugajski J, Gadek-Michalska A, Ołowska A, Borycz J, Głód R, Bugajski AJ (1995) Adrenergic regulation of the hypothalamic-pituitary-adrenal axis under basal and social stress conditions. J Physiol Pharmacol 46(3):297–312

Connolly AT, Jensen AL, Bello EM, Netoff TI, Baker KB, Johnson MD, Vitek JL (2015) Modulations in oscillatory frequency and coupling in globus pallidus with increasing parkinsonian severity. J Neurosci 35(15):6231–6240. doi:10.1523/JNEUROSCI.4137-14.2015

Dabrowska J, Nowak P, Brus R (2008) Reactivity of 5-HT1A receptor in adult rats after neonatal noradrenergic neurons' lesion–implications for antidepressant-like action. Brain Res 1239:66–76. doi:10.1016/j.brainres.2008.08.054

Date I, Notter MF, Felten SY, Felten DL (1990) MPTP-treated young mice but not aging mice show partial recovery of the nigrostriatal dopaminergic system by stereotaxic injection of acidic fibroblast growth factor (aFGF). Brain Res 526(1):156–160

de Lima MN, Presti-Torres J, Caldana F, Grazziotin MM, Scalco FS, Guimarães MR, Bromberg E, Franke SI, Henriques JA, Schröder N (2007) Desferoxamine reverses neonatal **iron**-induced recognition memory impairment in rats. Eur J Pharmacol 570(1–3):111–114

Dolphin A, Jenner P, Marsden CD (1976a) Noradrenaline synthesis from L-DOPA in rodents and its relationship to motor activity. Pharmacol Biochem Behav 5(4):431–439

Dolphin A, Elliott PN, Jenner P (1976b) The irritant properties of dopamine-beta-hydroxylase inhibitors in relation to their effects on L-dopa-induced locomotor activity. J Pharm Pharmacol 28(10):782–785

Dooley DJ, Bittiger H, Hauser KL, Bischoff SL, Waldmeier PC (1983a) Alteration of central alpha-2 and beta-adrenergic receptors in the rat after DSP4, a delective noradrenergic neurotoxin. Neuroscience 9:889–898

Dooley DJ, Mogilnicka E, Delini-Stula A, Waechter F, Truog A, Wood J (1983b) Functional supersensitivity to adrenergic agonists in the rat after DSP-4, a selective noradrenergic neurotoxin. Psychopharmacology 81:1–5

Dornelles AS, Garcia VA, de Lima MN, Vedana G, Alcalde LA, Bogo MR, Schröder N (2010) mRNA expression of proteins involved in iron homeostasis in brain regions is altered by age and by iron overloading in the neonatal period. Neurochem Res 35(4):564–571. doi:10.1007/s11064-009-0100-z

Fagherazzi EV, Garcia VA, Maurmann N, Bervanger T, Halmenschlager LH, Busato SB, Hallak JE, Zuardi AW, Crippa JA, Schröder N (2012) Memory-rescuing effects of cannabidiol in an animal model of cognitive impairment relevant to neurodegenerative disorders. Psychopharmacology 219(4):1133–1140. doi:10.1007/s00213-011-2449-3

Fornai F, Bassi L, Torracca MT, Alessandrì MG, Scalori V, Corsini GU (1996) Region-and neurotransmitter-dependent species and strain differences in DSP-4-induced monoamine depletion in rodents. Neurodegeneration. 5(3):241–249

Fornai F, Alessandrì MG, Torracca MT, Bassi L, Corsini GU (1997) Effects of noradrenergic lesions on MPTP/MPP + kinetics and MPTP nigrostriatel dopamine depletions. J Pharmacol Exp Ther 283:100–107

Fornai F, Giorgi FS, Gesi M, Chen K, Alessrì MG, Shih JC (2001) Biochemical effects of the monoamine neurotoxins DSP-4 and MDMA in specific brain regions of MAO-B-deficient mice. Synapse 39(3):213–221

Fowler CJ, Thorell G, Sundström E, Archer T (1988) Norepinephrine-stimulated inositol phospholipid breakdown in the rat cerebral cortex following serotoninergic lesion. J Neural Transm 73(3):205–215

Fox ME, Studebaker RI, Swofford NJ, Wightman RM (2015) Stress and drug dependence differentially modulate norepinephrine signaling in animals with varied HPA axis function. Neuropsychopharmacology 40(7):1752–1761. doi:10.1038/npp.2015.23

Fredriksson A, Archer T (1994) MPTP-induced behavioral and biochemical deficits: a parametric analysis. J Neural Transm 7:123–132

Fredriksson A, Archer T (2000) Effects of clonidine and α-adrenoceptor antagonists on motor activity in DSP4-treated mice. II. Interactions with apomorphine. Neurotox Res 1:249–259

Fredriksson A, Archer T (2003) Effect of postnatal iron administration on MPTP-induced behavioral deficits and neurotoxicity: behavioral enhancement by l-dopa-MK-801 co-administration. Behav Brain Res 139:31–46

Fredriksson A, Archer T (2006) Subchronic administration of haloperidol influences the functional deficits of postnatal iron administration in mice. Neurotox Res 9(4):305–312

Fredriksson A, Archer T (2007) Postnatal iron overload destroys NA-DA functional interactions. J Neural Transm. 114(2):195–203

Fredriksson A, Plaznik A, Sundström E, Jonsson G, Archer T (1990) MPTP-induced hyperactivity in mice: reversal by l-dopa. Pharmacol Toxicol 67:295–301

Fredriksson A, Palomo T, Chase TN, Archer T (1999) Tolerance to a suprathreshold dose of l-dopa in MPTP mice: effects of glutamate antagonists. J Neural Transm 106:283–300

Fredriksson A, Schröder N, Eriksson P, Izquierdo I, Archer T (2001) Neonatal iron potentiates adult MPTP-induced neurodegenerative and functional deficits. Parkinsonism Rel Disord 7:97–105

Giovanni A, Sieber BA, Heikkila RE, Sonsalla PK (1994a) Studies on species sensitivity to the dopaminergic neurotoxin 1-methyl-4-phenyl-1,2,3,6-tetrahydropyridine. Part 1: systemic administration. J Pharmacol Exp Ther 270(3):1000–1007

Giovanni A, Sonsalla PK, Heikkila RE (1994b) Studies on species sensitivity to the dopaminergic neurotoxin 1-methyl-4-phenyl-1,2,3,6-tetrahydropyridine. Part 2: central administration of 1-methyl-4-phenylpyridinium. J Pharmacol Exp Ther 270(3):1008–1014

Goldstein LB (2006) Neurotransmitters and motor activity: effects on functional **recovery** after brain injury. NeuroRx 3(4):451–457

Goshima Y, Misu Y, Arai N, Misugi K (1991) Nanomolar l-dopa facilitates release of dopamine via presynaptic beta-adrenoceptors: comparative studies on the actions in striatal slices from control and1-methyl-4-phenyl-1,2,3,6-tetrahydropyridine (MPTP)-treated C57 black mice, an animal model for Parkinson's disease. Jpn J Pharmacol 55(1):93–100

Hallman H, Jonsson G (1984) Pharmacological modifications of the neurotoxic action of the noradrenaline neurotoxin DSP4 on central noradrenaline neurons. Eur J Pharmacol 103(3–4):269–278

Hallman H, Sundström E, Jonsson G (1984) Effects of the noradrenaline neurotoxin DSP 4 on monoamine neurons and their transmitter turnover in rat CNS. J Neural Transm. 60(2):89–102

Hare DJ, Adlard PA, Doble PA, Finkelstein DI (2013) Metallobiology of 1-methyl-4-phenyl-1,2,3,6-tetrahydropyridine neurotoxicity. Metallomics 5(2):91–109. doi:10.1039/c2mt20164j

Heal DJ, Butler SA, Prow MR, Buckett WR (1993) Quantification of alpha 2-adrenoceptors in rat brain after short-term DSP-4 lesioning. Eur J Pharmacol 249:37–41

Heikkila RE, Sieber BA, Manzino L, Sonsalla PK (1989) Some features of the nigrostriatal dopaminergic neurotoxin 1-methyl-4-phenyl-1,2,3,6-tetrahydropyridine (MPTP) in the mouse. Mol Chem Neuropathol 10(3):171–183

Hornykiewizc O, Kish SJ (1987) Biochemical pathophysiology of Parkinsin's disease. Adv Neurol 45P:19–34

Hurko O, Boudonck K, Gonzales C, Hughes ZA, Jacobsen JS, Reinhart PH, Crowther D (2010) Ablation of the locus coeruleus increases oxidative stress in tg-2576 transgenic but not wild-type mice. Int J Alzheimers Dis 2010:864625. doi:10.4061/2010/864625

Jonsson G, Hallman H (1982) Response of central monoamine neurons following an early neurotoxic lesion. Bibl Anat 23:76–92

Jonsson G, Hallman H, Ponzio F, Ross SB (1981) (N-[2-chloroethyl]-N-ethyl-2-bromobenzylamine)—a useful denervation tool for central and peripheral noradrenaline neurons. Eur J Pharmacol 72:173–188

Jonsson G, Hallman H, Sundström E (1982) Effects of the noradrenaline neurotoxin DSP4 on the postnatal development of central noradrenaline neurons in the rat. Neuroscience 7(11):2895–2907

Jonsson G, Sundström E, Mefford I, Olson L, Johnson S, Freedman R, Hoffer B (1985) Electrophysiological and neurochemical correlates of the neurotoxic effects of 1-methyl-4-phenyl-1,2,3,6-tetrahydropyridine (MPTP) on central catecholamine neurons in the mouse. Naunyn Schmiedebergs Arch Pharmacol 331(1):1–6

Kostowski W (1979) Two noradrenergic systems in the brain and their interactions with other monoaminergic neurons. Pol J Pharmacol Pharm 31(4):425–436

Kostrzewa RM (2009) Evolution of neurotoxins: from research modalities to clinical realities (Chapter 1: Unit 1.18). Curr Protoc Neurosci. doi:10.1002/0471142301.ns0118s46

Kostrzewa RM, Kostrzewa JP, Kostrzewa RA, Kostrzewa FP, Brus R, Nowak P (2011) Stereotypic progressions in psychotic behavior. Neurotox Res 19(2):243–252. doi:10.1007/s12640-010-9192-0

Langston JW (1985) MPTP neurotoxicity: an overview and characterization of phases oftoxicity. Life Sci 36(3):201–206

Lategan AJ, Marien MR, Colpaert FC (1992) Suppression of nigrostriatal and mesolimbic dopamine release in vivo following noradrenaline depletion by DSP-4: a microdialysis study. Life Sci 50(14):995–999

Lavich IC, de Freitas BS, Kist LW, Falavigna L, Dargél VA, Köbe LM, Aguzzoli C, Piffero B, Florian PZ, Bogo MR, de Lima MN, Schröder N (2015) Sulforaphane rescues memory dysfunction and synaptic and mitochondrial alterations induced by brain iron accumulation. Neuroscience 301:542–552. doi:10.1016/j.neuroscience.2015.06.025

Liang KC, Chen LL, Huang TE (1995) The role of amygdala norepinephrine in memory formation: involvement in the memory enhancing effect of peripheral epinephrine. Chin J Physiol 38(2):81–91

Lindgren H, Demirbugen M, Bergqvist F, Lane EL, Dunnett SB (2014) The effect of additional noradrenergic and serotonergic depletion on a lateralised choice reaction time task in rats with nigral 6-OHDA lesions. Exp Neurol 253:52–62. doi:10.1016/j.expneurol.2013.11.015

Liu YP, Huang TS, Tung CS, Lin CC (2015a) Effects of atomoxetine on attention and impulsivity in the five-choice serial reaction time task in rats with lesions of dorsal noradrenergic ascending bundle. Prog Neuropsychopharmacol Biol Psychiatry 56:81–90. doi:10.1016/j.pnpbp.2014.08.007

Liu Y, Yue F, Tang R, Tao G, Pan X, Zhu L, Kung HF, Chan P (2015b) Progressive loss of striatal dopamine terminals in MPTP-induced acute parkinsonism in cynomolgus monkeys using vesicular monoamine transporter type 2 PET imaging ([(18)F]AV-133). Neurosci Bull 30(3):409–416. doi:10.1007/s12264-013-1374-3

Marien M, Briley M, Colpaert F (1993) Noradrenaline depletion exacerbates MPTP-induced striatal dopamine loss in mice. Eur J Pharmacol 236(3):487–489

Mathai A, Ma Y, Paré JF, Villalba RM, Wichmann T, Smith Y (2015) Reduced cortical innervation of the subthalamic nucleus in MPTP-treated parkinsonian monkeys. Brain 138(Pt 4):946–962. doi:10.1093/brain/awv018

Nagai Y, Minamimoto T, Ando K, Obayashi S, Ito H, Ito N, Suhara T (2012) Correlation between decreased motor activity and dopaminergic degeneration in the ventrolateral putamen in monkeys receiving repeated MPTP administrations: a positron emission tomography study. Neurosci Res 73(1):61–67. doi:10.1016/j.neures.2012.02.007

Nishi K, Kondo T, Narabayashi H (1991) Destruction of norepinephrine terminals in 1-methyl-4-phenyl-1,2,3,6-tetrahydropyridine (MPTP)-treated mice reduces locomotor activity induced by l-dopa. Neurosci Lett 123(2):244–247

Nowak P, Nitka D, Kwieciński A, Jośko J, Drab J, Pojda-Wilczek D, Kasperski J, Kostrzewa RM, Brus R (2009) Neonatal co-lesion by DSP-4 and 5,7-DHT produces adulthood behavioral sensitization to dopamine D(2) receptor agonists. Pharmacol Rep 61(2):311–318

O'Donohue TL, Crowley WR, Jacobowitz DM (1979) Biochemical mapping of the noradrenergic ventral bundle projection sites: evidence for a noradrenergic–dopaminergic interaction. Brain Res 172(1):87–100

Pain S, Gochard A, Bodard S, Gulhan Z, Prunier-Aesch C, Chalon S (2013) Toxicity of MPTP on neurotransmission in three mouse models of Parkinson's disease. Exp Toxicol Pathol 65(5):689–694. doi:10.1016/j.etp.2012.09.001

Persson T, Waldeck B (1970) Further studies on the possible interaction between dopamine and noradrenaline containing neurons in the brain. Eur J Pharmacol 11(3):315–320

Pifl C, Kish SJ, Hornykiewicz O (2012) Thalamic noradrenaline in Parkinson's disease: deficits suggest role in motor and non-motor symptoms. Mov Disord 27(13):1618–1624. doi:10.1002/mds.25109

Pifl C, Hornykiewicz O, Blesa J, Adánez R, Cavada C, Obeso JA (2013) Reduced noradrenaline, but not dopamine and serotonin in motor thalamus of the MPTP primate: relation to severity of parkinsonism. J Neurochem 125(5):657–662. doi:10.1111/jnc.12162

Ponzio F, Hallman H, Jonsson G (1981) Noradrenaline and dopamine interaction in rat brain during development. Med Bull 59:161–169

Post C, Persson ML, Archer T, Minor BG, Danysz W, Sundström E (1987) Increased antinociception by alpha-adrenoceptor drugs after spinal cord noradrenaline depletion. Eur J Pharmacol 137(1):107–116

Rényi L, Archer T, Minor BG, Tandberg B, Fredriksson A, Ross SB (1986) The inhibition of the cage-leaving response–a model for studies of the serotonergic neurotransmission in the rat. J Neural Transm. 65(3–4):193–210

Roczniak W, Wróbel J, Dolczak L, Nowak P (2013) Influence of central noradrenergic system lesion on the serotoninergic 5-HT3 receptor mediated analgesia in rats. Adv Clin Exp Med 22 (5):629–638

Roczniak W, Babuśka-Roczniak M, Kwapuliński J, Brodziak-Dopierała B, Widuchowski W, Cipora E, Nowak P, Oświęcimska JM (2015) The effect of central noradrenergic system lesion on dopamine (DA) and serotonin (5-HT) synthesis rate following administration of 5-HT3 receptor ligands in chosen parts of the rat brain. Pharmacol Rep 67(1):146–151. doi:10.1016/j. pharep.2014.08.018

Ross SB (1976) Long-term effects of N-2-chlororethyl-N-ethyl-2-bromobenzylamine hydrochloride on noradrenergic neurons in rat brain and heart. Br J Pharmacol 58:521–527

Ross SB, Renyi L (1976) On the long-lasting inhibitory effect of N-(2-chlororethyl)-N-ethyl-2-bromobenzylamine (DSP4) on the active uptake of noradrenaline. J Pharm Pharmacol 28:458–459

Ross SB, Stenfors C (2015) DSP4, a selective neurotoxin for the locus coeruleus noradrenergic system. A review of its mode of action. Neurotox Res 27(1):15–30. doi:10.1007/s12640-014-9482-z

Schendzielorz N, Oinas JP, Myöhänen TT, Reenilä I, Raasmaja A, Männisto PT (2013) Catechol-O-Methyltransferase (COMT) protein expression and activity after dopaminergic and noadrenergic lesions of rat brain. PLoS ONE: e61392. doi:10.1371/journal.pone.0061392

Schultz W, Studer A, Jonsson G, Sundström E, Mefford I (1985) Deficits in behavioral initiation and execution processes in monkeys with 1-methyl-4-phenyl-1,2,3,6-tetrahydropyridine-induced parkinsonism. Neurosci Lett 59(2):225–232

Shannak K, Rajput A, Rozdilsky B, Kish S, Gilbert J, Hornykiewicz O (1994) Noradrenaline, dopamine and serotonin levels and metabolism in the human hypothalamus: observations in Parkinson's disease and normal subjects. Brain Res 639(1):33–41

Shin E, Rogers JT, Devoto P, Björklund A, Carta M (2014) Noradrenaline neuron degeneration contributes to motor impairments and development of l-dopa-induced dyskinesia in a rat model of Parkinson's disease. Exp Neurol 257:25–38. doi:10.1016/j.expneurol.2014.04.011

Silva PF, Garcia VA, Dornelles Ada S, Silva VK, Maurmann N, Portal BC, Ferreira RD, Piazza FC, Roesler R, Schröder N (2012) Memory impairment induced by brain iron overload is accompanied by reduced H3K9 acetylation and ameliorated by sodium butyrate. Neuroscience 200:42–49. doi:10.1016/j.neuroscience.2011.10.038

Sonsalla PK, Heikkila RE (1986) The influence of dose and dosing interval on MPTP-induced dopaminergic neurotoxicity in mice. Eur J Pharmacol 129(3):339–345

Sundström E, Luthman J, Jonsson G, Goldstein M (1987) Determination of monoamines by use of liquid chromatography with electrochemical detection in the study of selective monoamine neurotoxins. Life Sci 41(7):857–860

Tassin JP, Trovero F, Hervé D, Blanc G, Glowinski J (1992) Biochemical and behavioural consequences of interactions between dopaminergic and noradrenergic systems in rat prefrontal cortex. Neurochem Int 20(Suppl):225S–230S

Tian L, Xia Y, Flores HP, Campbell MC, Moerlein SM, Perlmutter JS (2015) Neuroimaging analysis of the dopamine basis for apathetic behaviors in an MPTP-lesioned primate model. PLoS ONE 10(7):e0132064. doi:10.1371/journal.pone.0132064

Tong J, Hornykiewicz O, Kish SJ (2007) Inverse relationship between brain noradrenaline level and dopamine loss in Parkinson disease: a possible neuroprotective role for noradrenaline. Arch Neurol 63(12):1724–1728

Uehara S, Uno Y, Inoue T, Murayama N, Shimizu M, Sasaki E, Yamazaki H (2015) Activation and deactivation of 1-methyl-4-phenyl-1,2,3,6-tetrahydropyridine by cytochrome P450 enzymes and flavin-containing monooxygenases in common marmosets (*Callithrix jacchus*). Drug Metab Dispos 43(5):735–742. doi:10.1124/dmd.115.063594

Wang Y, Chen X, Wang T, Sun YN, Han LN, Li LB, Li Z, Wu ZH, Huang C, Liu J (2014) Additional noradrenergic depletion aggravates forelimb akinesia and abnormal subthalamic nucleus activity in a rat model of Parkinson's disease. Life Sci 119:18–27. doi:10.1016/j.lfs.2014.10.007

You LH, Li F, Wang L, Zhao SE, Wang SM, Zhang LL, Zhang LH, Duan XL, Yu P, Chang YZ (2015) Brain iron accumulation exacerbates the pathogenesis of MPTP-induced Parkinson's disease. Neuroscience 284:234–246. doi:10.1016/j.neuroscience.2014.09.071

Youdim MB, Arraf Z (2004) Prevention of MPTP (N-methyl-4-phenyl-1,2,3,6-tetrahydropyridine) dopaminergic neurotoxicity in mice by chronic lithium: involvements of Bcl-2 and Bax. Neuropharmacology 46(8):1130–1140

Zecca L, Stroppolo A, Gatti A, Tampellini D, Toscani M, Gallorini M, Giaveri G, Arosio P, Santambrogio P, Fariello RG, Karatekin E, Kleinman MH, Turro N, Hornykiewicz O, Zucca FA (2004) The role of iron and copper molecules in the neuronal vulnerability of locus coeruleus and substantia nigra during aging. Proc Natl Acad Sci USA 101(26):9843–9848

Zuddas A, Vaglini F, Fornai F, Corsini GU (1992) Selective lesion of the nigrostriatal dopaminergic pathway by MPTP and acetaldehyde or diethyldithiocarbamate. Neurochem Int 20(Suppl):287S–293S

Perinatal Domoic Acid
as a Neuroteratogen

Tracy A. Doucette and R. Andrew Tasker

Abstract In mammals, the period shortly before and shortly after birth is a time of massive brain growth, plasticity and maturation. It is also a time when the developing brain is exquisitely sensitive to insult, often with long-lasting consequences. Many of society's most debilitating neurological diseases arise, at least in part, from trauma around the time of birth but go undetected until later in life. For the past 15 years, we have been studying the consequences of exposure to the AMPA/kainate agonist domoic acid (DOM) on brain development in the rat. Domoic acid is a naturally occurring excitotoxin that enters the food chain and is known to produce severe neurotoxicity in humans and other adult wildlife. Our work, and that of others, however, has demonstrated that DOM is also toxic to the perinatal brain and that toxicity occurs at doses much lower than those required in adults. This raises concern about the current regulatory limit for DOM contamination that is based on data in adult animals, but has also allowed creation of a novel model of neurological disease progression. Herein, we review briefly the toxicity of DOM in adults, including humans, and describe features of the developing nervous system relevant to enhanced risk. We then review the data on DOM as a prenatal neuroteratogen and describe in detail the work of our respective laboratories to characterize the long-term behavioural and neuropathological consequences of exposure to low-dose DOM in the newborn rat.

Keywords Glutamate receptors · Amnesic shellfish toxin · Seizures · Epilepsy · Schizophrenia · Social interaction · Attentional processing · Cognition

T.A. Doucette
Department of Psychology, University of Prince Edward Island,
550 University Avenue, Charlottetown, PE C1A4P3, Canada
e-mail: tdoucette@upei.ca

R. Andrew Tasker (✉)
Department of Biomedical Sciences, University of Prince Edward Island,
550 University Avenue, Charlottetown, PE C1A4P3, Canada
e-mail: tasker@upei.ca

Curr Topics Behav Neurosci (2016) 29: 87–110
DOI 10.1007/7854_2015_417
Published Online: 23 December 2015

Contents

1 Introduction 88
2 The Developing Brain Is More Sensitive to Toxic Insult 90
3 Domoic Acid as a Neurotoxin in Both Humans and Wildlife 90
   3.1 The Current Regulatory Environment for Domoic Acid 91
4 The Neonatal Brain Is More Sensitive to Domoic Acid 92
5 Domoic Acid as a Prenatal Neuroteratogen 93
6 The Neonatal Domoate Model of Progressive Brain Dysfunction: A New Approach?.... 94
   6.1 The Need for Models of Disease not just Symptoms 94
   6.2 The Neonatal DOM Rat Protocol 95
   6.3 Translation in Reverse: Neonatal DOM as a Model of Epileptogenesis 98
   6.4 More Than Epilepsy: Effects on Measures of Psychotic Disease 101
   6.5 Changes in Cognitive Function and Attentional Processes 103
7 Conclusions 107
References 107

1 Introduction

Mental and neurological disorders are increasingly prevalent and constitute a major societal and economic burden worldwide, and with the ageing of the general population and ever-increasing life expectancy in many developed countries and emerging economies, it is estimated that mental and neurological diseases will account for 14.7 % of the global disease burden by the year 2020 (World Health Organization 2006). Further, many of society's most debilitating neurological diseases are now known to be progressive in nature. Alzheimer's disease (AD), Parkinson's disease (PD) and amyotrophic lateral sclerosis (ALS) are just three of many examples where the disease process begins long before the onset of symptoms that constitute the diagnostic criteria. For example, it is estimated that up to 60 % of the dopaminergic neurons in the substantia nigra must be lost before the first clinical signs of PD appear (Schulz and Falkenburger 2004). Further, even when diagnosed, the neurodegenerative process continues, resulting in escalating morbidity and ultimately death. This concept is depicted schematically in Fig. 1, and if those factors responsible for initiation and early progression of the disease process can be reliably identified, presymptomatic detection of such diseases is possible.

It is our contention, therefore, that presymptomatic detection of disease represents a largely unexplored opportunity for therapeutic intervention, albeit not an easy thing to implement. Even if presymptomatic biomarkers relevant to individual diseases could be identified, health budgets cannot possibly support sustainable screening of all members of the population, and the incidence of these diseases is too large to permit accurate presymptomatic risk analysis. Further, most neurological diseases are presumed to not result from a single precipitating event, but rather arise from a complex interaction between initiating stimuli, genetic predisposition,

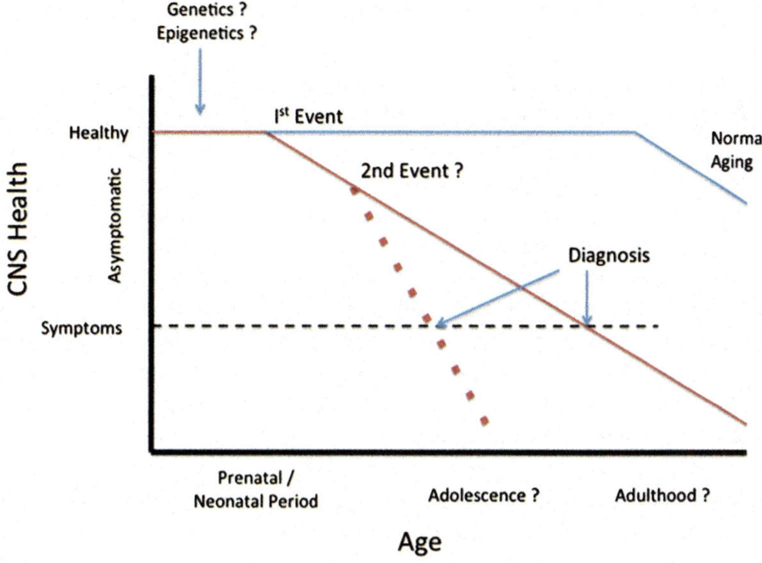

Fig. 1 Schematic representation of the concept of progressive neurological dysfunction relevant to disease. Normal decline of brain health begins in adulthood (*blue line*), but events that are often undetected may induce a degenerative process resulting in premature decline of central nervous system (*CNS*) health as illustrated by the *red lines*. Susceptibility may be enhanced by genetic/epigenetic predisposition, and degeneration may be accelerated by subsequent trauma. Initially, the patient remains asymptomatic until the process has progressed to the point of recognizable symptoms and subsequent diagnosis

environmental factors and lifestyle (Shaw and Hoglinger 2008). Nonetheless, the overall goal of identifying disease earlier is both socially and economically responsible and should be vigorously pursued.

As disturbing and emotionally challenging as diseases such as AD, PD and ALS are, however, with the exception of comparatively rare early-onset variants, diseases such as AD and PD are essentially diseases of elderly people who have often enjoyed full and productive lives before becoming recognizably ill. Of arguably greater consequence are a myriad of progressive neurological diseases that are often neurodevelopmental in origin, whereby the disease process initiates prior to, or shortly after, birth and where the affected patient suffers from reduced quality of life beginning in childhood, adolescence or early adulthood. Many of these neurodevelopmental diseases appear to originate from imbalances in the normal harmony of excitatory and inhibitory processes during brain development (Lujan et al. 2005). Consequently, accidental or deliberate exposure to drugs or toxins that alter excitatory or inhibitory processing at key developmental stages prior to or shortly after birth might initiate a disease process that culminates in recognizable deficits later in life. The current manuscript will discuss the neuroteratogenic potential of one such toxin, domoic acid, a naturally occurring excitatory amino acid that exists in the food chain of higher mammals including humans.

2 The Developing Brain Is More Sensitive to Toxic Insult

From an engineering standpoint, a system that is in rapid transition is necessarily and intrinsically unstable. This is true of the developing nervous system, wherein during "critical periods" of CNS development, the brain is undergoing rapid change and is especially vulnerable to exogenous stimuli (Rice and Barone 2000; Vorhees 1986; Colombo 1982). The notion of "critical period" makes reference to that period in time between the emergence (anatomically and/or functionally) of a particular biobehavioural system and when that system reaches its full maturational state. Changes effected during a "critical period" often have long-term consequences. We have known for some time that as well as producing transient pharmacological effects, receptor-acting compounds, particularly when exposure occurs during a so called "critical period", are capable of producing more-or-less permanent neurobehavioural effects. This irreversible insult can either be expressed as a permanent dysfunction of the neurochemical system involved, or may result in what has been termed "irreversible imprinting" of receptor densities, which in turn produces lasting functional and/or structural changes to the nervous system (reviewed in Kaufmann 2000). During the first three weeks of post-natal life in the rat, a developmental window roughly comparable to the third trimester of human foetal development, there are rapid and tremendous changes in brain development (Clancy et al. 2007; Vernadakis and Woodbury 1969). This period of development, generally referred to as the "brain growth spurt" (Dobbing and Sands 1979; Dobbing and Smart 1974), is a time of remarkable, structural and functional plasticity necessary to prepare the organism for life. It is also, however, a time when the developing brain is exquisitely sensitive to environmental disruption. Significantly, the immediate consequences of that disruption are often quite subtle and go unnoticed. However, in many cases, behavioural anomalies will be latently expressed—notable only when those systems that underlie the given behavioural profiles have reached full functional maturity.

3 Domoic Acid as a Neurotoxin in Both Humans and Wildlife

Domoic acid (DOM) ([2S,3S,4S]-2-carboxy-4-1-methyl-5[R]-carboxyl-l[Z]-3[E]-hexadienyl pyrrolidine-3-acetic acid) ($C_{15}H_{21}NO_6$) is a naturally occurring crystalline, water-soluble molecule with chemical properties similar to other acidic amino acids. The pharmacological actions of DOM in the brain are mediated primarily, but not exclusively, by the activation of the AMPA/kainate subclass of glutamate receptors with the "low affinity" kainate receptors (GluK1 and GluK2) showing selective affinity (Verdoorn et al. 1994; Tasker et al. 1996). In addition to the parent compound, several DOM isomers are known to occur either in nature or under certain laboratory conditions. Historically, DOM is a constituent of the red alga "doumoi" (*Chondria armata*) that is indigenous to certain Japanese archipelagos. In the late 1980s, however, DOM was found as a contaminant in cultured blue mussels (*Mytilus edulis*) (Wright et al. 1990) that once consumed resulted in a well-documented case of human intoxication (Perl et al. 1990; Teitelbaum et al. 1990). In that case (and subsequent cases of wildlife intoxication; see below), the source of DOM was a diatom, *Nitzschia pungens f. multiseries*, although subsequently a variety of planktonic diatoms of *Pseudo-Nitzschia* spp. have been identified worldwide as being capable of producing DOM under certain environmental conditions (for review, see Bates and Horner 1998; Bates and Trainer 2006). Details of the only documented case of human intoxication and the multiple well-documented cases of toxicity in wildlife species including various species of seabirds and marine mammals are documented elsewhere (for reviews, see Doucette and Tasker 2008; Tasker 2015 among others), but in brief, ingestion of DOM in sufficient quantity results in a dose-related sequence of human toxicity beginning with gastrointestinal distress and progressing through confusion, amnesia (hence the designation of "amnesic shellfish poisoning" (Can Dis Wkly Rep 1990) seizures, coma and death. Birds such as Brandt's cormorants (*Phalacrorax penicillatus*) (Work et al. 1993) and brown pelicans (*Pelecanus occidentalis*) (Sierra-Beltran et al. 1997) and marine mammals including California sea lions (*Zalophus californianus*) (Scholin et al. 2000; Gulland et al. 2002) exhibit lethargy, ataxia, seizures and death with increasing exposure. In all mammalian species examined to date, post-mortem histopathology reveals a reproducible pattern of cellular pathology with regions of the hippocampal formation and midbrain being particularly vulnerable (Pulido 2008).

3.1 The Current Regulatory Environment for Domoic Acid

Shortly after the 1987 outbreak of human toxicity following ingestion of DOM, the Canadian regulatory agencies convened a meeting (Can Dis Wkly Rep 1990) with three principal aims: (1) to define the clinical syndrome resulting from DOM intoxication,[1] (2) to discuss the current state of knowledge about DOM and related compounds and (3) to provide guidance on future regulatory limits for shellfish contamination. At that time, most of the available data on DOM toxicity in mammals were derived from studies of acute intoxication in adult rats and mice. Based largely on those data, and giving due consideration to a standard "margin of safety", a regulatory limit of 20 parts per million (ppm) in contaminated shellfish was established for Canada and subsequently adopted by other countries as DOM

---

[1]Due to the profound neurological consequences of DOM intoxication, including the unusual observation of anterograde amnesia in more severely affected individuals, the syndrome was subsequently termed "amnesic shellfish poisoning (ASP)".

contamination began to be detected globally. That regulatory limit still exists in jurisdictions with monitoring systems in place, although of note a report commissioned jointly by the FAO/IOC/WHO on toxins in marine bivalves did raise concerns that the limit might not be appropriate for sensitive human populations including the elderly, chronically ill and during pregnancy (Tasker et al. 2011).

4 The Neonatal Brain Is More Sensitive to Domoic Acid

In the case of DOM and related compounds, we and others have reported that the acute toxicity profile for excitatory amino acid agonists differs significantly between preweaning and adult rodents with respect to both potency and the displayed repertoire of toxicity-associated behaviours (Xi et al. 1997; Doucette et al. 2000). Administration of DOM to young rats produces a reproducible pattern of behavioural toxicity that is similar (although not identical) to that observed following the administration of the structurally similar toxin kainic acid. Although in adult rats and mice, DOM toxicity begins with lethargy and impaired mobility (Tasker et al. 1991), in neonatal rats, the first signs noted following systemic administration of DOM are "wet dog shakes", head tremors, purposeless mastication, Straub tail and circling behaviour. These are followed by a series of behaviours that manifest in a dose-related fashion and include forelimb tremors, hyperextension of the hindlimbs, scratching and hind-paw directed foot-biting. If the dosage of the toxin is sufficiently high, whole-body tremors become apparent, as do myoclonic and bilateral clonic jerks, a loss of the righting reflex, postural rigidity, hind limb "paddling" and extreme tail whipping. Those animals receiving a sufficient dosage of the toxin will next proceed to generalized clonic–tonic convulsions, which in our study (Doucette et al. 2000) served as an end point. By using a cumulative behavioural score to develop full dose–response curves for DOM at various developmental ages, we were able to determine that very young rats (PND 0 and PND 5) displayed signs of moderate toxicity at doses in the 100–200 µg/kg range, a finding that compared favourably with other reports in PND 5 rat pups (Xi et al. 1997), but is significantly below the 1.5–2.5 mg/kg dose required to produce toxicity in the adult rat (Tryphonas et al. 1990). Further, we reported in the same paper that the potency of DOM decreases progressively with age, such that rats at two weeks of age (PND 14) require approximately twice as much toxin to produce a comparable cumulative behavioural toxicity score (i.e. as compared to rats at PND 5), and between PND 14 and PND 22, a dramatic shift in potency is apparent with respect to behavioural toxicity scores (Doucette et al. 2000). What is particularly noteworthy, however, is that even at 22 days of age, rats are approximately twice as sensitive to systemically administered DOM compared to mature adults (Tryphonas et al. 1990). The reason (s) for this significantly enhanced vulnerability and progressive decrease in potency with age are as yet unclear, but are most likely the result of a complex interaction among factors such as incomplete BBB formation,[2] reduced renal clearance capacities [DOM is eliminated almost exclusively by excretion (Suzuki and Hierlihy 1993)], differences in the maturation of excitatory versus inhibitory CNS circuitry[3] and/or age-related changes in relative AMPA/kainate receptor expression in regions of the brain that mediate the behavioural signs of DOM toxicity. Regardless, it is abundantly clear that the neonatal rat, particularly during the "brain growth spurt" from PND 5/6–22, is considerably more sensitive to DOM toxicity than is the adult rat.

5   Domoic Acid as a Prenatal Neuroteratogen

Over 20 years ago, Dakshinamurti et al. (1993) reported that the offspring of CD-1 mice injected with 0.6 mg/kg DOM on gestational day (GD) 13 showed no immediate signs of toxicity but did have aberrant pathology of the hippocampus by PND 14 and an altered EEG and reduced seizure threshold from PND 10–30. Curiously, this issue lays dormant for many years until reports by Levin et al. (2005) in rats and Tanemura et al. (2009) in mice confirmed that prenatal exposure to subconvulsive doses of DOM produced subtle, but detectable changes in cognition and brain pathology. Taken in light of data demonstrating that DOM administered systemically to pregnant rats is detectable in both amniotic fluid and foetal brain (Maucher and Ramsdell 2007; Maucher-Fuquay et al. 2012), and given the conclusive data that the neonatal brain is exquisitely sensitive to DOM (see above), it would appear that exposure in utero is both possible and of consequence to the offspring. Moreover, the risk appears to be observable across vertebrate species in that neurotoxicity has been observed after ovo microinjection in zebra fish (Tiedeken et al. 2005; Tiedeken and Ramsdell 2007) and is detectable in the amniotic fluid of stranded sea lions (Brodie et al. 2006), giving rise to speculation that the increasing number of cases of neurotoxicity in young sea lions is due in part to prenatal exposure (Goldstein et al. 2008; Ramsdell and Zabka 2008). Clearly, further investigation of the negative consequences of prenatal DOM is warranted.

---

[2]In the rat, the blood–brain barrier develops progressively, commencing on around gestational day 15 or 20 and showing mature morphological characteristics and structural integrity on approximately postnatal day 24 (reviewed in Kaufmann 2000).

[3]Moshe et al. (1983) have argued that age differences in EAA-induced seizures may relate to the precocious development of excitatory systems within the limbic region which develop prenatally, with concurrently delayed inhibitory mechanisms—not functionally mature until after two weeks of age.

6 The Neonatal Domoate Model of Progressive Brain Dysfunction: A New Approach?

Certain forms of epilepsy and schizophrenia are diseases that are known to be frequently neurodevelopmental in origin. Often arising from a combination of genetic susceptibility and/or perinatal trauma, both epilepsy and schizophrenia are increasingly recognized as involving a presymptomatic period of progressive neurological dysfunction that lasts years or even decades before diagnosis. As discussed previously, the presymptomatic period represents a largely unexplored opportunity for therapeutic intervention. This requires a better understanding of the disease process from initiation to symptom onset. Such understanding, however, requires the use of well-validated models of disease aetiology. The remainder of this manuscript will describe such a model, the neonatal domoate rat model that was created, validated and continues to be characterized by our respective laboratories, but first we address the question of why such a model is needed.

6.1 The Need for Models of Disease not just Symptoms

Preclinical drug discovery and development in the neurosciences relies heavily on the use of animal models in a variety of species but primarily in rats and mice. At present, most of these models fall into three major categories: (1) genetically modified species including knockout, conditional knockout and transgenic lines, (2) surgical models including site-selective lesions and (3) acute drug administration models including those that use toxins or receptor-selective agonists and antagonists. A characteristic common to almost all of these models, however, is that they have been developed to replicate the major diagnostic symptoms of each individual disease. This is to be expected since most current drug development is geared towards identifying therapies to reduce or eliminate those symptoms of disease that are most detrimental to the patient's quality of life. For example, most drug development in PD is targeted to reducing or eliminating the motor deficits that compromise mobility and daily function. In the cases of epilepsy and schizophrenia, most of the widely used preclinical models involve induction of convulsive seizures or production of deficits in attentional and preattentional processing, respectively. Although there may be no direct connection to the underlying cause of these symptoms, such approaches have resulted in the development and commercialization of a number of drug candidates for symptomatic relief following diagnosis, although as depicted in Fig. 1 such therapies are largely palliative in nature because overall CNS health is already seriously compromised.

In contrast, we and others have argued that while more labour-intensive, what is needed are preclinical models of disease development rather than models of symptoms (signs). Rodent models in which the pathology of a disease can be measured in a progressive longitudinal sequence hold considerable potential for achieving the goal of presymptomatic detection and therapy. It could also be argued that such models have greater potential for identifying new drug candidates that will translate into clinical settings, but until preclinical models of disease become more widespread in the biotechnology and pharmaceutical industries, such a claim is premature.

In the ensuing sections, we will describe our attempts to produce such a pre-clinical model. Moreover, the creation of the neonatal domoate rat model has been guided largely by two *a priori* principles:

1. That the model be more clinically relevant. To increase the likelihood of relevancy, we have employed a strategy we term "translation in reverse". As described below, the rat model was an attempt to replicate a clinical case of delayed-onset neurological dysfunction and was therefore validated against documented behavioural and pathological data in human patients. Further, our more recent attempts to explore the molecular basis of these changes have employed an in vitro system that reproduces the changes seen in vivo.

2. That the model embrace rather than exclude elements of multiple diseases that often present as comorbidities in patients. As described below, the neonatal domoate rat model manifests signs of both temporal lobe epilepsy and schizophrenia. Because these conditions often appear as comorbidities in patients, a model that investigates elements of both should have a greater potential to identify common and/or causally related changes in brain development, maturation and function than would models investigated solely in the context of a single disease.

6.2   The Neonatal DOM Rat Protocol

Working from the notion of critical periods (see above) and guided by clinical reports of late-onset epilepsy in patients who had consumed DOM (see below), we sought to determine whether chronic low-dose exposure to DOM during critical periods of CNS maturation (i.e. with respect to the ontogeny of ionotropic glutamate receptor subunit expressions and the maturation state of particular forebrain structures) would produce latently expressed changes in the behavioural profile of the resulting adult animals. In a series of experiments that have spanned more than a decade and a half, we have presented an abundance of data which demonstrate that daily injections (s.c.) of a low dose of DOM (20 µg/kg, which is equivalent to 1/5 of the ED50 for PND 8 rats (Doucette et al. 2000)) throughout the second week of post-natal life (i.e. PND 8–14) result in a progressively dysfunctional phenotype relevant to both epilepsy and schizophrenia. The protocol produces no signs of toxicity in the newborn rats, but the dose is sufficient to be physiologically active in the CNS as evidenced by a slightly accelerated time of eye opening and an enhancement of conditioned odour preference relative to saline controls (Doucette et al. 2003; Tasker et al. 2005). With increasing age, however, DOM-treated rats show altered behavioural profiles (summarized in Table 1) as well as neurochemical

Table 1 Behavioural and neurodevelopmental changes relevant to psychiatric disease

Assessment	Age	Effect	References
Physical development	PND 8–16	Accelerated eye opening	Doucette et al. (2003), Burt et al. (2008)
Olfactory conditioning	PND 8–13	Development of CPP to DOM—paired olfactory stimulus	Doucette et al. (2003), Tasker et al. (2005)
Nicotine-induced conditioned place preference (CPP)	PND 40–52	Abolishment of CPP effect at a dosage (0.6 mg/kg) that produces a conditioned preference in untreated rats of this age	Burt et al. (2008a)
	6–7 months	CPP manifested to a dosage (0.6 mg/kg) that does not result in a conditioned preference in untreated rats of this age	Burt et al. (2008b)
Elevated plus maze	3–4 months	Alterations in emotionality	Gill et al. (2012)
	4–5 months	Alterations in emotionality	Doucette et al. (2007)
Playground maze	2 months	Increased behavioural response to novelty (e.g. exploration of objects)	Burt et al. (2008a, b)
Social interaction	4–5 months	Active social avoidance	Ryan et al. (2011)
8-Arm radial maze	1–1.5 months	Superior choice accuracy (indicative of improved working memory)	Adams et al. (2009)
Morris water maze	4–5 months	Increased susceptibility to proactive interference	Doucette et al. (2007)
	4–5 months	Superior performance during initial learning trials; but significant impairments when challenged with reversal tasks	Adams et al. (2009)
	3–5 months	Increased perseveration and impairments in re-learning new platform location when challenged with reversal task	Gill et al. (2012)
Water H maze	3–5 months	Increased perseveration and impairments in re-learning new platform location when challenged with reversal task	Gill et al. (2012)
Temporal memory task	8–9 months	Deficits in temporal memory	Robbins et al. (2013)
Latent inhibition	4–5 months	Impaired attentional processing evident in a conditioned emotional response task	Marriott et al. (2014)
	4–5 months	Impaired latent inhibition in a conditioned taste aversion task	Marriott et al. (2012)

(continued)

Table 1 (continued)

Assessment	Age	Effect	References
Prepulse inhibition	3–4 months	PPI deficits (magnitude of the deficit impacted upon by prepulse dB level, sex and light cycle)	Marriott et al. (2012)
	3–4 months	PPI deficits and altered baseline startle response	Marriott et al. (2008)
Sleep	3 months	Reductions in paradoxical sleep (PS) (no differences in total sleep time or stage shifts into PS)	Gill et al. (2009)
Seizure threshold	3 months	Enhanced sensitivity to PTZ and electrical stimulation in the absence of an increase in electrically induced seizure propagation	Gill et al. (2010a, b)
Stress-induced low-grade seizures	2–15 months	Manifestation of seizure-like behaviours in response to mild to moderate stress. Observed in various behavioural tasks	Doucette et al. (2004), Perry et al. (2009), Gill et al. (2012)

and neuropathological changes primarily in the hippocampus (Table 2). It is known
that there are dynamic and complex patterns of changing expression of ionotropic
glutamate receptor subunits within the hippocampus during neonatal development
with many undergoing transient periods of heightened expression commencing
around the second postnatal week of life (Ritter et al. 2002). Evidence of

Table 2 Molecular and morphological changes with relevant to psychiatric disease

Molecular/morphological change	Age	References
Hippocampus		
Mossy fibre sprouting	15 months	Doucette et al. (2004)
	3 months	Bernard et al. (2007)
Elevated BDNF mRNA in CA1	15 months	Doucette et al. (2004)
Elevated trkB receptor expression in hilus of DG and area CA3	3 months	Bernard et al. (2007)
Increases in hippocampal adrenergic and mineralocorticoid receptor expression	8 months	Gill et al. (2012)
Decreased number of parvalbumin containing cells	3 months	Gill et al. (2010a, b)
Decreased GAD65/67 immunoreactivity	3 months	Gill et al. (2010a, b)
mPFC		
Suppressed TH immunoreactivity	9 months	Robbins et al. (2013)
NAcc		
Increased TH immunoreactivity	9 months	Robbins et al. (2013)

Abbreviations Brain-derived neurotrophic factor (*BDNF*); cornu ammonis (*CA*); tropomyosin
receptor kinase B (*trkB*); dentate gyrus (*DG*); glutamic acid decarboxylase (*GAD*); medial
prefrontal cortex (*mPFC*); nucleus accumbens (*NAcc*); tyrosine hydroxylase (*TH*)

behavioural, neuropathological and neurochemical changes observed to date in this model is described below.

6.3 Translation in Reverse: Neonatal DOM as a Model of Epileptogenesis

6.3.1 It Started with Patients

In the only properly documented case of human poisoning with domoic acid, over 100 people in Montreal, Canada, became ill after consuming cultured blue mussels (*M. edulis*) that were contaminated with high concentrations of DOM. Details of the epidemiology, clinical course and post-mortem pathology are well described in a number of sources (Perl et al. 1990; Teitelbaum et al. 1990; reviewed in Pérez-Gómez and Tasker 2014a). Of the 107 persons that met the case definition (47 men and 60 women), however, 12 patients were sufficiently affected to require admission to intensive care and manifested seizures that were largely resistant to conventional anticonvulsant therapy (Teitelbaum et al. 1990). In a few cases, the seizures progressed to status epilepticus and sadly several of these patients died in hospital. Of particular relevance to the current work, however, was the interesting case of one of these patients who survived the initial poisoning event but presented with seizures about 1 year later and was subsequently diagnosed with temporal lobe epilepsy caused by DOM (Cendes et al. 1995). This patient died later of complications from pneumonia and on post-mortem examination was found to have extensive cellular pathology in the hippocampus (particularly regions H1 and H3 that correspond to CA1 and CA3 in the rat), amygdala and thalamus (Cendes et al. 1995). The case was quite remarkable largely for two reasons: (1) it was the first report of non-fatal glutamatergic excitotoxicity in humans and confirmed the selective pattern of neuropathology produced by DOM in the human brain, and (2) it demonstrated that exposure to DOM could induce a state of progressive neurological dysfunction that culminated in clinical epilepsy (i.e. epileptogenesis).

6.3.2 Rats Are a Clinically Relevant Surrogate

Based largely on this report, we initiated a series of experiments to investigate the immediate and long-term effects of DOM exposure during brain development in the rat (see Sect. 6.2), and in 2004, we reported that administration of very low doses of DOM during the second postnatal week resulted in an adult-onset syndrome characterized by stage 2/3 seizure behaviour (Racine 1972) induced by exposure to novel/stressful environments (Doucette et al. 2004). Further, and remarkably, the pattern of cellular pathology in the hippocampal formation of these rats was almost identical to that seen in the cases of human poisoning including cell loss in the hippocampal subfields CA1 and CA3 (Doucette et al. 2004; Bernard et al. 2007).

Further, these rats had extensive mossy fibre sprouting (aberrant growth of dentate granule cell axons) which is a hallmark of seizure states (Holmes et al. 1999) (see Fig. 2). Subsequently, we have demonstrated that the neonatal DOM rat model is characterized by many other signs of epileptiform activity including abnormal cortical EEG consistent with mild seizure states (Gill et al. 2010a) (see Fig. 3), a highly significant reduction in induced general and focal seizure threshold (Gill et al. 2010a), reductions in paradoxical (REM-like) sleep patterns when recorded by remote telemetry (Gill et al. 2009) and selective reductions in GABA neurons in the hippocampus (Gill et al. 2010b). Collectively, these data were sufficient to have the

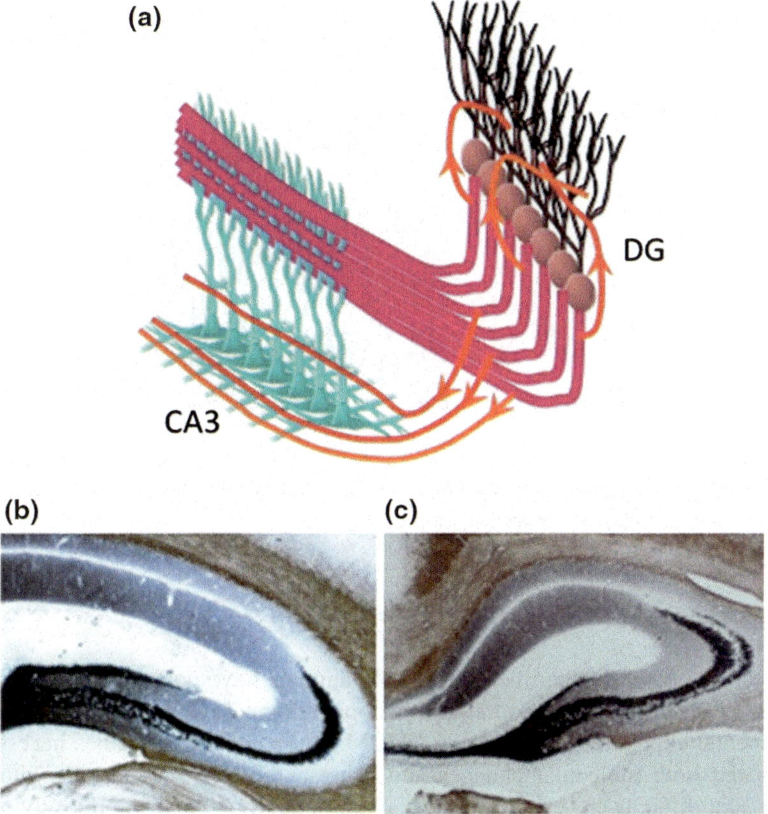

Fig. 2 **a** Cartoon depicting mossy fibre sprouting. Normally, granule cells in the dentate gyrus (*DG*) make synaptic contact with the apical dendrites of pyramidal cells in hippocampal subfield CA3 via mossy fibres. But in seizure states, mossy fibre axons can form aberrant connections with the basilar dendrites in CA3 and as well with dendrites of granule cells in DG. **b** MF axons making contact with apical dendrites in CA3 are visualized using Timm's stain and appear as a single black band in the hippocampus of a saline-treated rat. **c** A second band of staining indicating abnormal connections with basilar dendrites in CA3 is visible in the hippocampus of a DOM-treated rat

Fig. 3 EEG and EMG recordings via radiotelemetry from the cortex of a male SD rat treated with low-dose DOM as a neonate as described in Gill et al. (2010a, b). **a** Recordings from a normally behaving rat, **b** recordings during a stage 2/3 seizure induced by mild stress and **c** recordings following a low dose of pentylenetetrazol demonstrating a significantly reduced seizure threshold

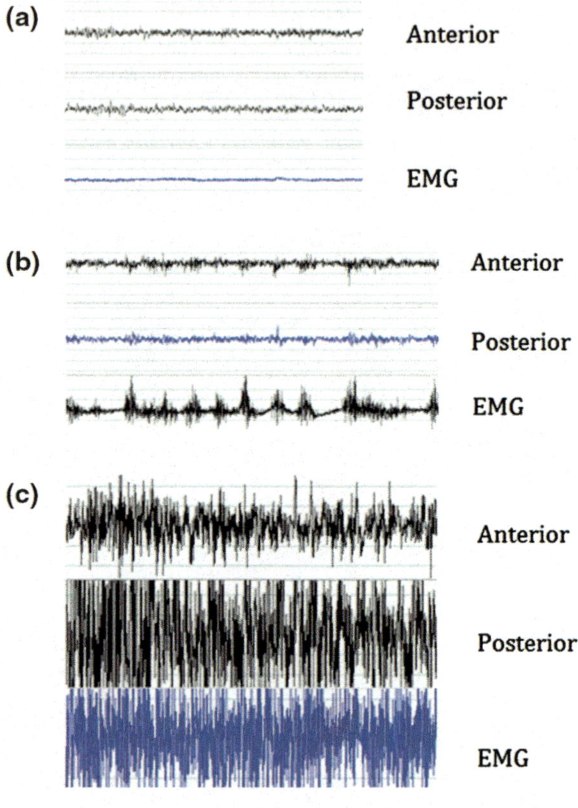

neonatal DOM rat recognized by both US and Canadian patent agencies as a novel and useful rat model of temporal lobe epilepsy.[4]

6.3.3 Organotypic Hippocampal Slice Cultures

As described above, experiments in whole animals have contributed much to our understanding of the behavioural and histopathological consequences of neurodevelopmental exposure to low doses of DOM. But mechanistic studies necessary to understand both molecular changes and cause-and-effect relationships within the model are difficult in whole animals. Consequently, we have recently further "reverse-translated" the DOM model into an in vitro system that retains the cellular composition and principal architecture of the hippocampus: organotypic hippocampal slice cultures (OHSCs) (Stoppini et al. 1991). Studies in OHSCs have demonstrated that DOM induces neurotoxicity in a concentration-dependent

---

[4]US Patents 7034201B2, 7521589 and 7622101 and Canadian Patent CA2448647.

manner with areas CA1 and CA3 being the most vulnerable as they are in whole rats (and humans) (Pérez-Gómez and Tasker 2012), but moreover, a concentration of DOM that induces only mild toxicity in CA1 in these cultures has been shown to cause neurogenesis in the subgranular zone (Pérez-Gómez and Tasker 2012) with accompanying mossy fibre sprouting and synaptogenesis in region CA3 (Pérez-Gómez and Tasker 2014b). The precise mechanisms responsible for these changes are still to be determined, but we do know that they are dependent on enhanced expression of brain-derived neurotrophic factor (BDNF) that is driven by the transcription factor CREB (Pérez-Gómez and Tasker 2013). The enhanced expression of BDNF in OHSCs is consistent with our previous observations in vivo of elevated BDNF mRNA (Doucette et al. 2004) and increases in the BDNF receptor trkB (Bernard et al. 2007) in the hippocampus of DOM-treated rats. Moreover, we have recently reported that DOM-induced CREB activation, increased BDNF, neurogenesis and synaptogenesis are blocked by AMPA/kainate antagonists but not by $Ca2^+$ channel antagonists in OHSCs (Pérez-Gómez and Tasker 2014b).

In summary, therefore, DOM is epileptogenic in adult humans at high doses, but is also epileptogenic in neonatal rats at doses considerably lower. Pathophysiological, pathological and pharmacological alterations known in humans can be replicated in neonatal rats (and cultures of neonatal brain), indicating clearly that low concentrations of DOM during neurodevelopment induce permanent changes in the adult brain of rats (and humans?) that are relevant to epilepsy.

6.4 More Than Epilepsy: Effects on Measures of Psychotic Disease

Most animals, rats included, show a well-structured and stable degree of social behaviour, and thus, measuring social behaviour is a relatively easy process. In one study, we evaluated dyadic encounters to determine whether early domoate treatment would result in rats who engaged actively in social avoidance. When tested as adults, those male rats who were exposed to DOM as neonates demonstrated a significant degree of social withdrawal, as was evidenced in a significant increase in the amount of time spent in active avoidance behaviour and a significant decrease in the amount of time spent engaged in direct social contact with a same-sex untreated conspecific (Ryan et al. 2011). Other evidence of abnormalities in social engagement comes from the finding that while early toxin treatment reliably produces deficits in PPI in animals who are raised using normal husbandry practices, DOM treatment appears to abolish deficits in isolation-induced PPI (Marriott et al. 2015). While the reason(s) for this somewhat discrepant finding remain to be elucidated, it is noteworthy that DOM-treated animals appear to "fair" better when housed in isolation which may suggest that DOM treatment, in conjunction with the presumed stress of forced social engagements, may be a part of our repeatedly reported PPI deficits.

DOM-treated rats consistently show altered affective responses in behavioural paradigms which are mildly to moderately stressful. For instance, when tested in the Elevated Plus Maze, DOM-treated female rats have been shown to perform significantly different from their saline-treated counterparts to this emotional challenge (Gill et al. 2012; Doucette et al. 2007), and when presented with a version of the MWM which significantly increases the complexity of the task by introducing repeated reversal trials, DOM-treated female rats responded to the uncertainty of the present situation by engaging in a significant degree of thigmotaxis (Adams et al. 2009). In general, thigmotaxis will decrease as the animal learns to navigate the maze, but when challenged, these animals regress to swimming around the edge of the pool. One interpretation of these results may be that the DOM-treated rats have an altered emotional response to novelty, and when the environment no longer met expectations, they had a greater deal of difficulty adapting to change. This interpretation is supported by other findings wherein rats treated neonatally with DOM displayed altered responses to novelty in the playground maze (Burt et al. 2008a). The playground maze consisted of an elevated circular platform, with 8 objects placed equidistant from each other around the circumference of the maze. Four familiarization trials, separated by 24 h, were administered to each rat, during which time rats were given 3 min to explore objects. The final trial was followed by a novelty trial, in which one object from previous trials was replaced by a new, previously unseen object. As expected, time spent exploring objects decreased over successive familiarization trials. However, while both experimental and control rats displayed exploration of the novel object, DOM-treated male rats also displayed a surprising and unusually increase in their exploration of all objects, new and old (Burt et al. 2008a).

Other behavioural anomalies are readily apparent in rats treated with domoic acid during critical periods of CNS maturation. One such anomaly is the manner in which these toxin-treated animals respond to drugs of abuse. We evaluated drug-seeking behaviour in both adolescent and adult rats using a nicotine-induced conditioned place preference task. It is first critical to note a few important considerations one must be aware of when attempting to use nicotine in a place conditioning task; these include dosage, procedures employed (i.e. biased vs. unbiased designs) and age of testing. For instance, dosages that are considerably "too high" will cause a conditioned place aversion, whereas doses that are "too low" will not induce any conditioned response (CR); a biased procedure is considered to be far more sensitive to conditioning than is an unbiased protocol; and juvenile rats are far more sensitive to developing a conditioned place preference than are adult rat. Thus, using what is considered to be a relatively low dose in adult rats (i.e. 0.6 mg/kg) and by employing an unbiased CPP procedure, such that CPP should not be observed in normal rats, we were able to show a nicotine-induced conditioned place preference in DOM-treated female rats which suggests an increase in the sensitivity of the rewarding properties of nicotine—an effect that was maintained for at least one month (Burt et al. 2008a).

This same dose, when administered to rats during late adolescence,[5] has been readily shown to produce a conditioned place response in rats. But when DOM-treated rats were tested in this paradigm, this CPP effect, readily apparent in saline-treated controls, was completely abolished (Burt et al. 2008b). It may be that with a heighted sensitivity to nicotine, in adolescent rats, this dose was sufficient to surpass the normal appetitive properties of the drug, shifting the experience instead to one that was mild to moderately aversive.

6.5 Changes in Cognitive Function and Attentional Processes

On first glance, when tested on what would be considered a relatively easy cognitive task, early DOM treatment seems to produce "smarter animals". For instance, over one week of testing in the RAM, DOM-treated adolescent rats outperformed their saline counterparts in measures reflecting working memory[6], an effect unlikely attributable to alterations in locomotor activity or motivation levels, as no main effects for (nor interactions with) the treatment variable were evident for either response latency (session latency/total arms entered) or the number of reward pellets consumed (Adams et al. 2009). Comparable effects are also found in adult rats tested in the MWM, wherein DOM-treated rats reliably outperform their saline-treated counterparts on many typical indices of "learning and memory"—at least during the initial learning phase of the test paradigm (Adams et al. 2009). However, as cognitive load increases and the tasks become more complex or challenging, significant impairments in behavioural indices which reflect learning and memory processes begin to manifest (Gill et al. 2012; Adams et al. 2009). For instance, although outperforming their saline counterparts on choice accuracy during RAM testing as adolescents (Adams et al. 2009), when retested again as adults, DOM-treated female rats showed significantly poorer choice accuracy scores than did saline-treated rats, suggesting impairments, at least to some degree, in reference memory[7] (unpublished observations). As adults in the MWM, while outperforming their saline-treated counterparts during the initial learning trials (Adams et al. 2009), once reversal tasks[8] were introduced, the performance of DOM-treated rats became significantly impaired, particularly for female rats.

---

[5]Adolescence is a developmental time period wherein the typical rat is exquisitely sensitive to the rewarding properties of nicotine.

[6]Assessed using entries to repeat (ERT); a reliable measure of choice accuracy in the RAM.

[7]Reference memory can be considered to be reflected as an inability to recall the overall "rules" of the test procedure (i.e. in these circumstances, each arm must be entered once and only once).

[8]A reversal task is one in which the platform is relocated to a new quadrant after a set number of trials. Its use in MWM testing has become commonplace as it tends to be more sensitive to subtle deficits produced by experimental intervention (McNamara and Skelton 1993).

Furthermore, while no group differences were found during the final probe trial,[9] analyses did reveal that DOM-treated rats spent significantly greater amounts of time in the quadrant which housed the platform in its first and original location (i.e. prior to the reversal task) and crossed into this quadrant significantly more often, both suggesting that these animals may be engaging in some form of perseveration. We also reported comparable results in the MWM and in the water H maze (Gill et al. 2012). In another study using a recency discrimination paradigm, DOM-treated female rats have been shown to demonstrate significant temporal memory dysfunction, evidenced in a significantly lower proportion of total exploratory behaviour directed towards a remote object. While to some this finding may initially seem unremarkable, it is important to note that temporal memory and timing behaviour are fundamental cognitive abilities that allow individuals to orient both themselves and events in time and to accurately perceive time, abilities which are compromised in certain neuropsychiatric conditions. Thus, an ability to demonstrate such dysfunction in laboratory animals following early low-dose toxin treatment suggests that this model may indeed have significant clinical relevance and importance.

An ongoing issue in the development of animal models of neuropsychiatric disorders is the difficulty that is obviously inherent in finding common symptoms displayed across species. For this reason, relatively simple behaviours such as reflexes and/or simple associative tasks provide invaluable information regarding the suitability of a particular animal model. Deficits in PPI and LI are commonly reported in various neuropsychiatric conditions within the clinical population, and fortuitously, PPI and LI are two such trans-species behaviours—readily assessable in both humans and rodents alike.

Latent inhibition is a normal cognitive process whereby repeated exposure to a non-reinforced stimulus will subsequently impair the ability of that stimulus to enter into new associations (see Fig. 4). In traditional learning theory language, this stimulus—termed the conditioned stimulus (CS)—upon repeated non-reinforced presentation loses its salience and thus its ability to serve as an effective CS when later paired with an unconditioned stimulus (US) and thus is ineffective in eliciting a CR. This "learned inattention" presents as an adaptive mechanism for an organism as it is believed important for the proper processing of incoming stimuli, the ability to ignore irrelevant information and to allocate appropriate mental resources. We have, on two separate occasions using two different paradigms, demonstrated that early domoate treatment will reliably produce deficits in LI (Marriott et al. 2012, 2014). For instance, in one experiment, by exploiting a typical conditioned taste aversion paradigm, we paired a novel stimulus (sucrose; CS) with the injection of a noxious stimulus (LiCl; US), thereby inducing nausea (UR) in test animals. When

---

[9]A probe trial is one in which after a period of time following the final swim trial, the platform is removed from the maze and rats are given a single swim trial in which they move freely in the maze for a set amount of time. Later analyses of this trial generally include time spent in target quadrant (i.e. area in which the platform was most recently located) and the number of annulus crossings; both assessments are used as indices of memory.

tested subsequently, and given a choice between sucrose and tap water, all rats manifested a CR—the avoidance of the 5 % sucrose solution. For some animals, prior to US/CS pairing, rats were pre-exposed to sucrose daily for 3 consecutive days prior to conditioning. As expected, control rats demonstrated latent inhibition, that is to say that sucrose consumption was not suppressed following the sucrose/LiCl pairing. Domoate-treated rats, on the other hand, did not show latent inhibition, such that the repeated pre-exposure to sucrose did nothing to diminish its salience as a CS, as sucrose consumption was significantly suppressed (Marriott et al. 2012) (see Fig. 4).

Prepulse inhibition (PPI) is a neurobiological phenomenon in which, when presented with a weaker prestimulus (termed a prepulse), the reaction of an

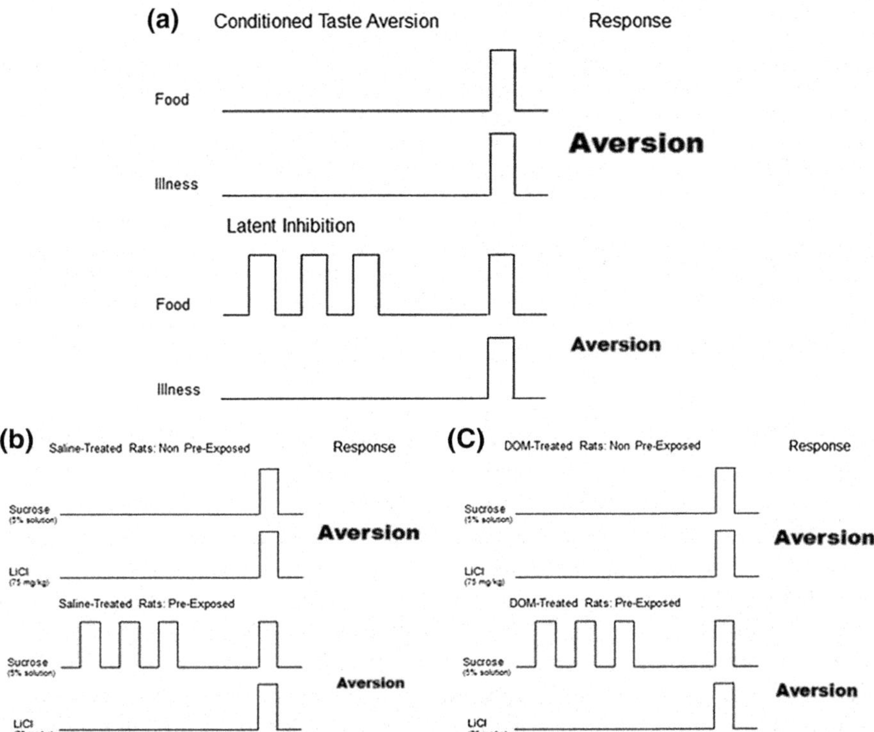

Fig. 4 a In conditioned taste aversion (*CTA*), a single pairing of food with the experience of nausea results in a strong association between the two stimuli and consequently a strong aversion to the food. In a latent inhibition task, pre-exposure to the food without consequence inhibits the later formation of an association between that food substance and nausea, resulting in a lesser degree of aversion to that food substance. **b** We have shown that early DOM exposure results in impaired LI in a CTA task. DOM-treated rats demonstrated a significant aversion (CR) to sucrose (CS), once paired with LiCl to induce nausea (US), even though they were pre-exposed to the sucrose solution for three consecutive days prior to CS/US pairing (Marriott et al. 2012). *Abbreviations* conditioned response (*CR*); conditioned stimulus (*CS*); unconditioned stimulus (*US*)

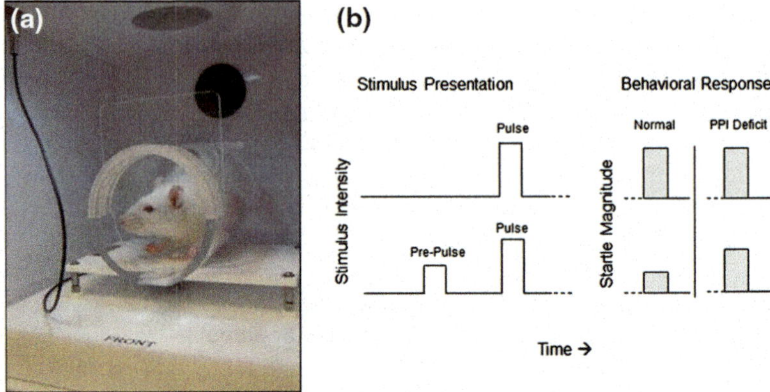

Fig. 5 **a** The startle apparatus used for all experiments described herein (*Photo Credit* A. Adams 2007). **b** Sudden onset acoustic stimuli (pulse) will elicit a startle response measured by the physical displacement of an accelerometer. Startle amplitude in response to the pulse can be suppressed by the administration of a lower-decibel, non-startling prepulse. This phenomenon is termed prepulse inhibition. For the experiments reported herein, animals received a 5-min acclimation period to the chamber before testing. Testing was comprised of 3 blocks of trials, with an intertrial interval for all trials averaging 15 s (ranging from 10 to 20 s) and a background white noise level maintained at 70 dBs. Startle amplitude (defined as the average of 100 readings) was obtained by measuring every 1 ms for 100 ms after the onset of the startle pulse. Blocks 1 and 3 consisted of 5–6 120 dBs white noise startle pulses (40 ms in length) and were used to normalize acoustic startle responses. Block 2 data were used to determine prepulse inhibition. Using prepulse levels of either 4, 8, 12 or 16 dBs above the background noise, %PPI was calculated by the following formula: PPI = 100 − (P/S) * 100, where P is the average startle amplitude for prepulse–pulse trials and S is the average startle amplitude on startle pulse-alone trials

organism to subsequent strong startling stimulus (termed a pulse) is inhibited. Deficits of PPI manifest in the inability to filter out the unnecessary information, and as such, they have been linked to abnormalities of sensorimotor gating which are believed to be controlled by structures located in the lower brainstem and mediated by input from the forebrain. PPI deficits are not typical to specific disease, but rather tell of disruptions in particular brain circuitry. As such, deficits in PPI are noted in patients suffering from a variety of psychiatric/neurological conditions. Using a standard acoustic startle apparatus (SR-Lab from San Diego Instruments, San Diego, CA, USA), we have repeatedly and reliably demonstrated that domoate treatment during the early post-natal period produces deficits in prepulse inhibition of the acoustic startle response[10](Marriott et al. 2012; Adams et al. 2008) (see Fig. 5).

---

[10]While PPI deficits are apparent in both male and female DOM-treated rats, it should be noted that the effects of early DOM treatment were in fact impacted upon by both sex and time of day.

7 Conclusions

The developing brain is exquisitely sensitive to insult, trauma and change, and disturbances of normal brain development often give rise to neurological disorders that may go undetected until later in life. In many cases, trauma appears to alter the excitatory/inhibitory balance in the brain leading to progressive neuropathology and consequent disturbances of normal behaviour. The environmental toxin domoic acid is known to produce severe neurotoxicity in adult humans and wildlife, but the experimental data presented herein support strongly that the developing brain is permanently altered by exposure to very low concentrations of DOM. This highlights an issue with the current regulatory limit for DOM contamination that is based on data in adult animals. However, the slowly progressing nature of neurological dysfunction in experimental animals exposed perinatally to DOM has also allowed for creation and characterization of new animal models of disease (rather than symptom) development. Such models are critical to understanding the presymptomatic processes underlying neurological disease progression and offer hope for identifying the molecular targets and biomarkers that are needed to develop new therapeutic interventions to arrest or slow disease before the onset of symptoms that adversely affect quality of life.

Acknowledgements Research attributed to the Doucette and Tasker laboratories was supported by the Natural Sciences and Engineering Research Council of Canada (NSERC), the Canadian Institutes of Health Research (CIHR), Atlantic Innovation Fund (AIF), Springboard Atlantic and Innovation PEI. T.A. Doucette holds the Jeanne and J Louis Levesque Research Professorship and receives partial salary support from Neurodyn Life Sciences Inc.

References

Adams AL, Doucette TA, Ryan CL (2008) Altered pre-pulse inhibition in adult rats treated neonatally with domoic acid. Amino Acids 35:157–160

Adams AL, Doucette TA, James R, Ryan CL (2009) Persistent changes in learning and memory in rats following neonatal treatment with domoic acid. Physiol Behav 96:505–512

Bates SS, Horner RA (1998) Bloom dynamics and physiology of domoic-acid-producing Pseudo-nitzschia species. In: Anderson DM, Cembella AD, Hallegraeff GM (eds) Physiological ecology of harmful algal blooms. Springer, Heidelberg, pp 267–292

Bates SS, Trainer VL (2006) The ecology of harmful diatoms. In: Granéli E, Turner J (eds) Ecology of harmful algae. Ecological studies, vol 189. Springer, Heidelberg, pp 81–93

Bernard PB, MacDonald DS, Gill DA, Ryan CL, Tasker RA (2007) Hippocampal mossy fiber sprouting and elevated trkB receptor expression following systemic administration of low dose domoic acid during neonatal development. Hippocampus 17:1121–1133

Brodie EC, Gulland FMD, Greig DJ, Hunter M, Jaakola J, Leger J, Leighfield TA, Van Dolan FM (2006) Domoic acid causes reproductive failure in California sea lions (*Zalophus californianus*). Mar Mammal Sci 22(3):700–707

Burt MA, Ryan CL, Doucette TA (2008a) Altered responses to novelty and drug reinforcement in adult rats treated neonatally with domoic acid. Physiol Behav 93:327–336

Burt MA, Ryan CL, Doucette TA (2008b) Low dose domoic acid in neonatal rats abolishes nicotine induced conditioned place preference during late adolescence. Amino Acids 35:247–249

Can Dis Wkly Rep (1990) Proceedings of a symposium on domoic acid. Can Dis Wkly Rep 16: Suppl. 1E

Cendes F, Andermann F, Carpenter S et al (1995) Temporal lobe epilepsy caused by domoic acid intoxication: evidence for glutamate receptor-mediated excitotoxicity in humans. Ann Neurol 37:123–126

Clancy B, Finlay BL, Darlington RB, Anand KJS (2007) Extrapolating brain development from experimental species to humans. Neurotoxicol 28(5):931–937

Colombo J (1982) The critical period concept: research, methodology, and theoretical issues. Psychol Bull 91(2):260–275

Dakshinamurti K, Sharma SK, Sundaram M, Watanabe T (1993) Hippocampal changes in developing postnatal mice following intrauterine exposure to domoic acid. J Neurosci 13 (10):4486–4495

Dobbing J (1976) Vulnerable periods in brain growth and somatic growth. In: Roberts DF, Thomson AM (eds) The biology of human fetal growth. Taylor & Francis, London, pp 137–147

Dobbing J, Sands J (1979) Comparative aspects of the brain growth spurt. Early Hum Dev 3:79–83

Dobbing J, Smart JL (1974) Vulnerability of developing brain and behaviour. Br Med Bull 30:164–168

Doucette TA, Tasker RA (2008) Domoic acid: detection methods, pharmacology and toxicology. In: Botana LM (ed) Seafood and freshwater toxins. Pharmacology, physiology and detection (2nd edn). CRC Press, Taylor Francis Group, New York, pp 398–429

Doucette TA, Strain SM, Allen GV, Ryan CL, Tasker RAR (2000) Comparative behavioural toxicity of domoic acid and kainic acid in neonatal rats. Neurotox Teratol 22:863–869

Doucette TA, Bernard PB, Yuill PC, Tasker RA, Ryan CL (2003) Low doses of non-NMDA glutamate receptor agonists alter neurobehavioral development in the rat. Neurotoxicol Teratol 25:473–479

Doucette TA, Bernard PB, Husum H, Perry MA, Ryan CL, Tasker RA (2004) Low doses of domoic acid during postnatal development produce permanent changes in rat behaviour and hippocampal morphology. Neurotox Res 6(7, 8):555–563

Doucette TA, Ryan CL, Tasker RA (2007) Gender-based changes in cognition and emotionality in a new rat model of epilepsy. Amino Acids 32:317–322

Gill DA, Bastlund JF, Anderson NJ, Tasker RA (2009) Reductions in paradoxical sleep time in adult rats treated neonatally with low dose domoic acid. Behav Brain Res 205:564–567

Gill DA, Bastlund JF, Watson WP, Ryan CL, Reynolds DS, Tasker RA (2010a) Neonatal exposure to low-dose domoic acid lowers seizure threshold in adult rats. Neuroscience 169:1789–1799

Gill DA, Ramsay R, Tasker RA (2010b) Selective reductions in subpopulations of GABAergic neurons in a developmental rat model of epilepsy. Brain Res 1331:114–123

Gill DA, Perry MA, McGuire EP, Perez-Gomez A, Tasker RA (2012) Low-dose neonatal domoic acid causes persistent changes in behavioural and molecular indicators of stress response in rats. Behav Brain Res 230:409–417

Goldstein T, Mazet JAK, Zabka TS et al (2008) Novel symptomatology and changing epidemiology of domoic acid toxicosis in California sea lions (*Zalophus californianus*): an increasing risk to marine mammal health. Proc R Soc B 275:267–276

Gulland FMD, Haulena M, Fauguier D et al (2002) Domoic acid toxicity in Californian sea lions (*Zalophus californianus*): clinical signs, treatment and survival. Vet Record 150:475–480

Holmes GL, Sarkisian M, Ben-Ari Y, Chevassus-Au-Louis N (1999) Mossy fiber sprouting after recurrent seizures during early development in rats. J Comp Neurol 22:537–553

Kaufmann W (2000) Developmental neurotoxicity. In: Krinke GJ (ed) The handbook of experimental animals: the laboratory rat. Academic Press, New York, pp 227–252

Levin ED, Pizarro K, Pang WG, Harrison J, Ramsdell JS (2005) Persisting behavioural consequences of prenatal domoic acid exposure in rats. Neurotox Teratol 27(5):719–725

Lujan R, Shigemoto R, Lopez-Bendito G (2005) Glutamate and GABA receptor signalling in the developing brain. Neuroscience 130:567–580

Marriott AL, Ryan CL, Doucette TA (2012) Neonatal domoic acid treatment produces alterations to prepulse inhibition and latent inhibition in adult rats. Pharm Biochem Behav 103:338–344

Marriott AL, Tasker RA, Ryan CL, Doucette TA (2014) Neonatal domoic acid abolishes latent inhibition in male but not female rats and has differential interactions with social isolation. Neurosci Lett 578:22–26

Marriott AL, Tasker RA, Ryan CL, Doucette TA (2015) Alterations to prepulse inhibition magnitude and latency in adult rats following neonatal treatment with domoic acid and social isolation rearing. Behav Brain Res doi:10.1016/j.bbr.2015.11.009 [Epub ahead of print]

Maucher JM, Ramsdell JS (2007) Maternal-fetal transfer of domoic acid in rats at two gestational time points. Environ Health Perspect 115(12):1743–1746

Maucher-Fuquay J, Muha N, Wang Z, Ramsdell JS (2012) Elimination kinetics of domoic acid from the brain and cerebrospinal fluid of the pregnant rat. Chem Res Toxicol 25(12):2805–2809

McNamara RK, Skelton RW (1993) The neuropharmacological and neurochemical basis of place learning in the Morris Water Maze. Brain Res Brain Res Rev 18:33–49

Moshe SL, Alabala BJ, Ackermann RF, Engel J Jr (1983) Increased seizure susceptibility of the immature brain. Brain Res 283:81–85

Pérez-Gómez A, Tasker RA (2012) Enhanced neurogenesis in organotypic cultures of rat hippocampus after transient subfield-selective excitotoxic insult induced by domoic acid. Neuroscience 208:97–108

Pérez-Gómez A, Tasker RA (2013) Transient domoic acid excitotoxicity increases BDNF expression and activates both MEK- and PKA-dependent neurogenesis in organotypic hippocampal slices. BMC Neurosci 14(1):72 (13 pages online only)

Pérez-Gómez A, Tasker RA (2014a) Domoic acid as a neurotoxin. In: Kostrewza R (ed) Handbook of neurotoxicity. Springer, New York, pp 399–419

Pérez-Gómez A, Tasker RA (2014b) Enhanced mossy fibre sprouting and synapse formation in organotypic hippocampal cultures following transient domoic acid excitotoxicity. Neurotox Res 25:402–410

Perl TM, Bedard L, Kosatsky T et al (1990) An outbreak of toxic encephalopathy caused by eating mussels contaminated with domoic acid. N Engl J Med 322:1775–1780

Perry MA, Ryan CL, Tasker RA (2009) Effects of low dose neonatal domoic acid administration on behavioural and physiological response to mild stress in adult rats. Physiol Behav 98:53–59

Pulido OM (2008) Domoic acid toxicologic pathology: a review. Mar Drugs 6:180–219

Racine RJ (1972) Modification of seizure activity by electrical stimulation. II. Motor seizure. Electroencephalogr Clin Neurophysiol 32:281–294

Ramsdell JS, Zabka TS (2008) In utero domoic acid toxicity: a fetal basis to adult diseases in the California sea lion (*Zalophus californianus*). Mar Drugs 6(2):262–290

Rice D, Barone S Jr (2000) Critical periods of vulnerability for the developing nervous system: evidence from humans and animal models. Environ Health Perspect 108(3):511–533

Ritter LM, Vazquez DM, Meador-Woodruff JH (2002) Ontogeny of ionotropic glutamate receptor subunit expression in the rat hippocampus. Dev Brain Res 139:227–236

Robbins MA, Ryan CL, Marriott AL, Doucette TA (2013) Temporal memory dysfunction and alterations in tyrosine hydroxylase immunoreactivity in adult rats following neonatal exposure to domoic acid. Neurosci Med 4:29–35

Ryan CL, Robbins MA, Smith MT, Gallant IC, Adams-Marriott AL, Doucette TA (2011) Altered social interaction in adult rats following neonatal treatment with domoic acid. Physiol Behav 102:291–295

Scholin CA, Gulland F, Doucette GJ et al (2000) Mortality of sea lions among the central California coast linked to a toxic diatom bloom. Nature 403:80–84

Schulz JB, Falkenburger BH (2004) Neuronal pathology in Parkinson's disease. Cell Tissue Res 318:135–137

Shaw CA, Hoglinger GU (2008) Neurodegenerative diseases: neurotoxins as sufficient etiological agents? Neuromolecular Med 10:1–9

Sierra-Beltran A, Palafox-Uribe M, Grajales-Montiel J et al (1997) Sea bird mortality at Cabo San Lucas, Mexico: evidence that toxic diatom blooms are spreading. Toxicon 35:447–453

Stoppini L, Buchs PA, Muller D (1991) A simple method for organotypic cultures of nervous tissue. J Neurosci Meth 37:173–182

Suzuki CAM, Hierlihy SL (1993) Renal clearance of domoic acid in the rat. Food Chem Toxicol 31(10):701–706

Tanemura K, Igarashi K, Matsugami TR, Aisaki K-I, Kitajima S, Kanno J (2009) Intrauterine environment-genome interaction and Children's development (2): brain structure impairment and behavioural disturbance induced in male mice offspring by a single intraperitoneal administration of domoic acid (DA) to their dams. J Toxicol Sci 34:SP279–SP286

Tasker RA (2015) Domoic acid and other amnesic toxins: toxicological profile. Mar Freshw Toxins 1–16

Tasker RA, Connell BJ, Strain SM (1991) Pharmacology of systemically administered domoic acid in mice. Can J Physiol Pharmacol 69:378–382

Tasker RA, Strain SM, Drejer J (1996) Selective reduction in domoic acid toxicity in vivo by a novel non-N-methyl-D-aspartate receptor antagonist. Can J Physiol Pharmacol 74:1047–1054

Tasker RAR, Perry MA, Doucette TA, Ryan CL (2005) NMDA receptor involvement in the effects of low dose domoic acid in neonatal rats. Amino Acids 28:193–196

Tasker RAR, Kuiper-Goodman T, Pulido O, Lawrence JF (2011) Domoic acid. In: Lawrence J, Loreal H, Toyofuku H, Karunasagar I, Ababouch L (eds) Assessment and management of biotoxin risks in bivalve molluscs. FAO Fisheries and Aquaculture Technical Paper No. 551. Rome, FAO

Teitelbaum J, Zatorre R, Carpenter S et al (1990) Neurological sequelae of domoic acid intoxication due to ingestion of contaminated mussels. N Engl J Med 322:1781–1787

Tiedeken JA, Ramsdell JS (2007) Embryonic exposure to domoic acid increases the susceptibility of zebrafish larvae to the chemical convulsant pentylenetetrazole. Environ Health Perspect 115 (11):1547–1552

Tiedeken JA, Ramsdell JS, Ramsdell AF (2005) Developmental toxicity of domoic acid in zebrafish (*Danio rerio*). Neurotoxicol Teratol 27(5):711–717

Tryphonas L, Truelove J, Nera E, Iverson F (1990) Acute neurotoxicity of domoic acid in the rat. Toxicol Pathol 18:1–9

Verdoorn TA, Johansen TH, Drejer J, Nielsen EO (1994) Selective block of recombinant GluR6 receptors by NS-102, a novel non-NMDA receptor antagonist. Eur J Pharmacol 269:43–49

Vernadakis A, Woodbury DM (1969) The developing animal as a model. Epilepsia 10:163–178

Vorhees CV (1986) Principles of behavioral teratology. In: Riley EP, Vorhees CV (eds) Handbook of behavioral teratology. Plenum Press, New York, pp 23–48

Work TM, Barr B, Beal AM, Fritz L, Quilliam MA, Wright JLC (1993) Epidemiology of domoic acid poisoning in brown pelicans (*Pelecanus occidentalis*) and Brant's cormorants (*Phalacrocorax penicillatus*) in California. J Zoo Wildl Med 24(1):54–62

World Health Organization (2006) Neurology and public health. http://www.who.int/mental_ health/neurology/en

Wright JLC, Bird CJ, deFreitas ASW, Hampson D, McDonald J, Quilliam MA (1990) Chemistry, biology, and toxicology of domoic acid and its isomers. Can Dis Wkly Rep 16(Suppl. 1E): 15–18

Xi D, Peng YG, Ramsdell JS (1997) Domoic acid is a potent neurotoxin to neonatal rats. Nat Toxins 5:74–79

Perinatal 192 IgG-Saporin as Neuroteratogen

Laura Petrosini, Paola De Bartolo, Debora Cutuli and Francesca Gelfo

Abstract The immunotoxin 192 IgG-saporin selectively destroys basal forebrain cholinergic neurons that provide cholinergic input to the hippocampus, entire cortical mantle, amygdala, and olfactory bulb. Perinatal immunotoxic lesions by 192 IgG-saporin induce long-lasting cholinergic depletion mimicking a number of developmental disorders reported in humans. The perinatal injection of 192 IgG-saporin induces several brain modifications, which are observed in neocortex and hippocampus at short and long term. These plastic changes involve both structural (alterations in brain volume, neuronal morphology, and neurogenesis) and molecular (modulations of the levels of neurotransmitters and other proteins related to neurodegeneration) levels. Moreover, the perinatal injection of 192 IgG-saporin may interact with the brain plastic capacity to react to other injuries. Perinatal 192 IgG-saporin lesions allowed investigating the role of the basal forebrain cholinergic system in modulating behavioral functions in developing as well as adult rats. After perinatal cholinergic depletion, rats display reduced ultrasonic

Debora Cutuli and Francesca Gelfo—Both the authors equally contributed.

L. Petrosini (✉) · P. De Bartolo · D. Cutuli · F. Gelfo
IRCCS Fondazione Santa Lucia, Via del Fosso di Fiorano 64,
00143 Rome, Italy
e-mail: laura.petrosini@uniroma1.it

L. Petrosini · D. Cutuli
Department of Psychology, University Sapienza of Rome, Via dei Marsi 78,
00185 Rome, Italy

P. De Bartolo
Department of Sociological and Psychopedagogical Studies,
University Guglielmo Marconi of Rome, Via Plinio 44, 00193 Rome, Italy

F. Gelfo
Department of Systemic Medicine, University of Tor Vergata,
Via Montpellier 1, 00133 Rome, Italy

© Springer International Publishing Switzerland 2015
Curr Topics Behav Neurosci (2016) 29: 111–124
DOI 10.1007/7854_2015_418
Published Online: 23 December 2015

vocalizations as neonates, learning and exploratory deficits as juveniles, altered discriminative abilities, impulsive and perseverative behaviors, and memory deficits as adults. Overall, these findings underline the importance of cholinergic system integrity for the development of specific structural and functional features.

Keywords 192 IgG-saporin · Perinatal lesion · Basal forebrain cholinergic system · Neuroplasticity · Cognitive functions

Contents

1 Introduction 112
2 Immunolesioning Techniques 113
3 192 IgG-Saporin 113
4 Development of the Rodent Cholinergic Basal Forebrain 114
5 Effectiveness of Perinatal 192 IgG-Saporin Lesions 114
6 Structural and Biochemical Effects of Perinatal 192 IgG-Saporin 115
7 Behavioral and Cognitive Outcomes of Perinatal 192 IgG-Saporin 118
   7.1 Short-Term Effects of Perinatal 192 IgG-Saporin 118
   7.2 Long-Term Effects of Perinatal 192 IgG-Saporin 119
8 Summary 121
References 121

1 Introduction

Numerous studies indicate that cholinergic system is critical for the normal development of neocortex and hippocampus. The nucleus basalis magnocellularis and the horizontal limb of the diagonal band of Broca (NBM/DBBh) provide the primary cholinergic afferents to neocortex, while the medial septum and the vertical limb of the diagonal band of Broca (MS/DBBv) provide the primary cholinergic afferents to hippocampus (Mesulam et al. 1983). These two neuronal populations, collectively referred to as cholinergic basal forebrain (CBF) system, are critically involved in the neuronal differentiation and synaptogenesis of their target regions. Developmental studies showed that CBF neurons are generated prenatally following a distinct spatiotemporal pattern and reach their complete maturation during the first postnatal month (see Sect. 4).

Although many works addressed the functional role of the CBF in adulthood, substantially fewer researches addressed its role during development. In general, the functional role of a neurotransmitter can be quite different during development as compared to adulthood. In fact, developmental lesions affect a forming system and alter its course of development, while adult lesions impact on an already functional network. Therefore, it is not surprising that developmental damage to a neurotransmitter system can have consequences very different from similar damage occurring in adulthood.

In this chapter, the consequences of the perinatally induced absence (or dramatic reduction) of CBF system on brain structure and function will be reviewed, taking into account studies that used the 192 IgG-saporin as neuroteratogen. 192 IgG-saporin is a highly efficient and selective immunotoxin for lesioning the CBF either in adulthood or in the early postnatal life. Understanding mechanisms by which cholinergic insults affect brain development is important since a lot of developmental disorders reported in humans, including Down syndrome, Rett syndrome, and hypothyroidism, as well as perinatal exposure to environmental contaminants, are reported to involve cholinergic impairment (Berger-Sweeney 1998).

2　Immunolesioning Techniques

Immunolesions use anti-neuronal monoclonal antibodies to selectively deliver citotoxins to only those neurons bearing surface moieties recognized by the antibodies. Immunolesioning can destroy a specific type of neurons locally or after axonal suicide transport from the injection site, thus combining selectivity for both anatomical region and cell type.

Immunotoxins are composed of monoclonal antibodies conjugated to various ribosome-inactivating proteins (RIPs). The monoclonal antibody component of an immunotoxin binds to specific cell-surface molecules to result in endocytosis of the antibody and its coupled RIP. Once internalized, the RIP component inhibits protein synthesis, catalytically inactivating ribosomes and resulting in cell death. If immunotoxin binds to nerve terminals or axons present at the injection site, it is internalized by receptor-mediated endocytosis and then by retrograde axonal transport travel to neuronal cell bodies, where the RIP acts to inhibit protein synthesis. If immunotoxin binds to antigen-bearing soma present at the injection site, then axonal transport is obviously not necessary to deliver the toxin.

The first immunotoxin reported to be effective in provoking neuronal lesions is OX7-saporin that consists of the monoclonal antibody OX7 coupled to saporin. Saporin is an often used RIP derived from the plant *Saponaria Officinalis* (soap wort). Given OX7 recognizes Thy-1, an abundant surface molecule present on all neurons, the resulting immunotoxin OX7-saporin is not selective for a specific type of neurons. For a review, see Wiley (1992).

3　192 IgG-Saporin

The first type-selective anti-neuronal active in vivo immunotoxin is the 192 IgG-saporin. It consists of the monoclonal antibody 192 IgG which is disulfide coupled to saporin. The antibody component is directed against low-affinity rat nerve growth factor (NGF) receptor (p75^{NGFR}). Targeting p75^{NGFR} serves as an ideal way to selectively lesion the CBF neurons because only the cholinergic neurons express p75^{NGFR} in the rat basal forebrain. Cholinergic neurons possess p75^{NGFR} while neurons containing other neurotransmitters of the same region do not express detectable levels of such a receptor. NGF plays a role in the maintenance of function of the CBF neurons which express p75^{NGFR} (Thomas et al. 1991). When injected, 192 IgG-saporin binds to the surface of p75^{NGFR} bearing neurons and is internalized by endocytosis. Once in the cytoplasm, the saporin moiety escapes endosomes and enzymatically inactivates the large ribosomal subunit halting protein synthesis and leading ultimately to cell death. For a review, see Wiley (1992).

4 Development of the Rodent Cholinergic Basal Forebrain

In the rodents, CBF neurons provide the major source of cholinergic innervation for the entire neocortex, hippocampus, amygdala, and olfactory bulb. CBF is a continuum of several overlapping nuclei with divergent efferent pathways such that a single projection field is not exclusive of a particular nucleus.

CBF neurons are generated during the embryonic days (E) (in rats between E12 and E17, and in mice between E11 and E16) (Bayer et al. 1993; Semba and Fibiger 1988). Just after their neurogenesis, in both rats and mice, the cholinergic neurons begin to send projections toward their cortical targets, entering neocortex in the first postnatal week (Coyle and Yamamura 1976; Berger-Sweeney 1998). Conversely, cholinergic innervation of hippocampus occurs a little bit later, in the second postnatal week (Linke and Frotscher 1993). The timing of the CBF innervation of its targets overlays the entire dynamic period of synaptogenesis and differentiation in these areas. So, it is reasonable to expect that the interruption of the CBF innervation during these critical periods would influence differentiation and morphogenesis in the neocortex and hippocampus (see Sect. 6).

5 Effectiveness of Perinatal 192 IgG-Saporin Lesions

To induce CBF degeneration in the perinatal period, 192 IgG-saporin is administrated by intracerebroventricular (i.c.v.) injection (Berger-Sweeney 1998; Leanza et al. 1996). No depletion of calbindin-immunoreactive Purkinje cells is found in the cerebellum of perinatally lesioned rats (Leanza et al. 1996). This finding is visibly in contrast to the extensive Purkinje cell loss described after the immunotoxin i.c.v. administration in adult rats (De Bartolo et al. 2009; Waite et al. 1995) and may reflect a lower expression of p75^{NGFR} in the Purkinje cells of immature rats, as compared to adult animals. This difference points to the perinatal immunolesion as an effective model to address the effects of CBF damage without the confounding presence of cerebellar damage.

Immunolesions performed in rats during the first postnatal week produce massive and selective loss of cholinergic neurons, both in the MS/DBBv and in the NBM/DBBh, which is paralleled by marked reduction of the catalytic enzyme CholineAcetylTrasferase (ChAT) activity levels in the hippocampus and neocortex. The 192 IgG-saporin effects are long-lasting. In fact, after 3–8 months from the perinatal lesion, the cholinergic cell loss in the MS/DBBv and NBM/DBBh, and the depletion of hippocampal and cortical ChAT activity are of a magnitude similar to, or even greater than, those observed in 4–5 weeks after the lesion (Leanza et al. 1996; Wenk et al. 1994). The damage extent is dose-dependent. The optimal dose of 192 IgG-saporin producing high reduction of ChAT and cholinergic neurons without nonspecific toxicity or animal death is reported to be 0.4 µg/ventricle (Leanza et al. 1995). The effects induced by perinatal immunolesion may be milder than those observed following i.c.v. administration in adult rats (Leanza et al. 1995). In fact, at the optimal dose for perinatal immunolesion (0.4 µg/ventricle), a 72 % loss of cholinergic neurons was observed in the MS/DBBv after 4–5 post-lesional weeks, compared to the 96 % depletion observed following injection of the optimal dose (5 µg/ventricle) in adult rats (Leanza et al. 1996). This does not apply to the NBM/DBBh, where the cell loss was comparable after both treatments (93 and 90 %, respectively). Similarly, the reduction in hippocampal ChAT activity levels showed by the perinatally lesioned animals (about 70 %) did not reach the level of reduction (about 80–90 %) showed by the rats lesioned when adults, while the extent of prefrontal ChAT depletion was similar after both treatments. A possible explanation of the lower cholinergic depletion observed in neonatally lesioned rat is that the CBF of developing rat exhibits low p75^{NGFR} expression during the first postnatal week and shows a p75^{NGFR} concentration peak only between the first and the second postnatal weeks (Leanza et al. 1996). Thus, perinatal 192 IgG-saporin lesions performed from the second postnatal week would allow inducing a damage in a neuronal system whose maturation stage is already comparable to that of adult animal, given that cholinergic fibers projecting to the hippocampus show a late development, as previously described (Berger-Sweeney 1998).

6 Structural and Biochemical Effects of Perinatal 192 IgG-Saporin

A number of studies were conducted with the aim to investigate the effects of the perinatal injection of 192 IgG-saporin on brain structure and biochemistry. The brain modifications that follow immunotoxin injection provide a powerful example of what the brain neuroplasticity is able to operate. Neuroplasticity is the nervous system ability to structurally and functionally change in response to virtually every experience, with the production of transient or enduring adjustments (Kolb and Teskey 2012). The perinatal injection of 192 IgG-saporin induces several short- and at long-term brain modifications either at structural (altered brain volume, neuronal morphology, and neurogenesis) or at molecular (modulated levels of neurotransmitters and neurodegeneration-related proteins) level.

Early studies on this topic regarded gross indices, such as brain weight and volume, and cortical thickness. Robertson et al. (1998) showed that the unilateral injection of 192 IgG-saporin (0.1–0.2 µg) executed at the postnatal day (pnd) 0–2 induced a 10 % reduction in neocortical thickness, detectable since pnd 8 and observable until 3 months of age. Similarly, Ricceri et al. (2002) showed that bilateral injections (0.5–1 µl of 192 IgG-saporin 0.2 µg/µl) at pnd 1–3 induced a reduction of the total cortical thickness at pnd 7. Conversely, Pappas et al. (1996) found that bilateral injections of 192 IgG-saporin (0.2 µg) at pnd 7 did not affect cortical thickness in adulthood (70–80 pnd). This bulk of findings seems to support the hypothesis that the CBF projections to neocortex are critical for its normal morphological development. When the typical developmental course is altered in the first postnatal week, these cortical neurons form abnormal synaptic contacts that may lead to malformations in the cortical networks (Berger-Sweeney 1998). On the other hand, cholinergic depletion at pnd 7 could have more negligible effect on cortical organization development and induce adult-like consequences on brain morphology (Ricceri et al. 1997). Nevertheless, in a more recent study, Koh et al. (2005) documented that even when executed at pnd 8, the unilateral injection of 192 IgG-saporin (2 µg) reduced brain weight at 10 days after the lesion. In accordance, at a microscope inspection, cortical and hippocampal neurons appeared of reduced size and more densely packed, suggesting that neuropil loss had occurred. Thus, the different effects of the injection of 192 IgG-saporin on brain regional volumes may not depend on (or not only on) the lesion day, but also on the dosage of the immunotoxin injected and/or on the time at which the effects are investigated.

Similar inconsistencies are provided by the studies that addressed more subtle changes in cortical and hippocampal neuronal morphology after perinatal injection of 192 IgG-saporin. Robertson et al. (1998) showed that unilateral injection of 192 IgG-saporin (0.1–0.2 µg), executed at pnd 0–2, induced a reduction of soma size, apical dendritic branching and spine density in the visual cortex layer-V pyramidal neurons, without changes in basal dendritic spine density. On the other hand, investigating the effects of the injection of 192 IgG-saporin at pnd 7, a number of studies obtained contrasting findings. Sherren and Pappas (2005) found at about 5.5 months of age that a pnd 7 dosage of 0.6 µg induced a reduction in apical and basal dendritic branching and spine density in layer-V prefrontal pyramidal neurons, as well as a reduction of dendritic branching without spine density modifications in layer-II/III retrosplenial neurons. Similar reductions were described by Fréchette et al. (2009) in hippocampal CA1 pyramidal neurons, after 8 weeks from the injection of 192 IgG-saporin (0.3 µg). Conversely, De Bartolo et al. (2010) reported that the injection of 0.2 µg of 192 IgG-saporin induced after 4.5 months the increase of the spine density in apical and basal dendrites in layer-III parietal pyramidal neurons. In this case, the increase in spine density was interpreted as a compensatory response to the cholinergic deafferentation. Actually, both increase and decrease of spine density are reported as outcome of an injury, such as a deafferentation (Fiala et al. 2002). In fact, both patterns may be the results of readjustments in brain networks to keep neuronal firing within a desirable range, and to produce compensatory reactions in the circuit excitability, thus re-establishing the baseline firing rate set point (Nahmani and Turrigiano 2014). However, the increase in spine density could also execute just a "mechanical" function finalized to prevent lesion-induced atrophy of dendritic branching (Harmon and Wellman 2003). The hypothesis that the spines might provide protection against dendritic atrophy is supported by evidence that in frontal pyramidal neurons of aged rats with forebrain cholinergic lesions spines did not increase (Harmon and Wellman 2003) and dendritic branching decreased (Works et al. 2004). Nevertheless, an increase in spines does not necessarily imply an increase in functional synapses. A study of slice preparations of hippocampal neurons showed that new spines form even when synapses are inactivated and persist for many hours in the absence of any functional activity (Kirov and Harris 1999). A recent in vivo study (Arellano et al. 2007) reported the existence of non-synaptic spines in the mice neocortex, indicating that dendritic spines might not be used to form synaptic contacts. In conclusion, following the perinatal injection of 192 IgG-saporin at pnd 7, the function of dendritic spine and branching controversial readjustments has still to be defined. The immunotoxic dosage, the time at which the effects are investigated, and the brain area analyzed may influence morphological outcomes and lead to conflicting findings.

Perinatal 192 IgG-saporin injection has been demonstrated to affect adult hippocampal neurogenesis, that is the generation of new cells that integrate themselves in pre-existing neural networks, highly contributing to the hippocampal function of acquiring new memories (Lledo et al. 2006; Deng et al. 2010). Pnd 7 injection of 192 IgG-saporin (0.3 µg) induced after 8 weeks a reduction in proliferation (Fréchette et al. 2009), but not in differentiation and survival (Rennie et al. 2011) of hippocampal neurons. The most plausible explanation for these results is that the perinatal lesion induces in adulthood an acceleration of the death of newborn cells, but no change on the overall proportion of the survived neurons or on the phenotypic differentiation. In fact, it has been advanced that the cholinergic system plays a survival-promoting role for neuronal progenitors and immature neurons, similar to the effects observed during brain development (Cooper-Kuhn et al. 2004).

As for molecular changes, the perinatal injection of 192 IgG-saporin affects also molecular processes that support the normal brain functioning in adulthood. Some studies demonstrated that a cholinergic deafferentation is able to induce alterations in the functioning of other neurotransmitter systems. 192 IgG-saporin injection (0.2, 0.4, or 0.8 µg) at pnd 4 after 5 weeks induced a general increase in brain noradrenalin and dopamine levels, with dose- and region-dependent differences. These changes remained even after 3 months from lesion (Leanza et al. 1996). Similarly, when the 192 IgG-saporin (0.2 µg) was injected at 7 pnd, the cortical and hippocampal noradrenalin levels increased at 70–80 pnd (Pappas et al. 1996).

Interestingly, perinatal cholinergic deafferentation induces alterations also in neurochemical markers reported to be relevant in the human decline from mild cognitive impairment to Alzheimer's Disease dementia (Sarter and Bruno 2004). The pnd 7 injection of 192 IgG-saporin (0.2 μg) decreased after 6 months the levels of hippocampal cyclooxygenase-2 protein and mRNA and of cortical and hippocampal presenilin 1 and 2 mRNA. Conversely, amyloid precursor protein levels were unaffected (Ricceri et al. 2004).

Finally, it is noteworthy that perinatal injection of 192 IgG-saporin may interact with the brain capacity to positively react to other injuries. As an example, pups exposed to unilateral 192 IgG-saporin injection (0.1 or 0.2 μg) at pnd 1 and successively to bilateral removal of the C-line whisker follicles exhibited reduced expansion of D row barrels and decreased D/C ratio after 7 days (Zhu and Waite 1998). Thus, perinatal cholinergic depletion can undermine the learning-dependent reorganization of cortical map representations and perhaps also the ability of the brain to keep itself in good repair in response to injury (Conner et al. 2003).

7 Behavioral and Cognitive Outcomes of Perinatal 192 IgG-Saporin

The short- and long-term effects of perinatal 192 IgG-saporin lesions have been extensively examined in a variety of behavioral paradigms. In the following, we review the results of the bulk of studies on how perinatal cholinergic depletion affects rats' behavioral repertoire earlier in development and later at adulthood.

7.1 Short-Term Effects of Perinatal 192 IgG-Saporin

The many studies on behavioral effects of perinatal cholinergic lesions before weaning (about pnd 21) provide new insights into the physiological role of the CBF in regulating cognitive and behavioral development.

Following pnd 7 192 IgG-saporin injections, learning impairment was evident on pnd 15 in a multi-trial version of passive avoidance (paradigm adapted to developing pups). In this task, the animal is placed in a white and lighted compartment and has to withhold a spontaneous escape response (entering a dark compartment) in order to avoid a mild aversive foot-shock. Pnd 7 192 IgG-saporin injection induced deficits no more evident on pnd 23–24, age in which the multi-trial paradigm of the passive avoidance has become too easy (Ricceri et al. 1999). Moreover, after pnd 1, 192 IgG-saporin injection no deficits were evident on pnd 15, given the neocortical cholinergic loss was too mild to alter behavioral responses (Ricceri et al. 1997). Interestingly, due to the larger extent of cholinergic damage, pnd 1 + 3 192 IgG-saporin lesions resulted in impaired passive avoidance learning even at pnd 15 (Ricceri et al. 2002).

Pnd 7 immunotoxic lesions induced alterations also in the open field test performed on pnd 19. Namely, 192 IgG-saporin injection markedly decreased wall rearings and time spent in exploring objects (Ricceri et al. 1997; Scattoni et al. 2003), suggesting that the perinatal cholinergic lesions altered the patterns of exploration of a novel environment. However, no significant deficits in spatial learning were found in pnd 7 192 IgG-saporin-injected rats tested in the Morris water maze 3 weeks after the lesion (Silveira et al. 2002).

The effects of perinatal 192 IgG-saporin injections have been evaluated also on ultrasonic vocalization (USV), a behavior strictly implied in mother–pup interaction and controlled by cholinergic system (Kehoe et al. 2001). Pnd 7 192 IgG-saporin injections markedly reduced the number of USV emitted on pnd 9–11 and 13, whereas the same effects were no longer evident on pnd 14, probably because of the normal dramatic decrease in the number of USV occurring in the third postnatal week (Branchi et al. 2001; Scattoni et al. 2005). A reduction of USV, not accompanied by any mnesic deficit, was also found in pnd 7 192 IgG-saporin-injected rats when tested in the fear conditioning task on pnd 18 (Ricceri et al. 2007). In contrast, using a paradigm of maternal potentiation consisting of pup isolation and subsequent reunion with the mother, no USV differences were found on pnd 13 in pnd 7 192 IgG-saporin-lesioned rats, since a brief reunion with the mother similarly induced a marked USV increase in both lesioned and control rats (Ricceri et al. 2007). It is noteworthy that cholinergic depletion at birth (pnd 0), even if not impacting on the general development of pups, is able to affect the acoustic structure of infant USV. Anyway, given this altered vocalization did not affect maternal care, it can be advanced that the subsequent behavioral deficits induced by 192 IgG-saporin are the direct result of cholinergic depletion rather than of abnormal mother–pup interactions (Krüger and Hanganu-Opatz 2013).

7.2 Long-Term Effects of Perinatal 192 IgG-Saporin

Most of the studies on pups injected with 192 IgG-saporin on pnd 4 and pnd 7 failed to reveal consistent impairments in spatial learning and memory in adult rats tested in the Morris water maze (De Bartolo et al. 2010; Leanza et al. 1996; Pappas et al. 1996; Ricceri et al. 1999; Sherren et al. 1999). However, in a few cases, mild effects consisting of initial longer latencies to find the hidden platform, less direct navigational trajectories, and increased swimming velocity in Morris water maze were reported (De Bartolo et al. 2010; Ricceri et al. 1999; Sherren et al. 1999). Only when tested in the Morris water maze at 22 months of age, pnd 7 IgG-saporin-injected rats exhibited a dramatic impairment in learning the hidden platform location (Pappas et al. 2005). This result indicates that the cognitive effects of perinatal cholinergic lesions may become more severe in aging, as a result of the added effects of aging on the cholinergic system functioning.

Studies using different, less stressful spatial tasks, such as the open field with objects, succeeded in showing behavioral deficits of perinatal 192 IgG-saporin lesions at adulthood (De Bartolo et al. 2010; Ricceri et al. 1999, 2002). The open field with objects is a non-aversive task in which spontaneous response to both a new spatial rearrangement of familiar objects and to a novel object can be measured. Interestingly, perinatal 192 IgG-saporin lesions on pnd 1 + 3 or pnd 7 did not interfere with locomotor activity or exploration of novel objects, but they selectively altered spatial discrimination making adult animals unable to recognize the occurrence of changes in the spatial relationships among the objects. In fact, when a spatial rearrangement of familiar objects was presented in the arena, perinatally 192 IgG-saporin-lesioned adult rats did not show any selective interest toward the displaced objects, as control rats did. Reinvestigation of the displaced objects by a rat implies that an internal representation of the topographical arrangement of the objects has been formed and then it is compared with the new arrangement. Perinatal cholinergic depletion selectively impairs spatial novelty discrimination. This impairment could be ascribed both to memory deficits in the spatial mapping ability and to attentional deficits in acquiring or encoding relevant spatial information from the environment. In any case, perinatal 192 IgG-saporin lesions seem to induce mild spatial deficits, limited to the detection of subtle modifications of the spatial relationships among the objects and not of greater magnitude modifications, such as the exposure to a novel object.

When tested in high-demanding tasks as adults, pnd 7 192 IgG-saporin-injected rats showed more severe deficits. Namely, during the operant conditioning test, 192 IgG-saporin-lesioned animals preferred small but immediate reinforces to large but delayed rewards (Scattoni et al. 2006). Such an elevated intolerance to delay is an index of increased impulsivity most likely linked to alterations in the modulatory influence of CBF on prefrontal cortex function (Dalley et al. 2004). In addition, pnd 7 192 IgG-saporin-lesioned animals were significantly impaired in accomplishing the radial arm maze and delayed spatial alternation tasks as they showed increased working and reference memory errors, as well as perseverative behaviors (De Bartolo et al. 2010; Pappas et al. 2000, 2005; Scattoni et al. 2006).

Finally, when examined at adulthood in a non-spatial associative memory task, such as social transmission food preference test, pnd 7 192 IgG-saporin-injected rats showed significant memory impairment (Ricceri et al. 2004). In this food preference task, after interacting with conspecifics previously fed on a particular cued food (demonstrators), observer rats exhibit enhanced cued-food preference that lasted up to 1 month (Galef and Whiskin 2003). Olfactory messages passed from demonstrators to observers through demonstrators' breath and this process required the integrity of hippocampal function (Galef 2002). In pnd 7 192 IgG-saporin-injected rats, the socially transmitted food preference extinguished dramatically earlier, lasting less than 4 h. A neophobia test conducted on the same animals demonstrated that the lack of preference toward the cued-food found in perinatally lesioned animals was a mnesic deficit, not due to a mere reduced motivation toward novel food stimuli (Ricceri et al. 2004).

8 Summary

The introduction of immunotoxic lesioning methods has proved to be a valuable tool in investigating the function of developing ascending cholinergic systems. The perinatally injected immunotoxin 192 IgG-saporin causes selective, long-lasting loss of the CBF neurons, inducing in turn a lot of effects on brain structure and function visible either at short and long term.

Several studies demonstrated that the perinatal injection of 192 IgG-saporin induces a number of acute and long-lasting brain modifications, which are observed in neocortex and hippocampus. These plastic changes involve both structural (alterations in brain volume, neuronal morphology, and neurogenesis) and molecular (modulations of the levels of neurotransmitters and neurodegeneration-related proteins) levels. A number of factors, such as lesion day, immunotoxin dosage and timing of analysis may induce differences in lesion outcomes.

The effects of perinatal selective immunotoxic lesions on the rats' behavioral repertoire analyzed earlier in development as well as later in adulthood have been examined by means of various behavioral paradigms. In particular, after perinatal cholinergic depletion, rats display reduced USV as neonates, and deficits in passive avoidance learning and exploration of novel environments as juveniles. Moreover, perinatally 192 IgG-saporin-lesioned rats exhibit altered reactions to spatial novelty, as well as impulsive, perseverative, and mnesic deficits when tested in high demanding tasks as adults. In conclusion, short- and long-term effects of perinatal cholinergic lesions appear to be task-specific and depend on the amount of cholinergic depletion due to the age at which pups are lesioned.

References

Arellano JI, Espinosa A, Fairèn A, Yuste R, DeFelipe J (2007) Non-synaptic dendritic spines in neocortex. Neuroscience 145(2):464–469

Bayer SA, Altman J, Russo RJ, Zhang X (1993) Timetables of neurogenesis in the human brain based on experimentally determined patterns in the rat. Neurotoxicology 14(1):83–144

Berger-Sweeney J (1998) The effects of neonatal basal forebrain lesions on cognition: towards understanding the developmental role of the cholinergic basal forebrain. Int J Dev Neurosci 16 (7–8):603–612

Branchi I, Santucci D, Alleva E (2001) Ultrasonic vocalisation emitted by infant rodents: a tool for assessment of neurobehavioural development. Behav Brain Res 125(1–2):49–56

Conner JM, Culberson A, Packowski C, Chiba AA, Tuszynski MH (2003) Lesion of the basal forebrain cholinergic system impair task acquisition and abolish cortical plasticity associated with motor skill learning. Neuron 38(5):819–829

Cooper-Kuhn CM, Winkler J, Kuhn HG (2004) Decreased neurogenesis after cholinergic forebrain lesion in the adult rat. J Neurosci Res 77(2):155–165

Coyle JT, Yamamura HI (1976) Neurochemical aspects of the ontogenesis of cholinergic neurons in the rat brain. Brain Res 118(3):429–440

Dalley JW, Theobald DE, Bouger P, Chudasama Y, Cardinal RN, Robbins TW (2004) Cortical cholinergic function and deficits in visual attentional performance in rats following 192 IgG-saporin-induced lesions of the medial prefrontal cortex. Cereb Cortex 14(8):922–932

De Bartolo P, Gelfo F, Mandolesi L, Foti F, Cutuli D, Petrosini L (2009) Effects of chronic donepezil treatment and cholinergic deafferentation on parietal pyramidal neuron morphology. J Alzheimers Dis 17(1):177–191

De Bartolo P, Cutuli D, Ricceri L, Gelfo F, Foti F, Laricchiuta D, Scattoni ML, Calamandrei G, Petrosini L (2010) Does age matter? Behavioral and neuro-anatomical effects of neonatal and adult basal forebrain cholinergic lesions. J Alzheimers Dis 20(1):207–227

Deng W, Aimone JB, Gage FH (2010) New neurons and new memories: how does adult hippocampal neurogenesis affect learning and memory? Nat Rev Neurosci 11(5):339–350

Fiala JC, Spacek J, Harris KM (2002) Dendritic spine pathology: cause or consequence of neurological disorders? Brain Res Brain Res Rev 39(1):29–54

Fréchette M, Rennie K, Pappas BA (2009) Developmental forebrain cholinergic lesion and environmental enrichment: behaviour, CA1 cytoarchitecture and neurogenesis. Brain Res 1252:172–182

Galef BG (2002) Social influences on food choices of Norway rats and mate choices of Japanese quail. Appetite 39(2):179–180

Galef BG, Whiskin EE (2003) Socially transmitted food preferences can be used to study long-term memory in rats. Learn Behav 31(2):160–164

Harmon KM, Wellman CL (2003) Differential effects of cholinergic lesions on dendritic spines in frontal cortex of young adult and aging rats. Brain Res 992(1):60–68

Kehoe P, Callahan M, Daigle A, Mallinson K, Brudzynski S (2001) The effect of cholinergic stimulation on rat pup ultrasonic vocalizations. Dev Psychobiol 38(2):92–100

Kirov SA, Harris KM (1999) Dendrites are more spiny on mature hippocampal neurons when synapses are inactivated. Nat Neurosci 2(10):878–883

Koh S, Santos TC, Cole AJ (2005) Susceptibility to seizure-induced injury and acquired microencephaly following intraventricular injection of saporin-conjugated 192 IgG in developing rat brain. Exp Neurol 194(2):457–466

Kolb B, Teskey GC (2012) Age, experience, injury, and the changing brain. Dev Psychobiol 54 (3):311–325

Krüger HS, Hanganu-Opatz IL (2013) Neonatal cholinergic lesion alters the acoustic structure of infant rat vocalization but not the early cognitive development. Dev Psychobiol 55(3):294–308

Leanza G, Nilsson OG, Wiley RG, Björklund A (1995) Selective lesioning of the basal forebrain cholinergic system by intraventricular 192 IgG-saporin: behavioural, biochemical and stereological studies in the rat. Eur J Neurosci 7(2):329–343

Leanza G, Nilsson OG, Nikkhah G, Wiley RG, Björklund A (1996) Effects of neonatal lesions of the basal forebrain cholinergic system by 192 immunoglobulin G-saporin: biochemical, behavioural and morphological characterization. Neuroscience 74(1):119–141

Linke R, Frotscher M (1993) Development of the rat septohippocampal projection: tracing with DiI and electron microscopy of identified growth cones. J Comp Neurol 332(1):69–88

Lledo PM, Alonso M, Grubb MS (2006) Adult neurogenesis and functional plasticity in neuronal circuits. Nat Rev Neurosci 7(3):179–193

Mesulam MM, Mufson EJ, Wainer BH, Levey AI (1983) Central cholinergic pathways in the rat: an overview based on an alternative nomenclature (Ch1-Ch6). Neuroscience 10:1185–1201

Nahmani M, Turrigiano GG (2014) Adult cortical plasticity following injury: recapitulation of critical period mechanisms? Neuroscience 283:4–16

Pappas BA, Davidson CM, Fortin T, Nallathamby S, Park GA, Mohr E, Wiley RG (1996) 192 IgG-saporin lesion of basal forebrain cholinergic neurons in neonatal rats. Brain Res Dev Brain Res 96(1–2):52–61

Pappas BA, Nguyen T, Brownlee B, Tanasoiu D, Fortin T, Sherren N (2000) Ectopic noradrenergic hyperinnervation does not functionally compensate for neonatal forebrain acetylcholine lesion. Brain Res 867(1–2):90–99

Pappas BA, Payne KB, Fortin T, Sherren N (2005) Neonatal lesion of forebrain cholinergic neurons: further characterization of behavioral effects and permanency. Neuroscience 133(2):485–492

Rennie K, Fréchette M, Pappas BA (2011) The effects of neonatal forebrain cholinergic lesion on adult hippocampal neurogenesis. Brain Res 1373:79–90

Ricceri L, Calamandrei G, Berger-Sweeney J (1997) Different effects of postnatal day 1 versus 7 192 immunoglobulin G-saporin lesions on learning, exploratory behaviors, and neurochemistry in juvenile rats. Behav Neurosci 111(6):1292–1302

Ricceri L, Usiello A, Valanzano A, Calamandrei G, Frick K, Berger-Sweeney J (1999) Neonatal 192 IgG-saporin lesions of basal forebrain cholinergic neurons selectively impair response to spatial novelty in adult rats. Behav Neurosci 113(6):1204–1215

Ricceri L, Hohmann C, Berger-Sweeney J (2002) Early neonatal 192 IgG saporin induces learning impairments and disrupts cortical morphogenesis in rats. Brain Res 954(2):160–172

Ricceri L, Minghetti L, Moles A, Popoli P, Confaloni A, De Simone R, Piscopo P, Scattoni ML, di Luca M, Calamandrei G (2004) Cognitive and neurological deficits induced by early and prolonged basal forebrain cholinergic hypofunction in rats. Exp Neurol 189(1):162–172

Ricceri L, Cutuli D, Venerosi A, Scattoni ML, Calamandrei G (2007) Neonatal basal forebrain cholinergic hypofunction affects ultrasonic vocalizations and fear conditioning responses in preweaning rats. Behav Brain Res 183(1):111–117

Robertson RT, Gallardo KA, Claytor KJ, Ha DH, Ku KH, Yu BP, Lauterborn JC, Wiley RG, Yu J, Gall CM, Leslie FM (1998) Neonatal treatment with 192 IgG-saporin produces long-term forebrain cholinergic deficits and reduces dendritic branching and spine density of neocortical pyramidal neurons. Cereb Cortex 8(2):142–155

Sarter M, Bruno JP (2004) Developmental origins of the age-related decline in cortical cholinergic function and associated cognitive abilities. Neurobiol Aging 25(9):1127–1139

Scattoni ML, Calamandrei G, Ricceri L (2003) Neonatal cholinergic lesions and development of exploration upon administration of the GABAa receptor agonist muscimol in preweaning rats. Pharmacol Biochem Behav 76(2):213–221

Scattoni ML, Puopolo M, Calamandrei G, Ricceri L (2005) Basal forebrain cholinergic lesions in 7-day-old rats alter ultrasound vocalisations and homing behaviour. Behav Brain Res 161(1):169–172

Scattoni ML, Adriani W, Calamandrei G, Laviola G, Ricceri L (2006) Long-term effects of neonatal basal forebrain cholinergic lesions on radial maze learning and impulsivity in rats. Behav Pharmacol 17(5–6):517–524

Semba K, Fibiger HC (1988) Time of origin of cholinergic neurons in the rat basal forebrain. J Comp Neurol 269(1):87–95

Sherren N, Pappas BA (2005) Selective acetylcholine and dopamine lesions in neonatal rats produce distinct patterns of cortical dendritic atrophy in adulthood. Neuroscience 136(2):445–456

Sherren N, Pappas BA, Fortin T (1999) Neural and behavioral effects of intracranial 192 IgG-saporin in neonatal rats: sexually dimorphic effects? Brain Res Dev Brain Res 114(1):49–62

Silveira DC, Cha BH, Holmes GL (2002) Effects of lesions of basal forebrain cholinergic neurons in newborn rats on susceptibility to seizures. Brain Res Dev Brain Res 139(2):277–283

Thomas LB, Book AA, Schweitzer JB (1991) Immunohistochemical detection of a monoclonal antibody directed against the NGF receptor in basal forebrain neurons following intraventricular injection. J Neurosci Methods 37:37–45

Waite JJ, Chen AD, Wardlow ML, Wiley RG, Lappi DA, Thal LJ (1995) 192 immunoglobulin G-saporin produces graded behavioral and biochemical changes accompanying the loss of cholinergic neurons of the basal forebrain and cerebellar Purkinje cells. Neuroscience 65(2):463–476

Wenk GL, Stoehr JD, Quintana G, Mobley S, Wiley RG (1994) Behavioral, biochemical, histological, and electrophysiological effects of 192 IgG-saporin injections into the basal forebrain of rats. J Neurosci 14(10):5986–5995

Wiley RG (1992) Neural lesioning with ribosome-inactivating proteins: suicide transport and immunolesioning. Trends Neurosci 15:285–290

Works SJ, Wilson RE, Wellman CL (2004) Age-dependent effect of cholinergic lesion on dendritic morphology in rat frontal cortex. Neurobiol Aging 25(7):963–974

Zhu XO, Waite PM (1998) Cholinergic depletion reduces plasticity of barrel field cortex. Cereb Cortex 8(1):63–72

NGF in Early Embryogenesis, Differentiation, and Pathology in the Nervous and Immune Systems

Luisa Bracci-Laudiero and Maria Egle De Stefano

Abstract The physiology of NGF is extremely complex, and although the study of this neurotrophin began more than 60 years ago, it is far from being concluded. NGF, its precursor molecule pro-NGF, and their different receptor systems (i.e., TrkA, p75NTR, and sortilin) have key roles in the development and adult physiology of both the nervous and immune systems. Although the NGF receptor system and the pathways activated are similar for all types of cells sensitive to NGF, the effects exerted during embryonic differentiation and in committed mature cells are strikingly different and sometimes opposite. Bearing in mind the pleiotropic effects of NGF, alterations in its expression and synthesis, as well as variations in the types of receptor available and in their respective levels of expression, may have profound effects and play multiple roles in the development and progression of several diseases. In recent years, the use of NGF or of inhibitors of its receptors has been prospected as a therapeutic tool in a variety of neurological diseases and injuries. In this review, we outline the different roles played by the NGF system in various moments of nervous and immune system differentiation and physiology, from

---

We wish to dedicate this review to the late Prof. Rita Levi-Montalcini, a splendid mentor and physiologist.

L. Bracci-Laudiero (✉)
Institute of Translational Pharmacology, Consiglio Nazionale Delle Ricerche (CNR),
Via Fosso Del Cavaliere 100, 00133 Rome, Italy
e-mail: luisa.braccilaudiero@ift.cnr.it

L. Bracci-Laudiero
Bambino Gesù Children Hospital, IRCCS, Rome, Italy

M.E. De Stefano
Istituto Pasteur-Fondazione Cenci Bolognetti, Department of Biology
and Biotechnology "Charles Darwin", University of Rome La Sapienza,
Piazzale a. Moro 5, 00185 Rome, Italy
e-mail: egle.destefano@uniroma1.it

M.E. De Stefano
Center for Research in Neurobiology "Daniel Bovet", Università Di Roma La Sapienza,
Rome, Italy

© Springer International Publishing Switzerland 2015
Curr Topics Behav Neurosci (2016) 29: 125–152
DOI 10.1007/7854_2015_420
Published Online: 23 December 2015

embryonic development to aging. The data collected over the past decades indicate that NGF activities are highly integrated among systems and are necessary for the maintenance of homeostasis. Further, more integrated and multidisciplinary studies should take into consideration these multiple and interactive aspects of NGF physiology in order to design new therapeutic strategies based on the manipulation of NGF and its intracellular pathways.

Keyword Neurotrophin · proNGF · NGF receptors · Neuronal degeneration · Inflammation

Abbreviations

CGRP	Calcitonin gene-related peptide
CIPA	Congenital insensitivity to pain with anhidrosis
CNS	Central nervous system
CREB	CRE-binding protein
DRG	Dorsal root ganglion
EAE	Experimental autoimmune encephalomyelitis
ERK	Extracellular signal-regulated kinase
IL	Interleukin
MAPK	Mitogen-activated protein kinase
NGF	Nerve growth factor
NPY	Neuropeptide Y
NT	Neurotrophin
p75NTR	p75neurotrophin receptor
PI3K	Phosphatidylinositol 3-kinase
PKC	Protein kinase C
PLC	Phospholipase C
PNS	Peripheral nervous system
SOS	Son of sevenless
SP	Substance P
TH	Tyrosine hydroxylase
TrkA	Tropomyosin-related kinase A

Contents

1 The Discovery of NGF 127
2 The NGF System: NGF Precursor, Mature Forms, and Their Receptors 128
3 NGF and Early Embryo Development 132
4 NGF in Health and Disease of the Peripheral and Central Nervous Systems 134
5 NGF in Immune System Differentiation and the Immune Response 139
6 Conclusion 141
References 142

1   The Discovery of NGF

The discovery of nerve growth factor (NGF) is intimately linked to the development of the nervous system. More than sixty years ago, Levi-Montalcini's pioneer studies on nervous system differentiation, using Cajal's silver staining technique in chick embryos, lead to the identification of NGF (Levi-Montalcini 1987). Levi-Montalcini's scientific background as a neuroanatomist prompted her to re-investigate the effects of limb extirpation on the nervous system in chick embryos and to hypothesize that contrary to the prevailing opinion of the time, limb ablation resulted in a massive neuronal degeneration due to the impossibility of the neurons to form synapsis with their end organs. The massive death of peripheral neurons was also observed in normal chick embryos between E6 and E12, suggesting that there were more neurons in the developing nervous system than the periphery could support. This intuition was confirmed in experiments with supernumerary limbs, while an unexpected clue for the interpretation of this phenomenon was provided by the transplantation of solid tumors in chick embryos (Levi-Montalcini and Hamburger 1951). Here, the enlarged size of sensory and sympathetic ganglia, the increased number of surviving neurons, the abnormal distribution of fibers, that not only penetrated the sarcoma but also invaded the embryo viscera and the lumen of the veins, suggested that the tumor was releasing an unknown factor able to control the growth and survival of sensory and sympathetic neurons. The possibility to purify the newly discovered factor in huge amounts from male mouse submaxillary glands (Levi-Montalcini and Cohen 1960) and to produce antibodies against NGF (Levi-Montalcini and Booker 1960) was instrumental in understanding that the target organ, by producing limited amounts of NGF, was responsible for the survival of peripheral neurons during development. In vivo NGF deprivation, using neutralizing NGF antibodies, demonstrated, in a number of animal models, a massive destruction of immature sympathetic cells and a marked reduction in sensory neurons in dorsal root ganglia (DRG) (Levi Montalcini and Angeletti 1968; Aloe et al. 1981). Even though the technical approach at that time was limited to morphological and biochemical analysis, these findings, demonstrating the existence and effects of NGF, still represent a milestone in neurobiology. This was the beginning of what has been called "The NGF saga," since the spectrum of NGF activities is much broader and more complex than that was initially imagined (Levi-Montalcini 1997). More than 60 years after the discovery of NGF, our knowledge of its properties and functions is far from complete, as is evident from the astonishing number of papers published every year on NGF.

The study of NGF is indeed still a fertile terrain of discovery, as shown by the recent findings of a variety of biological activities of the immature form of NGF (pro-NGF) and the critical balance between pro-NGF and NGF concentrations within tissues for their neurotrophic activity, mediated by different receptors (i.e., p75NTR versus TrkA) and intracellular signaling pathways.

2 The NGF System: NGF Precursor, Mature Forms, and Their Receptors

NGF, isolated from mouse salivary glands, is a complex comprising three subunits: α, β, and γ, where only the β subunit has a biological effect (Bax et al. 1997). This high molecular weight complex, also known as 7S NGF, because of its sedimentation velocity, is present only in the mouse. In other species, NGF is present only as the β subunit, a dimer of two identical monomers of 118 amino acids, held together by non-covalent bindings. The crystallographic analysis of murine β-NGF shows that each monomer has an elongated shape and comprises antiparallel pairs of β-strands, forming a flat surface and four β-hairpin loop regions (McDonald et al. 1991). The monomer is also characterized by three disulfide bridges arranged in a peculiar ring structure, known as a "cysteine knot" that, first described for NGF, is also common to other growth factors (McDonald and Hendrickson 1993). The two monomers associate through the flat region (which is rich in hydrophobic residues) that gives stability to the homodimer, which shows a high association constant. NGF is synthesized in the endoplasmic reticulum as an immature form of 241 amino acids—the proNGF—which is either processed in the Golgi network within secretory vesicles by furin and other convertases (Seidah et al. 1996), or released in immature forms of different molecular weights, due to alternative splice variants and different levels of glycosylation (Bierl et al. 2005; Bierl and Isaacson 2007). In the extracellular space, pro-NGF is either processed into mature NGF by plasmin or rapidly degraded by metalloproteinases (Bruno and Cuello 2006). ProNGF has recently received attention because it appears to have biological effects (Lee et al. 2001) that are distinct from, or even opposite, to those of mature NGF in certain cell types (D'Onofrio et al. 2011). In vivo studies have shown that the prevalent form present in the brain (Fanhestock et al. 2001) and peripheral nervous tissues (Bierl et al. 2005) is proNGF and not the mature NGF. Levels of proNGF in the brain are also increased during aging (Bierl and Isaacson. 2007), in neurodegenerative diseases (i.e., Alzheimer's and Parkinson's diseases) (Fanhestock et al. 2001; Xia et al. 2013) and diabetic encephalopathy (Soligo et al. 2015).

Studies of the sequence of NGF led to the discovery of other molecules, sharing a high level of sequence homology with NGF and together constituting the neurotrophin family (Ebendal 1992). Neurotrophins are structurally related proteins, and all are key factors in the development and physiology of the nervous system. In addition to NGF, the neurotrophin family includes brain-derived neurotrophic factor (BDNF), neurotrophin-3 (NT-3), and neurotrophin 4/5 (NT-4/5). These neurotrophins have been found in all vertebrates, while two additional neurotrophins, closely related to NGF, neurotrophin-6, and neurotrophin-7, have been found only in fish, and Lf-NT and Mg-NT have been isolated only in lamprey and Atlantic hagfish (Agnatha) (Hallböök 1999; Lanave et al. 2007). Molecular evolution studies have shown that neurotrophins evolved early in vertebrate history, and the phylogenetic tree suggests the duplication of an ancestor *NGF* gene in two clusters of genes, one including *NGF*, *NT-3*, and *NT-6/7* and the other comprising

BDNF and *NT-4/5*. A second duplication separated the *BDNF* gene from that of *NT-4/5* and the *NGF* gene from that of *NT-3*, leading to the formation of a variety of proteins that have acquired multiple functions during the evolution of the vertebrate nervous system with its increasing complexity (Hallböök 1999). The rate of evolution of the *NGF* gene has been higher than that of other neurotrophin genes (Lanave et al. 2007), and this could possibly explain the greater diversity and complexity of functions requiring NGF, both within and without the nervous system, in comparison with the more conserved functions that require BDNF and NT-3 (Lanave et al. 2007).

The biological effects of NGF are directly dependent on its initial binding to specific cell surface receptors: p75NTR, a 75 kD glycoprotein, that belongs to the TNF receptor superfamily, and TrkA, a transmembrane tyrosine kinase of 140 kD. p75NTR is also known as a pan-neurotrophin receptor because it also binds the other neurotrophins with similar affinity, while the transmembrane tyrosine kinases, TrkA, TrkB, and TrkC, bind more specifically NGF, BDNF, and NT-3, respectively (Rodríguez-Tébar et al. 1991). Phylogenetic analysis of the *trk* genes shows that they originate from duplications of an ancestral gene during vertebrate evolution. These multiple *trk* genes coevolved in parallel with the neurotrophin genes, allowing preferred ligand–receptor interactions and the development of specific functions in different cell populations (Hallböök 1999).

The structure of TrkA is characterized by an extracellular domain that contains cysteine repeats, leucine-rich repeats, and two immunoglobulin domains, the second of which is the site that interacts with NGF. The binding of NGF to TrkA induces receptor dimerization and phosphorylation of specific tyrosine residues (Kaplan et al. 1991), which act as docking sites for cytoplasmic adaptor proteins that activate intracellular pathways (Fig. 1). Numerous studies have demonstrated that the activation of TrkA signaling regulates the survival and differentiation of neuronal cells. The best known pathways activated by NGF trough TrkA are the phosphatidylinositol 3-kinase (PI3K), MAPK, and PLC-γ (Reichardt 2006). The binding of Shc to the phosphorylated tyrosine residue Y490 leads to the activation of PI3K and the phosphorylation of Akt. Activation of Akt results in the inhibition of the forkhead transcription factor FKHRL1, which controls the transcription of pro-apoptotic genes such as Bim and Fas ligand (Fukunaga et al 2005), and in inhibition of GSK-3β, a point of convergence of many signaling pathways that influence the transcription of CREB, AP-1, and NF-κB (Beurel et al. 2010). The activation of PI3K pathway in neurons is involved in the maintenance of survival and in promoting axonal elongation.

The binding of Shc to residue Y490 also allows the recruitment of Grb2 and Sos, which leads to the activation of Ras and the downstream MAPK pathway, which results in Erk activation and CREB transcription. This pathway has a key role in regulating neuronal survival and differentiation.

Phosphorylation of the TrkA residue Y785 induces activation of PLC-γ and the formation of diacylglycerol (DAG) and inositol triphosphate (IP3). The increase in IP3 releases Ca^{++} from the intracellular stores, thus activating calcineurin and other enzymes sensitive to Ca^{++} levels. The accumulation of DAG induces activation of

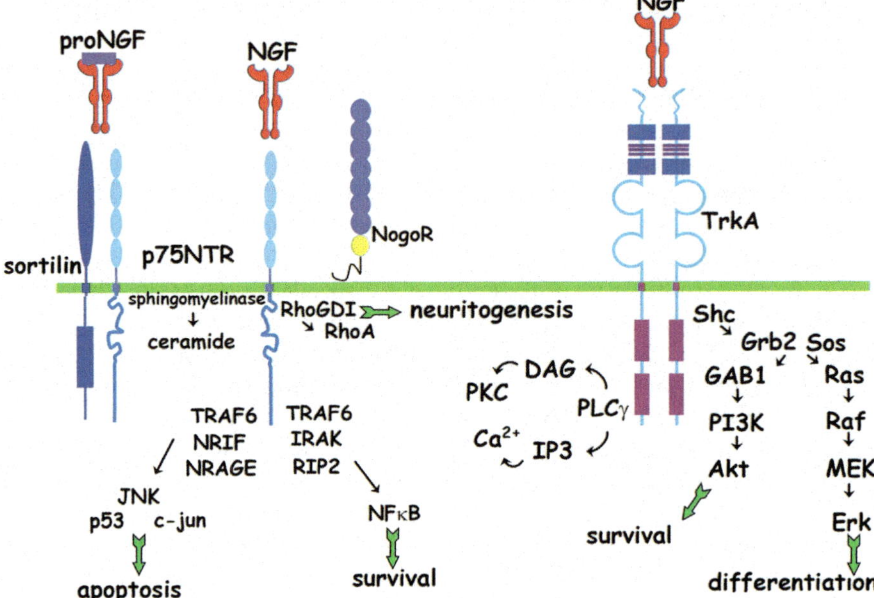

Fig. 1 The binding of NGF to TrkA dimers induces phosphorylation of tyrosine residues in the intracellular domain that results in activation of specific pathways. The presence of p75NTR increases the affinity and specificity of TrkA for NGF. p75NTR can interact with other co-receptors (i.e., sortilin, NogoR) and activates different intracellular adaptors and signaling pathways regulating a broad spectrum of diverse cellular responses

protein kinase C (PKC), which is involved in neurite outgrowth (for more detailed reviews on TrkA signaling see Reichardt 2006; Uren and Turnley 2014).

The majority of studies on TrkA signaling have been performed using either a cellular model of mouse pheochromocytoma, the PC12 cell line (Greene 1978), which in the presence of NGF differentiates in sympathetic neurons, or primary neuronal cells derived from different areas of the nervous system. The intracellular pathways activated by NGF binding to TrkA in non-neuronal cells are similar to those described in neuronal models, although they regulate other cell functions (Barouch et al. 2001; Torcia et al. 2001). For example, in masts cells, NGF acts as a chemotactic factor by binding to TrkA and activating MAPK and PI3K signaling (Sawada et al. 2000). However, in monocytes, NGF binding to TrkA plays a key anti-inflammatory role by activating the PI3K pathway and inhibiting the synthesis of inflammatory cytokines (Prencipe et al. 2014).

The expression of TrkA is considered necessary and sufficient to elicit NGF biological responses, but the role and contribution of p75NTR to NGF signaling and function is still debated and far from being univocally defined. For a long time, the role of p75NTR was considered limited to facilitating TrkA functions and enhancing its specificity. Indeed, the co-expression of p75NTR together with TrkA increases the specificity and affinity of NGF binding. Both receptors have a similar affinity for NGF with a Kd in the range of 10^{-9} M; however, when they are co-expressed, the Scatchard analysis reveals the presence of a high-affinity-binding site with a Kd of 10^{-11} M. Different biochemical models, describing the possible interaction between p75NTR and TrkA, have been proposed to explain the increased binding affinity: p75NTR can either act as TrkA co-receptor, by concentrating NGF and presenting it to TrkA, or directly interacts with TrkA, inducing allosteric modifications that enhance its affinity for NGF (Chao and Hempstead 1995). A novel hypothesis is that there is no direct extracellular interaction between TrkA and p75NTR and that they interact through common downstream signaling pathways and shared adaptor molecules (Wehrman et al. 2007). What is now emerging as a critical factor in the regulation of NGF/proNGF activities is the relative levels of expression of p75NTR in comparison with the TrkA, since their ratio plays a role in determining which of the intracellular pathways and biological activities are triggered (Masoudi et al. 2009).

For a long time, p75NTR was considered unable to convey any signals by itself because, in common with the other members of the TNF receptor family, it lacks any intrinsic enzymatic activity in the intracellular domain. As studies on TNF-R signaling progressed (Cabal-Hierro and Lazo 2012), it also became clear that p75NTR, by interacting with specific co-receptors and intracellular adaptors, could activate intracellular pathways very distinct from those activated by TrkA. The intracellular region of p75NTR has a death domain, although it is not identical to the death domain of other receptors of the TNFR1 group. This p75NTR intracellular motif is not able to self-assemble and does not recruit signaling proteins containing death domains such as FADD or TRADD (Wang et al. 2001), but instead binds to TRAF6 (Ye et al. 1999; Yeiser et al. 2004) and to other adaptor molecules, which do not have a death domain, such as IRAK (Mamidipudi et al. 2002) RhoA (Yamashita et al. 1999), NRAGE (Salehi et al. 2000), and NogoR (Wang et al. 2002).

In common with other receptors of the TNF-R superfamily (Hehlgans and Pfeffer 2005), it is now clear that p75NTR can regulate a broad spectrum of extremely diverse biological responses, such as neurite outgrowth, proliferation, and cell differentiation, as well as cell survival and death. This depends on the cell types in which p75NTR is expressed, on their differentiation state and on co-expression with TrkA and other co-receptors, such as sortilin (Nykjaer and Willnow 2012). This variety of effects indicates an extremely complex regulation of p75NTR activity, which as yet is only partially understood. Some of the pathways now known to be regulated by p75NTR are the NF-kB, c-Jun kinase, RhoA, and ceramide pathways (for more extensive reviews on p75NTR signaling see Roux and Barker 2002; Kraemer et al. 2014).

Recently, much attention has been directed to sortilin, one of the p75NTR co-receptors belonging to the VPS10P-domain receptor family (Willnow et al. 2008), because it can significantly increase the affinity of p75NTR for proNGF (Nykjaer et al. 2004). A number of studies have shown that in the presence of low levels of TrkA, p75NTR is structurally and functionally coupled with the transmembrane protein sortilin (Nykjaer et al. 2004; Teng et al. 2005; Domeniconi et al. 2007). This transforms p75NTR into a "death-promoting" receptor, whose activation drives neurons to apoptosis by binding proNGF with high affinity and activating caspase 6 (Lee et al. 2001; Majdan et al. 2001; Huang and Reichardt 2003; Fahnestock et al. 2004; Reichardt 2006).

3   NGF and Early Embryo Development

NGF and its receptors are expressed in the embryo well before the formation of the neural tube and the differentiation of neuronal cells (Fig. 2). This is not surprising, considering that NGF mRNA is found in oocytes, being present in small pre-vitellogenic as well as in fully grown oocytes (Carriero et al. 1991; Abir et al. 2005; Dissen et al. 2009). Although NGF and its receptors are involved in follicle formation and oocyte maturation (Dissen 1995; Dissen et al. 2009; Kerr et al. 2009; Linher-Melville and Li 2013), and NGF is expressed and accumulated in oocytes (Carriero et al. 1991), there are few indications of its possible role in early embryogenesis. The expression of p75NTR and TrkA appears in mouse embryos at the blastocyst stage, with the expression of p75NTR preceding that of TrkA, and both being confined within the inner mass of the blastocyst and absent from the trophoblast (Moscatelli et al. 2009). Mouse embryonic stem cell lines derived from blastocysts retain the expression of NGF receptors in vitro, and the addition of NGF, while increasing their number, maintains the expression of staminal markers of pluripotency such as Oct4 and Nanog (Moscatelli et al. 2009). Human embryonic stem cells (hESCs) also express neurotrophin receptors (Pyle et al. 2006; Schuldiner et al. 2000), and the addition of NGF results in an increase in cell survival (Pyle et al. 2006). In studies of the differentiation patterns of hESCs in a 3D culture environment, it was found that NGF, together with Retinoic Acid (RA), preferentially favors the differentiation of hESCs toward ectodermal and mesodermal lineages (Inanç et al. 2008).

In embryos, the expression of NGF receptors persists in specific areas during gastrulation and neurulation (Zhang et al. 1996), and their localization is not always identical (Wheeler et al. 1998), suggesting different functions of TrkA and p75NTR in cell differentiation and morphogenesis. During early embryogenesis, NGF expression is modulated in the ectoderm and seems to be involved in body shaping, but not in early neural differentiation (Bhargava and Modak 2002). Thus, NGF and its receptor are expressed well before the onset of neurogenesis, and their expression characterizes specifically non-neuronal tissues such as somites, notochord, and neural crest (Yao et al. 1994; Bhargava 2007; Tomellini et al. 2014). Indeed, neutralization of NGF in chick embryos in the first stages of development causes alterations of the embryo's axial rotation and influences a number of genes involved in developmental processes, cell movements in the notochord and somite formation,

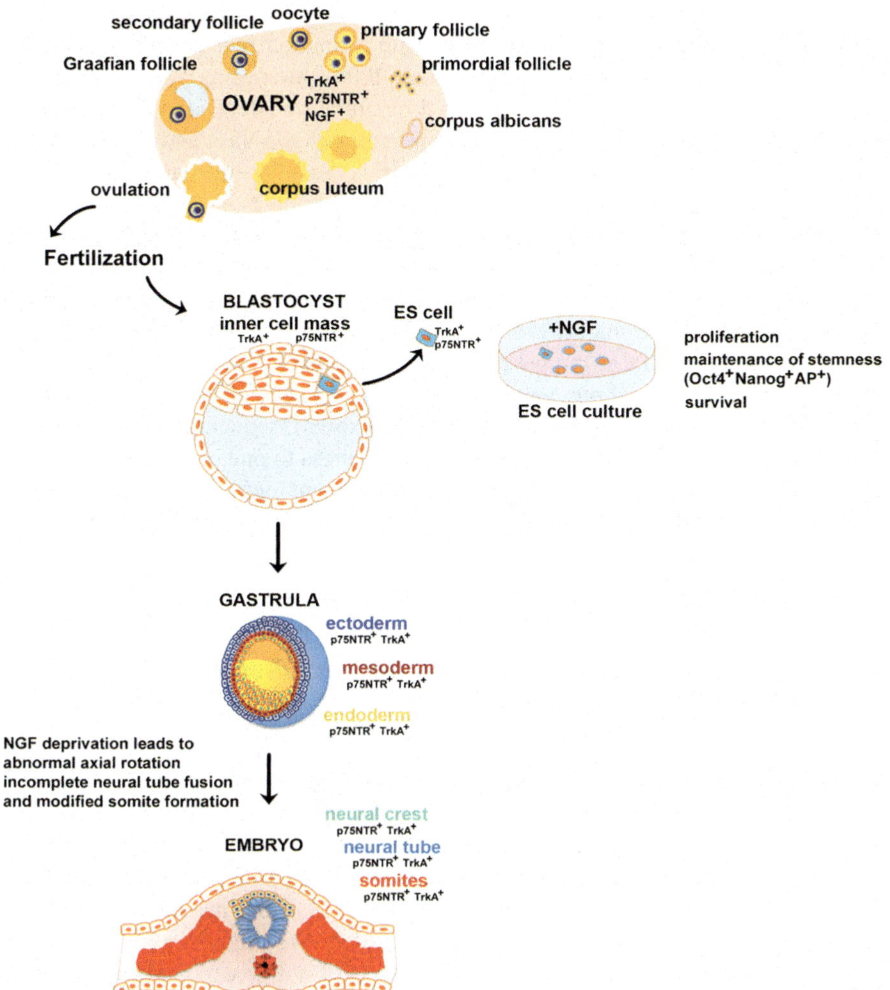

Fig. 2 NGF, p75NTR, and TrkA are expressed in the ovary during fetal life and contribute to follicle formation and at puberty onset they are involved in oocytes maturation. NGF mRNA is accumulated in the oocytes so that NGF mRNA is transferred to the embryo at fertilization. The novo synthesis of NGF and expression of TrkA and p75NTR are detected in the blastula. The embryonic stem cells that derive from the inner mass retain the expression of p75NTR and TrkA, and addition of NGF to the culture medium increases their proliferation and survival but does not alter the stemness potential as the unchanged expression of Nanog and Oct4 indicates. In the gastrula, the expression of NGF receptors characterizes all three germ layers. During morphogenesis and neurogenesis, there is a well-defined spatiotemporal expression of NGF, p75NTR, and TrkA that influences differentiation of both non-neuronal and neuronal cells in specific areas

cell cycle regulation, and proliferation (Manca et al. 2012). Neutralization of NGF or of p75NTR in chick embryos during somite formation reduces apoptosis of the sclerotome and dermomyotome (Cotrina et al. 2000). Interestingly, sonic hedgehog (Shh), a key factor regulating sclerotome differentiation (Resende et al. 2010) and survival (Britto et al. 2000), reduces the expression of p75NTR and NGF in somite explants, suggesting that the regulation of NGF is essential for maintaining control of programmed cell death in non-neuronal cells (Cotrina et al. 2000).

Neural crest cells begin to express p75 from E10.5, when they start to migrate (Wilson et al. 2004). These cells, although of ectodermal origin, acquire mesenchymal cell properties, migrating throughout the embryo to generate multiple tissues that include the majority of the cranial connective tissue and skeletal elements, neurons and glia of the peripheral nervous system, endocrine cells, smooth muscle cells, tendons, and pigment cells (Mayor and Theveneau 2013). During their migration and regulated differentiation, they express sequentially different Trk receptors (Davies 1997), thus acquiring responsiveness to one or other neurotrophin (Vogel 1993; Rifkin et al. 2000) that canalizes and defines the differentiative program of peripheral neuronal populations (for a more detailed review see Marmigère and Carroll 2014).

4 NGF in Health and Disease of the Peripheral and Central Nervous Systems

A large body of studies has discussed and demonstrated the crucial role of NGF in the development, maintenance, and regeneration of mammalian sympathetic and sensory neurons of the peripheral nervous system (PNS). NGF-sensitive neurons express both TrkA and p75NTR receptors, the activities of which are exquisitely balanced. The sensitivity and affinity of TrkA for NGF are increased by its functional interaction with p75NTR (Nykjaer et al. 2005; Wehrman et al. 2007) and are modulated by the changes in the expression levels of both receptors (Esposito et al. 2001).

During the development of neural circuits, intermediate and final neuron targets release gradients of NGF and support neuron survival via a main long-distance signaling initiating within distal axons (Levi-Montalcini 1987), which implies the retrograde transport of internalized NGF-TrkA receptors complexes. Once they reached neuronal cell bodies, these so-called signaling endosomes will regulate genes expression, promoting neuron survival, axon growth and pathfinding, and synaptogenesis (Miller and Kaplan 2001; Barker et al. 2002; Harrington and Ginty 2013; Howe and Mobley 2005; Reichardt 2006; Cosker et al. 2008; Pazyra-Murphy et al. 2009; Sharma et al. 2010), while enhancing sensitization to NGF and protecting from p75NTR-mediated apoptosis (Deppmann et al. 2008). Although another neurotrophin, NT-3, can also bind to TrkA receptors, only the TrkA/NGF complex is internalized and forms signaling endosomes supporting neuronal survival. This mechanism relies on the inability of NT-3 to activate the intracellular signaling cascades regulated by Rac1, GTP, and cofilin proteins that promote F-actin depolymerization, a process essential to initiate the trafficking of signaling endosomes (Harrington et al. 2011). Fundamental evidence on the pro-survival role of NGF during development derives from pioneering in vivo studies on DRG and sympathetic ganglion neurons of NGF- and Trk-knockout mice (Snider 1994). In vitro studies, on the other hand, were decisive for identifying the molecular mechanisms through which NGF, proNGF, and their receptors operate (Campenot 1977; Ye et al. 2003; Mok and Campenot 2007; for review see Harrington and Ginty 2013).

Along with neuron survival, target-released NGF also regulates axon growth and retraction (Campenot 1977), synapse and neural circuit formation (Ladle et al. 2007; Sharma et al. 2010), and expression of neurotransmitters (Luo et al. 2007; Patel et al. 2003). In vitro studies, which use compartmentalized chambers, have shown that NGF applied directly to distal axons acts locally to support their extension; however, when only cell bodies are exposed to NGF, neurons fail to extend axons into a compartment that lacks NGF (Campenot 1977). In sympathetic neurons, NGF is also required for dendritic arborization, the degree of which strictly correlates with the size of the neuron's peripheral targets (Voyvodic 1989). Interruption of retrograde signaling by axotomy causes dendritic retraction, which persists as long as axons require regeneration (Purves 1975). Finally, retrograde NGF/TrkA signaling also drives synapse establishment and subsequent maintenance and plasticity in adults. In sympathetic neurons, appropriate coupling of pre- and postsynaptic specializations has been validated by several studies, in which retrograde signaling was abolished by either axotomy or the administration of NGF-blocking antibodies (Purves 1975; Mandai et al. 2009; Sharma et al. 2010).

Along with the "classic" and well-described mechanism of retrograde signaling, NGF/TrkA activity can also be purely local, through the tightly regulated activation of intracellular signaling cascades. This local signaling has been extensively studied during axon outgrowth, guidance, and regulation of preterminal branching. In general, neurotrophin-stimulated axon growth requires the activation of transcription factors, which regulate gene expression and subsequent synthesis of the proteins necessary for axon growth. However, in addition to gene expression, local signaling pathways are also required for the control of cytoskeletal dynamics, as elegantly described for the NGF by Campenot (1982a, b). The activation of a spatially controlled signal transduction based on PI3K-Akt activation at the growth cones, the major sites where neurons receive and integrate extracellular signals to direct axonal cytoskeletal dynamics (Baas and Luo 2001), has been demonstrated for filopodia formation and axon growth (Kuruvilla et al. 2000; Zhou et al. 2004; Ketschek and Gallo 2010). Fig. 3 shows the three major intracellular domains of sympathetic ganglionic neurons, in which NGF-TrkA exerts its activity: nucleus, for the modulation of gene expression by NGF signaling endosomes (control of neuronal survival, axon growth, and synaptogenesis); postsynaptic specializations, for a more local control by NGF signaling endosomes of synaptogenesis and

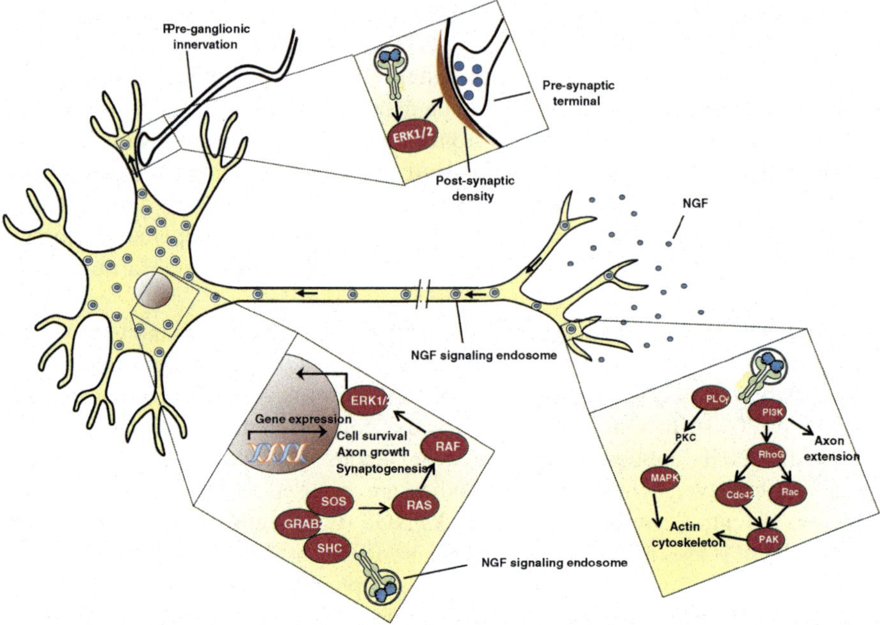

Fig. 3 In sympathetic neurons, NGF activates different intracellular signaling pathways. At the growth cones, NGF released by peripheral targets binds to TrkA receptors and elicits two main types of responses. In the first, the NGF-TrkA complex is internalized in signaling endosomes. These are retrogradely transported along the axon to the cell body, where they trigger an intracellular signaling cascade ending up in Erk1/2 activation and consequent modulation of gene expression. NGF-regulated genes encode proteins important for cell survival and differentiation, axon growth, and synaptogenesis. Signaling endosomes, after reaching cell bodies, can move into dendrites, where they will regulate processes related to synaptogenesis and local control of synaptic plasticity. Furthermore, the NGF-TrkA complex can also remain within the growth cone plasma membrane, initiating a local signaling cascade, involving MAPK and PAK activation, directed at promoting growth cone mobility and advancement through cytoskeletal remodeling, a process important during both development and axon regeneration

synaptic activity; and growth cones, for a local control of axon growth, which does not require NGF endocytosis.

The role of NGF in adult life is not only confined to the maintenance of neuronal survival and structure, but also includes the fine modulation of the chemical phenotype of neurons. It has been shown that in adult dorsal root ganglion neurons in vitro, the expression of mRNAs encoding the precursors of both substance P (SP) and calcitonin gene-related peptide (CGRP) is regulated by NGF (Lindsay and Harmar 1989). In vivo studies have also shown that the infusion of NGF after sciatic nerve transection reactivated α-CGRP, β-CGRP, and SP expression in DRG neurons (Verge et al. 1995). Similarly, in sympathetic neurons, the expression of tyrosine hydroxylase, the rate-limiting enzyme for catecholamine synthesis, is strictly NGF-dependent (Thoenen et al. 1971). Altogether, these findings provide evidences of a continuous and dynamic regulation of peptide neurotransmitter/neuromodulator levels in adult neurons by NGF.

As with the PNS, a specific role for NGF has also been proposed for the cholinergic neuron population of the central nervous system (CNS). Cholinergic neurons are situated largely in different areas of the basal forebrain (BF), which are the major source of cholinergic innervation to the cerebral cortex, hippocampus, amygdala, and remaining portions of the cortical mantle (Niewiadomska et al. 2009). They are highly and critically dependent on NGF during both development and adulthood, including aging. When NGF is synthesized by target neurons, it is retrogradely transported to the cell bodies of the BF neurons, contributing to the maintenance of cell morphology (i.e., cell body size and extent of terminal arborization (Higgins et al. 1989), and physiology, i.e., up-regulation of choline acetyltransferase (ChAT) gene expression, protein levels and activity, synthesis and release of acetylcholine, and expression of the vesicular acetylcholine transporter (VAChT) (see Niewiadomska et al. 2011 for review). NGF expression is highly modulated by neuron activity (i.e., in the hippocampus, it is increased by gluta-matergic and cholinergic neurotransmission, and decreased by GABAergic neuro-transmission) (Huh et al. 2008), and by pathological events (i.e., it is up-regulated by seizures, forebrain ischemia, marked hypoglycemia, and tissue injury) (Lindvall et al. 1994). NGF is also synthesized by subpopulations of hippocampal and BF GABAergic interneurons (Zhang et al. 2007) and by astrocytes and microglia, the CNS glial cells. In both these cell types, NGF expression is significantly up-regulated by several factors, such as inflammation, cytokines, and the bacterial lipopolysaccharide (LPS) (Tonchev et al. 2008), indicating a role in brain protection and regeneration from injuries.

In the past, a reduction in NGF trophic support, which could derive from a decrease in the correct cleavage of the proNGF to form mature NGF, has been indicated as one of the causes of age-related cholinergic neuron atrophy and neu-rodegenerative diseases, i.e., Alzheimer's disease (AD) (Tuszynski and Blesch 2004). In effect, high levels of proNGF and sortilin in adult and AD patient brains have been described (Fahnestock et al. 2001; Nykjaer et al. 2004). However, the aging-related cholinergic atrophy and cell loss in normal brains are not always accompanied by the reductions in the levels of NGF (Katoh-Semba et al. 1998). This has led to the hypothesis that aging cholinergic neurons fail to respond to NGF. Indeed, NGF retrograde transport in aged rat BF cholinergic neurons is significantly reduced, with consequent NGF signaling impairment, cell atrophy, and changes in gene expression (Cooper et al. 1994; DeLacalle et al. 1996; Sofroniew et al. 2001). Similarly, mild cognitive impairment (a prodromal stage of AD) and early forms of AD are characterized by the loss of cholinergic function more than by neurodegeneration, as neither a loss of NGF receptor mRNA levels (Goedert et al. 1989) nor a failure of NGF synthesis (Scott et al. 1995) has been observed. This marks an important difference between the early stages of AD, and the massive cholinergic neuron death observed in advanced stages of late-onset AD and other pathologies associated with cognitive deficits, such as Parkinson's disease (PD) and Down syndrome, among others (for review see Schliebs and Arendt 2010; Iulita and Cuello 2014). It is possible that the early formation and deposition of β-amyloid, the hallmark of AD, play a role in inducing an initial cholinergic dysfunction, which later evolves to become neuronal loss. This hypothesis is supported by a number of experimental findings showing that β-amyloid deposits may trigger cholinergic dysfunction in several ways: (i) by binding to α7-containing nicotinic acetylcholine receptors (a7nAChRs), thereby affecting their activity; (ii) by affecting NGF signaling; (iii) by mediating tau phosphorylation; (iv) by interacting with acetylcholinesterase (AChE); or (v) by specifically affecting the cholinergic neuron proteome (for review see Schliebs and Arendt 2011). Several in vivo studies have shown that age-related dysfunction of the cholinergic system may be ameliorated by treatment with NGF (for review see Niewiadomska et al. 2011). Another important aspect to be taken into consideration is the presence of prominent neuroinflammatory events triggered by the early deposition of intracellular Aβ oligomers (Ferretti et al. 2012). As demonstrated in transgenic animal models for AD, characterized by extracellular plaque deposition, neuroinflammation induces an increase in levels of metalloproteinase-9, one of the enzymes involved in NGF degradation in the extracellular space (Bruno et al. 2009). Amplified NGF degradation, together with a concomitant decreased in the conversion of proNGF into its mature form, results in a signaling deficiency of NGF and atrophy of basal forebrain cholinergic neurons (Cuello et al. 2012).

NGF may affect a variety of additional CNS neurons since its binding sites are present during early development in many neuronal systems, not necessarily of cholinergic origin, including the visual system (Vantini et al. 1989; Yan and Johnson 1988, 1989). Here, NGF and TrkA are expressed by almost all eye components. NGF released into the aqueous humor (Lambiase et al. 2002) and in the retina, the neural part of the eye, is produced and utilized by retinal ganglion cells, bipolar cells, and glial cells (Frade et al. 1999; Wang et al. 2014). During visual system development, NGF, TrkA, and p75 are highly expressed along the entire visual pathway, where NGF influences neuronal outgrowth, survival, apoptosis, and physiology (Roberti et al. 2014). Interestingly, topically applied NGF eye drops are able to reach the retina and the optic nerve (Ferrari et al. 2014), greatly enhancing the expectations regarding the clinical use of this neurotrophin for a number of ocular neurodegenerative diseases, such as glaucoma (Wang et al. 2014).

Finally, the presence of NGF in those limbic areas of the CNS involved in mood and cognition (i.e., the amygdala) and in the orchestration of neuroendocrine responses and circadian activities (i.e., the hypothalamus) indicates a much wider role for this NT than previously hypothesized. Several experimental studies currently suggest that NGF may also function as an intercellular messenger or humoral factor to help regulate endocrine responses to stress (for review see Berry et al. 2012).

5 NGF in Immune System Differentiation and the Immune Response

During embryo development, the expression of NGF receptors is finely modulated in primary and secondary lymphoid organs (Ernfors et al. 1988; Lomen-Hoerth and Shooter 1995; Ciriaco et al. 1996; Aloe et al. 1997) and declines during postnatal life (Ernfors et al. 1988; Ciriaco et al. 1996). In bone marrow and in the thymus, NGF receptor expression has been found in stromal cells, which release specific growth factors and signals (Pezzati et al. 1992; Cattoretti et al. 1993; Caneva et al. 1995; Ciriaco et al. 1996; Rezaee et al. 2010; Lee et al. 2008), that regulate the correct differentiation of myeloid and lymphoid precursors. In the thymus, the expression of NGF and of both its receptors is very high in the final stages of embryo development and in the early postnatal period and then decreases with age (Laurenzi et al. 1994; Aloe et al. 1997; Garcia-Suárez et al. 2001). In the Bursa of Fabricius, the lymphoid organ in which B cells differentiate in birds, the expression of NGF receptors (Ciriaco et al. 1996) is elevated during B-precursor differentiation in the epithelial follicles. The administration of NGF to chick embryos accelerates follicle formation (Bracci-Laudiero et al. 1991).

In immune organs, there is a local production of NGF (Laurenzi et al. 1994) that regulates sympathetic and sensory innervation, and neuropeptide and neurotransmitter synthesis in the embryo and in adult life (Madden and Felten 1995, Elenkov et al. 2000) These effects are directly and actively modulated by the amount of NGF available, as demonstrated in transgenic mice overexpressing NGF, which shows enhanced fiber density and modified innervation patterns in the spleen and lymph nodes (Carlson et al. 1995). The constant production of NGF in lymphoid organs (Laurenzi et al. 1994; Yamamoto et al. 1996; Aloe et al. 1997) also seems to be important for regulating the differentiation of hematopoietic stem cells. These hematopoietic stem cells express TrkA (Chevalier et al. 1994; Bracci-Laudiero et al. 2003), and its expression is at its highest levels in the more undifferentiated cells, declining during lineage differentiation. This modulated expression of TrkA suggests that NGF may play different functions depending on the state of differentiation of immune cell and functional activity. The possibility that NGF may be a key factor for haemopoiesis is supported by the fact that hematopoietic stem cells produce their own NGF in an autocrine fashion, probably regulating anti-apoptotic genes. In vitro studies have shown that the administration of NGF in semisolid cultures increases long-term survival of human hematopoietic cells (Bracci-Laudiero et al. 1993, Auffray et al 1996) and promote the commitment toward specific lineages in human and murine myeloid progenitor cells (Matsuda et al. 1988, 1991; Tsuda et al. 1991; Welker et al. 2000). Support for the possible pro-survival role of NGF in hematopoiesis is also provided by studies on leukemia showing a constitutive activation of TrkA in blasts from patients with de novo or secondary acute leukemia that affects survival of the leukemic cells in patients and in animal models (Li et al. 2009). A high expression of TrkA also characterizes Hodgkin–Reed/Sternberg cell lines, in which the constitutive activation of TrkA and the Akt pathway promotes cell survival, which can be strongly reduced by using TrkA inhibitors (Renné et al. 2008).

Differentiation of B cells is also influenced by NGF. Studies of mature lymphoid cells have shown that NGF has a proliferative effect on both B and T cells (Brodie and Gelfand 1992) and causes the differentiation of B cells into immunoglobulin-secreting plasma cells (Kimata et al. 1991). NGF seems also to influence the survival and, consequently, the antibody production of pulmonary plasma cells via regulation of the Ire1/XBP-1 pathway (Abram et al. 2009). NGF stimulates B-lymphocytes to produce IgM, IgA, and IgG (Brodie and Gelfand 1994) and regulates the survival of B-memory cells (Torcia et al. 1996), which rapidly and efficiently counteract pathogens when re-encountering antigens during the secondary response, a key feature of immunological memory.

The effects of NGF do not seem to be restricted only to differentiation processes, and a growing body of data is accumulating to support the hypothesis that NGF and its receptors are also involved in modulating functions and activity of mature immune cells. Mature immune cells express much lower TrkA levels than hematopoietic stem cells (Bracci-Laudiero et al. 2003; Antonelli et al. 2003), but after antigenic or inflammatory stimulation, when strong functional activity is necessary, TrkA expression is strongly up-regulated (Ehrhard et al. 1993; Caroleo et al. 2001; Ralainirina et al. 2010). The activity of NGF on cells of the myeloid lineage is to enhance effector functions, such as the release of inflammatory mediators, chemotaxis, and proliferation (Kannan et al. 1991; Noga et al. 2002; Gibbs et al. 2005; Samah et al. 2008). In lymphocytes and monocytes, in addition to proliferation and survival, NGF influences cytokine release (Susaki et al. 1996; Bayas et al. 2003; Shi et al. 2012; Prencipe et al. 2014) and the production of neuropeptides that have an immunological role (Bracci-Laudiero et al. 1996, 2005). In differentiated cells of both myeloid and lymphoid origin, NGF appears to regulate their survival (Kawamoto et al. 1995; Bullock and Johnson 1996; Hamada et al. 1996; la Sala et al. 2000;) by inducing the expression of anti-apoptotic genes such as bcl-2 (Bullock et al. 1996; Torcia et al.1996; la Sala et al. 2000).

Altogether these data suggest that NGF has a specific function in the differentiation and activation of immune cells and can thus play an important role in regulating the immune response in vivo. Confirmation of this hypothesis has come from studies on patients with congenital insensitivity to pain with anhidrosis (CIPA), a rare autosomal peripheral sensory neuropathy caused by mutations in the gene encoding TrkA (Indo et al. 1996). These patients, in addition to neurological alterations and loss of pain and sensation, are also prone to recurrent infections, slow wound healing, and inflammatory complications that are related to the altered functions of certain immune cell populations. The chemotactic activity of neutrophils is significantly impaired (Beigelman et al. 2009), and B-lymphocytes show changes in the intracellular pathways (Melamed et al. 2004) and impaired anti-apoptotic activity (Sato et al. 2004).

Studies on inflammatory and autoimmune diseases have clearly demonstrated that NGF and its receptor expression are highly modulated during inflammatory response and disease progression. Inflammatory mediators and cytokines can greatly enhance the synthesis of NGF in cells (Torcia et al. 1996; Caroleo et al. 2001; Kobayashi et al. 2002) and in tissues, and NGF concentrations appear to be correlated with clinical severity (Aloe et al. 2001). The biological meaning of this enhanced in vivo production of NGF at the site of inflammation and the ways in which NGF can affect inflammatory pathways, immune responses, and healing processes is far from being understood.

An example of the complexity of NGF activities in vivo can be extrapolated from studies on multiple sclerosis (MS). In MS patients, there is an increase in NGF concentration in the cerebrospinal fluid that correlates with the inflammatory state of the patients (Bracci-Laudiero et al. 1992; Caggiula et al. 2005). Administration of NGF in animal models of experimental autoimmune encephalomyelitis (EAE) delays the onset of clinical symptoms and prevents the full development of EAE lesions by directly reducing immune cell infiltrates (Villoslada et al. 2000; Parvaneh Tafreshi 2006). The administration of NGF exerts its effects directly on the activity of immune cells, by regulating T-lymphocyte response, monocyte infiltration, and the release of anti-inflammatory cytokines (Villoslada et al. 2000; Arredondo et al. 2001; Flügel et al. 2001; Parvaneh Tafreshi 2006). In EAE models, a local production of NGF (De Simone et al. 1996; Acosta et al. 2015) has also been observed in different brain areas during the acute phase of the disease. In addition to the immunological effects, this release of NGF also protects neuronal population (Linker et al. 2009), induces myelin repair (Øren et al. 2004; Acosta et al. 2013), and efficiently suppresses the formation of new lesions (Villoslada et al. 2000). This neuroprotective effect of immune cell-produced NGF also appears to be common to other neuropathologies characterized by different etiologies, as AD, PD, and spinal cord injury (Ebadi et al. 1997; Schulte-Herbrüggen et al. 2007; Saab et al. 2009; Colafrancesco and Villoslada 2011). This further confirms the integrative role that NGF has in human physiology and how the local increase of NGF in vivo can simultaneously have multiple effects on different cell types.

6 Conclusion

Much still remains to be elucidated regarding the highly complex physiology of NGF and its key role not only in the nervous system but in other systems as well. The complexity of NGF signaling, the number of receptors involved, and the diverse biological activities of immature and mature NGF are directly correlated with the need for the differential regulation of a variety of cell types in the various states of differentiation and activation. This intricate control of a multitude of activities, some of which have opposite effects, is necessary for a correct home-ostasis. The pleiotropic effects of NGF are thus a demonstration of its major integrative role, and alterations in the NGF-NGF receptor axis can affect many cell types, tissues, organs, and systems. Manipulation of the production of NGF, its receptors, and intracellular pathways at different times and in selected cell types could become a powerful tool for the treatment of many neurological and immune diseases, although more integrative and interdisciplinary studies are needed. As aptly described by Dr. Ibáñez in an editorial on the discovery of the biological effects of pro-NGF (Ibáñez 2002), the story of NGF resembles a "roller-coaster," still full of discoveries and surprises and, unusually for such an "old molecule," still unfolding.

References

Due to the vastness and complexity of the subjects treated in the present review, the list of bibliographic references is far from exhaustive and should be considered as merely representative. The authors apologize for not including the original work of many researchers who have helped to define the physiology of NGF and refer readers to other reviews for further informations.

Abir R, Fisch B, Jin S, Barnnet M, Ben-Haroush A, Felz C, Kessler-Icekson G, Feldberg D, Nitke S, Ao A (2005) Presence of NGF and its receptors in ovaries from human fetuses and adults. Mol Hum Reprod 11:229–236

Abram M, Wegmann M, Fokuhl V, Sonar S, Luger EO, Kerzel S, Radbruch A, Renz H, Zemlin M (2009) Nerve growth factor and neurotrophin-3 mediate survival of pulmonary plasma cells during the allergic airway inflammation. J Immunol 182:4705–4712

Acosta CM, Cortes C, MacPhee H, Namaka MP (2013) Exploring the role of nerve growth factor in multiple sclerosis: implications in myelin repair. CNS Neurol Disord: Drug Targets 12:1242–1256

Acosta C, Cortes C, Altaweel K, MacPhee H, Hoogervorst B, Bhullar H, MacNeil B, Torabi M, Burczynski F, Namaka MP (2015) Immune system induction of nerve growth factor in an animal model of multiple sclerosis: implications in re-myelination and myelin repair. CNS Neurol Disord Drug Targets. PMID: 25801841 (in press)

Aloe L, Cozzari C, Calissano P, Levi-Montalcini R (1981) Somatic and behavioral postnatal effects of fetal injections of nerve growth factor antibodies in the rat. Nature 291:413–415

Aloe L, Micera A, Bracci-Laudiero L, Vigneti E, Turrini P (1997) Presence of nerve growth factor in the thymus of prenatal and pregnant rats. Thymus 24:221–231

Aloe L, Tirassa P, Bracci-Laudiero L (2001) Nerve growth factor in neurological and non-neurological diseases: basic findings and emerging pharmacological prospectives. Curr Pharm Des 7:113–123

Antonelli A, Bracci-Laudiero L, Aloe L (2003) Altered plasma nerve growth factor-like immunoreactivity and nerve growth factor-receptor expression in human old age. Gerontology 49:185–190

Arredondo LR, Deng C, Ratts RB, Lovett-Racke AE, Holtzman DM, Racke MK (2001) Role of nerve growth factor in experimental autoimmune encephalomyelitis. Eur J Immunol 31:625–633

Auffray I, Chevalier S, Froger J, Izac B, Vainchenker W, Gascan H, Coulombel L (1996) Nerve growth factor is involved in the supportive effect by bone marrow derived stromal cells of the factor-dependent human cell line UT-7. Blood 88:1608–1618

Baas PW, Luo L (2001) Signaling at the growth cone: the scientific progeny of Cajal meet in Madrid. Neuron 32:981–984

Bhargava S (2007) Role of nerve growth factor and its receptor in the morphogenesis of neural tube in early chick embryo. Gen Comp Endocrinol 153:141–146

Bhargava S, Modak SP (2002) Expression of nerve growth factor during the development of nervous system in early chick embryo. Brain Res Dev Brain Res 136:43–49

Barker PA, Hussain NK, McPherson PS (2002) Retrograde signaling by the neurotrophins follows a well-worn trk. Trends Neurosci 25:379–381

Barouch R, Kazimirsky G, Appel E, Brodie C (2001) Nerve growth factor regulates TNF-alpha production in mouse macrophages via MAPkinase activation. J Leukoc Biol 69:1019–1026

Bax B, Blundell TL, Murray-Rust J, McDonald NQ (1997) Structure of mouse 7S NGF: a complex of nerve growth factor with four binding proteins. Structure 5:1275–1285

Bayas A, Kruse N, Moriabadi NF, Weber F, Hummel V, Wohleben G, Gold R, Toyka KV, Rieckmann P (2003) Modulation of cytokine mRNA expression by brain-derived neurotrophic factor and nerve growth factor in human immune cells. Neurosci Lett 335:155–158

Beigelman A, Levy J, Hadad N, Pinsk V, Haim A, Fruchtman Y, Levy R (2009) Abnormal neutrophil chemotactic activity in children with congenital insensitivity to pain with anhidrosis (CIPA): the role of nerve growth factor. Clin Immunol 130:365–372

Berry A, Bindocci E, Alleva E (2012) NGF, brain and behavioral plasticity. Neural Plast 2012:784040. doi:10.1155/2012/784040

Beurel E, Michalek SM, Jope RS (2010) Innate and adaptive immune responses regulated by glycogen synthase kinase-3 (GSK3). Trends Immunol 31:24–31

Bierl MA, Jones EE, Crutcher KA, Isaacson LG (2005) 'Mature' nerve growth factor is a minor species in most peripheral tissues. Neurosci Lett 380:133–137

Bierl MA, Isaacson LG (2007) Increased NGF proforms in aged sympathetic neurons and their targets. Neurobiol Aging 28:122–134

Bracci-Laudiero L, Vigneti E, Aloe L (1991) In vivo and in vitro effect of NGF on bursa of Fabricius cells during chick embryo development. Int J Neurosci 59:189–198

Bracci-Laudiero L, Aloe L, Levi-Montalcini R, Buttinelli C, Schilter D, Gillessen S, Otten U (1992) Multiple sclerosis patients express increased levels of b-nerve growth factor in cerebrospinal fluid. Neurosci Lett 147:9–12

Bracci-Laudiero L, Vigneti E, Iannicola C, Aloe L (1993) NGF retards apoptosis in chick embryo bursal cell in vitro. Differentiation 53:61–66

Bracci-Laudiero L, Aloe L, Stenfors C, Tirassa P, Theodorsson E, Lundberg T (1996) Nerve growth factor stimulates production of neuropeptide Y in human lymphocytes. NeuroReport 7:485–488

Bracci-Laudiero L, Celestino D, Starace G, Antonelli A, Lambiase A, Procoli A, Rumi C, Lai M, Picardi A, Ballatore G, Bonini S, Aloe L (2003) CD34-positive cells in human umbilical cord blood express nerve growth factor and its specific receptor TrkA. J Neuroimmunol 136:130–139

Bracci-Laudiero L, Aloe L, Caroleo MC, Buanne P, Costa N, Starace G, Lundeberg T (2005) Endogenous NGF regulates CGRP expression in human monocytes, and affects HLA-DR and CD86 expression and IL-10 production. Blood 106:3507–3514

Britto JM, Tannahill D, Keynes RJ (2000) Life, death and Sonic hedgehog. BioEssays 22:499–502

Brodie C, Gelfand EW (1992) Functional nerve growth factor receptors on human B-lymphocytes. Interaction with IL-2. J Immunol 148:3492–3497

Brodie C, Gelfand EW (1994) Regulation of immunoglobulin production by nerve growth factor: comparison with anti-CD40. J Neuroimmunol 52:87–96

Bruno MA, Cuello AC (2006) Activity-dependent release of precursor nerve growth factor, conversion to mature nerve growth factor, and its degradation by a protease cascade. Proc Natl Acad Sci USA 103:6735–6740

Bruno MA, Leon WC, Fragoso G, Mushynski WE, Almazan G, Cuello AC (2009) Amyloid beta-induced nerve growth factor dysmetabolism in Alzheimer disease. J Neuropathol Exp Neurol 68:857–869

Bullock ED, Johnson EM (1996) Nerve growth factor induces the expression of certain cytokines genes and bcl2 in mast cells: potential role in survival promotion. J Biol Chem 271:27500–27508

Cabal-Hierro L, Lazo PS (2012) Signal transduction by tumor necrosis factor receptors. Cell Signal 24:1297–1305

Caggiula M, Batocchi AP, Frisullo G, Angelucci F, Patanella AK, Sancricca C, Nociti V, Tonali PA, Mirabella M (2005) Neurotrophic factors and clinical recovery in relapsing-remitting multiple sclerosis. Scand J Immunol 62:176–182

Campenot RB (1977) Local control of neurite development by nerve growth factor. Proc Natl Acad Sci USA 74:4516–4519

Campenot RB (1982a) Development of sympathetic neurons in compartmentalized cultures. II. Local control of neurite survival by nerve growth factor. Dev Biol 93:13–21

Campenot RB (1982b) Development of sympathetic neurons in nerve growth factor. Dev Biol 93:1–12

Caneva L, Soligo D, Cattoretti G, De Harven E, Deliliers GL (1995) Immuno-electron microscopy characterization of human bone marrow stromal cells with anti-NGFR antibodies. Blood Cells Mol Dis 21:73–85

Carlson SL, Albers KM, Beiting DJ, Parish M, Conner JM, Davis BM (1995) NGF modulates sympathetic innervation of lymphoid tissues. J Neurosci 15:5892–5899

Caroleo MC, Costa N, Bracci-Laudiero L, Aloe L (2001) Human monocyte/macrophages activate by exposure to LPS overexpress NGF and NGF receptors. J Neuroimmunol 113:193–201

Carriero F, Campioni N, Cardinali B, Pierandrei-Amaldi P (1991) Structure and expression of the nerve growth factor gene in Xenopus oocytes and embryos. Mol Reprod Dev 29:313–322

Cattoretti G, Schirò R, Orazi A, Soligo D, Colombo MP (1993) Bone marrow stroma in humans: anti-nerve growth factor receptor antibodies selectively stain reticular cells in vivo and in vitro. Blood 81:1726–1738

Chao MV, Hempstead BL (1995) p75 and Trk: a two-receptor system. Trends Neurosci 18:321–326

Chevalier S, Praloran V, Smith C, Mac Grogan D, Ip NY, Yancopoulos GD, Brachet P, Pouplard A, Gascan H (1994) Expression and functionality of the trkA proto-oncogene product/NGF receptor in undifferentiated hematopoietic cells. Blood 83:1479–1485

Ciriaco E, Dall'Aglio C, Hannestad J, Huerta JJ, Laurà R, Germanà G, Vega JA (1996) Localization of TrK neurotrophin receptor-like proteins in avian primary lymphoid organs: thymus and bursa of Fabricius. J Neuroimmunol 69:73–83

Colafrancesco V, Villoslada P (2011) Targeting NGF pathway for developing neuroprotective therapies for multiple sclerosis and other neurological diseases. Arch Ital Biol 14:183–192

Cooper JD, Lindholm D, Sofroniew MV (1994) Reduced transport of 125I-NGF by cholinergic neurons and downregulated TrkA expression in the medial septum of aged rats. Neuroscience 62:625–629

Cosker KE, Courchesne SL, Segal RA (2008) Action in the axon: generation and transport of signaling endosomes. Curr Opin Neurobiol 18:270–275

Cotrina ML, González-Hoyuela M, Barbas JA, Rodríguez-Tébar A (2000) Programmed cell death in the developing somites is promoted by nerve growth factor via its p75(NTR) receptor. Dev Biol 228:326–336

Cuello AC, Ferretti MT, Iulita MF (2012) Preplaque ('preclinical') Aβ-induced inflammation and nerve growth factor deregulation in transgenic models of Alzheimer's disease-like amyloid pathology. Neurodegener Dis 10:104–107

Davies AM (1997) Neurotrophin switching: where does it stand? Curr Opin Neurobiol 7:110–118

DeLacalle S, Cooper JD, Svendsen CN, Dunnett SB, Sofroniew MV (1996) Reduced retrograde labeling with fluorescent tracer accompanies neuronal atrophy of basal forebrain cholinergic neurons in aged rats. Neuroscience 75:19–27

Deppmann CD, Mihalas S, Sharma N, Lonze BE, Niebur E, Ginty DD (2008) A model for neuronal competition during development. Science 320:369–373

De Simone R, Micera A, Tirassa P, Aloe L (1996) mRNA for NGF and p75 in the central nervous system of rats affected by experimental allergic encephalomyelitis. Neuropathol Appl Neurobiol 22:54–59

Dissen GA, Garcia-Rudaz C, Ojeda SR (2009) Role of neurotrophic factors in early ovarian development. Semin Reprod Med 27:24–31

Dissen GA, Hirshfield AN, Malamed S, Ojeda SR (1995) Expression of neurotrophins and their receptors in the mammalian ovary is developmentally regulated: changes at the time of folliculogensis. Endocrinology 136:4681–4692

Domeniconi M, Hempstead BL, Chao MV (2007) Pro-NGF secreted by astrocytes promotes motor neuron cell death. Mol Cell Neurosci 34:271–279

D'Onofrio M, Paoletti F, Arisi I, Brandi R, Malerba F, Fasulo L, Cattaneo A (2011) NGF and proNGF regulate functionally distinct mRNAs in PC12 cells: an early gene expression profiling. PLoS ONE 6:e20839. doi:10.1371/journal.pone.0020839

Ebadi M, Bashir RM, Heidrick ML, Hamada FM, Refaey HE, Hamed A, Helal G, Baxi MD, Cerutis DR, Lassi NK (1997) Neurotrophins and their receptors in nerve injury and repair. Neurochem Int 30:347–374

Ebendal T (1992) Function and evolution in the NGF family and its receptors. J Neurosci Res 32:461–470

Ehrhard PB, Erb P, Graumann U, Otten U (1993) Expression of nerve growth factor and nerve growth factor receptor tyrosine kinase Trk in activated CD4-positive T-cell clones. Proc Natl Acad Sci USA 90:10984–10988

Elenkov IJ, Wilder RL, Chrousos GP, Vizi ES (2000) The sympathetic nerve—an integrative interface between two supersystems: the brain and the immune system. Pharmacol Rev 52:595–638

Ernfors P, Hallböök F, Ebendal T, Shooter EM, Radeke MJ, Misko TP, Persson H (1988) Developmental and regional expression of beta-nerve growth factor receptor mRNAin the chick and rat. Neuron 1:983–996

Esposito D, Patel P, Stephens RM, Perez P, Chao MV, Kaplan DR, Hempstead BL (2001) The cytoplasmic and transmembrane domain of the p75 and TrkA receptors regulate high affinity binding to nerve growth factor. J Biol Chem 276:32687–32695

Fahnestock M, Michalski B, Xu B, Coughlin MD (2001) The precursor pro-nerve growth factor is the predominant form of nerve growth factor in brain and is increased in Alzheimer's disease. Mol Cell Neurosci 18:210–220

Fahnestock M, Yu G, Coughlin MD (2004) ProNGF: a neurotrophic or an apoptotic molecule? Prog Brain Res 146:101–110

Ferrari MP, Mantelli F, Sacchetti M, Antonangeli MI, Cattani F, D'Anniballe G, Sinigaglia F, Ruffini PA, Lambiase A (2014) Safety and pharmacokinetics of escalating doses of human recombinant nerve growth factor. BioDrugs 28:275–283

Ferretti MT, Bruno MA, Ducatenzeiler A, Klein WL, Cuello AC (2012) Intracellular Aβ-oligomers and early inflammation in a model of Alzheimer's disease. Neurobiol Aging 33:1329–1342

Flügel A, Matsumuro K, Neumann H, Klinkert WE, Birnbacher R, Lassmann H, Otten U, Wekerle H (2001) Anti-inflammatory activity of nerve growth factor in experimental autoimmune encephalomyelitis: inhibition of monocyte transendothelial migration. Eur J Immunol 31:11–22

Frade JM, Bovolenta P, Rodríguez-Tébar A (1999) Neurotrophins and other growth factors in the generation of retinal neurons. Microsc Res Tech 45:243–251

Fukunaga K, Ishigami T, Kawano T (2005) Transcriptional regulation of neuronal genes and its effect on neural functions: expression and function of forkhead transcription factors in neurons. J Pharmacol Sci 98:205–211

Garcia-Suárez O, Germanà A, Hannestad J, Ciriaco E, Silos-Santiago I, Germanà G, Vega JA (2001) Involvement of the NGF receptors (Trka and p75lngfr) in the development and maintenance of the thymus. Ital J Anat Embryol 106:279–285

Gibbs BF, Zillikens D, Grabbe J (2005) Nerve growth factor influences IgE-mediated human basophil activation: functional properties and intracellular mechanisms compared with IL-3. Int Immunopharmacol 5:735–747

Goedert M, Fine A, Dawbarn D, Wilcock GK, Chao MV (1989) Nerve growth factor receptor mRNA distribution in human brain: normal levels in basal forebrain in Alzheimer's disease. Brain Res Mol Brain Res 5:1–7

Greene LA (1978) Nerve growth factor prevents the death and stimulates the neuronal differentiation of clonal PC12 pheochromocytoma cells in serum-free medium. J Cell Biol 78:747–755

Hallböök F (1999) Evolution of the vertebrate neurotrophin and Trk receptor gene families. Curr Opin Neurobiol 9:616–621

Hamada A, Watanabe N, Ohtomo H, Matsuda H (1996) Nerve growth factor enhances survival and cytotoxic activity of human eosinophils. Br J Haematol 93:299–302

Harrington AW, St. Hillaire C, Zweifel LS, Glebova NO, Philippidou P, Halegoua S, Ginty DD (2011) Recruitment of actin modifiers to TrkA endosomes governs retrograde NGF signaling and survival. Cell 146:421–434

Harrington AW, Ginty DD (2013) Long-distance retrograde neurotrophic factor signalling in neurons. Nat Rev Neurosci 14:177–187

Hehlgans T, Pfeffer K (2005) The intriguing biology of the tumour necrosis factor/tumour necrosis factor receptor superfamily: players, rules and the games. Immunology 115:1–20

Higgins GA, Koh S, Chen KS, Gage FH (1989) NGF induction of NGF receptor gene expression and cholinergic neuronal hypertrophy within the basal forebrain of the adult rat. Neuron 3:247–256

Howe CL, Mobley WC (2005) Long-distance retrograde neurotrophic signaling. Curr Opin Neurobiol 15:40–48

Huang EJ, Reichardt LF (2003) Trk receptors: roles in neuronal signal transduction. Annu Rev Biochem 72:609–642

Huh CY, Danik M, Manseau F, Trudeau LE, Williams S (2008) Chronic exposure to nerve growth factor increases acetylcholine and glutamate release from cholinergic neurons of the rat medial septum and diagonal band of Broca via mechanisms mediated by p75NTR. J Neurosci 28:1404–1409

Ibáñez CF (2002) Jekyll-Hyde neurotrophins: the story of proNGF. Trends Neurosci 25:284–286

Inanç B, Elçin AE, Elçin YM (2008) Human embryonic stem cell differentiation on tissue engineering scaffolds: effects of NGF and retinoic acid induction. Tissue Eng Part A 14:955–964

Indo Y, Tsuruta M, Hayashida Y, Karim MA, Ohta K, Kawano T, Mitsubuchi H, Tonoki H, Awaya Y, Matsuda I (1996) Mutations in the TRKA/NGF receptor gene in patients with congenital insensitivity to pain with anhidrosis. Nat Genet 13:485–488

Iulita MF, Cuello AC (2014) Nerve growth factor metabolic dysfunction in Alzheimer'sdisease and Down syndrome. Trends Pharmacol Sci 35:338–348

Kannan Y, Ushio H, Koyama H, Okada M, Oikawa M, Yoshihara T, Kaneko M, Matsuda H (1991) Nerve growth factor enhances survival, phagocytosis, and superoxide production of murine neutrophils. Blood 77:1320–1325

Kaplan DR, Martin-Zanca D, Parada LF (1991) Tyrosine phosphorylation and tyrosine kinase activity of the trk proto-oncogene product induced by NGF. Nature 350:158–160

Katoh-Semba R, Semba R, Takeuchi IK, Kato K (1998) Age-related changes in levels of brain-derived neurotrophic factor in selected brain regions of rats, normal mice and senescence-accelerated mice: a comparison to those of nerve growth factor and neurotrophin-3. Neurosci Res 31:227–234

Kawamoto K, Okada T, Kannan Y, Ushio H, Matsumoto M, Matsuda H (1995) Nerve growth factor prevents apoptosis of rat peritoneal mast cells through the trk proto-oncogene receptor. Blood 86:4638–4644

Kerr B, Garcia-Rudaz C, Dorfman M, Paredes A, Ojeda SR (2009) NTRK1 and NTRK2 receptors facilitate follicle assembly and early follicular development in the mouse ovary. Reproduction 138:131–140

Ketschek A, Gallo G (2010) Nerve growth factor induces axonal filopodia through localized microdomains of phosphoinositide 3-kinase activity that drive the formation of cytoskeletal precursors to filopodia. J Neurosci 30:12185–12197

Kimata H, Yoshida A, Ishioka C, Kusunoki T, Hosoi S, Mikawa H (1991) Nerve growth factor specifically induces human IgG4 production. Eur J Immunol 21:137–141

Kobayashi H, Gleich GJ, Butterfield JH, Kita H (2002) Human eosinophils produce neurotrophins and secrete nerve growth factor on immunologic stimuli. Blood 99:2214–2220

Kraemer BR, Yoon SO, Carter BD (2014) The biological functions and signaling mechanisms of the p75 neurotrophin receptor. Handb Exp Pharmacol 220:121–164

Kuruvilla R, Ye H, Ginty DD (2000) Spatially and functionally distinct roles of the PI3-K effector pathway during NGF signaling in sympathetic neurons. Neuron 27:499–512

la Sala A, Corinti S, Federici M, Saragovi HU, Girolomoni G (2000) Ligand activation of nerve growth factor receptor TrkA protects monocytes from apoptosis. J Leukoc Biol 68:104–110

Ladle DR, Pecho-Vrieseling E, Arber S (2007) Assembly of motor circuits in the spinal cord: driven to function by genetic and experience-dependent mechanisms. Neuron 56:270–283

Lambiase A, Bonini S, Manni L, Ghinelli E, Tirassa P, Rama P, Aloe L (2002) Intraocular production and release of nerve growth factor after iridectomy. Invest Ophthalmol Vis Sci 43:2334–2340

Lanave C, Colangelo AM, Saccone C, Alberghina L (2007) Molecular evolution of the neurotrophin family members and their Trk receptors. Gene 394:1–12

Laurenzi MA, Barbany G, Timmusk T, Lindgren JA, Persson H (1994) Expression of mRNA encoding neurotrophins and neurotrophin receptors in rat thymus, spleen tissue and immunocompetent cells. Eur J Biochem 223:733–741

Lee HW, Na YJ, Jung PK, Kim MN, Kim SM, Chung JS, Kim BS, Kim JB, Moon JO, Yoon S (2008) Nerve growth factor stimulates proliferation, adhesion and thymopoietic cytokine expression in mouse thymic epithelial cells in vitro. Regul Pept 147:72–81

Lee R, Kermani P, Teng KK, Hempstead BL (2001) Regulation of cell survival by secreted proneurotrophins. Science 294:1945–1948

Levi-Montalcini R, Hamburger V (1951) Selective growth stimulating effects of mouse sarcoma on the sensory and sympathetic nervous system of the chick embryo. J Exp Zool 116:321–361

Levi-Montalcini R, Booker B (1960) Destruction of the sympathetic ganglia in mammals by an antiserum to a nerve-growth protein. Proc Natl Acad Sci USA 46:384–391

Levi-Montalcini R, Cohen S (1960) Effects of the extract of the mouse submaxillary salivary glands on the sympathetic system of mammals. Ann N Y Acad Sci 85:324–341

Levi-Montalcini R, Angeletti PU (1968) Nerve growth factor. Physiol Rev 48:534–569

Levi-Montalcini R (1987) The nerve growth factor 35 years later. Science 237:1154–1162

Levi-Montalcini R (1997) The Saga of the nerve growth factor, preliminary studies, discovery, further development. In: World Scientific Series in 20th Century Biology (vol 3). World Scientific, Singapore

Li Z, Beutel G, Rhein M, Meyer J, Koenecke C, Neumann T, Yang M, Krauter J, von Neuhoff N, Heuser M, Diedrich H, Göhring G, Wilkens L, Schlegelberger B, Ganser A, Baum C (2009) High-affinity neurotrophin receptors and ligands promote leukemogenesis. Blood 113:2028–2037

Lindsay RM, Harmar AJ (1989) Nerve growth factor regulates expression of neuropeptide genes in adult sensory neurons. Nature 337:362–364

Lindvall O, Kokaia Z, Bengzon J, Elmr E, Kokaia M (1994) Neurotrophins and brain insults. Trends Neurosci 17:490–496

Linher-Melville K, Li J (2013) The roles of glial cell line-derived neurotrophic factor, brain-derived neurotrophic factor and nerve growth factor during the final stage of folliculogenesis: a focus on oocyte maturation. Reproduction 145:R43–R54

Linker R, Gold R, Luhder F (2009) Function of neurotrophic factors beyond the nervous system: inflammation and autoimmune demyelination. Crit Rev Immunol 29:43–68

Lomen-Hoerth C, Shooter EM (1995) Widespread neurotrophin receptor expression in the immune system and other nonneuronal rat tissues. J Neurochem 64:1780–1789

Luo W, Wickramasinghe SR, Savitt JM, Griffin JW, Dawson TM, Ginty DD (2007) A hierarchical NGF signaling cascade controls Ret-dependent and Ret-independent events during development of nonpeptidergic DRG neurons. Neuron 54:739–754

Madden KS, Felten DL (1995) Experimental basis for neuronal-immune interactions. Physiol Rev 75:77–106

Majdan M, Walsh GS, Aloyz R, Miller FD (2001) TrkA mediates developmental sympathetic neuron survival in vivo by silencing an ongoing p75NTR-mediated death signal. J Cell Biol 155:1275–1285

Mamidipudi V, Li X, Wooten MW (2002) Identification of interleukin 1 receptor-associated kinase as a conserved component in the p75-neurotrophin receptor activation of nuclear factor-kappa B. J Biol Chem 277:28010–28018

Mandai K, Guo T, St Hillaire C, Meabon JS, Kanning KC, Bothwell M, Ginty DD (2009) LIG family receptor tyrosine kinase-associated proteins modulate growth factor signals during neural development. Neuron 63:614–627

Manca A, Capsoni S, Di Luzio A, Vignone D, Malerba F, Paoletti F, Brandi R, Arisi I, Cattaneo A, Levi-Montalcini R (2012) Nerve growth factor regulates axial rotation during early stages of chick embryo development. Proc Natl Acad Sci USA 109:2009–2014

Marmigère F, Carroll P (2014) Neurotrophin signalling and transcription programmes interactions in the development of somatosensory neurons. Handb Exp Pharmacol 220:329–353

Masoudi R, Ioannou MS, Coughlin MD, Pagadala P, Neet KE, Clewes O, Allen SJ, Dawbarn D, Fahnestock M (2009) Biological activity of nerve growth factor precursor is dependent upon relative levels of its receptors. J Biol Chem 284:18424–18433

Matsuda H, Coughlin MD, Bienenstock J, Denburg JA (1988) Nerve growth factor promotes human hemapoietic colony growth and differentiation. Proc Natl Acad Sci USA 85:6508–6512

Matsuda H, Kannan Y, Ushio H, Kiso Y, Kanemoto T, Suzuki H, Kitamura Y (1991) Nerve growth factor induces development of connective tissue-type mast cells in vitro from murine bone marrow cells. J Exp Med 174:7–14

Mayor R, Theveneau E (2013) The neural crest. Development 140:2247–2251

McDonald NQ, Hendrickson WA (1993) A structural superfamily of growth factors containing a cystine knot motif. Cell 73:421–424

McDonald NQ, Lapatto R, Murray-Rust J, Gunning J, Wlodawer A, Blundell TL (1991) New protein fold revealed by a 2.3-Å resolution crystal structure of nerve growth factor. Nature 354:411–414

Melamed I, Levy J, Parvari R, Gelfand EW (2004) A novel lymphocyte signaling defect: trk A mutation in the syndrome of congenital insensitivity to pain and anhidrosis (CIPA). J Clin Immunol 24:441–448

Miller FD, Kaplan DR (2001) Neurotrophin signalling pathways regulating neuronal apoptosis. Cell Mol Life Sci 58:1045–1053

Mok SA, Campenot RB (2007) A nerve growth factor-induced retrograde survival signal mediated by mechanisms downstream of TrkA. Neuropharmacology 52:270–278

Moscatelli I, Pierantozzi E, Camaioni A, Siracusa G, Campagnolo L (2009) p75 neurotrophin receptor is involved in proliferation of undifferentiated mouse embryonic stem cells. Exp Cell Res 315:3220–3232

Niewiadomska G, Baksalerska-Pazera M, Riedel G (2009) The septo-hippocampal system, learning and recovery of function. Prog Neuropsychopharmacol Biol Psychiatry 33:791–805

Niewiadomska G, Mietelska-Porowska A, Mazurkiewicz M (2011) The cholinergic system, nerve growth factor and the cytoskeleton. Behav Brain Res 221:515–526

Noga O, Englmann C, Hanf G, Grützkau A, Guhl S, Kunkel G (2002) Activation of the specific neurotrophin receptors TrkA, TrkB and TrkC influences the function of eosinophils. Clin Exp Allergy 32:1348–1354

Nykjaer A, Lee R, Teng KK, Jansen P, Madsen P, Nielsen MS, Jacobsen C, Kliemannel M, Schwarz E, Willnow TE, Hempstead BL, Petersen CM (2004) Sortilin is essential for proNGF induced neuronal cell death. Nature 427:843–848

Nykjaer A, Willnow TE (2012) Sortilin: a receptor to regulate neuronal viability and function. Trends Neurosci 35:261–270

Nykjaer A, Willnow TE, Petersen CM (2005) p75NTR—live or let die. Curr Opin Neurobiol 15:49–57

Øren A, Falk K, Rötzschke O, Bechmann I, Nitsch R, Gimsa U (2004) Production of neuroprotective NGF in astrocyte-T helper cell cocultures is upregulated following antigen recognition. J Neuroimmunol 149:59–65

Parvaneh Tafreshi A (2006) Nerve growth factor prevents demyelination, cell death and progression of the disease in experimental allergic encephalomyelitis. Iran J Allergy Asthma Immunol 5:177–181

Patel TD, Kramer I, Kucera J, Niederkofler V, TM Jesell, Arber S, Snider WD (2003) Peripheral NT3 signaling is required for ETS protein expression and central patterning of proprioceptive sensory afferents. Neuron 38:403–416

Pazyra-Murphy MF, Hans A, Courchesne SL, Karch C, Cosker KE, Heerssen HM, Watson FL, Kim T, Greenberg ME, Segal RA (2009) A retrograde neuronal survival response: target-derived neurotrophins regulate MEF2D and bcl-w. J Neurosci 29:6700–6709

Pezzati P, Stanisz AM, Marshall JS, Bienenstock J, Stead RH (1992) Expression of nerve growth factor receptor immunoreactivity on follicular dendritic cells from human mucosa associated lymphoid tissues. Immunology 76:485–490

Prencipe G, Minnone G, Strippoli R, De Pasquale L, Petrini S, Caiello I, Manni L, De Benedetti F, Bracci-Laudiero L (2014) Nerve growth factor downregulates inflammatory response in human monocytes through TrkA. J Immunol 192:3345–3354

Purves D (1975) Functional and structural changes in mammalian sympathetic neurones following interruption of their axons. J Physiol 252:429–463

Pyle AD, Lock LF, Donovan PJ (2006) Neurotrophins mediate human embryonic stem cell survival. Nat Biotechnol 24:344–350

Ralainirina N, Brons NH, Ammerlaan W, Hoffmann C, Hentges F, Zimmer J (2010) Mouse natural killer (NK) cells express the nerve growth factor receptor TrkA, which is dynamically regulated. PLoS ONE 5:e15053

Reichardt LF (2006) Neurotrophin-regulated signalling pathways. Philos Trans R Soc Lond B Biol Sci 361:1545–1564

Renné C, Minner S, Küppers R, Hansmann ML, Bräuninger A (2008) Autocrine NGFbeta/TRKA signalling is an important survival factor for Hodgkin lymphoma derived cell lines. Leuk Res 32:163–167

Resende TP, Ferreira M, Teillet MA, Tavares AT, Andrade RP, Palmeirim I (2010) Sonic hedgehog in temporal control of somite formation. Proc Natl Acad Sci USA 107:12907–12912

Rezaee F, Rellick SL, Piedimonte G, Akers SM, O'Leary HA, Martin K, Craig MD, Gibson LF (2010) Neurotrophins regulate bone marrow stromal cell IL-6 expression through the MAPK pathway. PLoS ONE 5(3):e9690. doi:10.1371/journal.pone.0009690

Rifkin JT, Todd VJ, Anderson LW, Lefcort F (2000) Dynamic expression of neurotrophin receptors during sensory neuron genesis and differentiation. Dev Biol 227:465–480

Roberti G, Mantelli F, Macchi I, Massaro-Giordano M, Centofanti M (2014) Nerve growth factor modulation of retinal ganglion cell physiology. J Cell Physiol 229:1130–1133

Rodríguez-Tébar A, Dechant G, Barde YA (1991) Neurotrophins: structural relatedness and receptor interactions. Philos Trans R Soc Lond B Biol Sci 331:255–258

Roux PP, Barker PA (2002) Neurotrophin signaling through the p75 neurotrophin receptor. Prog Neurobiol 67:203–233

Saab CY, Shamaa F, El Sabban ME, Safieh-Garabedian B, Jabbur SJ, Saadé NE (2009) Transient increase in cytokines and nerve growth factor in the rat dorsal root ganglia after nerve lesion and peripheral inflammation. J Neuroimmunol 208:94–103

Salehi AH, Roux PP, Kubu CJ, Zeindler C, Bhakar A, Tannis LL, Verdi JM, Barker PA (2000) NRAGE, a novel MAGE protein, interacts with the p75 neurotrophin receptor and facilitates nerve growth factor-dependent apoptosis. Neuron 27:279–288

Samah B, Porcheray F, Gras G (2008) Neurotrophins modulate monocyte chemotaxis without affecting macrophage function. Clin Exp Immunol 151:476–486

Sato Y, Tsuboi Y, Kurosawa H, Sugita K, Eguchi M (2004) Anti-apoptotic effect of nerve growth factor is lost in congenital insensitivity to pain with anhidrosis (CIPA) B lymphocytes. J Clin Immunol 24:302–308

Sawada J, Itakura A, Tanaka A, Furusaka T, Matsuda H (2000) Nerve growth factor functions as a chemoattractant for mast cells through both mitogen-activated protein kinase and phosphatidylinositol 3-kinase signalling pathways. Blood 95:2052–2058

Schliebs R, Arendt T (2011) The cholinergic system in aging and neuronal degeneration. Behav Brain Res 221:555–563

Schuldiner M, Yanuka O, Itskovitz-Eldor J, Melton DA, Benvenisty N (2000) Effects of eight growth factors on the differentiation of cells derived from human embryonic stem cells. Proc Natl Acad Sci USA 97:11307–11312

Schulte-Herbrüggen O, Braun A, Rochlitzer S, Jockers-Scherübl MC, Hellweg R (2007) Neurotrophic factors–a tool for therapeutic strategies in neurological, neuropsychiatric and neuroimmunological diseases? Curr Med Chem 14:2318–2329

Scott SA, Mufson EJ, Weingartner J, Skau KA, Crutcher KA (1995) Nerve growth factor in Alzheimer's disease: increased levels throughout the brain coupled with declines in nucleus basalis. J Neurosci 15:6213–6221

Seidah NG, Benjannet S, Pareek S, Savaria D, Hamelin J, Goulet B, Laliberte J, Lazure C, Chrétien M, Murphy RA (1996) Cellular processing of the nerve growth factor precursor by the mammalian pro-protein convertases. Biochem J 314:951–960

Sharma N, Deppmann CD, Harrington AW, St Hillaire C, Chen ZY, Lee FS, Ginty DD (2010) Long-distance control of synapse assembly by target-derived NGF. Neuron 67:422–434

Shi Y, Jin Y, Guo W, Chen L, Liu C, Lv X (2012) Blockage of nerve growth factor modulates T cell responses and inhibits allergic inflammation in a mouse model of asthma. Inflamm Res 61:1369–1378

Snider WD (1994) Functions of the neurotrophins during nervous system development: what the knockouts are teaching us. Cell 77:627–638

Sofroniew MV, Howe CL, Mobley WC (2001) Nerve growth factor signaling, neuroprotection, and neural repair. Annu Rev Neurosci 24:1217–1281

Soligo M, Protto V, Florenzano F, Bracci-Laudiero L, De Benedetti F, Chiaretti A, Manni L (2015) The mature/pro nerve growth factor ratio is decreased in the brain of diabetic rats: Analysis by ELISA methods. Brain Res. doi:10.1016/j.brainres.2015.08.005

Susaki Y, Shimizu S, Katakura K, Watanabe N, Kawamoto K, Matsumoto M, Tsudzuki M, Furusaka T, Kitamura Y, Matsuda H (1996) Functional properties of murine macrophages promoted by nerve growth factor. Blood 88:4630–4637

Teng HK, Teng KK, Lee R, Wright S, Tevar S, Almeida RD, Kermani P, Torkin R, Chen ZY, Lee FS, Kraemer RT, Nykjaer A, Hempstead BL (2005) ProBDNF induces neuronal apoptosis via activation of a receptor complex of p75NTR and sortilin. J Neurosci 25:5455–5463

Thoenen H, Angeletti PU, Levi-Montalcini R, Kettler R (1971) Selective induction by nerve growth factor of tyrosine hydroxylase and dopamine- -hydroxylase in the rat superior cervical ganglia. Proc Natl Acad Sci USA 68:1598–1602

Tomellini E, Lagadec C, Polakowska R, Le Bourhis X (2014) Role of p75 neurotrophin receptor in stem cell biology: more than just a marker. Cell Mol Life Sci 71:2467–2481

Tonchev AB, Boneva NB, Kaplamadzhiev DB, Kikuchi M, Mori Y, Sahara S, Yamashima T (2008) Expression of neurotrophin receptors by proliferating glia in postischemic hippocampal CA1 sector of adult monkeys. J Neuroimmunol 205:20–24

Torcia M, Bracci-Laudiero L, Lucibello M, Nencioni L, Labardi D, Rubartelli A, Cozzolino F, Aloe L, Garaci E (1996) Nerve growth factor is an autocrine survival factor for memory B lymphocytes. Cell 85:345–356

Torcia M, De Chiara G, Nencioni L, Ammendola S, Labardi D, Lucibello M, Rosini P, Marlier LN, Bonini P, Dello Sbarba P, Palamara AT, Zambrano N, Russo T, Garaci E, Cozzolino F (2001) Nerve growth factor inhibits apoptosis in memory B lymphocytes via inactivation of p38 MAPK, prevention of Bcl-2 phosphorylation, and cytochrome c release. J Biol Chem 276:39027–39036

Tsuda T, Wong D, Dolovich J, Bienenstock J, Marshall J, Denburg JA (1991) Synergistic effects of nerve growth factor and granulocytes-macrophage colony-stimulating factor on human basophilic cell differentiation. Blood 77:971–979

Tuszynski MH, Blesch A (2004) Nerve growth factor: from animal models of cholinergic neuronal degeneration to gene therapy in Alzheimer's disease. Prog Brain Res 146:441–449

Uren RT, Turnley AM (2014) Regulation of neurotrophin receptor (Trk) signaling: suppressor of cytokine signaling 2 (SOCS2) is a new player. Front Mol Neurosci 7:39. doi:10.3389/fnmol.2014.00039

Vantini G, Schiavo N, Di Martino A, Polato P, Triban C, Callegaro L, Toffano G, Leon A (1989) Evidence for a physiological role of nerve growth factor in the central nervous system. Neuron 3:267–273

Verge VM, Richardson PM, Wiesenfeld-Hallin Z, Hökfelt T (1995) Differential influence of nerve growth factor on neuropeptide expression in vivo: a novel role in peptide suppression in adult sensory neurons. J Neurosci 15:2081–2096

Villoslada P, Hauser SL, Bartke I, Unger J, Heald N, Rosenberg D, Cheung SW, Mobley WC, Fisher S, Genain CP (2000) Human nerve growth factor protects common marmosets against autoimmune encephalomyelitis by switching the balance of T helper cell type 1 and 2 cytokines within the central nervous system. J Exp Med 2000(191):1799–1806

Vogel KS (1993) Development of trophic interactions in the vertebrate peripheral nervous system. Mol Neurobiol 7:363–382

Voyvodic JT (1989) Peripheral target regulation of dendritic geometry in the rat superior cervical ganglion. J Neurosci 9:1997–2010

Wang H, Wang R, Thrimawithana T, Little PJ, Xu J, Feng ZP, Zheng W (2014) The nerve growth factor signaling and its potential as therapeutic target for glaucoma. Biomed Res Int 2014:759473. doi:10.1155/2014/759473

Wang KC, Kim JA, Sivasankaran R, Segal R, He Z (2002) P75 interacts with the Nogo receptor as a co-receptor for Nogo, MAG and OMgp. Nature 420:74–78

Wang X, Bauer JH, Li Y, Shao Z, Zetoune FS, Cattaneo E, Vincenz C (2001) Characterization of a p75(NTR) apoptotic signaling pathway using a novel cellular model. J Biol Chem 276:33812–33820

Wehrman T, He X, Raab B, Dukipatti A, Blau H, Garcia KC (2007) Structural and mechanistic insights into nerve growth factor interactions with the TrkA and p75 receptors. Neuron 5:325–338

Welker P, Grabbe J, Gibbs B, Zuberbier T, Henz BM (2000) Nerve growth factor-beta induces mast-cell marker expression during in vitro culture of human umbilical cord blood cells. Immunology 99:418–426

Wheeler EF, Gong H, Grimes R, Benoit D, Vazquez L (1998) p75NTR and Trk receptors are expressed in reciprocal patterns in a wide variety of non-neural tissues during rat embryonic development, indicating independent receptor functions. J Comp Neurol 391:407–428

Willnow TE, Petersen CM, Nykjaer A (2008) VPS10P-domain receptors—regulators of neuronal viability and function. Nat Rev Neurosci 9:899–909

Wilson YM, Richards KL, Ford-Perriss ML, Panthier JJ, Murphy M (2004) Neural crest cell lineage segregation in the mouse neural tube. Development 131:6153–6162

Xia Y, Chen BY, Sun XL, Duan L, Gao GD, Wang JJ, Yung KK, Chen LW (2013) Presence of proNGF-sortilin signaling complex in nigral dopamine neurons and its variation in relation to aging, lactacystin and 6-OHDA insults. Int J Mol Sci 14:14085–14104

Yamamoto M, Sobue G, Yamamoto K, Terao S, Mitsuma T (1996) Expression of mRNAs for neurotrophic factors (NGF, BDNF, NT-3, and GDNF) and their receptors (p75NGFR, trkA, trkB, and trkC) in the adult human peripheral nervous system and nonneural tissues. Neurochem Res 21:929–938

Yamashita T, Tucker KL, Barde YA (1999) Neurotrophin binding to the p75 receptor modulates Rho activity and axonal outgrowth. Neuron 24:585–593

Yan Q, Johnson EM Jr (1988) An immunohistochemical study of the nerve growth factor receptor in developing rats. J Neurosci 8:3481–3498

Yan Q, Johnson EM Jr (1989) Immunoistochemical localization and biochemical characterization of nerve growth receptor in adult rat brain. J Comp Neurol 290:585–598

Yao L, Zhang D, Bernd P (1994) The onset of neurotrophin and trk mRNA expression in early embryonic tissues of the quail. Dev Biol 165:727–730

Ye H, Kuruvilla R, Zweifel LS, Ginty DD (2003) Evidence in support of signaling endosome-based retrograde survival of sympathetic neurons. Neuron 39:57–68

Ye X, Mehlen P, Rabizadeh S, VanArsdale T, Zhang H, Shin H, Wang JJ, Leo E, Zapata J, Hauser CA, Reed JC, Bredesen DE (1999) TRAF family proteins interact with the common neurotrophin receptor and modulate apoptosis induction. J Biol Chem 274:30202–30208

Yeiser EC, Rutkoski NJ, Naito A, Inoue J, Carter BD (2004) Neurotrophin signaling through the p75 receptor is deficient in traf6-/- mice. J Neurosci 24:10521–10529

Zhang D, Yao L, Bernd P (1996) Expression of neurotrophin trk and p75 receptors in quail embryos undergoing gastrulation and neurulation. Dev Dyn 205:150–161

Zhang HT, Li LY, Zou XL, Song XB, Hu YL, Feng ZT, Wang TT (2007) Immunohistochemical distribution of NGF, BDNF, NT-3, and NT-4 in adult rhesus monkey brains. J Histochem Cytochem 55:1–19

Zhou FQ, Zhou J, Dedhar S, Wu YH, Snider WD (2004) NGF-Induced axon growth is mediated by localized inactivation of GSK-3b and functions of the microtubule plus end binding protein APC. Neuron 42:897–912

Part II
Neuroteratogens to Model Human Disorders

Disrupted Circadian Rhythm as a Common Player in Developmental Models of Neuropsychiatric Disorders

Eva M. Marco, Elena Velarde, Ricardo Llorente
and Giovanni Laviola

Abstract The environment in which individuals develop and mature is critical for their physiological and psychological outcome; in particular, the intrauterine environment has reached far more clinical relevance given its potential influence on shaping brain function and thus mental health. Gestational stress and/or maternal infection during pregnancy has been related with an increased incidence of neuropsychiatric disorders, including depression and schizophrenia. In this framework, the use of animal models has allowed a formal and deep investigation of causal determinants. Despite disruption of circadian clocks often represents a hallmark of several neuropsychiatric disorders, the relationship between disruption of brain development and the circadian system has been scarcely investigated. Nowadays, there is an increasing amount of studies suggesting a link between circadian system malfunction, early-life insults and the appearance of neuropsychiatric diseases at adulthood. Here, we briefly review evidence from clinical literature and animal models suggesting that the exposure to prenatal insults, i.e. severe gestational stress or maternal immune activation, changes the foetal hormonal milieu increasing the circulating levels of both glucocorticoids and pro-inflammatory cytokines. These two biological events have been reported to affect genes expression in experimental models and critically interfere with brain development triggering and/or exacerbating behavioural anomalies in the offspring. Herein, we highlight the importance

Eva M Marco and Elena Velarde: These authors contributed equally to the present manuscript.

E.M. Marco (✉)
Department Physiology (Animal Physiology II), Faculty of Biological Sciences,
Universidad Complutense de Madrid (UCM), 28040 Madrid, Spain
e-mail: emmarco@bio.ucm.es

E. Velarde · R. Llorente
Department Basic Biomedical Sciences, Faculty of Biomedical Sciences,
Universidad Europea (UE), Villaviciosa de Odón, Madrid, Spain

G. Laviola (✉)
Section of Behavioral Neuroscience, Department Cell Biology and Neurosciences,
Istituto Superiore di Sanità, Viale Regina Elena, 299, 00161 Rome, Italy
e-mail: giovanni.laviola@iss.it

© Springer International Publishing Switzerland 2015
Curr Topics Behav Neurosci (2016) 29: 155–182
DOI 10.1007/7854_2015_419
Published Online: 05 January 2016

to unravel the individual components of the body circadian system that might also be altered by prenatal insults and that may be causally associated with the disruption of neural and endocrine developmental programming.

Keywords Early-life stress · Prenatal · Immune activation · Animal models · Clock genes

Contents

1 Changes in the Prenatal Environment to Model Neuropsychiatric Disorders 157
   1.1 Brief Notes on the Maternal Immune Activation (MIA) Model 157
   1.2 Prenatal Restraint Stress (PRS) 159
2 Relevance of the Circadian Rhythms in Neuropsychiatric Disorders 160
   2.1 Circadian Rhythms: Brief Note on Neuroanatomy and Molecular Machinery 161
   2.2 The Circadian System During the Perinatal Period 161
   2.3 Disruption of Circadian Rhythms from Postnatal Periods to Adulthood 163
   2.4 Chronodisruption as a Perinatal Insult 165
3 Possible Molecular Players Involved in the Long-Term Consequences of Prenatal
   Insults on Behaviour and Circadian Rhythm Disruption 167
   3.1 Glucocorticoids and the Hypothalamic–Pituitary–Adrenal (HPA) Axis 167
   3.2 Immune System Activation Through Inflammation and Cytokine Production 170
   3.3 Crosstalk Between Glucocorticoids and Cytokines 171
4 Conclusions and Future Remarks 172
References 174

Despite genetics are critical for neuropsychiatric disorders' vulnerability, an increasing body of clinical and experimental evidence attests to the relevance of the environment during the developmental period. The time immediately before and after birth is a highly sensitive period characterized by a rapid and continuous neural maturation. During this prenatal and early postnatal period, the brain is particularly susceptible to insults and increasing literature gives support to the impact environmental changes have in the adult individual. Actually, there is now compelling evidence that exposure to suboptimal environments during perinatal life alters brain development and increases the risk for suffering several neuropsychiatric disorders, including schizophrenia, mood and anxiety disorders (Cirulli et al. 2009; Marco et al. 2011). In this regard, experimental animal models do represent an essential and valuable tool to investigate the causal relationship between a given environmental manipulation during development and the observation of persistent neurobehavioral anomalies reminiscent in the model of those frequently observed in mental health problems. Herein, we will briefly review current literature on animal models based upon in utero environmental challenges—prenatal insults—that have been extensively used in the investigation of neuropsychiatric disorders with a neurodevelopmental origin. In particular, we will first focus on models dealing with, i.e. maternal infection or severe stress conditions during gestation. Then, we will consider how disruption of circadian sleep cycle and expression/function of clock genes are generated as a function of severe environmental insults during gestation. Despite the time of birth and early postnatal period are also time windows of enhanced sensitivity to environmental changes, analysing the diversity of animal models that consider perinatal and postnatal insults remains far beyond the scope of the present manuscript (for reviews on animal models of early-life stress consult (Cirulli et al. 2009; Laviola et al. 2009; Marco et al. 2011, 2015).

1 Changes in the Prenatal Environment to Model Neuropsychiatric Disorders

1.1 Brief Notes on the Maternal Immune Activation (MIA) Model

Maternal infection during pregnancy has consistently been associated with increased risk of developing symptoms related to schizophrenia in the offspring (Boksa 2010; Brown 2006; Brown and Susser 2002; Meyer and Feldon 2009; Patterson 2007). Epidemiological studies have mostly investigated an association between maternal immune activation and schizophrenia, while some other neuropsychiatric disorders have been disregarded. Maternal infection has also been suggested to play a role in the pathogenesis of autism spectrum disorders (ASD) (Hyman et al. 2006), although more epidemiology is needed here. Failed support for the hypothesis that prenatal exposure to a viral infection might be associated with risk of subsequent depression has derived from a human study with over 6000 subjects (Pang et al. 2009). However, more recently, a systematic analysis of the literature explored the potential association between depression, metabolic syndrome, inflammation and hypothalamic–pituitary–adrenal (HPA) axis; although the study did not achieve clear conclusions the importance of the immune system function, possibly through the release of cytokines, to induce and/or maintain depressive symptoms became of critical biological relevance (Martinac et al. 2014). Therefore, a role for prenatal activation of the immune system in depression should not be completely disregarded.

Several rodent models that mimic prenatal infection of bacterial or viral origin have been developed and a multitude of infectious agents have been tested, i.e. influenza, poliovirus, rubella, measles, varicella–zoster, retrovirus and several bacterial agents such as bacterial endotoxin lipopolysaccharide (LPS) or viral mimic polyinosinic: polycytidylic acid (poly I:C). In general, animal models of maternal immune activation have demonstrated alterations in behaviours relevant to both schizophrenia and depression [see recent reviews (Boksa 2010; Meyer and Feldon 2009; Samsom and Wong 2015)]. In this framework, changes in the prepulse inhibition of startle response (PPI), a measure of sensorimotor gating that reflects the ability of an organism to attain information and process it correctly, have been reported. Loss of normal PPI is widely accepted as an endophenotype of schizophrenia with high translational validity since it can be assessed in both human subjects and animal models. Exposure to LPS during prenatal life has been reported to induce a deficit in PPI. Interestingly, the alteration emerges at "puberty" and seems to persist throughout adult life in male rats (Borrell et al. 2002; Fortier et al. 2007; Romero et al. 2010; Wischhof et al. 2015). With respect to sex-related vulnerability, however, a similar PPI impairment was only found at adolescence (not persisting in adulthood) in prenatally LPS-treated females (Wischhof et al. 2015). Worth mentioning, the administration of antipsychotics, i.e. haloperidol, successfully reversed the PPI impairment induced by prenatal LPS administration, thus providing pharmacological validity to this animal model of schizophrenia (Borrell et al. 2002; Romero et al. 2007). In contrast, other studies have reported no effects of gestational LPS on PPI response at adulthood, although an enhanced acoustic startle response was observed (Fortier et al. 2004). Maternal LPS administration also induced a marked increase in amphetamine-induced hyperactivity, compared to the offspring from control dams (Fortier et al. 2004), and more recently, locomotor hyperactivity has been reported following prenatal LPS administration in both male and female rats offspring (Wischhof et al. 2015).

Important cognitive deficits have also been reported following maternal immune activation. LPS injection to pregnant mice induced an enhancement in recognition memory, a deficit in associative learning and memory but no alterations in spatial memory as measured in the Morris water maze (MWM) (Golan et al. 2005). Impaired object recognition memory has been recently reported in both sexes following prenatal LPS administration, although males appeared to be more severely affected (Wischhof et al. 2015). Similarly, poly I:C administration to near-term mouse dams led to an impairment in reversal learning in the adult offspring (Meyer et al. 2006b). Notwithstanding, a deficit in the reversal phase is indicative of perseverative behaviour, also implicated in schizophrenia, autism, obsessive compulsive disorders, and addictive behaviour (Ridley 1994). Thus, prenatal poly I:C administration has been associated with the exhibition this symptom of altered mental health, i.e. perseverative behaviour. Prenatal poly I:C administration also reduced sucrose preference, indicative of anhedonia, but only if pregnant dams lost weight following MIA (Missault et al. 2014). Taken together, MIA has been related to psychotic symptoms, cognitive impairments and depression-like responses [consult (Meyer et al. 2009; Samsom and Wong 2015) for an extended review]. Different factors associated with the maternal infection process, including the time of maternal immune challenge (Meyer et al. 2006a, b) also seem to be critical for the behavioural outcome in offspring: whether or not dams displayed a febrile response (Lowe et al. 2008) and/or a loss in body weight (Missault et al. 2014) in response to the infectious agent have been identified as factors that should be taken into account when employing animal models of maternal immune activation.

1.2 Prenatal Restraint Stress (PRS)

In general, in both clinical reports and rodent studies, gestational stress has been related with an increased incidence of anxiety, depression and attention deficits in the offspring, together with major neurobehavioral disturbances relevant to ASD and schizophrenia (Beydoun and Saftlas 2008; Talge et al. 2007; Weinstock 2008). However, severe stress during gestation not only positively affects the probability to suffer from several neuropsychiatric disorders, but seems also to facilitate the onset of these disorders as well as potentiate comorbidity problems (Mill and Petronis 2008; Mittal et al. 2008; Rice et al. 2007).

Experimental research has employed different animal models of gestational stress, including altered maternal nutrition (Budge et al. 2007; MacLaughlin and McMillen 2007), administration of glucocorticoid hormones (Catalani et al. 2011; Macri et al. 2011), and obstetric complications mimicking reduced placental perfusion such as episodes of neonatal hypoxia (Boksa 2004; Laviola et al. 2004a). Among the available animal models of gestational stress, we have explored the consequences of exposure to intermittent restraint which also includes an important psychological component in the rat. In particular, we focused on the consequences in the offspring of exposure of pregnant dams to physical restrain immobilization during the last week of gestation.

A prominent reduction in social play behaviour has been reported as a consequence of prenatal restraint stress in adolescent male rats, which came in the absence of changes in environment exploration (Morley-Fletcher et al. 2003b); indeed, our observations confirmed other studies reporting impairments in affiliative behaviour as well as a reduced amount of age-typical rough-and-tumble play in the PRS offspring (Koenig et al. 2005; Lee et al. 2007; Ward and Stehm 1991). The offspring of stressed dams, as both adolescents and adults, has been reported to exhibit increased anxiety levels. In particular, an increased latency to approach a novel object in an open field (Laviola et al. 2004b), as well as increased anxiety-like responses in the elevated plus maze (EPM) (Darnaudery and Maccari 2008; Zuena et al. 2008). Depressive-like responses have also been observed following PRS since an increase in immobility time was registered in the forced swim test, a behavioural paradigm extensively used to validate the effectiveness and potency of antidepressant drugs (Morley-Fletcher et al. 2003a). Learning and attention deficits are also common in the offspring of PRS rats: impairments in spatial learning in the MWM were observed in PRS male rats (Lemaire et al. 2000), as well as impaired recognition memory and altered working memory (Maccari et al. 2003); notably, sex differences have been described with cognitive deficits being more evident in female PRS rats than in males (Zuena et al. 2008). Taken together, PRS seems to increase anxiety levels, induce a depressive-like phenotype and impair cognitive function in the offspring.

Further, profound disruption of circadian rhythmicity of several physiological responses, i.e. heart rate, body temperature, and physical activity (Mastorci et al. 2009), has been reported in rodent models following maternal exposure to prenatal intermittent restraint. Long-term consequences of PRS significantly depend on the sex of the offspring, on the behavioural parameter being considered, and most importantly on the intensity and timing of the maternal stress [see (Darnaudery and Maccari 2008; Weinstock 2008) for review]. Great efforts have been devoted to understand the underlying mechanisms of the deleterious outcomes observed in the offspring of stressed pregnancies (Darnaudery and Maccari 2008; Maccari et al. 2003; Weinstock 2002, 2008); however, research in this field is still needed.

2 Relevance of the Circadian Rhythms in Neuropsychiatric Disorders

Alterations in circadian rhythm have a profound impact on the physical and psychological homeostasis of an individual (Atcheson and Tyler 1975). Indeed, a disruption of circadian clocks has been found in several neuropsychiatric disorders such as depression, post-traumatic stress disorder, mania and schizophrenia [(Agorastos et al. 2014; Mendlewicz 2009; Novakova et al. 2015) for a review, see (Karatsoreos 2014)]. The neurobehavioral deficits, such as changes in mood, affect or cognitive impairments, derived from altered circadian patterns have long been established. For instance, animal models relevant for jet-lag syndrome have provided evidence for negative long-lasting effects on hippocampal neurogenesis, deficient performance in hippocampal-dependent learning and memory and underlying depression as a consequence of altered circadian conditions (Gibson et al. 2010). However, it is yet to be determined if the alterations of the circadian system observed in neuropsychiatric disorders are mere symptoms or contributing factors.

In the case of depressive syndrome, this dual role of the circadian system is even more pronounced. On the one hand, patients with major depression exhibit direct disturbances of the circadian system, including changes in daily mood variation, brain activity, core body temperature, hormone secretion, sleep–wake cycle, motor activity and seasonal mood variation (Monteleone et al. 2011). On the other hand, disruption of the circadian system has been linked to the pathophysiology of depression, as it has been shown in transgenic mice for clock and clock-controlled genes (Mukherjee et al. 2010; Roybal et al. 2007). Moreover, in clinical studies, single-nucleotide polymorphisms in CLOCK, BMAL1, PER3 genes or in the circadian regulator glycogen synthase-kinase-3β (GSK-3β) have been related to sensitivity to antidepressant treatment (Benedetti et al. 2005; Serretti et al. 2003), increased recurrent rate of affective episodes (Benedetti et al. 2003), development of bipolar disorder (Nievergelt et al. 2006) and risk for seasonal affective disorder (Partonen et al. 2007). A more recent study also delves into the role of the circadian system in the pathophysiology of depression. By using a validated mouse model of depression, i.e. exposure to chronic mild stress, authors provide support for an association between mice depressive-like phenotype and a desynchronization of the core components of the molecular clock within the amygdala. Actually, the rhythmicity exhibited by the basolateral amygdala of control mice was completely abolished in anhedonic mice, thus suggesting that the observed disruption in normal daily oscillations might be a consequence of the illness (Savalli et al. 2014).

2.1 Circadian Rhythms: Brief Note on Neuroanatomy and Molecular Machinery

All organisms have the ability to adapt to their environment by anticipating periodic changes such as the alternation of light and dark periods. This is possible due to the existence of an endogenous circadian system that works as a clock that can be synchronized by environmental cues, setting a period of around 24 h (thus the name circadian). In mammals, this system consists of a hierarchical structure, where the central or master clock is located in the suprachiasmatic nucleus (SCN) of the hypothalamus, which controls a network of peripheral clocks in every tissue (Hastings et al. 2007). This master clock—SCN—is mainly entrained to the light/dark cycle, receiving the light information directly from the retinohypothalamic tract and driving an output of metabolic (i.e. glucose homeostasis, gene expression) and hormonal rhythms (i.e. melatonin, cortisol/corticosterone) that keep the rest of the body oscillators in synchrony (Hastings et al. 2007; Iuvone et al. 2005). However, these peripheral clocks can be entrained by external cues different from the light/dark cycle and therefore remain functional without the control of the SCN. The strongest cue that can entrain peripheral clocks without affecting the phase of the SCN is food (in terms of feeding schedule and/or caloric consumption) despite other clocks related to the reward system have also been described for methamphetamine (Honma and Honma 2009; Mendoza 2007).

The molecular machinery responsible for circadian rhythm generation can be found in every single body cell and consists of transcriptional–translational feedback loops that involve a highly conserved set of "clock genes". In mammals, these loops are formed by the so-called positive elements, CLOCK and BMAL1, which heterodimerize and enhance the transcription of the negative components *Period* (*Per1*, *Per2* and *Per3*) and *Cryptochrome* (*Cry1* and *Cry2*) genes. The PER and CRY proteins form complexes that inhibit their own transcription by binding to the CLOCK:BMAL1 complex blocking its function. This negative loop allows a daily rhythm in expression of *Per* and *Cry* transcripts and their protein products (Iuvone et al. 2005; Okamura et al. 2002).

2.2 The Circadian System During the Perinatal Period

Considering that a hypothalamic structure—SCN—is central for the circadian system functioning, it seems plausible that an aberrant development of the related hypothalamic circuitries may lead to a malfunctioning of this master clock as well as of extra-SCN non-photic oscillators. Accordingly, it seems reasonable to deep into the importance of an appropriate development of the circadian system during early stages and also to investigate the possible consequences of circadian rhythm disruption during the prenatal time window that might be related to the emergence of neuropsychiatric disorders later in life. It is well known that the SCN is fully formed and innervated by mid-gestation in humans and rhesus monkeys, and it presents oscillations in utero. Indeed, clear-cut circadian rhythms of heart rate, respiratory movements, foetal movements and hormones are already displayed by the foetus (Davis and Reppert 2001; Seron-Ferre et al. 2007; Weirnert 2005). In contrast, these rhythms seem to appear at a postnatal age in rodents; neurogenesis of SCN is completely close to birth, by gestational day (GD) 17 (Davis and Reppert 2001), although there are evidences of foetal rhythms in metabolic activity (Davis and Gorski 1985), vasopressin mRNA levels (Reppert and Uhl 1987) and spontaneous neural activity (Shibata and Moore 1987). However, in every mammalian specie, foetal circadian system depends on maternal signals (core body temperature, metabolic cues and hormonal milieu) that allow the immature offspring to synchronize to the external environment (Seron-Ferre et al. 2007). Thus, from a molecular perspective, the morphological development of the SCN parallels the gradual development of molecular clock robustness as shown by an increase in the clock gene expression rhythm amplitudes from no rhythmicity at GD 19 to highly developed rhythms at postnatal day (PND) 10 (Houdek and Sumova 2014). In rats, an analysis of clock gene transcripts at GD 19 showed that no rhythmic expression could be detected for *Per2* and *Bmal1* although some other genes related to cellular activity showed rhythmic, thus providing evidence for the maternal circadian system to drive these oscillations (Houdek and Sumova 2014). Likewise, a study in which pregnant dams were exposed to a 6-h delay of the dark period and then released into constant darkness at different stages of the foetal development showed that the expression of clock genes *Per1* and *Per2* in the SCN of newborn pups was shifted differently, according to the day the mother was housed in constant darkness (El-Hennamy et al. 2008). Besides, give that rat pups are blind until PND 15, studies with rats highlight the relevance of non-photic entrainment during perinatal stages. In this regard, it has been shown that keeping pregnant rats under a scheduled feeding regime—that is, food availability for a limited period of time without caloric restriction—is able to phase advance the expression of several clock genes at birth (PND 0), with less pronounced effects at PND 10 when maternal feeding schedules are not as important as nursing time for the offspring (Olejnikova et al. 2015). Therefore, after birth, maternal care becomes crucial for the generation and maintenance of pup's circadian rhythms, as the mother will nurse and take care of the pups until weaning. During lactation, nocturnal animals will provide nursing and maternal care during the light phase, shifting the clocks of the pups opposite to adults. Meanwhile, the SCN becomes a light-driven oscillator and peripheral clocks are entrained by competing signals from the SCN and the feeding regime (Polidarova et al. 2014).

However, both in foetal and postnatal stages, the main environmental factor controlling for the adequate development of the circadian system is the light–dark cycle. In studies with animals raised at different photoperiods, it has been shown that SCN photic sensitivity develops gradually, with clock genes like *Per1* or *Per2* being rhythmic from PND 10 but others like *Cry1* from PND 20 (Kovacikova et al. 2005). Before birth, foetuses do not receive light directly, but photic entrainment is set by maternal cues via melatonin secretion. This hormone is mainly produced by the pineal gland keeping a high-at-night secretion pattern that contributes to photic entrainment of physiological and behavioural processes. By contrast, glucocorticoids are produced in the adrenal glands controlled by the HPA axis and are also released in a rhythmic manner but opposite in phase to melatonin, peaking during the daytime. Brain structures controlling the secretion of these hormones, the pineal gland and the parvocellular neurosecretory cells of the hypothalamus, respectively, are under regulation of the SCN which maintains melatonin and glucocorticoids in opposite phases as direct outputs of the clock. Both hormones will affect the development of the circadian rhythms in pups. It has been shown that suppressing maternal melatonin rhythmic secretion by exposing animals to constant light at mid-gestation leads to intrauterine growth retardation, to an altered pattern of clock gene expression in foetal adrenal glands with suppression of glucocorticoid rhythm together with an altered response to corticotrophin-releasing hormone (ACTH) effects that were counteracted by melatonin injection during the subjective night (Mendez et al. 2012). Likewise, it has been shown that constant light during the last third of pregnancy in primates affected the entrainment of temperature rhythms in the newborns, which were re-synchronized with melatonin injections to the pregnant mothers highlighting the physiological importance of the maternal melatonin rhythm during pregnancy (Seron-Ferre et al. 2013).

2.3 Disruption of Circadian Rhythms from Postnatal Periods to Adulthood

Chronodisruption has been used in several ways in adult and gestational animal models, and it can be defined as an alteration of the normal 12-h light/12-h darkness (12L:12D) photoperiod. In humans, the most common circadian disruptors include environmental lighting (including both the exposure to low light during daytime or to electric light sources during night time), shift work (which implies the exposure to abnormal light cycles and the interference of work hours with sleep timing), jet lag from transmeridian travel (which requires the adaptation of body clocks to a new time zone), social jet lag (temporal differences between the endogenous clock and the social clock) and sleep disorders (Bedrosian et al. 2015 in press). Several animal models have been developed to study these conditions including long photoperiods (16L:8D, to mimic summer conditions), short photoperiods (8L:16D to mimic winter conditions), phase advances (i.e. 12L:12D photoperiod but lights on 5 h earlier to mimic jet lag) or constant conditions (i.e. 24 h of light to mimic shift work). As a conclusion from those studies, darkness is needed for the normal development of the circadian system. This consideration is of special relevance for neonatal intensive care units (NICUs) as early newborn babies born with low weight are kept in facilities with artificial light conditions during 24 h rather than to the 24-h darkness they get inside the womb. The importance of the amount of light received during the perinatal stages became evident in studies performed with mice exposed to different seasonal photoperiods until weaning and then changed to the opposite seasonal photoperiod for four weeks; this study demonstrates that perinatal photoperiod has long-term effects on the rhythmicity of clock neurons and on animal behaviour. Perinatal seasonal lighting seems to imprint individual clock neurons. Thus, long-photoperiod-raised pups display low-amplitude rhythms of *Per1*, a free running period of behavioural activity and a stability of these conditions when season shifted. In contrast, short-photoperiod-exposed pups showed high-amplitude rhythms, and waveform changes when season light was changed (Ciarleglio et al. 2011a). A similar study has measured the ability to develop circadian rhythmicity under constant light in rats that received different quantity and quality of light during suckling. Normal circadian patterns under constant light, that is becoming arrhythmic, were only developed by those animals which received darkness during suckling, whereas the animals that were kept under light during suckling remained rhythmic with a stability dependent on light intensity received prior to weaning (Cambras et al. 2015). This imprinting of the circadian system by perinatal seasonal light has proven to have implications in the later development of neurobehavioral disorders. Early exposure to short photoperiods in rodents has been linked to an increase in depressive and anxiety-like behaviours (Pyter and Nelson 2006). In humans, light levels during the night may play an increasingly evident role in regulating behaviour and mood in adults (Bedrosian and Nelson 2013), and therefore, light conditions may play a yet unrecognized role in controlling physiological functions, sleep–wake cycles, alertness and cognitive functions in preterm and neonatal infants. For instance, the rate and severity of seasonal affective disorder are elevated in winter-born humans, as is the risk for developing schizophrenia or bipolar disorder (Castrogiovanni et al. 1998; Foster and Roenneberg 2008). Moreover, studies performed at the NICUs indicated continuous light as a stressful condition for preterm babies that could be related to later impaired academic performance, attention-deficit, general hyperactivity and psychiatric disorders at puberty (Perlman 2001).

The basis for this relationship between seasonal photoperiod imprinting and neurobehavioral disorders may lay, at least in part, in the link between circadian and serotonergic systems in the brain. The SCN is innervated by serotonin projections from the median raphe nucleus and also receives indirect information via neuropeptide-Y (NPY) input from the intergeniculate leaflet driven by the dorsal raphe (Deurveilher and Semba 2005). Serotonin regulates SCN response to light acting both presynaptically on retinal afferent terminals and postsynaptically on SCN neurons to inhibit retinal input to the central biological clock (Smith et al. 2001). Likewise, the dorsal raphe nuclei receive direct input from the circadian visual system and indirect input from the SCN via the dorsomedial hypothalamic nuclei (Morin 2013). Besides, serotonergic raphe neurons express the elements of the molecular clock and rhythms in key serotonergic genes, as tryptophan hydrolase, and in serotonin secretion (Malek et al. 2007). Therefore, both systems are anatomically and genetically intertwined and together regulate affective behaviours, and any malfunction in their relationship might be associated with a range of mood disorders (Ciarleglio et al. 2011b). Recent studies have demonstrated that photoperiod received during perinatal period (both pre- and postnatal) imprints the serotonergic neurons of the dorsal raphe programming their firing rate, responsiveness to noradrenergic stimulation, intrinsic electrical properties, serotonin and norepinephrine content in the midbrain, and depression-/anxiety-related behaviour in a melatonin receptor 1 (MT1)-dependent manner, and these features remain later in life even after several subsequent photoperiod shifting (Green et al. 2015).

There is growing clinical evidence highlighting the importance of early-life photoperiod exposure in the development of a certain chronotype, referred to the phase of the endogenous sleep–wake cycle. The so-called *evening type* is normally born during spring and summer (long-photoperiod seasons), while the "*morning type*" is more frequently born during autumn and winter (short-photoperiod seasons) (Natale and Di Milia 2011; Takao et al. 2009; Tonetti et al. 2012). Such an association becomes more evident when seasonality is more clearly marked, like in residents of higher latitudes (Mongrain et al. 2006) or in rural areas with less amount of artificial light (Borisenkov et al. 2012). A person's chronotype influences the individual ability to adapt to circadian disruption and also to their future health and well-being. Moreover, chronotype and the amount of light received during childhood and adolescence may affect the susceptibility of a person to develop mood disorders later in life (Erren et al. 2012). Notwithstanding, puberty has been recently described as a period of increased sensitivity to light, in particular to evening light exposure. In the study, participants received 1 h of light prior to bedtime and the amounts of sleep as well as the melatonin salivary content were assessed. The suppression of melatonin shown by prepubertal participants was higher than the suppression observed among the more mature adolescents suggesting an increased sensitivity to evening light in early pubertal children (Crowley et al. 2015). The suppression in melatonin production seems to depend upon light intensity, and the study suggests that exposure to low light intensity, as that provided by the use of electronic devices as tablets or computers in the hour prior to sleep, may not only have short-term effects in adolescents' sleep quality but also long-term effects imprinting serotonergic and circadian systems as previously stated, thus increasing the risk for the development of mood disorders later in life.

2.4 Chronodisruption as a Perinatal Insult

In spite of the fact that the investigation of the relationship between brain developmental disruption and circadian system is a considerable novel topic, there is an increasing amount of studies, both in animal models and humans, linking circadian system malfunctioning, early-life insults and the appearance of neuropsychiatric diseases at adulthood. Brain developmental disruption and circadian system seem to interact in a bidirectional way; not only exposure to a prenatal insult can provoke a misalignment in the biological rhythms of the offspring, but a disruption of the maternal circadian rhythm during gestation can also induce an aberrant development in neural circuitries that may render the offspring more vulnerable to undergo neuropsychiatric disorders later in life. Although this relationship between the circadian system and the development of neurobehavioral disorders has been described, not many studies have considered the use of chronodisruption as a perinatal insult. Conditions such as shift work are increasingly becoming frequent in modern society, and there is no exception for pregnant women. In a work developed by Roman and Karlsson (2013), authors recall that a technical problem at the animal facilities during their ongoing experiment led to seven days of constant light for a group of pregnant rats, from GD20 to PND 4; when these animals reached adulthood, a battery of behavioural tests was carried out, i.e. open field, object recognition and water maze. Animals exposed to constant light during the perinatal period showed intact recognition memory and no deficits in spatial learning or memory; however, these animals exhibited increased thigmotaxis in the open field and the water maze, less ambulatory behaviour in the open field, as well as lower exploration times in the object recognition test. Taken together, in this study, authors suggest an increased anxiety-like phenotype as a long-term effect of perinatal chronodisruption. This impact of light can be produced with dim light, mimicking the conditions of artificial light during the night for a pregnant woman. In this regard, pregnant mice were raised on either a 14L:10D photoperiod or a 14L:10dim-light condition. After weaning, the offspring was all maintained under constant darkness and at adulthood several behavioural and molecular measures were assessed. Mice early exposed to constant light (dim light) showed reduced growth rates, displayed an anxiety-like behaviour in the elevated plus maze and increased fear responses in the passive avoidance test (Borniger et al. 2014). Interestingly, other studies exposing pregnant rats to chronodisruption showed also impairment of hippocampal spatial memory in the adult offspring. Hippocampal clock gene rhythms in the foetuses from dams exposed to constant light were completely abolished and no detectable differences in plasma melatonin or corticosterone were reported although animals were reared under 12L:12D photoperiod. Notably, a significant deficit of spatial memory was also observed (Vilches et al. 2014). Considering the importance of light in current society, with artificial lights on almost all the time, further studies on the impact of shift work and social jet lag during gestation on mental health are needed. It is critical to better understand the impact of developmental chronodisruption on the adult circadian system, and its influence on the development of additional brain circuitries controlling for cognition, emotionality and energetic homeostasis. The investigation of a possible association between gestational circadian rhythm disruption and the risk for the development of neurobehavioral disorders during adulthood is of great clinical relevance.

3 Possible Molecular Players Involved in the Long-Term Consequences of Prenatal Insults on Behaviour and Circadian Rhythm Disruption

The interactions between central nervous system, endocrine and immune system are profuse and intricate (Bilbo and Schwarz 2012; Eskandari and Sternberg 2002) and initiate very early in life. The developing brain is exquisitely sensitive to both endogenous and exogenous signals. The prenatal insults here described include endocrine signals (glucocorticoids), immunological molecules (cytokines) and additional external factors comprised in the experimental procedures (injections, animal manipulation, etc.). All together are able to critically affect brain development so that the adult individual will exhibit abnormal emotional responses, cognitive deficits, and, possibly, an altered circadian activity (see previous sections). In this section, we will briefly delve into the particular contribution of glucocorticoids and cytokines to the hypothesized circadian rhythm disruption and the long-term behavioural outcomes.

3.1 Glucocorticoids and the Hypothalamic–Pituitary–Adrenal (HPA) Axis

Dysfunctions in the HPA axis activity have been reported in many psychiatric disorders [e.g. (Jacobson 2014; Wingenfeld and Wolf 2011)]. The HPA axis and the autonomic nervous system are responsible for the elaborated multi-directional communication pathway designed to restore homeostasis upon a change in the environmental conditions, i.e. a stressful challenge. Both systems become activated to cope with adverse environmental situations that might be considered as physical and/or psychological challenge, e.g. low temperatures, undernutrition, inflammation, restraint immobilization. HPA actions are mediated by the glucocorticoids (cortisol/corticosterone) released from the adrenal glands act through their ubiquitously distributed intracellular receptors, glucocorticoid and mineralocorticoid receptors (GR and MR, respectively) (de Kloet et al. 2008). The importance of glucocorticoid receptors not only consists in their role in stress response, but also in their participation in processes related with synaptic plasticity and memory formation (Brinks et al. 2007; Oitzl et al. 1997). Both receptors are located in brain areas involved in emotion, learning and memory (de Kloet 2003). A balanced MR: GR activation in the limbic brain appears to be critical for the emotional and cognitive functioning required for optimal performance in a changing environment and thus beneficial for mental health (Oitzl et al. 2010). The development of HPA system is not uniform; different components of the system have different ontogenetic patterns. Actually, HPA axis circadian rhythmicity and feedback regulation in the rat are not yet fully developed until late in development (Levine 1994). Moreover, during brain development, the brain is highly sensitive to the effects of glucocorticoids; therefore, maintaining glucocorticoid levels within a physiological range during the different stages of the nervous system development seems to be critical for a correct organization of brain. Worth noting, the existence of a time window during postnatal life (PND 4–14) of diminished circulating corticosterone levels and reduced adrenal sensitivity—the so-called stress hyporesponsive period, SHRP—that seems to be an adaptive mechanism to protect the brain developmental processes from the deleterious impact of heightened glucocorticoid levels during a critical time window (Levine 1994).

Early-life environmental factors seem to be critical for the development and functionality of the HPA axis (Karrow 2006; Levine 1994, 2000; Pryce et al. 2005). Indeed, a dysregulation of HPA activity has been consistently described in animal models of prenatal insults. Heightened stress during pregnancy increases plasma levels of glucocorticoids and ACTH in the mother and foetus and may consequently interfere with the required adaptive mechanisms triggering to a persistent dysregulation of the HPA axis (Laviola et al. 2004b; Lazinski et al. 2008; Mastorci et al. 2009; Morley-Fletcher et al. 2003b; Weinstock 2002). The offspring of PRS dams showed a prolonged corticosterone stress response together with a reduction in the hippocampal expression of both MR and GR. In animal studies, PRS-induced impairments in behaviour and HPA axis responsiveness are prevented by maternal adrenalectomy. However, maternal injection of corticosterone only reverses the increased anxiety and HPA alterations in the offspring (Weinstock 2008). Similar effects have been reported following MIA. Animals prenatally exposed to LPS showed, as adults, augmented corticosterone levels at baseline, a blunted stress response, as well as a decreased expression of MR and GR within the hippocampus (Basta-Kaim et al. 2011; Lin et al. 2012; Reul et al. 1994). Indeed, since hippocampal GR seems to participate in an inhibitory feedback mechanism, the decrease in the levels of GR in this structure might involve HPA axis hyperactivity. Remarkably, the reported HPA axis disturbances reported following gestational LPS administration were reversed by the chronic administration of antipsychotic drugs, clozapine and to a lesser extent chlorpromazine (Basta-Kaim et al. 2011). Notably, the reported changes in hippocampal glucocorticoid receptors (GR and MR) may not only affect HPA functioning, but also emotional processing and cognitive function; therefore, the cognitive deficits described in these animal models may rely, at least in part, on the changes in glucocorticoid receptor expression already described within the hippocampus. Circulating maternal glucocorticoid levels may constitute one of the critical factors mediating the foetal programming of neural circuitries involved in the control of emotion, cognition and stress response. Last but not least, an affection of the placental function as a consequence of prenatal stress cannot be discharged and may also be considered as a critical factor mediating some of the deleterious outcomes described in the offspring (O'Donnell et al. 2009). Taken together, an elevation in glucocorticoid levels within the foetus—with foetal, maternal or a placental origin—may serve as one hormonal pathway that could mediate the brain development disruption that in the long term may provoke the behavioural anomalies already described.

The activity of the HPA axis follows a circadian rhythm, with peak levels of glucocorticoids during the active phase (daytime in humans and night in nocturnal animals, such as rats or mice). The circadian rhythm of the HPA axis is characterized by a pulsatile release of glucocorticoids from the adrenal gland that results in rapid ultradian oscillations of hormone levels both in the blood and within target tissues, including the brain. This rhythm is under control of the central clock system, in particular the hypothalamic SCN, throughout synapses from the SCN to the hypothalamic PVN (see Sect. 2.1). However, this control also affects other structures of the HPA axis such as adrenal glands, due to the activation of the autonomic nervous system and/or to the presence of an adrenal peripheral clock. In animal studies, glucocorticoids have been reported to be able to phase shift many clock-related genes, such as *Per1* and *Per2*, in peripheral tissues such as liver, heart and kidney. The HPA axis also influences the circadian system. Glucocorticoids seem to be able to change the circadian expression of different molecules providing an adaptive mechanism through which respond to stressors. The stress system, through the HPA axis, communicates with the clock system; therefore, any uncoupling or dysregulation could potentially cause several disorders, such as metabolic, autoimmune, and mood disorders [for recent review consult: (Nicolaides et al. 2014; Spiga et al. 2011)].

Neural connections between the HPA axis and the circadian system are established early during brain development; thus, any environmental impact during a critical time frame could drive to a mismatch between these systems. In this regard, exposure to severe stress during gestation has been reported to alter the circadian system, and more particularly, the temporal functioning of the HPA axis (Koehl et al. 1999; Maccari et al. 2003; Maccari and Morley-Fletcher 2007). PRS induced higher levels of corticosterone secretion at the end of the light period in both males and females and hypercorticism over the entire diurnal cycle in females (Koehl et al. 1999). In addition, PRS induced a phase advance in the circadian rhythm of locomotor activity as well as an increase in the paradoxical sleep in adult rats (Maccari et al. 2003; Maccari and Morley-Fletcher 2007). These two behavioural effects might be mediated, at least in part, by the specific change in the temporal pattern of hippocampal GR expression induced by prenatal stress; PRS induced a reduction in hippocampal GR expression both at the beginning of the light period and at the end of the light period times at which total corticosterone levels are increased in PRS rats. Data from the prenatal immune activation model also demonstrated a disruption in the circadian system. Indeed, prenatal LPS administration (GD17) was reported to alter sleep architecture in mice by using continuous quantitative video/electroencephalogram/electromyogram analyses; changes that seem to be circadian cycle and activity state dependent (Adler et al. 2014). Further research is still needed to control for the consequences of prenatal insults on the temporal profile of the HPA functioning.

3.2 Immune System Activation Through Inflammation and Cytokine Production

Nowadays, extensive literature gives support to the fact that maternal infection during pregnancy is associated with increased risk of developing schizophrenia (Brown 2006; Meyer and Feldon 2009; Patterson 2007); however, the causal link underlying this observation is far from clear. The specific mechanisms by which maternal infection may lead to psychopathology include direct infection of the developing foetus and subsequent abnormal neural development, the generation of autoantibodies by the mother that subsequently react with foetal neural tissue and alterations in cytokine production, which may be an underlying component of all three mechanisms (Pearce 2001). Literature from animal models suggests a causal relationship between maternal immune activation and the altered behavioural traits observed in the adult offspring. Among the diversity of immunological events that can be triggered by infectious agents, the cytokine-associated inflammatory response may be of particular relevance. Cytokines are soluble bioactive mediators released by diverse immune cell types; these include interleukins (ILs), interferons, tumour necrosis factors (TNFs), chemokines and growth factors. Cytokines act within a complex network, either synergistically or antagonistically, and are generally associated with inflammation, immune activation and cell differentiation or death (Allan and Rothwell 2003). Cytokines can be produced by brain immune cells, also by neurons, and, more interestingly, cytokines can cross the blood–brain barrier. Thus, the central nervous system can be affected not only by cytokines produced within the brain, but also through the actions of mediators originating from the periphery (Lucas et al. 2006). Several studies have demonstrated an increase in pro-inflammatory cytokines in the adult offspring of dams exposed to gestational infection. In particular, prenatal LPS exposure induced an increase in chemokines and cytokines expression (Borrell et al. 2002). In the same line, the neonatal administration of pro-inflammatory cytokines (Samuelsson et al. 2006; Tohmi et al. 2004) or leukaemia inhibitory factors (Watanabe et al. 2004) to rats has been reported to induce behavioural changes, including cognitive deficits, together with neurobiological changes in specific brain regions such as the hippocampus. Following prenatal LPS administration, an unbalanced inflammatory reaction in the foetal environment that activates the foetal stress axis has also been suggested (Gayle et al. 2004). Remarkably, most of the cytokines altered by LPS administration have an important influence in different synaptic plasticity processes, and in learning and memory processes. Consequently, the increase in the expression of these cytokines may mediate, at least partially, the long-lasting cognitive deficits observed in this animal model (Bilbo and Schwarz 2012). In conclusion, the maternal immune response, possibly through the increase of pro-inflammatory cytokines, may represent one of the key events interfering with foetal brain development and maturation at critical time windows; the disruption of the balance between pro-inflammatory and anti-inflammatory cytokine during prenatal life may trigger the debut of the behavioural anomalies already described as a consequence of prenatal immune activation and extensively related to neuropsychiatric disorders.

Severe stress condition during gestation has been reported to negatively affect the immune system and may possibly contribute to the maladaptive immune responses to stress that occur later in life. Exposure to psychosocial stress early in life—during early stages of pre- and postnatal life—seems to aggravate the effects of immunotoxicants or immune-mediated diseases in infants (Bellinger et al. 2008; Meerlo et al. 2008). In the long term, PRS produced important alterations in a number of peripheral and central immunological parameters, as previously described for several animal models (Coe et al. 2002; Gotz and Stefanski 2007; Llorente et al. 2002; Tuchscherer et al. 2002). Specifically, a decrease in blood cell populations devoted to immune competence, especially the $CD4^+$ T-lymphocytes and T4/T8 ratio were observed in response to PRS (Laviola et al. 2004b). PRS also produced an elevation in pro-inflammatory IL-1β concentration, both in the rats' spleen and frontal cortex (Laviola et al. 2004b). This increment in IL-1β production among the PRS offspring resembles human studies suggesting that stress and psychopathology can be indexed by a hypersecretion of pro-inflammatory cytokines, although other intervening variables have to be considered (Dabkowska and Rybakowski 1994; Elenkov and Chrousos 2002; Fan et al. 2007; Schiepers et al. 2005).

The immune system is also subjected to circadian rhythms. Most immune cells express circadian clock genes and present a wide array of genes expressed with a 24-h rhythm (Labrecque and Cermakian 2015). However, knowledge on the biological relevance of the immunological circadian clock is scarce, and few data are available regarding the ontogeny of the rhythmicity within this system. Cytokines have also a crucial role in brain maturation given their biological relevance in developmental processes such as neurogenesis, neuronal and glia cell migration, proliferation, differentiation, and synaptic maturation and pruning. The levels of several cytokines fluctuate during development based upon the neurodevelopmental processes occurring in the brain in a region-dependent manner; as an example, the content of IL-1β, TNFα or IL-6 seems to be higher during early phases of brain development than in adult brains (Bilbo and Schwarz 2012). Further research needs to focus on the ontogeny of the rhythmicity within the production of cytokines, and on its possible relationship with developmental events critical for brain maturation.

3.3 Crosstalk Between Glucocorticoids and Cytokines

On the one hand, glucocorticoids play a modulatory role on the expression of numerous cytokines, adhesion molecules and other inflammatory molecules (Eskandari and Sternberg 2002). Classically glucocorticoids are considered as anti-inflammatory molecules, although some studies indicate that their immunomodulatory properties are tissue specific. Actually, stress has been considered to induce inflammation in the brain. Stress, depending of the age, nature, intensity and length of the exposure, may activate the pro-inflammatory pathway that triggers the release of cytokines and promotes cell damage that may underlie some of the detrimental behavioural outcomes of stress exposure (Garcia-Bueno et al. 2008). On the other hand, interleukins such as IL-1, IL-1β and IL-6 induce the activation of the HPA axis, increasing the levels of glucocorticoids (Karrow 2006). The bidirectional influence between glucocorticoids and immune molecules may initiate at early stages of development, such as during foetal and neonatal developmental windows, as it might represent an adaptive feature that contributes to the survival of the offspring in its new environment. However, if the individual's environment is drastically changed such that neuroendocrine–immune programming becomes maladaptive, it may trigger or exacerbate certain diseases, including neuropsychiatric disorders (Karrow 2006). Further research is still needed to distinguish the temporal sequence of events triggered by prenatal insults, whether glucocorticoids induce pro-inflammatory release or *viceversa*, and to identify the critical time windows that may enable a modulation and/or regional and temporal restriction in brain damage by pharmacological or environmental interventions.

4 Conclusions and Future Remarks

The neurodevelopmental hypothesis suggests that the disruption of early brain development may increase the risk for the appearance of several neurobehavioural disorders later in life. Although genetics plays a clear crucial role, the maternal–foetal environment arises as a critical factor that needs to be taken into account. Neonatal neural programming seems to be highly sensitive to the content of glucocorticoids and to several immune signalling molecules, i.e. cytokines. Therefore, prenatal insults through the activation of the HPA axis and/or the release of pro-inflammatory cytokines may trigger a dysfunctional brain network and connectivity responsible for the emergence of behavioural symptoms that have been described in animal models of neuropsychiatric disorders.

The activity of the HPA axis and the immune system is subjected to circadian rhythms, and, in the recent years, a disruption of circadian clocks has been extensively described in several neuropsychiatric disorders. Moreover, optimal brain development has been related to intact synchrony between circadian and diurnal rhythms (Powell and LaSalle 2015). The role of altered expression and/or function of circadian genes in neuropsychiatric disorders is particularly compelling. Actually, gene mutations and single-nucleotide polymorphisms and haplotypes in several circadian genes have been recently associated with susceptibility to mood disorders (Mendlewicz 2009) (Fig. 1).

Changes in environmental contingencies have been shown to affect circadian rhythms. In particular, exposure to social stress—repeated social defeat—during the dark/active phase seems to induce long-lasting consequences for the functional output of the biological clock, i.e. general motor activity and core body temperature, effects that seem to depend, at least in part, on the clock genes *Per1* and *Per2*

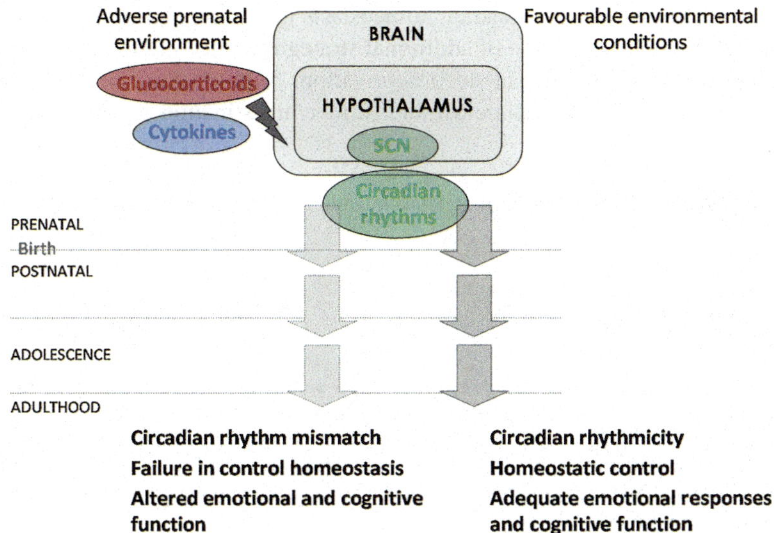

Fig. 1 Gestational stress and/or maternal infection during pregnancy has been related with an increased incidence of neuropsychiatric disorders, including depression and schizophrenia. It is of critical importance to unravel the individual components of the body circadian system that might also be altered by prenatal insults since they might be related with the disruption of neural and endocrine developmental programming

(Bartlang et al. 2015). If stress can affect the circadian system at adulthood, exposure to stress at critical developmental windows (gestation) may more critically affect the circadian clock. Whether clock genes are modified following exposure to prenatal stress deserves further investigation, as well as the analyses of similar changes in response to maternal immune activation.

Epigenetic regulation provides a mechanism for cells to integrate genetic programs with environmental signals in order to generate an adaptive and consistent output. DNA methylation is one epigenetic mechanism that entrains the circadian clock to a diurnal environment (Powell and LaSalle 2015) and has also been implicated in the pathophysiology of neurodevelopmental disorders including schizophrenia and ASD. Moreover, epigenetic changes have been described following maternal immune activation (Basil et al. 2014). Hence, epigenetic changes generated by prenatal insults have become an emergent and important target of investigation as a common molecular mechanism which may underlie changes in the circadian system, the HPA axis and the immune system.

Revealing a common pattern of circadian disruption in developmental models of neuropsychiatric disorders will open new avenues in the investigation of the biological bases of neuropsychiatric disorders as well as for the development of innovative strategies for therapy in mental health. In this framework, new antidepressants acting on melatonin receptors have been successfully developed. Interestingly, their efficacy seems to rely in the capacity to restore the internal clock

(Mendlewicz 2009) and the circadian homeostasis in rats exposed to PRS (Mairesse et al. 2013). Future application of additional strategies aimed at the manipulation of the circadian timing system via sleep deprivation, bright light or pharmacological therapy will become of great interest and further investigation in this field will be needed.

Acknowledgments The editorial help of Stella Falsini is acknowledged.

References

Adler DA et al (2014) Circadian cycle-dependent EEG biomarkers of pathogenicity in adult mice following prenatal exposure to in utero inflammation. Neuroscience 275:305–313

Agorastos A, Kellner M, Baker DG, Otte C (2014) When time stands still: an integrative review on the role of chronodisruption in posttraumatic stress disorder. Curr Opin Psychiatry 27:385–392

Allan SM, Rothwell NJ (2003) Inflammation in central nervous system injury. Philos Trans R Soc Lond B Biol Sci 358:1669–1677

Atcheson JB, Tyler FH (1975) Circadian rhythm: man and animals. In: Greep RO, Astwood EB (eds) Handbook of physiology. American Physiological Society, Washington, pp 127–134

Bartlang MS, Oster H, Helfrich-Forster C (2015) Repeated psychosocial stress at night affects the circadian activity rhythm of male mice. J Biol Rhythms 30:228–241

Basil P, Li Q, Dempster EL, Mill J, Sham PC, Wong CC, McAlonan GM (2014) Prenatal maternal immune activation causes epigenetic differences in adolescent mouse brain. Transl Psychiatry 4:e434

Basta-Kaim A et al (2011) Hyperactivity of the hypothalamus-pituitary-adrenal axis in lipopolysaccharide-induced neurodevelopmental model of schizophrenia in rats: effects of antipsychotic drugs. Eur J Pharmacol 650:586–595

Bedrosian TA, Fonken LK, Nelson RJ (2015) Endocrine effects of circadian disruption. Annu Rev Physiol (In press)

Bedrosian TA, Nelson RJ (2013) Influence of the modern light environment on mood. Mol Psychiatry 18:751–757

Bellinger DL, Lubahn C, Lorton D (2008) Maternal and early life stress effects on immune function: relevance to immunotoxicology. J immunotoxicol 5:419–444

Benedetti F, Serretti A, Colombo C, Barbini B, Lorenzi C, Campori E, Smeraldi E (2003) Influence of CLOCK gene polymorphism on circadian mood fluctuation and illness recurrence in bipolar depression. Am J Med Genet B Neuropsychiatr Genet 123B:23–26

Benedetti F, Serretti A, Pontiggia A, Bernasconi A, Lorenzi C, Colombo C, Smeraldi E (2005) Long-term response to lithium salts in bipolar illness is influenced by the glycogen synthase kinase 3-beta -50 T/C SNP. Neurosci Lett 376:51–55

Beydoun H, Saftlas AF (2008) Physical and mental health outcomes of prenatal maternal stress in human and animal studies: a review of recent evidence. Paediatr Perinat Epidemiol 22:438–466

Bilbo SD, Schwarz JM (2012) The immune system and developmental programming of brain and behavior. Front Neuroendocrinol 33:267–286

Boksa P (2004) Animal models of obstetric complications in relation to schizophrenia. Brain Res Brain Res Rev 45:1–17

Boksa P (2010) Effects of prenatal infection on brain development and behavior: a review of findings from animal models. Brain Behav Immun 24:881–897

Borisenkov MF, Kosova AL, Kasyanova ON (2012) Impact of perinatal photoperiod on the chronotype of 11- to 18-year-olds in northern European Russia. Chronobiol Int 29:305–310

Borniger JC, McHenry ZD, Abi Salloum BA, Nelson RJ (2014) Exposure to dim light at night during early development increases adult anxiety-like responses. Physiol Behav 133:99–106

Borrell J, Vela JM, Arevalo-Martin A, Molina-Holgado E, Guaza C (2002) Prenatal immune challenge disrupts sensorimotor gating in adult rats Implications for the etiopathogenesis of schizophrenia. Neuropsychopharmacology 26:204–215

Brinks V, van der Mark MH, de Kloet ER, Oitzl MS (2007) Differential MR/GR activation in mice results in emotional states beneficial or impairing for cognition. Neural plasticity 2007:90163

Brown AS (2006) Prenatal infection as a risk factor for schizophrenia. Schizophr Bull 32:200–202

Brown AS, Susser ES (2002) In utero infection and adult schizophrenia. Ment Retard Dev Disabil Res Rev 8:51–57

Budge H, Stephenson T, Symonds ME (2007) Maternal nutrient restriction is not equivalent to maternal biological stress. Curr Drug Targets 8:888–893

Cambras T, Canal MM, Cernuda-Cernuda R, Garcia-Fernandez JM, Diez-Noguera A (2015) Darkness during early postnatal development is required for normal circadian patterns in the adult rat. Chronobiol Int 32:178–186

Castrogiovanni P, Iapichino S, Pacchierotti C, Pieraccini F (1998) Season of birth in psychiatry. A review Neuropsychobiology 37:175–181

Catalani A, Alema GS, Cinque C, Zuena AR, Casolini P (2011) Maternal corticosterone effects on hypothalamus-pituitary-adrenal axis regulation and behavior of the offspring in rodents. Neurosci Biobehav Rev 35:1502–1517

Ciarleglio CM, Axley JC, Strauss BR, Gamble KL, McMahon DG (2011a) Perinatal photoperiod imprints the circadian clock. Nat Neurosci 14:25–27

Ciarleglio CM, Resuehr HE, McMahon DG (2011b) Interactions of the serotonin and circadian systems: nature and nurture in rhythms and blues. Neuroscience 197:8–16

Cirulli F, Francia N, Berry A, Aloe L, Alleva E, Suomi SJ (2009) Early life stress as a risk factor for mental health: role of neurotrophins from rodents to non-human primates. Neurosci Biobehav Rev 33:573–585

Coe CL, Kramer M, Kirschbaum C, Netter P, Fuchs E (2002) Prenatal stress diminishes the cytokine response of leukocytes to endotoxin stimulation in juvenile rhesus monkeys. J Clin Endocrinol Metab 87:675–681

Crowley SJ, Cain SW, Burns AC, Acebo C, Carskadon MA (2015) Increased sensitivity of the circadian system to light in early/mid puberty. J Clin Endocrinol Metab jc20152775

Dabkowska M, Rybakowski J (1994) Stress, depression and schizophrenia in view of psychoimmunology. Psychiatr Pol 28:23–32

Darnaudery M, Maccari S (2008) Epigenetic programming of the stress response in male and female rats by prenatal restraint stress. Brain Res Rev 57:571–585

Davis FC, Gorski RA (1985) Development of hamster circadian rhythms. I. Within-litter synchrony of mother and pup activity rhythms at weaning. Biol Reprod 33:353–362

Davis FC, Reppert SM (2001) Development of mammalian circadian rhythms. In: Takahashi JS, Turek FW, Moore RY (eds) Handbook of behavioral neurobiology, vol 12., Circadian ClocksKluwer Academic/Plenum Publishers, New York, pp 247–290

de Kloet ER (2003) Hormones, brain and stress. Endocr Regul 37:51–68

de Kloet ER, Karst H, Joëls M (2008) Corticosteroid hormones in the central stress response: Quick-and-slow. Front Neuroendocrinol 29:268–272

Deurveilher S, Semba K (2005) Indirect projections from the suprachiasmatic nucleus to major arousal-promoting cell groups in rat: implications for the circadian control of behavioural state. Neuroscience 130:165–183

El-Hennamy R, Mateju K, Bendova Z, Sosniyenko S, Sumova A (2008) Maternal control of the fetal and neonatal rat suprachiasmatic nucleus. J Biol Rhythms 23:435–444

Elenkov IJ, Chrousos GP (2002) Stress hormones, proinflammatory and antiinflammatory cytokines, and autoimmunity. Ann N Y Acad Sci 966:290–303

Erren TC, Koch MS, Gross JV, Reiter RJ, Meyer-Rochow VB (2012) A possible role of perinatal light in mood disorders and internal cancers: reconciliation of instability and latitude concepts. Neuro Endocrinol Lett 33:314–317

Eskandari F, Sternberg EM (2002) Neural-immune interactions in health and disease. Ann N Y Acad Sci 966:20–27

Fan X, Goff DC, Henderson DC (2007) Inflammation and schizophrenia. Expert Rev Neurother 7:789–796

Fortier ME, Joober R, Luheshi GN, Boksa P (2004) Maternal exposure to bacterial endotoxin during pregnancy enhances amphetamine-induced locomotion and startle responses in adult rat offspring. J Psychiatr Res 38:335–345

Fortier ME, Luheshi GN, Boksa P (2007) Effects of prenatal infection on prepulse inhibition in the rat depend on the nature of the infectious agent and the stage of pregnancy. Behav Brain Res 181:270–277

Foster RG, Roenneberg T (2008) Human responses to the geophysical daily, annual and lunar cycles. Curr Biol 18:R784–R794

Garcia-Bueno B, Caso JR, Leza JC (2008) Stress as a neuroinflammatory condition in brain: damaging and protective mechanisms. Neurosci Biobehav Rev 32:1136–1151

Gayle DA, Beloosesky R, Desai M, Amidi F, Nunez SE, Ross MG (2004) Maternal LPS induces cytokines in the amniotic fluid and corticotropin releasing hormone in the fetal rat brain. Am J Physiol Regul Integr Comp Physiol 286:R1024–R1029

Gibson EM, Wang C, Tjho S, Khattar N, Kriegsfeld LJ (2010) Experimental 'jet lag' inhibits adult neurogenesis and produces long-term cognitive deficits in female hamsters. PLoS ONE 5:e15267

Golan HM, Lev V, Hallak M, Sorokin Y, Huleihel M (2005) Specific neurodevelopmental damage in mice offspring following maternal inflammation during pregnancy. Neuropharmacology 48:903–917

Gotz AA, Stefanski V (2007) Psychosocial maternal stress during pregnancy affects serum corticosterone, blood immune parameters and anxiety behaviour in adult male rat offspring. Physiol Behav 90:108–115

Green NH, Jackson CR, Iwamoto H, Tackenberg MC, McMahon DG (2015) Photoperiod programs dorsal raphe serotonergic neurons and affective behaviors. Curr Biol 25:1389–1394

Hastings M, O'Neill JS, Maywood ES (2007) Circadian clocks: regulators of endocrine and metabolic rhythms. J Endocrinol 195:187–198

Honma K, Honma S (2009) The SCN-independent clocks, methamphetamine and food restriction. Eur J Neurosci 30:1707–1717

Houdek P, Sumova A (2014) In vivo initiation of clock gene expression rhythmicity in fetal rat suprachiasmatic nuclei. PLoS ONE 9:e107360

Hyman SL, Arndt TL, Rodier PM (2006) Environmental agents and autism: once and future associations Intern Rev Res. Ment Retard 30:171–194

Iuvone PM, Tosini G, Pozdeyev N, Haque R, Klein DC, Chaurasia SS (2005) Circadian clocks, clock networks, arylalkylamine N-acetyltransferase, and melatonin in the retina. Prog Retin Eye Res 24:433–456

Jacobson L (2014) Hypothalamic-pituitary-adrenocortical axis: neuropsychiatric aspects. Compr Physiol 4:715–738

Karatsoreos IN (2014) Links between circadian rhythms and psychiatric disease. Front Behav Neurosci 8:162

Karrow NA (2006) Activation of the hypothalamic–pituitary–adrenal axis and autonomic nervous system during inflammation and altered programming of the neuroendocrine–immune axis during fetal and neonatal development: Lessons learned from the model inflammagen, lipopolysaccharide. Brain Behav Immun 20:144–158

Koehl M, Darnaudéry M, Dulluc J, Van Reeth O, Moal ML, Maccari S (1999) Prenatal stress alters circadian activity of hypothalamo–pituitary–adrenal axis and hippocampal corticosteroid receptors in adult rats of both gender. J Neurobiol 40:302–315

Koenig JI et al (2005) Prenatal exposure to a repeated variable stress paradigm elicits behavioral and neuroendocrinological changes in the adult offspring: potential relevance to schizophrenia. Behav Brain Res 156:251–261

Kovacikova Z, Sladek M, Laurinova K, Bendova Z, Illnerova H, Sumova A (2005) Ontogenesis of photoperiodic entrainment of the molecular core clockwork in the rat suprachiasmatic nucleus. Brain Res 1064:83–89

Labrecque N, Cermakian N (2015) Circadian clocks in the immune system. J Biol Rhythms 30:277–290

Laviola G, Adriani W, Rea M, Aloe L, Alleva E (2004a) Social withdrawal, neophobia, and stereotyped behavior in developing rats exposed to neonatal asphyxia. Psychopharmacology 175:196–205

Laviola G, Ognibene E, Romano E, Adriani W, Keller F (2009) Gene-environment interaction during early development in the heterozygous reeler mouse: clues for modelling of major neurobehavioral syndromes. Neurosci Biobehav Rev 33:560–572

Laviola G et al (2004b) Beneficial effects of enriched environment on adolescent rats from stressed pregnancies. Eur J Neurosci 20:1655–1664

Lazinski MJ, Shea AK, Steiner M (2008) Effects of maternal prenatal stress on offspring development: a commentary. Arch Women's Ment Health 11:363–375

Lee PR, Brady DL, Shapiro RA, Dorsa DM, Koenig JI (2007) Prenatal stress generates deficits in rat social behavior: Reversal by oxytocin. Brain Res 1156:152–167

Lemaire V, Koehl M, Le Moal M, Abrous DN (2000) Prenatal stress produces learning deficits associated with an inhibition of neurogenesis in the hippocampus. Proc Natl Acad Sci USA 97:11032–11037

Levine S (1994) The ontogeny of the hypothalamic-pituitary-adrenal axis. Influence Matern Factors Ann NY Acad Sci 746:275–288; discussion 289–293

Levine S (2000) Influence of psychological variables on the activity of the hypothalamic-pituitary-adrenal axis. Eur J Pharmacol 405:149–160

Lin Y-L, Lin S-Y, Wang S (2012) Prenatal lipopolysaccharide exposure increases anxiety-like behaviors and enhances stress-induced corticosterone responses in adult rats. Brain Behav Immun 26:459–468

Lowe GC, Luheshi GN, Williams S (2008) Maternal infection and fever during late gestation are associated with altered synaptic transmission in the hippocampus of juvenile offspring rats. Am J Physiol Regul Integr Comp Physiol 295:R1563–R1571

Lucas SM, Rothwell NJ, Gibson RM (2006) The role of inflammation in CNS injury and disease. Br J Pharmacol 147(1):S232–S240

Llorente E, Brito ML, Machado P, Gonzalez MC (2002) Effect of prenatal stress on the hormonal response to acute and chronic stress and on immune parameters in the offspring. J Physiol Biochem 58:143–149

Maccari S, Darnaudery M, Morley-Fletcher S, Zuena AR, Cinque C, Van Reeth O (2003) Prenatal stress and long-term consequences: implications of glucocorticoid hormones. Neurosci Biobehav Rev 27:119–127

Maccari S, Morley-Fletcher S (2007) Effects of prenatal restraint stress on the hypothalamus-pituitary-adrenal axis and related behavioural and neurobiological alterations Psychoneuroendocrinology 32(1):S10–S15

MacLaughlin SM, McMillen IC (2007) Impact of periconceptional undernutrition on the development of the hypothalamo-pituitary-adrenal axis: does the timing of parturition start at conception? Curr Drug Targets 8:880–887

Macri S, Zoratto F, Laviola G (2011) Early-stress regulates resilience, vulnerability and experimental validity in laboratory rodents through mother-offspring hormonal transfer. Neurosci Biobehav Rev 35:1534–1543

Mairesse J et al (2013) Chronic agomelatine treatment corrects the abnormalities in the circadian rhythm of motor activity and sleep/wake cycle induced by prenatal restraint stress in adult rats. Int J Neuropsychopharmacol 16:323–338

Malek ZS, Sage D, Pevet P, Raison S (2007) Daily rhythm of tryptophan hydroxylase-2 messenger ribonucleic acid within raphe neurons is induced by corticoid daily surge and modulated by enhanced locomotor activity. Endocrinology 148:5165–5172

Marco EM, Llorente R, Lopez-Gallardo M, Mela V, Llorente-Berzal A, Prada C, Viveros MP (2015) The maternal deprivation animal model revisited. Neurosci Biobehav Rev 51:151–163

Marco EM, Macri S, Laviola G (2011) Critical age windows for neurodevelopmental psychiatric disorders: evidence from animal models. Neurotox Res 19:286–307

Martinac M, Pehar D, Karlovic D, Babic D, Marcinko D, Jakovljevic M (2014) Metabolic syndrome, activity of the hypothalamic-pituitary-adrenal axis and inflammatory mediators in depressive disorder. Acta Clin Croat 53:55–71

Mastorci F et al (2009) Long-term effects of prenatal stress: changes in adult cardiovascular regulation and sensitivity to stress. Neurosci Biobehav Rev 33:191–203

Meerlo P, Sgoifo A, Suchecki D (2008) Restricted and disrupted sleep: effects on autonomic function, neuroendocrine stress systems and stress responsivity. Sleep Med Rev 12:197–210

Mendez N et al (2012) Timed maternal melatonin treatment reverses circadian disruption of the fetal adrenal clock imposed by exposure to constant light. PLoS ONE 7:e42713

Mendlewicz J (2009) Disruption of the circadian timing systems: molecular mechanisms in mood disorders. CNS Drugs 23(2):15–26

Mendoza J (2007) Circadian clocks: setting time by food. J Neuroendocrinol 19:127–137

Meyer U, Feldon J (2009) Neural basis of psychosis-related behaviour in the infection model of schizophrenia. Behav Brain Res 204:322–334

Meyer U, Feldon J, Fatemi SH (2009) In-vivo rodent models for the experimental investigation of prenatal immune activation effects in neurodevelopmental brain disorders. Neurosci Biobehav Rev 33:1061–1079

Meyer U, Feldon J, Schedlowski M, Yee BK (2006a) Immunological stress at the maternal-foetal interface: a link between neurodevelopment and adult psychopathology. Brain Behav Immun 20:378–388

Meyer U et al (2006b) The time of prenatal immune challenge determines the specificity of inflammation-mediated brain and behavioral pathology. J Neurosci 26:4752–4762

Mill J, Petronis A (2008) Pre- and peri-natal environmental risks for attention-deficit hyperactivity disorder (ADHD): the potential role of epigenetic processes in mediating susceptibility. J Child Psychol Psychiatry 49:1020–1030

Missault S et al (2014) The risk for behavioural deficits is determined by the maternal immune response to prenatal immune challenge in a neurodevelopmental model. Brain Behav Immun 42:138–146

Mittal VA, Ellman LM, Cannon TD (2008) Gene-environment interaction and covariation in schizophrenia: the role of obstetric complications. Schizophr Bull 34:1083–1094

Mongrain V, Paquet J, Dumont M (2006) Contribution of the photoperiod at birth to the association between season of birth and diurnal preference. Neurosci Lett 406:113–116

Monteleone P, Martiadis V, Maj M (2011) Circadian rhythms and treatment implications in depression. Prog Neuropsychopharmacol Biol Psychiatry 35:1569–1574

Morin LP (2013) Neuroanatomy of the extended circadian rhythm system. Exp Neurol 243:4–20

Morley-Fletcher S, Darnaudery M, Koehl M, Casolini P, Van Reeth O, Maccari S (2003a) Prenatal stress in rats predicts immobility behavior in the forced swim test. Effects of a chronic treatment with tianeptine. Brain Res 989:246–251

Morley-Fletcher S, Rea M, Maccari S, Laviola G (2003b) Environmental enrichment during adolescence reverses the effects of prenatal stress on play behaviour and HPA axis reactivity in rats. Euro J Neurosci 18:3367–3374

Mukherjee S et al (2010) Knockdown of clock in the ventral tegmental area through RNA interference results in a mixed state of mania and depression-like behavior. Biol Psychiatry 68:503–511

Natale V, Di Milia L (2011) Season of birth and morningness: comparison between the northern and southern hemispheres. Chronobiol Int 28:727–730

Nicolaides NC, Charmandari E, Chrousos GP, Kino T (2014) Circadian endocrine rhythms: the hypothalamic–pituitary–adrenal axis and its actions. Ann NY Acad Sci 1318:71–80

Nievergelt CM et al (2006) Suggestive evidence for association of the circadian genes PERIOD3 and ARNTL with bipolar disorder. Am J Med Genet B Neuropsychiatr Genet 141B:234–241

Novakova M, Prasko J, Latalova K, Sladek M, Sumova A (2015) The circadian system of patients with bipolar disorder differs in episodes of mania and depression. Bipolar Disord 17:303–314

O'Donnell K, O'Connor TG, Glover V (2009) Prenatal stress and neurodevelopment of the child: focus on the HPA axis and role of the placenta. Dev Neurosci 31:285–292

Oitzl MS, Champagne DL, van der Veen R, de Kloet ER (2010) Brain development under stress: hypotheses of glucocorticoid actions revisited. Neurosci Biobehav Rev 34:853–866

Oitzl MS, de Kloet ER, Joels M, Schmid W, Cole TJ (1997) Spatial learning deficits in mice with a targeted glucocorticoid receptor gene disruption. Euro J Neurosci 9:2284–2296

Okamura H, Yamaguchi S, Yagita K (2002) Molecular machinery of the circadian clock in mammals. Cell Tissue Res 309:47–56

Olejnikova L, Polidarova L, Pauslyova L, Sladek M, Sumova A (2015) Diverse development and higher sensitivity of the circadian clocks to changes in maternal-feeding regime in a rat model of cardio-metabolic disease. Chronobiol Int 32:531–547

Pang D, Syed S, Fine P, Jones PB (2009) No association between prenatal viral infection and depression in later life–a long-term cohort study of 6152 subjects. Can J Psychiatry 54:565–570

Partonen T et al (2007) Three circadian clock genes Per2, Arntl, and Npas2 contribute to winter depression. Ann Med 39:229–238

Patterson PH (2007) Neuroscience. Maternal effects on schizophrenia risk. Science 318:576–577

Pearce BD (2001) Schizophrenia and viral infection during neurodevelopment: a focus on mechanisms. Mol Psychiatry 6:634–646

Perlman JM (2001) Neurobehavioral deficits in premature graduates of intensive care–potential medical and neonatal environmental risk factors. Pediatrics 108:1339–1348

Polidarova L, Olejnikova L, Pauslyova L, Sladek M, Sotak M, Pacha J, Sumova A (2014) Development and entrainment of the colonic circadian clock during ontogenesis. Am J Physiol Gastrointest Liver Physiol 306:G346–G356

Powell WT, LaSalle JM (2015) Epigenetic mechanisms in diurnal cycles of metabolism and neurodevelopment. Hum Mol Genet R1:R1–R9

Pryce CR, Ruedi-Bettschen D, Dettling AC, Weston A, Russig H, Ferger B, Feldon J (2005) Long-term effects of early-life environmental manipulations in rodents and primates: potential animal models in depression research. Neurosci Biobehav Rev 29:649–674

Pyter LM, Nelson RJ (2006) Enduring effects of photoperiod on affective behaviors in Siberian hamsters (Phodopus sungorus). Behav Neurosci 120:125–134

Reppert SM, Uhl GR (1987) Vasopressin messenger ribonucleic acid in supraoptic and suprachiasmatic nuclei: appearance and circadian regulation during development. Endocrinology 120:2483–2487

Reul JM, Stec I, Wiegers GJ, Labeur MS, Linthorst AC, Arzt E, Holsboer F (1994) Prenatal immune challenge alters the hypothalamic-pituitary-adrenocortical axis in adult rats. J Clin Investig 93:2600–2607

Rice F, Jones I, Thapar A (2007) The impact of gestational stress and prenatal growth on emotional problems in offspring: a review. Acta Psychiatr Scand 115:171–183

Ridley RM (1994) The psychology of perserverative and stereotyped behaviour. Prog Neurobiol 44:221–231

Roman E, Karlsson O (2013) Increased anxiety-like behavior but no cognitive impairments in adult rats exposed to constant light conditions during perinatal development. Ups J Med Sci 118:222–227

Romero E, Ali C, Molina-Holgado E, Castellano B, Guaza C, Borrell J (2007) Neurobehavioral and immunological consequences of prenatal immune activation in rats. Influence Antipsychotics Neuropsychopharmacol 32:1791–1804

Romero E, Guaza C, Castellano B, Borrell J (2010) Ontogeny of sensorimotor gating and immune impairment induced by prenatal immune challenge in rats: implications for the etiopathology of schizophrenia. Mol Psychiatry 15:372–383

Roybal K et al (2007) Mania-like behavior induced by disruption of CLOCK. Proc Nat Acad Sci USA 104:6406–6411

Samsom JN, Wong AH (2015) Schizophrenia and depression co-morbidity: what we have learned from animal models. Front Psychiatry 6:13

Samuelsson AM, Jennische E, Hansson HA, Holmang A (2006) Prenatal exposure to interleukin-6 results in inflammatory neurodegeneration in hippocampus with NMDA/GABA(A) dysregulation and impaired spatial learning. Am J Physiol Regul Integr Comp Physiol 290:R1345–R1356

Savalli G, Diao W, Schulz S, Todtova K, Pollak DD (2014) Diurnal oscillation of amygdala clock gene expression and loss of synchrony in a mouse model of depression. Int J Neuropsychopharmacol 18

Schiepers OJ, Wichers MC, Maes M (2005) Cytokines and major depression. Prog Neuropsychopharmacol Biol Psychiatry 29:201–217

Seron-Ferre M et al (2013) Impact of chronodisruption during primate pregnancy on the maternal and newborn temperature rhythms. PLoS ONE 8:e57710

Seron-Ferre M, Valenzuela GJ, Torres-Farfan C (2007) Circadian clocks during embryonic and fetal development. Birth Defects Res C Embryo Today 81:204–214

Serretti A, Benedetti F, Mandelli L, Lorenzi C, Pirovano A, Colombo C, Smeraldi E (2003) Genetic dissection of psychopathological symptoms: insomnia in mood disorders and CLOCK gene polymorphism. Am J Med Genet B Neuropsychiatr Genet 121B:35–38

Shibata S, Moore RY (1987) Development of neuronal activity in the rat suprachiasmatic nucleus. Brain Res 431:311–315

Smith BN, Sollars PJ, Dudek FE, Pickard GE (2001) Serotonergic modulation of retinal input to the mouse suprachiasmatic nucleus mediated by 5-HT1B and 5-HT7 receptors. J Biol Rhythms 16:25–38

Spiga F, Walker JJ, Terry JR, Lightman SL (2011) HPA Axis-Rhythms. In: Comprehensive Physiology. Wiley, Hoboken

Takao M, Kurachi T, Kato H (2009) Photoperiod at birth does not modulate the diurnal preference in asian population. Chronobiol Int 26:1470–1477

Talge NM, Neal C, Glover V (2007) Antenatal maternal stress and long-term effects on child neurodevelopment: how and why? J Child Psychol Psychiatry 48:245–261

Tohmi M, Tsuda N, Watanabe Y, Kakita A, Nawa H (2004) Perinatal inflammatory cytokine challenge results in distinct neurobehavioral alterations in rats: implication in psychiatric disorders of developmental origin. Neurosci Res 50:67–75

Tonetti L, Fabbri M, Martoni M, Natale V (2012) Season of birth and mood seasonality in late childhood and adolescence. Psychiatry Res 195:66–68

Tuchscherer M, Kanitz E, Otten W, Tuchscherer A (2002) Effects of prenatal stress on cellular and humoral immune responses in neonatal pigs. Vet Immunol Immunopathol 86:195–203

Vilches N et al (2014) Gestational chronodisruption impairs hippocampal expression of NMDA receptor subunits Grin1b/Grin3a and spatial memory in the adult offspring. PLoS ONE 9:e91313

Ward IL, Stehm KE (1991) Prenatal stress feminizes juvenile play patterns in male rats. Physiol Behav 50:601–605

Watanabe Y et al (2004) Neonatal impact of leukemia inhibitory factor on neurobehavioral development in rats. Neurosci Res 48:345–353

Weinstock M (2002) Can the behaviour abnormalities induced by gestational stress in rats be prevented or reversed? Stress 5:167–176

Weinstock M (2008) The long-term behavioural consequences of prenatal stress. Neurosci Biobehav Rev 32:1073–1086

Weirnert D (2005) Ontogenic development of the mammalian circadian system. Chronobiol Int 22:179–205

Wingenfeld K, Wolf OT (2011) HPA axis alterations in mental disorders: impact on memory and its relevance for therapeutic interventions. CNS Neurosci Ther 17:714–722

Wischhof L, Irrsack E, Osorio C, Koch M (2015) Prenatal LPS-exposure–a neurodevelopmental rat model of schizophrenia–differentially affects cognitive functions, myelination and parvalbumin expression in male and female offspring. Prog Neuropsychopharmacol Biol Psychiatry 57:17–30

Zuena AR et al (2008) Prenatal restraint stress generates two distinct behavioral and neurochemical profiles in male and female rats. PLoS ONE 3:e2170

Neurobehavioral Effects from Developmental Methamphetamine Exposure

Sarah A. Jablonski, Michael T. Williams and Charles V. Vorhees

Abstract Intrauterine methamphetamine exposure adversely affects the neurofunctional profile of exposed children, leading to a variety of higher order cognitive deficits, such as decreased attention, reduced working-memory capability, behavioral dysregulation, and spatial memory impairments (Kiblawi et al. in J Dev Behav Pediatr 34:31–37, 2013; Piper et al. in Pharmacol Biochem Behav 98:432–439 2011; Roussotte et al. in Neuroimage 54:3067–3075, 2011; Twomey et al. in Am J Orthopsychiatry 83:64–72, 2013). In animal models of developmental methamphetamine, both neuroanatomical and behavioral outcomes critically depend on the timing of methamphetamine administration. Methamphetamine exposure during the third trimester human equivalent period of brain development results in well-defined and persistent wayfinding and spatial navigation deficits in rodents (Vorhees et al. in Neurotoxicol Teratol 27:117–134, 2005, Vorhees et al. in Int J Dev Neurosci 26:599–610, 2008; Vorhees et al. in Int J Dev Neurosci 27:289–298, 2009; Williams et al. in Psychopharmacology (Berl) 168:329–338, 2003b), whereas drug delivery during the first and second trimester equivalents produces no such effect (Acuff-Smith et al. in Neurotoxicol Teratol 18:199–215, 1996; Schutova et al. in Physiol Res 58:741–750, 2009a; Slamberova et al. in Naunyn Schmiedebergs Arch Pharmacol 380:109–114, 2009, Slamberova et al. in Physiol Res 63:S547–S558, 2014b). In this review, we examine the impact of developmental metham-

S.A. Jablonski (✉)
Division of Neurology, Cincinnati Children's Hospital Medical Center,
3333 Burnet Ave., R-1447, Cincinnati, OH 45229-3039, USA
e-mail: sarah.jablonski@cchmc.org

M.T. Williams
Division of Neurology, Cincinnati Children's Hospital Medical Center and Department
of Pediatrics, University of Cincinnati College of Medicine,
45229 Cincinnati, OH, USA
e-mail: michael.williams@cchmc.org

C.V. Vorhees
Division of Neurology, Cincinnati Children's Hospital Medical Center and Department
of Pediatrics, University of Cincinnati College of Medicine,
3333 Burnet Ave., MLC 7044, 45229-3039 Cincinnati, OH, USA
e-mail: charles.vorhees@cchmc.org

© Springer International Publishing Switzerland 2015
Curr Topics Behav Neurosci (2016) 29: 183–230
DOI 10.1007/7854_2015_405
Published Online: 01 November 2015

phetamine on emerging neural circuitry, neurotransmission, receptor changes, and behavioral outcomes in animal models. The review is organized by type of effects and timing of drug exposure (prenatal only, pre- and neonatal, and neonatal only). The findings elucidate functional patterns of interconnected brain structures (e.g., frontal cortex and striatum) and neurotransmitters (e.g., dopamine and serotonin) involved in methamphetamine-induced developmental neurotoxicity.

Keywords Methamphetamine · Prenatal · Postnatal · Spatial learning · Egocentric learning · Allocentric learning · Dopamine · Serotonin · Glutamate

Contents

1 Introduction.. 184
   1.1 Neuroimaging Studies.. 185
2 Neurobehavioral Effects of Prenatal Methamphetamine Exposure............................ 187
   2.1 Introduction.. 187
   2.2 Long-Term Neurobehavioral Effects... 187
   2.3 Neurotransmitter Effects.. 191
   2.4 Prenatal Methamphetamine and Evoked Seizures... 194
3 Neurobehavioral Effects of Pre- and Postnatal Methamphetamine Exposure............ 199
   3.1 Maternal Behavior and Sensorimotor Function 199
   3.2 Spatial Learning Effects... 200
4 Neurobehavioral Effects of Neonatal Methamphetamine Exposure........................... 201
   4.1 Introduction.. 201
   4.2 Intrinsic Drug Effects... 202
   4.3 Long-Term Neurobehavioral Effects 203
   4.4 Neurotransmitter Effects 205
   4.5 Stress Effects.. 210
5 Experimental Design: Review by McDonnell-Dowling and Kelly (2015).................. 212
6 Conclusions.. 220
References.. 221

1 Introduction

Methamphetamine use among women of childbearing age is a persistent concern throughout the world. From 2009 to 2012, methamphetamine use among women seeking drug treatment rose from 18 to 27 % (Substance Abuse and Mental Health Services Administration 2013). Methamphetamine is a potent psychostimulant that blocks dopamine, serotonin, and norepinephrine reuptake from synapses as well as into synaptic vesicles (Sekine et al. 2006; Sulzer et al. 2005; Volkow et al. 2001a). Methamphetamine abuse in adults has been linked to a range of negative central nervous system effects, including tachycardia, respiratory problems, confusion, increased wakefulness, hyperactivity, hypertension, decreased appetite, and extreme weight loss (Plessinger 1998; Wijetunga et al. 2003). Chronic methamphetamine users exhibit alterations in regional brain volume (particularly in the striatum), abnormal balance of brain neurotransmitters, decreased dopamine receptors, and alterations in glucose metabolism (for review, see Chang et al. 2007). In addition to its direct properties, methamphetamine use during pregnancy exerts significant transplacental effects through direct and indirect mechanisms; indirect mechanisms include vasoconstriction and restriction of nutrients and oxygen to the fetus (Bartu et al. 2009). Because methamphetamine can cross the placenta (Rambousek et al. 2014), the fetus is also directly exposed to the drug, significantly increasing the likelihood of adverse effects on brain development.

The infant development, environment, and lifestyle (IDEAL) study is a comprehensive, prospective cohort study, which addresses the long-term effects of prenatal methamphetamine exposure on affected populations in the USA and New Zealand. Ongoing reports describe that methamphetamine exposure is associated with newborn neurobehavioral patterns of decreased arousal, increased stress and cortisol reactivity, and poor quality of movement (Kirlic et al. 2013; LaGasse et al. 2011; Smith et al. 2008). Increased prematurity, fetal loss, and incidence of small gestational age are also evident (Brecht and Herbeck 2014; Chomchai et al. 2004; Nguyen et al. 2010). Prenatal methamphetamine exposure results in increased neonatal intensive care unit admissions and poor infant sucking (Shah et al. 2012), smaller head circumference, and low birth weight (Abar et al. 2014; Chomchai et al. 2004; Smith et al. 2006), as well as reduced height and motor deficits during early childhood (Roos et al. 2014, 2015; Smith et al. 2011; Wouldes et al. 2014; Zabaneh et al. 2012). The IDEAL study elucidates specific endophenotypes (e.g., attention, memory) that provide behavioral correlates to structural and functional brain abnormalities obtained from imaging data on prenatal methamphetamine-exposed children (Derauf et al. 2009; Smith et al. 2015; see next section).

1.1 Neuroimaging Studies

Magnetic resonance (MR)-based brain imaging methods provide visualization of brain structures and functional activity patterns that are most vulnerable to the neuroteratogenic effects of drugs of abuse. For example, diffusion tensor imaging (DTI) studies reveal abnormalities in frontal and parietal white matter microstructure, on measures of fractional anisotropy (FA) and axial diffusivity (AD) in children after prenatal methamphetamine exposure (Cloak et al. 2009; Colby et al. 2012; Roos et al. 2015). These findings suggest white matter integrity and myelination and/or axonal density are disrupted following prenatal methamphetamine. In addition, methamphetamine-induced changes in cellular metabolism are evident from brain proton MR spectroscopy, in which increased creatine levels were reported in both the striatum and frontal lobe and increased glutamate/glutamine (GLX) concentrations in the frontal lobe (Chang et al. 2009; Smith et al. 2001).

Other studies have demonstrated neuroanatomical and physiological changes that correlate with functional differences among groups. For example, quantitative morphological analysis showed smaller putamen, globus pallidus, and hippocampal volumes in children exposed to prenatal methamphetamine compared with those without drug exposure. These regional volume reductions correlated with poorer performance in attention and verbal memory (Chang et al. 2004). Reduction in striatal volumes were also linked with lower full-scale intelligence quotient (FSIQ) scores in affected children (Sowell et al. 2010), and reduced caudate nucleus volumes and cortical thickness were associated with attention and reaction time deficits (Derauf et al. 2012a).

Localized brain abnormalities resulting from alterations and/or circuit rewiring in corticostriatal networks are evident following prenatal methamphetamine. Using functional magnetic resonance imaging (fMRI), two studies (Roussotte et al. 2011, 2012) examined blood oxygenation level-dependent (BOLD) fluctuations during a working-memory task. Compared with controls, methamphetamine-exposed subjects displayed increased connectivity between the putamen and the frontal brain regions, while caudate showed decreased connectivity within the same areas. Similarly, Lu et al. (2009) demonstrated more diffuse brain activation in medial temporal structures during verbal learning in children with prenatal methamphetamine exposure compared with unexposed controls (Lu et al. 2009). These studies suggest that compensatory and/or alternate brain regions/systems are activated in children prenatally exposed to methamphetamine.

Alterations in several brain networks are likely the cause of performance deficits and behavioral difficulties in children exposed to methamphetamine in utero. For example, drug-exposed children exhibited impaired learning in a spatial memory task, requiring spatial navigation to a hidden target in a virtual environment (Piper et al. 2011). Executive functional deficits, as well as a variety of outcomes predictive of ADHD, including problems with behavioral disinhibition and attention, have been reported in a number of studies (Abar et al. 2013; Derauf et al. 2012b; Himes et al. 2014; Kiblawi et al. 2013; LaGasse et al. 2011, 2012; Roos et al. 2015).

In summary, brain imaging studies on prenatal methamphetamine exposure in humans show significant effects on fronto-striatal networks. These dopamine-rich areas (e.g., frontal cortex, basal ganglia) are known to be affected in adult methamphetamine users, as well as in animal models of developmental methamphetamine exposure (see Sect. 4).

This review describes experiments concerning developmental methamphetamine treatment in animal models conducted to enhance an understanding of the pathophysiology involved in the neurobehavioral effects of methamphetamine-induced neurotoxicity. In an examination characterizing methamphetamine usage during pregnancy, Della Grotta et al. (2010) reported that 84.3 % of drug-abusing women used methamphetamine in the first trimester, 56.0 % in the second trimester, and 42.4 % continued to use methamphetamine during the third trimester. Because the timing of methamphetamine exposure significantly influences the degree of effects, the review is organized by outcomes of methamphetamine on brain and behavior in rodent progeny after drug administration during the prenatal period (Sect. 2), throughout both gestation and lactation (Sect. 3), and during the postnatal period only (Sect. 4).

2 Neurobehavioral Effects of Prenatal Methamphetamine Exposure

2.1 Introduction

Prenatal methamphetamine exposure, as reviewed in the following discussion, refers to drug administration during a subset of or throughout the entire prenatal period. This period of brain development, from embryonic day (E) 1–21, corresponds to the first (E1–9) and second (E10–21) human trimester equivalents, during which organogenesis, neurulation, and histogenesis occur and during which neuronal proliferation, migration, differentiation, synaptogenesis, apoptosis, and gliogenesis begin and expand rapidly (Rice and Barone 2000). Methamphetamine exposure during this time leads to a range of physiological and anatomical abnormalities, such as reduced offspring body weight gain and growth, cleft palate, optic nerve damage, and delays in eye opening, that are generally dose-dependent (Acuff-Smith et al. 1996; Cho et al. 1991; McDonnell-Dowling et al. 2014; Melo et al. 2006, 2008; Mirjalili et al. 2013; Yamamoto et al. 1992). Regional neurogenesis in rats occurs during this time period in the medulla, pons, thalamus, hypothalamus, hippocampal sub-regions, and some parts of the cerebellum (Bayer et al. 1993; Rice and Barone 2000). Targeting of one or many of these brain regions can cause behavioral deficits, such as impairments in surface and air-righting, inclined plane test, locomotor activity, and motor coordination in methamphetamine-exposed progeny (Cho et al. 1991; Motz and Alberts 2005; Slamberova et al. 2006). Importantly, methamphetamine administration to gravid animals rapidly exposes the embryo and fetus to the drug, leading to high methamphetamine concentrations in brain regions that are important for learning and memory (e.g., Burchfield et al. 1991; Won et al. 2001). The following section concerns the neurobehavioral impact of methamphetamine exposure in the first and second human trimester equivalents.

2.2 Long-Term Neurobehavioral Effects

2.2.1 Locomotor Activity

Prenatal methamphetamine exposure induces a reduction in spontaneous locomotor activity when examined during adolescence. For example, one study (Weissman and Caldecott-Hazard 1993) showed that methamphetamine given throughout gestation (10 mg/kg/day) to rats resulted in a significant decrease in square crossings and rearing in an open-field when measured at P30. Similarly, another study (Sato and Fujiwara 1986) reported a decrease in motor activity, but an increase in vertical activity, in P24–27 rats after a lower dose of methamphetamine (2 mg/kg/day) administered throughout gestation. This reduction in locomotor activity was also seen in P21 offspring following drug exposure during a narrower time window (E7–20) and at lower methamphetamine doses (1–4.5 mg/kg/day; Cho et al. 1991). Thus, gestational methamphetamine exerts an overall decrease in activity in open-field testing in young rats.

Other studies explored the effect of prenatal methamphetamine (10 mg/kg/day, throughout gestation) on locomotor activity, in combination with acute application of the same drug during adulthood. Slamberova and colleagues used the Laboras apparatus (an alternate version of the open-field), during which adult animals are placed in a cage that resembles the home cage (filled with bedding material and ad libitum food and water). The cage is placed on a sensor platform, and the movement (speed, distance travelled, time spent immobile) is automatically measured (Schutova et al. 2013; Slamberova et al. 2011, 2013). In one of these studies, the authors (Schutova et al. 2013) showed a reduction in movement speed in methamphetamine-exposed females, but not in males, whereas acute methamphetamine administered in adulthood (1 mg/kg) increased locomotor activity in both sexes, irrespective of prenatal drug exposure history. In a similar experiment, these authors (Slamberova et al. 2013) reported that, regardless of acute methamphetamine exposure in adulthood, female offspring exposed to methamphetamine prenatally were faster, spent more time in motion, and traveled longer distances compared with prenatally saline-exposed females. It is possible that the discrepancies in locomotor findings in female rats are dependent on the stage of the estrous cycle during testing. Indeed, naïve females in proestrus and estrous (when ovarian hormone levels are high) are more active on measures of speed, time in locomotion, and distance travelled, compared with females in diestrus (when hormones are low). In an experiment in adult males, prenatal methamphetamine-exposed rats spent significantly less time immobile compared with prenatal saline-treated rats. Additionally, acute methamphetamine (1 mg/kg) increased the time spent in locomotion in prenatal methamphetamine-exposed, but not prenatal saline-exposed, males (Slamberova et al. 2011). In general, prenatal methamphetamine exposure leads to sex-/hormone-dependent increases in locomotor activity during adulthood.

In summary, findings involving the degree of spontaneous locomotor activity observed in prenatal methamphetamine-exposed versus saline-exposed rats shows that (1) the exposure to a novel versus familiar environment influences activity measures; (2) the effect of prenatal methamphetamine depends on whether animals are tested during the preweanling or weanling periods (suppression of activity) versus during adulthood (increases in activity); and (3) once animals reach maturity, activity becomes sex-dependent. Indeed, it has been reported (Sato and Fujiwara 1986) that there is no effect of sex on activity measures at P30; however, sex differences are evident once the progeny reach adulthood (Macuchova et al. 2013; Schutova et al. 2008; Slamberova et al. 2011, 2013). (4) Adult females may be more susceptible to the effects of prenatal methamphetamine compared with adult males, and (5) acute methamphetamine administration during adulthood increases locomotor activity, but the influence of prenatal methamphetamine on this response may be sex-dependent.

2.2.2 Spatial Navigation

Learning in the Morris water maze (MWM) is unaffected by prenatal methamphetamine. This lack of effect is apparent in both cued learning and spatial memory variants of the procedure. During place navigation, rats learn the location of a fixed platform after placement in pseudorandom start positions between trials. Data published by Slamberova and colleagues have shown that methamphetamine exposure (5 mg/kg/day, throughout gestation) does not influence place navigation, in either sex (Macuchova et al. 2013; Schutova et al. 2008, 2009a, b; Slamberova et al. 2014b). Others have found that prenatal methamphetamine does not alter place navigational learning following longer exposure periods (one week prior to mating and throughout gestation) or using lower or higher doses of methamphetamine (0.625–10 mg/kg/day; McDonnell-Dowling et al. 2014; Weissman and Caldecott-Hazard 1993).

Rats with prenatal methamphetamine exposure (5 mg/kg/day) surprisingly had shorter latencies to locate a fixed platform during a retention test (retention memory) that occurred 7 days after completion of the acquisition navigation test (Schutova et al. 2008, 2009b; Slamberova et al. 2014b). Such improved learning in methamphetamine-exposed rats was later attributed to an increased use of spatial search strategies (focal search in target quadrant, spatial search with direct swim path in the target quadrant) in methamphetamine-exposed animals, compared with those prenatally exposed to saline (Schutova et al. 2009a). Prenatal methamphetamine had no effect on "new learning" in which the escape platform was located in a different quadrant on each of 4 consecutive days (Slamberova et al. 2005c). Using a similar testing procedure, one study (Acuff-Smith et al. 1996) tested offspring of pregnant dams treated with saline, 5, 10, 15, or 20 mg/kg/b.i.d. methamphetamine from either E7–12 or E13–18. In this experiment, rats were required to learn new locations of the platform after 4 days of stable platform testing. Unlike (Slamberova et al. 2005c), offspring exposed to the highest doses (15 and 20 mg/kg) and during the early treatment period (E7–12) displayed increases in latency on reversal trials. However, in-line with the studies described above, no effect of this treatment regimen, at any methamphetamine dose or exposure time window, affected acquisition learning (Acuff-Smith et al. 1996). Thus, methamphetamine, at higher doses, when administered during early gestation, affects a cognitively demanding aspect of MWM learning (reversal) that requires new learning (i.e., the forfeiting of searching in the previous location and adapting to a different contingency).

Findings from probe trials, during which the escape platform is removed following the completion of acquisition learning, follow the same general pattern. In adult male offspring, 5–10 mg/kg/b.i.d. methamphetamine administered throughout gestation did not influence performance in the probe test (Acuff-Smith et al. 1996; Macuchova et al. 2014; Schutova et al. 2008, 2009b). However, in some, but not all instances, adult females prenatally exposed to methamphetamine swam less often across the quadrant where the platform was previously located compared with control offspring (Macuchova et al. 2013, 2014; McDonnell-Dowling et al. 2014).

These findings corroborate others, which suggest females may be more sensitive to the effects of prenatal methamphetamine (see above, Sect. 2.2.1).

In sum, there is no effect on any phase of the MWM after prenatal methamphetamine doses up to 10 mg/kg/day. Thus, exposure to methamphetamine during the prenatal period results in spatial learning and memory impairments that are only evident after high doses during reversal learning in the MWM.

Acute Methamphetamine During Adulthood

Even though prenatal methamphetamine itself does not influence spatial learning during the acquisition phase of the MWM, it is possible that prenatal exposure, in combination with acute methamphetamine during adult testing, may impact learning. In a series of studies, Slamberova and colleagues tested whether adult male rats prenatally treated with methamphetamine responded differently to acute methamphetamine administration during MWM testing in adulthood, compared with males prenatally treated with saline. In one study, the authors (Schutova et al. 2008, 2009a) administered methamphetamine (1 mg/kg) to half the animals from each prenatal exposure group (5 mg/kg/day methamphetamine, saline, untreated controls) following each day of testing in acquisition, probe, and retention memory phases of the MWM (see above). Even though acute methamphetamine increased the trajectory length during acquisition, this effect was not influenced by prenatal methamphetamine exposure (Schutova et al. 2008, 2009a). More recent work replicated this finding (increased place trajectory after acute MA exposure) in females. They also observed detrimental effects of acute methamphetamine during acquisition by showing increased latency to find the hidden platform in adult methamphetamine-exposed rats compared with controls. As noted before, prenatal exposure to the drug did not affect methamphetamine-induced cognitive impairments to acute doses (Macuchova et al. 2013).

The possible sensitizing effects of prenatal methamphetamine exposure to acute amphetamine (5 mg/kg) were also explored. Consistent with acute methamphetamine effects, prenatal drug exposure did not affect spatial learning when adult amphetamine (5 mg/kg) was given after each testing phase. Specifically, female, but not male, rats given amphetamine in adulthood showed increased latency, distance traveled, and search errors in the acquisition and retention testing phases, compared with females given saline in adulthood, regardless of prenatal exposure (Macuchova et al. 2014; Slamberova et al. 2014a, b). Taken together, exposure to methamphetamine during the first two human trimester equivalents did not affect spatial learning and memory during MWM acquisition. Likewise, methamphetamine exposure during this period did not sensitize animals to the drug during adulthood.

2.3 Neurotransmitter Effects

2.3.1 Dopamine

Prenatal methamphetamine exposure during brain development has only a mild effect on dopamine systems. Sato and Fujiwara (1986) administered methamphetamine (2 mg/kg/day) throughout gestation and later examined regional dopamine concentration and receptor-specific binding. At P35, there was a decrease in dopamine concentration, but not dopamine-specific binding, in the frontal cortex. Neither dopamine concentration nor receptor binding was altered in the striatum. Additionally, Weissman and Caldecott-Hazard (1993) administered methamphetamine (2 or 10 mg/kg/day) throughout gestation, and in comparison with Sato and Fujiwara (1986), the low dose produced a decrease in striatal dopamine uptake, whereas the high dose resulted in an increase in uptake in the midbrain, pons, and hippocampus at P48. Thus, the impact of prenatal methamphetamine on dopamine concentration and specific binding in the young adult is not settled.

Similarly, selective dopamine effects after prenatal methamphetamine are not apparent in the offspring as adults. For example, following prenatal treatment with 0, 5, 10, 15, or 20 mg/kg/b.i.d. methamphetamine from E7–12 or E13–18, Acuff-Smith et al. (1996) examined dopamine and 3,4-dihydroxyphenylacetic acid (DOPAC) levels on P70. No change was evident after early gestational exposure in any brain region, and no differences were found in the medial frontal cortex or hippocampus after later gestational exposure. In the caudate, only females in the late-exposed, highest dose group (E13–18; 20 mg/kg) showed elevated dopamine concentrations compared with controls. No differences were evident among males for any measure, and the DOPAC/dopamine ratio was increased proportionately among all groups for both sexes. Together, these data suggest a minimal effect of prenatal methamphetamine exposure on dopamine and its metabolites. Additionally, Won et al. (2002) examined reaggregated tissue cultures of cells prepared from E14 mesencephalic and striatal regions in mouse offspring subjected to twice daily methamphetamine (40 mg/kg) or saline from E6–13. Dopamine and DOPAC levels were monitored after 14, 29, 43, and 64 days in culture. Tissue derived from methamphetamine-exposed embryos did not significantly affect dopaminergic neurons at any culture age, suggesting insensitivity of developing dopaminergic nigrostriatal projections to methamphetamine exposure in utero. However, Bubenikova-Valesova et al. (2009) reported that prenatal methamphetamine (5 mg/kg/day throughout gestation) increased dopamine levels ($\sim$288 %), DOPAC ($\sim$67 %), and homovanillic acid (HVA) ($\sim$74 %) in the nucleus accumbens of adult males. Thus, the long-term effects on dopamine concentration and its metabolites are only slightly affected by prenatal methamphetamine exposure, and these effects are region-specific.

The effect of prenatal methamphetamine on tyrosine hydroxylase (TH) activity has also been examined. For example, Gomes-da-Silva et al. (2002) investigated whether TH gene expression was altered following prenatal methamphetamine

(5 mg/kg/day from E8–22). TH mRNA levels were decreased in the ventral tegmental area of methamphetamine-exposed females on P7 and P14. However, this effect disappeared by P30 and was not evident in males at any age. Similarly, Jeng et al. (2005) treated pregnant mice with a single dose of methamphetamine (20 or 40 mg/kg) on E14 or 17. One hour following exposure, methamphetamine enhanced DNA oxidation, quantified by 8-oxoguanine (8-oxoG) formation, by at least 100 %; however, there was no evidence of degeneration of striatal dopaminergic nerve terminals at P84, as determined by TH staining. Interestingly, adult 8-oxoG glycosylase (OGG1, which repairs 8-oxoG) in KO mice showed no striatal dopaminergic terminal degeneration following exposure to methamphetamine in utero, suggesting that protection by OGG1 against neurodevelopmental deficits (enhanced motor coordination deficits in OGG1 KO mice) does not occur by preventing the degeneration of dopaminergic nerve terminals (Wong et al. 2008). Thus, some evidence points to dopaminergic markers being affected following prenatal methamphetamine, but these effects, particularly with TH, may be transient. In comparison with the prenatal drug exposure, methamphetamine treatment of mature animals exerts a profound effect on the monoamine system via the same mechanisms outlined here (e.g., for review see, Yu et al. 2015). Thus, unlike adulthood, the immature dopaminergic system may compensate for prenatal insult or respond less to methamphetamine, thereby producing little to no dopaminergic neurotoxicity.

2.3.2 Serotonin

Prenatal methamphetamine influences the serotonergic system both early and late in postnatal life. Following high-dose methamphetamine exposure (20 mg/kg/b.i.d. from E13–17), Acuff-Smith et al. (1996) demonstrated a significant reduction in serotonin (5-HT) concentrations in the nucleus accumbens in P70 females compared with controls. However, similar to its effects on dopamine, no change in serotonin concentrations was evident in any brain region after early gestational methamphetamine (E7–12) exposure or in the medial frontal cortex or hippocampus after exposure later in gestation (E13–17). Males were unaffected in all measures at doses up to 15 mg/kg (Acuff-Smith et al. 1996).

Sato and Fujiwara (1986) showed that the B_{max} of ^{3}H-spiperone (specific to 5-HT$_2$ receptors) receptor binding was significantly lower in the frontal cortex at P35 of methamphetamine-exposed offspring (2 mg/kg/day, throughout gestation), suggesting a decrease in the number of frontal cortex 5-HT receptors in the prenatal methamphetamine-exposed adolescent brain (Sato and Fujiwara 1986). However, Weissman and Caldecott-Hazard (1993) found that low-dose methamphetamine treatment (2 mg/kg/day) throughout gestation produced a significant decrease in ^{3}H-paroxetine binding (serotonin uptake sites) from 18 % in the hypothalamus, to 32 % in the striatum, and 42 % in the hippocampus. Conversely, a higher dose (10 mg/kg) produced significant increases in ^{3}H-paroxetine binding in the hippocampus (25 %) and the pons (33 %). This opposing effect parallels that of dopamine in this same experiment.

Cabrera et al. (1993) examined functional alterations of the serotonergic system by measuring changes in serotonin-mediated plasma hormones after a single injection of the serotonin releaser p-chloroamphetamine (PCA, 8 mg/kg) in P70 male and P30 female offspring exposed to methamphetamine (5 mg/kg/day) from E13–20. Prenatal methamphetamine produced significant attenuation of plasma renin responses to PCA at both ages without altering basal levels of hormones. In contrast to Weissman and Caldecott-Hazard (1993), the density of cortical and hypothalamic 5-HT uptake sites also did not change in response to prenatal methamphetamine. However, the long-term postnatal deficits in 5-HT-mediated renin secretion suggest selective functional alterations of the serotonergic system throughout ontogeny, in both sexes (Cabrera et al. 1993).

Finally, in an examination of the impact of methamphetamine (40 mg/kg/day, from E6–13) on serotonergic markers assessed shortly after drug administration, Won et al. (2002) reported that reaggregates prepared from E14 striatal cells in methamphetamine-exposed embryos showed a significant elevation in 5-HT levels at all culture ages (P14, 29, 43, and 64), compared with reaggregates prepared from saline-treated embryos. Additionally, 5-hydroxyindolacetic acid (5-HIAA) levels were also elevated in 14- and 29-day-old cultures derived from methamphe-tamine-exposed embryos, and this elevation persisted for up to 2 months (Won et al. 2002).

Thus, targeting of the serotoninergic system by prenatal methamphetamine varies as a function of brain region, gestational window of exposure (early vs. late), sex, and dose. The available data suggest that the developing serotoninergic system may be more sensitive to prenatal methamphetamine compared with the dopaminergic system.

2.3.3 Other Neurotransmitters

Norepinephrine and acetylcholine, as measured in a variety of brain regions, as well as in both sexes, are unaffected by prenatal methamphetamine exposure (Cabrera et al. 1993; Sato and Fujiwara 1986; Weissman and Caldecott-Hazard 1993). Conversely, the long-term impact of prenatal methamphetamine on glutamate release has been reported to occur (Vrajova et al. 2014). Slamberova and colleagues showed an increase in both the NR1 and NR2B subunits of the N-methyl-D-as-partate receptor (NMDAr) in the hippocampus of adult, but not adolescent off-spring, prenatally exposed to methamphetamine (5 mg/kg/day) throughout gestation (Vrajova et al. 2014). In a related experiment, Slamberova and colleagues suggested that prenatal methamphetamine effects on NMDArs may correlate with behavioral performance during object recognition (OR) testing. Results showed impaired novel object, but not novel location, recognition in prenatally methamphetamine-exposed progeny, at the same dose and period of exposure as before (Slamberova et al. 2014b). It is possible that prenatal methamphetamine may lead to deficits in non-spatial, rather than spatial, memory (Fialova et al. 2015). In fact, the enhancement of spatial memory in methamphetamine-treated offspring during the MWM (see above, Sect. 2.3.2) may correspond to the increase in NMDAr subtype isoforms required for spatial learning and memory.

2.4 Prenatal Methamphetamine and Evoked Seizures

Prenatal exposure to psychostimulants, including methamphetamine, is often associated with an increase in seizure susceptibility and/or induction of seizures. The following section summarizes the work of Slamberova and colleagues in determining the impact of prenatal methamphetamine (5 mg/kg/day, throughout gestation) on seizure induction, maintenance, and type.

2.4.1 Chemically Induced Seizures

GABA and NMDA

Inhibitory GABA and excitatory NMDA systems play an important role in the effect of methamphetamine exposure on primary and secondary generalized seizures. Using chemically induced seizure models, increased seizure susceptibility in prenatally methamphetamine exposed animals has been reported (Slamberova 2005; Slamberova and Rokyta 2005a; Slamberova et al. 2008, 2009). For example, Slamberova and Rokyta (2005b) showed differences in susceptibility to bicuculline (GABA$_A$ antagonist) and NMDA-induced seizures between prenatal treatment groups. Specifically, adult male prenatally methamphetamine-exposed progeny were administered either bicuculline (7.5 mg/kg), NMDA (NMDA agonist, 250 mg/kg), or kainic acid (non-NMDA receptor agonist, 15 mg/kg) and observed for 30–60 min, during which the incidence and the latency to the onset of tonic–clonic seizures were recorded. Clonic seizures are characterized by a rapid contraction and relaxation of the muscles of the head and forelimbs, whereas tonic–clonic seizures begin with running, followed by a tensing of the muscles (tonus), and then long-lasting clonus (Slamberova and Rokyta 2005a). Prenatal methamphetamine exposure shortened the latency to tonic–clonic seizures after bicuculline and shortened the latency to clonic–tonic seizures after NMDA. However, no differences in kainic acid-induced seizure susceptibility were found (Slamberova and Rokyta 2005a).

Flurothyl (pro-convulsant) interferes with GABA$_A$ receptor-mediated neurotransmission (Slamberova 2005), but the effect of flurothyl on seizure susceptibility in prenatal methamphetamine-exposed animals does not follow the pattern seen with bicuculline on seizures. Following flurothyl exposure to adult prenatally methamphetamine-exposed males and females (in the diestrus phase of the estrous cycle), drug-exposed males, but not diestrus females, showed a decrease in clonic seizure threshold, but the tonic–clonic seizure threshold was not altered by prenatal methamphetamine exposure for either sex (Slamberova 2005; Slamberova et al. 2008). The authors conclude that the dissimilarities in findings between the effects of these two GABA-related stimulants (bicuculline and flurothyl) may be due to the effect of flurothyl on more than one neurotransmitter (Slamberova et al. 2008). In any event, enhancement of seizure induction in adulthood by prenatal methamphetamine appears specific to GABA and NMDA receptor-mediated effects.

Acute Methamphetamine During Adulthood

Additional studies examined the possibility that prenatal methamphetamine exposure induces long-term changes in the sensitivity (alteration in seizure threshold) to the same drug administered in adulthood. This approach was used for evaluating the combined effects of prenatal and acute methamphetamine on behavioral output (see Section "Acute Methamphetamine During Adulthood"). Thirty minutes prior to seizure induction by flurothyl, Slamberova et al. (2008) administered a single injection of methamphetamine (1 mg/kg) or saline to prenatal methamphetamine-exposed or saline-exposed rats. The acute methamphetamine increased the threshold to both the first focal clonus and clonic seizure compared with animals that received acute saline. Interestingly, this increase in seizure threshold was higher in prenatally methamphetamine-exposed animals compared with prenatally saline-exposed animals. As such, these results suggest that methamphetamine may provide a protective effect against flurothyl-induced seizures and prenatal exposure (Slamberova et al. 2008).

In response to NMDA-induced seizures, though, acute methamphetamine decreased the latency to the onset of stereotypy and increased the length of seizures (Slamberova et al. 2009). However, the decrease in the latency to stereotypy onset induced by acute methamphetamine was smaller in methamphetamine-exposed and saline-exposed rats when compared with untreated rats (Slamberova et al. 2009). Even though methamphetamine-induced differences are apparent in comparison with one control group and other factors may contribute to this effect (see below), the response to acute methamphetamine prior to chemically induced seizure induction may be dependent on prenatal drug exposure history. Consistent with seizure induction via the GABA system, prenatal methamphetamine may operate by mitigating the detrimental effects of adult methamphetamine by altering the sensitivity of NMDA-type seizures.

Even though non-NMDA receptors do not contribute to the effect of prenatal methamphetamine on seizure induction (Slamberova and Rokyta 2005a), it is possible that acute methamphetamine in adulthood may differentially influence seizure activity across prenatal treatment groups. In this case, Slamberova et al. (2010a, b) administered a single injection of methamphetamine (1 mg/kg) prior to seizure induction using kainic acid (15 mg/kg). Both latency to onset and duration of clonic seizures was significantly lower in prenatal methamphetamine-exposed females given acute methamphetamine during proestrus/estrus, but not in those given the drug during diestrus. In males given acute methamphetamine, the latency to the onset of clonic seizures was increased in both prenatally saline- and methamphetamine-exposed males relative to undisturbed controls. Thus, even though prenatal methamphetamine exposure alone does not have a long-term effect on seizure susceptibility induced with kainic acid in adult progeny (Slamberova and Rokyta 2005b), the sensitivity to the acute dose of methamphetamine differed among prenatal exposure groups, in some instances (Slamberova et al. 2010b). Behavioral evidence, however, does not support an enhanced sensitivity/tolerance effect of prenatal methamphetamine. For example, in the conditioned place preference (CPP), test animals receive a drug in one of two conditioning chambers. During testing, animals are allowed free access to either side of the chamber, and active drug-seeking behavior is recorded if the animal spends more time in the chamber associated with the drug. Numerous reports reveal that prenatal methamphetamine exposure does not affect CPP conditioning with adult methamphetamine (Slamberova et al. 2011, 2012a, b, 2014a).

Overall, findings concerning the response to acute methamphetamine administration prior to seizure induction in adulthood emphasize the long-lasting influence of prenatal methamphetamine exposure on the response following adult exposure to the same drug. GABA and NMDA systems appear to play some role in such sensitization/tolerance processes.

2.4.2 Non-chemically Induced Seizures

As described above, chemically induced seizures generally lead to an increase in seizure susceptibility after methamphetamine treatment during the prenatal period (Slamberova 2005; Slamberova and Rokyta 2005a; Slamberova et al. 2008, 2009). However, prenatal methamphetamine, as well as acute methamphetamine in adulthood, decreased focal epileptiform activity (lowered seizure susceptibility) after electrical stimulation of sensorimotor cortex in male rats (Bernaskova et al. 2011). On the other hand, in a similar study with female rats (Matejovska et al. 2014), prenatal methamphetamine exposure decreased the seizure threshold in females in the diestrus phase of the estrous cycle, but not in females in the proestrus/estrus phases. The authors note that chemically induced seizure models use systemic delivery of convulsant drugs (i.e., NMDA, bicuculline, or flurothyl) that operate through glutamatergic and GABAergic receptors and spread rapidly to induce generalized seizures. Yet, electrical stimulation provides localization, allowing for control over the site of seizure origin (Bernaskova et al. 2011). Thus, female rats that undergo seizure induction via electrical stimulation respond more similar to males and females following chemically induced stimulation (e.g., Slamberova and Rokyta 2005b; Slamberova 2005), and increased seizure susceptibility following prenatal methamphetamine may be apparent during diestrus, when female gonadal hormones are low. As previously mentioned, sexual dimorphism is evident following chemically induced seizures in methamphetamine-exposed animals (Slamberova and Rokyta 2005a, b; Slamberova 2005), in which gonadal hormone specificity directly influenced the effect. In addition, prenatal methamphetamine may differentially influence seizure activity depending on the nature of the seizure induction with focal (electrically stimulated) versus generalized (chemically induced) seizure mechanisms.

2.4.3 Conclusions

Estrous Cycle

The effect of prenatal methamphetamine on seizure induction during adulthood differs as a function of multiple factors. As mentioned, seizure susceptibility following prenatal methamphetamine exposure in adult female rats is dependent on the level of ovarian hormones (Slamberova and Rokyta 2005a, b; Slamberova 2005; Slamberova et al. 2008; Matejovska et al. 2014). For example, prenatal methamphetamine-induced reductions in the first fasciculation after stimulation via flurothyl and increases in the duration of NMDA-induced seizure activity are evident in only those females in diestrus (Slamberova et al. 2005). It seems that the mutual effect of low estrogen and progesterone levels in the diestrus phase, combined with prenatal methamphetamine, may result in increased seizure susceptibility. However, females in proestrus/estrous are more affected when acute methamphetamine is delivered during adulthood (Slamberova et al. 2010a, b). Thus, the degree of seizure susceptibility after (1) prenatal methamphetamine exposure, (2) methamphetamine application during adulthood, and (3) the combined effect of both occurrences is largely influenced by the stage of the estrous cycle.

Stress

For many of the studies summarized above, prenatal methamphetamine-exposed rats differ in their response to seizure induction from one control group but not another (i.e., untreated controls vs. saline-treated controls). For example, Slamberova and Rokyta (2005a) reported an increase in seizure susceptibility for methamphetamine-exposed offspring following induction via bicuculline and NMDA compared with untreated controls. However, there was no difference in seizure susceptibility between prenatal saline- and methamphetamine-exposed groups. Importantly, the authors concluded that daily saline injections to pregnant dams may serve as a prenatal stressor (cf., Peters 1984) that alone may induce long-term changes in seizure susceptibility (Slamberova and Rokyta 2005a). In addition, significant differences between methamphetamine-exposed offspring and the untreated, but not the saline-treated, control group are evident in seizures induced by NMDA (Slamberova et al. 2009) and flurothyl (Slamberova et al. 2008; Slamberova 2005), as well as kainic acid and NMDA induction with acute methamphetamine during adulthood (Slamberova et al. 2009, 2010b; Bernaskova et al. 2011). Thus, it is difficult to rule out the effect of injection stress on the impact of prenatal methamphetamine on seizures induced by different convulsant drugs.

Sex

Another aspect of methamphetamine-induced seizure susceptibility is the numerous sex-related differences in responding to chemical stimulation that cannot be explained by the phase of the estrous cycle. For example, after seizure induction by NMDA, males demonstrated an increase in latency to stereotypy and a shorter length of clonic-tonic seizures compared with females (Slamberova et al. 2009). Additionally, in bicuculline-induced seizures (GABA), Slamberova and Rokyta (2005b) reported a decrease in the latency to the onset of tonic–clonic seizures in male rats (Slamberova and Rokyta 2005a); however, female rats showed no difference in the latency to the onset of this seizure type after bicuculline (Slamberova and Rokyta 2005b). Sex differences are also evident in seizure sensitivity induced by flurothyl (GABA). In this case, compared with females, males had a lower threshold of induction on all measures: first focal clonus, clonic seizures, and tonic–clonic seizures (Slamberova et al. 2008). In those studies that included a comparison between prenatal exposure group and sex (others only included one sex or analyzed males and females separately), there was no interaction between the two factors (Slamberova et al. 2008, 2009). However, sex differences are apparent in every type of chemically induced seizure model. Thus, variation in seizure susceptibility between males and females, independent of prenatal methamphetamine exposure history and estrous cycle classification, should be accounted for when interpreting these findings.

In summary, prenatal methamphetamine exposure increases seizure susceptibility in adult male and female rats, and NMDA and GABA, but not NMDA-independent systems, play a role in such effects (Slamberova 2005; Slamberova and Rokyta 2005a). This was shown by results indicating increased seizure susceptibility from GABA receptor inhibiting or NMDA receptor activating drugs (Slamberova et al. 2009; Slamberova 2005; Slamberova and Rokyta 2005a, b). Findings from acute methamphetamine administration during adulthood suggest sensitivity of prenatal methamphetamine treatment to later exposure to the same drug, in which prenatal methamphetamine may offer protection. Slamberova and colleagues suggest that prenatal methamphetamine exposure may alter the development of neurotransmitter systems involved in seizure initiation and propagation (GABAergic, glutamatergic systems) and/or others such as dopaminergic, serotoninergic, and noradrenergic systems that are associated with methamphetamine exposure during the prenatal period (Slamberova et al. 2008; see above, Sect. 2.3).

3 Neurobehavioral Effects of Pre- and Postnatal Methamphetamine Exposure

A limited number of studies have incorporated a dosing regimen in which methamphetamine is administered prior to gestation, throughout the entire gestational period, and during the preweaning period. The studies reviewed in this section comprise additional work of Slamberova and colleagues in which methamphetamine (5 mg/kg/day) was administered for 9 weeks (about three weeks prior to impregnation), throughout all of gestation, and for 23 days of lactation (until weaning).

3.1 Maternal Behavior and Sensorimotor Function

Using the aforementioned dosing regimen, Slamberova et al. (2005a) reported that, compared with control dams, methamphetamine-treated dams were slower in a pup retrieval test to return the first pup and all pups back to the nest. This delay in offspring recovery was also seen when methamphetamine treatment was restricted to the prenatal period (Malinova-Sevcikova et al. 2014; Slamberova et al. 2005b). Similarly, methamphetamine-exposed dams exhibit less time in the blanket position of active nursing and an increase in passive nursing, regardless of whether exposure was pre- and postnatal versus prenatal only exposure. Thus, methamphetamine before, during, and after gestation has a negative effect on maternal behaviors toward pups.

Prenatal, along with postnatal methamphetamine administration, negatively affects pups' sensorimotor function (Hruba et al. 2008; Slamberova et al. 2007). For example, methamphetamine-exposed pups showed a delay in achieving righting reflexes and had more falls in the rotorod and bar-holding tests than subjects from both untreated and saline-exposed mothers (Hruba et al. 2008). Slamberova et al. (2007) extended these observations to a second generation of rats that were not exposed to the drug. In this case, adult females that were exposed to methamphetamine during the prenatal and preweaning periods were mated with unexposed males. They found that pups from mothers exposed to methamphetamine during prenatal and preweaning periods had impaired sensorimotor coordination. Thus, pre- and postnatal methamphetamine treatment may affect two generations of offspring, emphasizing the long-term effect of developmental methamphetamine on sensorimotor processes.

Further analyses on the effect of cross-fostering in these studies revealed that methamphetamine given during either the prenatal or postnatal period alone, or during both, led to deficits in sensorimotor function. Additionally, postnatal care of control mothers partially suppressed the negative effect of prenatal methamphetamine, suggesting that cross-fostering may improve postnatal development of pups (Hruba et al. 2008). The authors conclude that it is possible that variation in rat maternal care could serve as a mechanism for a non-genomic behavioral mode of transmission of traits, and cross-fostering may mitigate the detrimental effects of methamphetamine on sensorimotor development (Hruba et al. 2008).

3.2 Spatial Learning Effects

Even though the timing of methamphetamine exposure (pre- vs. postnatal) does not differentially influence methamphetamine-induced changes in pup sensorimotor function, the drug exposure window greatly determines the impact of methamphetamine on cognitive processes, particularly spatial learning. In another fostering/cross-fostering study, all animals, regardless of prenatal exposure treatment (methamphetamine, saline, untreated controls), that were fostered to methamphetamine-treated dams had longer latencies, more search errors, and used fewer spatial strategies during acquisition in the MWM compared with animals fostered to control or saline-treated mothers (Hruba et al. 2009). In addition to replicating the insensitivity of MWM learning to prenatal methamphetamine (see Sect. 2.2.2), these studies showed that postnatal, but not prenatal, methamphetamine exposure (P1–21) affects MWM learning during adulthood. Follow-up studies confirmed that these effects are completely due to neonatal exposure (Hruba et al. 2010). Taken together, prenatal exposure to methamphetamine (5 mg/kg/day) does not impair learning in the MWM, while postnatal exposure to methamphetamine is effective at inducing learning deficits.

In these studies, "better" maternal care of control mothers does not affect the learning and memory of rat pups prenatally exposed to methamphetamine. Thus, even though cross-fostering influenced the learning of adult male rats (postnatal methamphetamine exposure impaired spatial learning, regardless of prenatal exposure group), in contrast to its effect on sensorimotor function (Hruba et al. 2008), the role of maternal care on spatial learning abilities during adulthood is minimal at best.

Developmental methamphetamine also increases anxiety-like behavior during adulthood, and the anxiogenic effect of the drug may be more evident when exposure occurs during the neonatal period (Hruba et al. 2012; Slamberova et al. 2015).

Additionally, it appears that prenatal, but not postnatal exposure, increases seizure susceptibility, as initiated by both NMDA and GABA systems (Slamberova et al. 2010a). However, it is unclear to what extent cross-fostering and variations in maternal behavior influence both of these findings (Hruba et al. 2008, 2009; Slamberova et al. 2005a, b, 2010a).

In conclusion, by incorporating a fostering/cross-fostering approach, Slamberova and colleagues delineated the postnatal exposure period as a heightened time of sensitivity to the detrimental effect of methamphetamine on MWM learning, with no detectable change from prenatal exposure.

4 Neurobehavioral Effects of Neonatal Methamphetamine Exposure

4.1 Introduction

The neonatal period of methamphetamine administration covered in the remainder of this review is from P1 to P21 or shorter windows of exposure within this range. The preweaning period of rat brain development is analogous to human brain development during the second half of gestation, dependent upon the region (Rice and Barone 2000). Importantly, brain regions that mediate higher cognitive function are developing during this time, and it is a sensitive period during which the "brain growth-spurt" occurs in which synaptic refinement is evolving. There is a substantial increase in the proliferation of glial cells, hippocampal neurogenesis, synaptogenesis, and dendritic arborization, and rapid increases in neurotransmitter and associated receptor expression (Bayer 1982; Gil-Mohapel et al. 2010; Herlenius and Lagercrantz 2004). Methamphetamine administration during this time disrupts neuronal populations and potentially leads to long-term behavioral changes.

We (and others) utilize a range of methamphetamine doses that are applicable to human patterns of drug abuse. On a mg/kg basis, the dose used in most of the experiments discussed below (10 mg/kg; 4×/day, every 2 h) approaches the relatively high doses described in the human literature in chronic methamphetamine users (e.g., Melega et al. 2007). However, when repeated exposure occurs in humans (human dose–frequency estimates are approximately every 6 h), accumulated plasma methamphetamine concentrations are higher due to the longer drug half-life in humans ($T_{1/2}$ = 10–12 h), compared to adult rats ($T_{1/2}$ = 1–1.5 h; Cho et al. 2001; Simon et al. 2000). To partially compensate for the rapid metabolism of methamphetamine in the rat, we deliver methamphetamine every 2 h. In the only pharmacokinetic study performed on neonatal rats (methamphetamine, 10 mg/kg; 4×/day), we determined that the plasma and brain T_{max} occur $\sim$25 min after s.c. injection in P11 rats, and the plasma and brain $T_{1/2}$ values are 2.8 and 2.3 h, respectively (Cappon and Vorhees 2001). In this case, the 4×/day dosing regimen produces some accumulation for up to 8 h, but not from one day to the next, as can occur in humans. Following a discussion of physiological effects associated with developmental methamphetamine, the majority of this section focuses on the use of the 10 mg/kg, 4×/day doses, in examining cognitive and other deficits after neonatal exposure during critical developmental stages.

4.2 Intrinsic Drug Effects

4.2.1 Body Weight

Similar to body weight variability described in humans following prenatal methamphetamine exposure (Dixon and Bejar 1989; Little et al. 1988; Oro and Dixon 1987; Smith et al. 2003), reduced body weight during and after treatment is a consistent observation from neonatal methamphetamine administration in the rodent model (e.g., Vorhees et al. 1994a; Williams et al. 2000, 2003a, b, c, 2004, 2006). Decline in body weight is dose-dependent, and decreases are apparent following a relatively low daily dose from P11–20 (2.5 mg/kg; 4x/day; Williams et al. 2004a). Methamphetamine-exposed pups (10 mg/kg; 4×/day; P11–20) quickly recover to 90–95 % body weight of control animals once treatment ends, but they remain 5 to $\sim$7 % lighter even at P60 (Vorhees et al. 2009). Importantly, though, long-term body weight changes are not predictive of long-term learning and memory changes. For example, methamphetamine (10 mg/kg; 4×/day) from P1–10 produces small but persistent body weight decreases, whereas administration from P11–20 or P6–15 (that induces the greatest MWM and Cincinnati water maze (CWM) deficits; see below) produces only transient reductions (Vorhees et al. 1994a, 2009). In addition, distributing a 40 mg/kg dose into 4, 10 mg/kg injections produces more severe cognitive impairment in the MWM compared with administering the same dose in 2, 20 mg/kg injections, even though both groups show the same body weight change (Vorhees et al. 2000). Accordingly, body weight change is a sensitive component of neonatal methamphetamine; yet, behavioral effects are reliably dissociated from drug-induced weight change.

4.2.2 Other Factors

Methamphetamine-induced effects on cognition are distinct from non-specific nutritional and maternal-pup disruption effects (Vorhees et al. 2000; Williams et al. 2004a, 2006). As discussed in Vorhees et al. (2000), drug treatments that cause protracted and more severe weight changes than those observed following a typical methamphetamine regimen have consistently shown that postnatal undernutrition has no effect on spatial learning (e.g., Campbell and Bedi 1989; Goodlett et al. 1986). Additionally, all groups, regardless of drug and delivery schedule, are matched for maternal-pup disruption during treatment (typically 4 times per day); thus, brief maternal–pup separation is unlikely to contribute to group differences in learning and memory performance (Vorhees et al. 2000). Even with intrinsic experimental variables, such as variation in body weight and nutrient consumption and early handling, the preponderance of data demonstrate a clear dissociation between the effect of neonatal methamphetamine and potentially confounding factors.

4.3 Long-Term Neurobehavioral Effects

4.3.1 Locomotor and Startle Response Activity

Neonatal methamphetamine consistently leads to decreased locomotor activity when examined during adolescence and adulthood (Vorhees et al. 1994b, 2009). Vorhees et al. (1994b) showed no difference in hypoactivity between methamphetamine-treated animals that were dosed from P1–10 or P11–20. While these effects endured for males up to P60, sustained age effects were transient for females. By 30 min into activity testing, methamphetamine-altered activity patterns converge with those of controls and habituation is relatively complete by 40 min for both drug treatment groups compared with controls. Vorhees et al. (2009) replicated this non-exposure period-dependent effect of methamphetamine treatment on lowering spontaneous locomotor activity. Again, similar findings were evident in the exploratory and habituation phases of locomotion in a novel environment, and no treatment effects were seen once baseline activity was reached (Vorhees et al. 2009). Thus, neonatal methamphetamine suppresses adult activity; however, these effects are not dependent upon the dosing period (Vorhees et al. 1994b) whereas cognitive effects are.

Similar to locomotor activity, augmented acoustic startle responsiveness (voltage changes by platform deflections to a startle stimulus) is evident following P1–10 or P11–20 methamphetamine (30 mg/kg; 2×/day), with no variation in startle response amplitude between these two exposure windows (Vorhees et al. 1994a). Vorhees et al. (2009) later reported a reduction of acoustic startle following a variety of exposure periods (P1–10, P6–15, and P11–20) with different methamphetamine doses, ranging from 10–25 mg/kg; 4x/day. No regimen-specific effects of methamphetamine were obtained for acoustic startle reactivity on pre-pulse or non-pre-pulse trials. Thus, methamphetamine at either of the 10-day intervals between P1 and P20 caused reliable reductions in acoustic startle, confirming previous findings for the P1–10 and P11–20 exposure regimens (Vorhees et al. 1994a, 1996, 1998).

Hence, developmental methamphetamine suppresses acoustic startle reactivity and reduces spontaneous locomotor activity, effects that span all treatment regimens during the first 20 days of life. Accordingly, methamphetamine has potent adverse developmental effects across development from P1–20. However, compared to locomotor activity and startle effects, the impact of neonatal methamphetamine on egocentric learning in the CWM and spatial learning in the MWM (all phases) can be affected by narrower windows of drug exposure within the P1–20 developmental period (see below).

4.3.2 Spatial and Non-spatial Navigation

In the following discussion, we describe findings from studies concerning the impact of neonatal methamphetamine on later learning and memory. We show that the developmental window of exposure is a crucial predictor of methamphetamine's effect and demonstrate that it exerts a long-term, robust influence on cognitive function.

In an initial set of experiments, Vorhees et al. (1994a, b) reported for the first time selective spatial learning and memory impairment in response to methamphetamine administered from P11–20, but not from P1–10. Compared with controls, adult animals exposed to methamphetamine (30 mg/kg; 2×/day) from P11–20 showed increased latencies to find both the hidden and moved platform in the hippocampal-dependent MWM (Vorhees et al. 1994b). The most sensitive periods of methamphetamine-induced impairment for both the MWM and for another form of learning in the CWM occur from either P6–15 or P11–20, compared with P1–10, P21–30, or later 10-day exposure periods (10–25 mg/kg; 4×/day; Vorhees et al. 2008, 2004, 2005, 2009; Williams et al. 2003c). In comparison with allocentric learning assessed in the MWM (as described in Sect. 1), the CWM is a 9-unit T-maze used as a test of route-based egocentric learning. More recent studies using the CWM (and those discussed below) eliminate reliance on distal cues by testing animals under infrared light, consequently requiring the use of internally generated, self-movement cues (Vorhees and Williams 2014). Dorsal striatal dopamine reductions via 6-hydroxydopamine (6-OHDA) lesions severely impair CWM and MWM navigation, but if only the dorsal medial or dorsal lateral striatal regions are lesioned, there is no deficit in MWM performance while CWM deficits remain (Braun et al. 2012, 2014). Hence, narrow 5- to 10-day neonatal methamphetamine exposure periods are sufficient to disrupt two types of directional learning that rely on the hippocampus (MWM) and striatum (CWM) for allocentric and egocentric learning, respectively.

These findings, in combination with others showing that P11–20 methamphetamine-induced spatial learning deficits occur in multiple strains of rats, in both males and females, in juvenile and adult animals, along with morphological changes in brain regions associated with these types of learning, in multiple phases of the MWM (acquisition, reversal, and shift), with and without experience in non-spatial learning tasks (fixed, cued MWM learning, acoustic startle response, passive avoidance), without compromising swim speed or swimming ability and other swimming-related performance factors (e.g., platform climbing, swim-overs), and in the absence of anxiety-related effects, emphasize the specificity of neonatal methamphetamine's effects on navigational learning and memory (Acevedo et al. 2007; Skelton et al. 2007; Vorhees et al. 1994a, b, 1999, Vorhees et al. 2000, 2007, 2008, 2009; Williams et al. 2002, 2003a, c, 2004a, b).

Despite numerous experimental manipulations that could conceivably diminish the persistent impact of developmental drug treatment, we reliably demonstrate *long-term* effects of neonatal methamphetamine on learning and memory. For example, decreasing the methamphetamine dose to as low as 0.625 mg/kg (4×/day, from P11–20) still results in reference memory acquisition impairments in adulthood for measures such as latency to reach the platform, path length, and cumulative distance from the platform in the MWM (Williams et al. 2004a). Additional findings reveal similar damaging effects following low-dose administration (5 mg/kg; 4×/day from P11–20) on both acquisition and reversal phases (Williams et al. 2003c, 2004a) that persist until at least 1 year of age (Vorhees et al. 2007).

Moreover, Williams et al. (2003b) examined whether extensive experience in other behavioral tests would eliminate and/or attenuate these methamphetamine-induced spatial deficits. Pups received methamphetamine (5, 10, or 15 mg/kg; 4/day) from P11–20. Prior to spatial learning during adulthood, animals were first tested for swimming ability in the straight swimming channel, followed by learning in the CWM under lighted conditions, and cued learning in the MWM. Next, the same animals were tested in the MWM in both an acquisition and reversal phase. Finally, animals were trained in a trial-dependent, match-to-sample, working-memory version of the MWM. There were no group differences in straight channel, CWM under lighted conditions, or cued performance in the MWM. However, all methamphetamine-exposed animals were impaired in both spatial learning phases, but not in the working-memory version of the MWM. As well as highlighting neonatal methamphetamine dose sensitivity to spatial learning, there was no benefit of extensive pre-training in other non-spatial swimming tasks for ameliorating drug-induced spatial learning and memory impairments (see also, Williams et al. 2002). Partial environmental enrichment (stainless steel enclosure) is also ineffective in overcoming learning deficits in both the MWM and CWM, even with the use of the most narrow effective methamphetamine window of administration (P11–15; Vorhees et al. 2008). Taken together, the learning and memory deficits summarized here are (1) specific to neonatal methamphetamine treatment and (2) persistent despite altering a range of experimental parameters.

4.4 Neurotransmitter Effects

Methamphetamine targets a number of neurotransmitter systems, including dopamine and serotonin especially during adulthood. Single high-dose exposure (Cappon et al. 2000) or repetitive moderate doses of methamphetamine (5-10 mg/kg; 4/day) reduce monoamine content and their metabolites and alter neurotransmitter release, transporter binding, and TH and tryptophan hydroxylase activity (Bowyer et al. 1998; Kokoshka et al. 1998; Ricaurte et al. 1982). Unlike methamphetamine administration during the neonatal period, changes in the dopamine system can last for at least 6 weeks in the adult rat (e.g., Ricaurte et al. 1980; Wagner et al. 1980).

4.4.1 Dopamine

The effect of methamphetamine on mature dopaminergic markers suggests that developing neurons may also be sensitive to drug-induced dopaminergic neurotoxicity (Frost and Cadet 2000). However, in contrast to exposure during adulthood, neonatal methamphetamine results in only mild influences on dopaminergic neurodevelopment. Biochemical analyses on developing striatal regions at time points shortly after methamphetamine administration reveal relative insensitivity of dopamine to neonatal methamphetamine. For example, multiple injections of methamphetamine (10 or 20 mg/kg; 4×/day) on P20 do not alter striatal dopamine or glial fibrillary acid protein (GFAP) levels when assessed on P20 or P23 (Cappon et al. 1997; Pu and Vorhees 1993). Additionally, Gomes-da-Silva et al. (2004) administered methamphetamine (10 mg/kg; 2×/day) from P1 to the time of sacrifice on P7, P14, or P30 and found no differences in dopamine content, DOPAC levels or in the DOPAC/dopamine ratio in the substantia nigra, ventral tegmental area, caudate-putamen, nucleus accumbens, or medial prefrontal cortex on P7 or P14. The only significant finding was a decrease in the DOPAC/dopamine ratio in the substantia nigra at P30 in methamphetamine-exposed males that were given the lengthiest dosing regimen (P1–30). Similarly, methamphetamine from P11–20 (10 mg/kg; 4×/day) increases striatal DOPAC and the DOPAC/dopamine ratio when assessed on P21. However, this effect does not persist in animals as young as P30, and dopamine itself is unaffected at P11, P21, and P30 (Schaefer et al. 2006, 2008). Single-day administration of the same dose as used in these studies (10 mg/kg) and even lower doses (5 mg/kg) to an adult rat produce over a 50 % decrease in dopamine content that continues for up to 6 months (Cass and Manning 1999; Wallace et al. 1999). Taken together, in contrast to the adult profile, dopaminergic markers appear relatively unaffected following up to 14 days of methamphetamine exposure and dopamine content is unchanged for at least 30 days after treatment, even when levels are measured shortly after exposure.

Catecholamine synthesis (particularly dopamine) depends on the rate-limiting enzyme TH. TH protein and mRNA expression levels measured shortly after neonatal methamphetamine also suggest a limited role of dopamine in drug-induced neurotoxicity. For example, Kaewsuk et al. (2009) administered methamphetamine from P4–10 (5 or 10 mg/kg; 1×/day) and analyzed TH enzyme levels on P10. Methamphetamine significantly decreased TH levels in the striatum, prefrontal cortex, nucleus accumbens, and substantia nigra. However, in comparison with the decreased TH protein levels by as much as ~65 % of controls in adult animals (e.g., Bowyer et al. 1998), the largest reduction at P10 was 30 % of control values in all regions analyzed. Further, in order to encompass the entire period of postnatal development of the dopaminergic system, Gomes-da-Silva et al. (2000) administered methamphetamine (10 mg/kg; 2×/day) from P1–29 and sacrificed animals on P30. Drug exposure increased TH activity (ng L-DOPA formed/mg of protein/h) in the caudate-putamen and substantia nigra; however, TH mRNA was elevated only in the substantia nigra, and all significant effects were observed in males only. Thus, even increasing the exposure period to include the entire first month of postnatal life does not lead to clear evidence of dopamine-specific targeting by methamphetamine after early exposure.

Long-term consequences of neonatal methamphetamine exposure, measured later during ontogeny or in adulthood, are consistent with the drug's limited impact on dopaminergic markers analyzed shortly after treatment. For example, a high dose of methamphetamine (100 mg/kg/day) delivered on P7–10 or P17–20 induces only modest reductions in striatal dopamine content on P40. In fact, the methamphetamine-induced depletions do not approach the reductions produced by the lowest effective dose of 6-OHDA (a known catecholaminergic neurotoxin; Lucot et al. 1982). Postnatal methamphetamine exposure from P11–20 (10 mg/kg; 4×/day) results in significant long-term reductions in dopamine and DOPAC content (%), striatal D_2 receptor binding sites, as well as reductions in protein kinase A (PKA) activity (a critical signal transduction enzyme of dopamine D_1- and D_2-like receptors, Crawford et al. 2003). As noted before, the decreases in dopamine and DOPAC in adulthood following developmental methamphetamine administration were less than those observed when exposure occurs during adulthood (e.g., Cappon et al. 1997; Sabol et al. 2001). Thus, the association between indices of neurotoxicity (altered dopamine and DOPAC content) and behavioral deficits in rats given methamphetamine during the preweanling period do not appear commensurate with one another, i.e., that the magnitude of the dopaminergic changes does not appear sufficient to account for the magnitude of learning and memory deficits (see above).

As discussed in Sect. 4.3.1, long-term changes in locomotor activity, in particular, are evident in response to developmental methamphetamine, and drug challenge approaches can be used to examine enduring neurochemical changes using behavioral outcomes. For instance, Vorhees et al. (1994b) administered a drug challenge during adulthood to test the role of dopamine receptors in animals given neonatal methamphetamine (30 mg/kg; 2×/day) from P1–10 or P11–20. The question of interest was whether exposure to methamphetamine in adulthood would change the response sensitivity to its locomotor stimulating effects, as a function of neonatal methamphetamine treatment. Compared with saline controls, the acute drug challenge slightly suppressed locomotor activity for the late-exposed methamphetamine group and slightly increased locomotor activity in the early exposure group, providing indirect evidence that dopaminergic functionality rather than content may be affected by neonatal methamphetamine (Vorhees et al. 1994b; Williams et al. 2003a).

Graham et al. (2013) provide the first evidence of long-lasting alterations to dopamine receptor function after neonatal methamphetamine administration. Pups were exposed to methamphetamine (10 mg/kg; 4×/day) from P11–20 and were tested during adulthood following selective pharmacological challenge with two dopamine receptor agonists, SKF-82958 (D_1) and quinpirole (D_2). Quinpirole induced an attenuated hyperactivity in methamphetamine-treated rats. On the other hand, SKF-82958 significantly increased hyperactivity in methamphetamine-exposed rats at multiple dose levels, possibly because of up-regulation of D_1 receptors (Graham et al. 2013). These data, along with the finding of long-term changes to D_2 and PKA activity (Crawford et al. 2003), suggest that persistent changes in D_1, D_2 receptors (Williams et al. 2003a) may play a role in the cognitive deficits induced by neonatal methamphetamine.

4.4.2   Serotonin

In addition to functioning as a neurotransmitter, serotonin can directly and indirectly regulate a variety of ontogenetic processes, including neuronal differentiation, synaptogenesis, and axon and dendritic growth (Lauder 1990). Serotonin disruption during early development (P10–20) produces spatial memory deficits when animals are tested in adulthood (Mazer et al. 1997). Similar to its effects on the dopaminergic system, methamphetamine administration during adulthood results in significant striatal serotonin depletion, which persists for days, weeks, and even months (Broening et al. 2005; Cappon et al. 2000; Fukumura et al. 1998; Richards et al. 1993).

The impact of repeated neonatal methamphetamine on serotonergic neurotransmission during the neonatal period, though, is evident only shortly after treatment and diminishes steadily throughout ontogeny, and is essentially absent during adulthood. When animals were given methamphetamine (10 mg/kg; 4×) on P11, there was no decrease in striatal or hippocampal 5-HT content 24 h later on P12, indicating a refractory effect at this young age (Schaefer et al. 2006). However, because it is possible that multiple days, rather than a single-day administration, may influence 5-HT concentrations, Schaefer et al. (2008) gave methamphetamine (10 mg/kg; 4×/day) for either 5 or 10 days (P11–15 or P11–20) and collected regional brain tissue on P16, P21, or P30. On P16, 5-HT was significantly decreased in the hippocampus of methamphetamine-treated pups compared with saline controls. Serotonin content in both the neostriatum and hippocampus was lower in methamphetamine-treated pups on P21, indicating a greater effect following the longer, 10-day exposure period. However, no brain 5-HT changes were found on P30 (Schaefer et al. 2008), indicating that the effects on 5-HT are transient. Similarly, Grace et al. (2010a) found 5-HT decreases 30 min following the first methamphetamine treatment (10 mg/kg/dose × 4/day) on each day between P11 and P20. In the same study, hippocampal 5-HIAA levels were also reduced in methamphetamine-exposed pups on most days of treatment (P13–20). Thus, neonatal methamphetamine-induced serotonergic changes, measured shortly after treatment, are more pronounced following drug treatment during later exposure windows (P11–20) rather than earlier time points (prior to P16), but do not persist past 30 days of age.

When the long-term effects of neonatal methamphetamine (10 mg/kg; 4×/day; P11–20) were examined at P60–70, Graham et al. (2013) reported no change in locomotor response following a drug challenge with the 5-HT releaser p-chloroamphetamine or the 5-HT$_{2/3}$ receptor agonist, quipazine. Grace et al. (2010b) showed reduced striatal and hippocampal 5-HT and 5-HIAA in adulthood following identical methamphetamine treatment, but these reductions do not appear to correlate with changes in egocentric learning and are smaller than those induced during the period of drug administration (see above; Grace et al. 2010b). Vorhees et al. (1994b) conducted a similar experiment using a higher dose (30 mg/kg; 2×/day) from P1–10 or P11–20. Suppression of locomotor activity during adulthood with a 5-HT reuptake inhibitor (fluoxetine) was only evident in the early exposure groups. Further, following a challenge dose of methamphetamine during adulthood, striatal 5-HT levels were depleted in all groups of rats that received the challenge dose, but there were no differences in the degree of depletion as a function of neonatal treatment (5 mg/kg; 4×/day from P11–20; Williams et al. (2003a). Taken together, these studies suggest a minimal effect of neonatal methamphetamine on serotonin function on locomotor function.

4.4.3   Other Neurotransmitters

During adulthood, following exposure to methamphetamine (10 mg/kg; 4×/day) from P11–20, there were marked attenuations in hyperactivity in animals given a pharmacological challenge with a range of doses of the NMDAr antagonist, MK-801 (Graham et al. 2013). In adult animals, the combined effect of dopamine and glutamate release contributes to increases in oxidative stress and excitotoxicity following acute methamphetamine administration (Braren et al. 2014; Sonsalla et al. 1989). Indeed, activation of NMDA-Rs produces an increase in plasma membrane insertion of D_1 receptors in cultured hippocampal neurons (Pei et al. 2004), and methamphetamine increases glutamate release via a D_1 regulatory mechanism in the substantia nigra (Mark et al. 2004). While NMDAr-mediated effects of methamphetamine are present with drug exposure during the neonatal period (Slamberova et al. 2009; Smith et al. 2007), it is unknown whether a similar association between dopamine and glutamate is responsible for the attenuated response to NMDA antagonism in locomotor activity and/or long-term cognitive deficits following developmental methamphetamine treatment (Graham et al. 2013).

Additional neurotransmitters and their receptors are altered by neonatal methamphetamine. For instance, in a series of experiments, Raber and colleagues examined the role of histamine (H) in methamphetamine-induced long-term cognitive deficits in mice. P11–20 administration of the H_3/H_4 receptor antagonist thioperamide, that increases histamine release, mimicked the effects of methamphetamine (5 mg/kg; 1×/day from P11–20) and co-administration of methamphetamine with the H_3 receptor agonist immepip, that inhibits histamine release, antagonized the object recognition and spatial learning deficits in adulthood (Acevedo et al. 2007) and adolescence compared with methamphetamine-treated offspring that did not receive immepip (Eastwood et al. 2012). In a later experiment, mice treated with methamphetamine or thioperamide, at the same dose levels, showed reductions in the dendritic marker microtubule-associated protein 2 (MAP2) in the CA3 region of the hippocampus and entorhinal cortex compared with animals treated with methamphetamine along with immepip, who did not show cognitive impairments (Acevedo et al. 2008).

Additionally, P11–20 methamphetamine (5 mg/kg; 1×/day) increases the number of muscarinic acetylcholine (ACh) receptors in the hippocampus and cortex during adulthood in mice that are impaired in novel object recognition (Siegel et al. 2010), thus identifying a role for the cholinergic system in long-term methamphetamine-induced behavioral deficits. Because the cholinergic system is modulated by the histaminergic system, increases in neonatal histamine release and/or neonatal ACh receptors may lead to long-term changes in both of these systems that may relate to methamphetamine-induced behavioral impairments (Siegel et al. 2010).

4.5 Stress Effects

The mechanisms by which neonatal methamphetamine cause enduring learning and memory deficits are unknown. In addition to neurotransmitter signaling, changes in endocrine system function have been suggested as an hypothesis for the observed methamphetamine-induced behavioral impairments. Alteration in glucocorticoid levels early in ontogeny interferes with normal brain development and may be responsible for learning and memory deficits later in life (e.g., Moriceau et al. 2004), and methamphetamine administration increases the release of corticosterone (CORT) in the neonatal rat (Schaefer et al. 2006, 2008, 2010; Williams et al. 2006). P11, P11–15, and P11–20 methamphetamine administration increases plasma CORT that lasts for least 24 h, and increases in adrenocorticotrophin hormone (ACTH) occur after methamphetamine on P11 (Schaefer et al. 2006, 2008; Williams et al. 2000). Indeed, CORT release after developmental methamphetamine administration is more potent than CORT release in response to stressors such as forced swim or isolation at early ages (Grace et al. 2008). Neonatal methamphetamine also alters the adrenal response to a forced swim stressor during adulthood (Williams et al. 2003a). Because early stress induces changes in brain development and behavior (Chapillon et al. 2002; Fenoglio et al. 2006), and prolonged activity of the HPA axis can have neurotoxic effects (Gould et al. 1991; McEwen et al. 1992; Meaney et al. 1988), studies were begun investigating dysregulated CORT release as a potential mediator of long-lasting methamphetamine-induced spatial learning deficits.

Fluctuation in adrenal response to methamphetamine, particularly CORT release, is evident between the two periods of drug administration in animals that display distinct behavioral effects (i.e., P1–10 and P11–20). As described in Sect. 4.3.2, P11–20, but not P1–10, methamphetamine-treated rats exhibit long-term spatial learning and memory impairment in the MWM. While 10-day exposure windows result in more severe spatial learning impairment (Vorhees et al. 2008), there are disproportionately more MWM deficits after P11–15 than after P16–20 exposure (Williams et al. 2003b). Conversely, methamphetamine exposure during P11–15 or P11–20 results in a reliable decrement in CWM performance

(Vorhees et al. 2008). P1–10 methamphetamine treatment, however, induces no behavioral impairment in either task (Vorhees et al. 1994a; 2009).

An age-dependent effect is also evident in the degree of CORT release following neonatal methamphetamine. For example, a U-shaped response function exists when methamphetamine (10 mg/kg/day) is given and measured every other day from P1-19 (Williams et al. 2006). Augmented CORT release occurs on P1, P3, and P5 compared with any other day between P7 and P13. CORT returns to elevated levels again on P15, P17, and P19. However, when repeated rather than single doses were examined (4 doses/day for 4 days with an acute dose on the 5th day), rather than displaying a U-shaped function, CORT increased progressively with age (Williams et al. 2006). Thus, in addition to characterizing methamphetamine as a robust activator of CORT release during development, these data showed a differential CORT response depending on the dosing regimen, as well as the age at which treatment occurred.

The blunted CORT release during intermediate developmental ages (P7–13) led us to postulate that the attenuated adrenal response to stress associated with the stress hyporesponsive period (SHRP) may explain the differential age effects we observed in methamphetamine-induced CORT production and learning and memory performance. The SHRP (P4–14) is an interval during which basal CORT is low and CORT elevation to stressors is attenuated compared with earlier or later periods of development and adulthood (Sapolsky and Meaney 1986). Diminished CORT following acute methamphetamine on P7–13 aligns with the SHRP and chronic methamphetamine not only inhibits the early CORT response, but amplifies the later response, particularly after P15 when the SHRP is over (Williams et al. 2006). Given that the SHRP is hypothesized to protect against stress-induced overstimulation of glucocorticoid receptors (Sapolsky and Meaney 1986; Sapolsky 1996) and methamphetamine treatment induces an HPA axis over-response, we hypothesized that methamphetamine treatment during the SHRP would lead to more severe long-term behavioral effects than drug administration outside of the SHRP (Vorhees et al. 2009). Vorhees et al. (2009) exposed pups to methamphetamine following one of three treatment windows: P11–20 (associated with cognitive deficits), P1–10 (not associated with cognitive deficits), and P6–15 (aligned with the SHRP). As predicted, in the CWM, the number of errors and latency to escape were increased in the P6-15 group compared with the P11–20 group, with no effect following P1–10 drug treatment. Thus, egocentric learning deficits from neonatal methamphetamine are associated with the SHRP and may result from HPA axis dysregulation during a critical period of brain development (Vorhees et al. 2009).

The next step in determining if changes in plasma CORT production were responsible for the cognitive impairments observed was to diminish the methamphetamine-induced CORT response prior to drug delivery via surgical adrenal autotransplantation (ADXA). Grace et al. (2010b) showed that ADXA on P9, two days before the start of methamphetamine treatment from P11–20 (10 mg/kg; 4/day), attenuated the methamphetamine-induced increase in CORT from P12–20. On P11, ADXA reduced the increase in CORT to ~30 % of methamphetamine-exposed animals that received sham surgery. This reduction in CORT increase rose to ~ 75 % by P20 (Grace et al. 2010b). Thus, neonatal ADXA is an effective method for partially reducing methamphetamine-induced CORT release, while preserving sufficient levels to support normal growth and development. However, the attenuation of the CORT response to methamphetamine by ADXA did not alter the long-lasting MWM or CWM deficits, despite ADXA having no effect on the long-term methamphetamine-induced 5-HT reductions in the neostriatum or entorhinal cortex or on 5-HIAA reductions in the hippocampus (Schaefer et al. 2010; Grace et al. 2010a, b). Therefore, it appears that reducing the extent of the methamphetamine-induced release of CORT during a sensitive period of neuronal development (SHRP) is not protective against later behavioral impairments, although it should be pointed out that the methamphetamine-treated animals still had higher levels of CORT than the control animals. While the mechanism of methamphetamine-induced neurotoxicity during the neonatal period is unknown, an alternate hypothesis may be that overstimulation of dopamine receptors caused by dopamine overflow and increased generation of reactive oxygen species (ROS) and reactive nitrogen species (RNS) may influence receptor function through adulthood, contributing to the long-term behavioral effects we describe.

5 Experimental Design: Review by McDonnell-Dowling and Kelly (2015)

A recent review of developmental amphetamine studies addresses differences in the design of developmental studies. McDonnell-Dowling and Kelly (2015) state that there is little consistency among maternal and offspring experimental variables used in these studies and that the findings are therefore not relevant to the human situation. It is true that there is variability among the developmental substituted amphetamine studies, but these are no greater (and in fact are less) than in most developmental drug studies, such as those on cocaine, ethanol, opiates, or any other drug whether it be a drug of abuse or not. This is because two laboratories have generated most of the data: our laboratory and that of Dr. Slamberova. This has resulted in more consistency and more replication of findings than for any other developmental drug study area. Of course, consistency is a significant issue, and McDonnell-Dowling and Kelly (2015) make some valid points about how some types of design inconsistencies can impede aspects of progress in understanding how methamphetamine (and also amphetamine and MDMA in their review) affect neurodevelopment. Unfortunately, their review contains mischaracterizations of several studies, apparent misunderstandings of experimental design procedures, and despite tabularizing 112 studies on procedural factors, does not review the effects of developmental exposure to amphetamines at all. Not a single finding from these 112 studies is presented or discussed in any detail, and no generalization or interpretations are provided; instead, the article describes how the authors' believe developmental amphetamine studies should be designed based on ideas not in accord with the field of developmental drug studies that has evolved over the last 40 years. In fact, they often reject experimental design features that the field, through much research, has identified as invaluable. For example, they reject such control procedures as fostering–cross-fostering, pair-feeding, direct treatment of pups, taking species differences in brain development into account, taking phar- macokinetics into account, and others (see below). Rather than providing an analysis of the data available, identifying gaps in our knowledge, and pointing to potentially fruitful directions, they provide a list of complaints not in tune with accepted standards for developmental drug studies.

Following the literature search that they report generated 833 articles associated with the effects of amphetamine-type stimulants during pregnancy and/or lactation, McDonnell-Dowling and Kelly (2015) state that they reviewed 112 in their article based on a criterion of "relevance," which they never define. They suggest that the use of experimental variables should be held constant across developmental studies. For example, they argue in favor of using a single rat strain in all experiments. In so doing, they fail to acknowledge the benefit of data derived from different strains (not to mention different species) in answering questions related to outcomes, as well as the benefit of the converging lines of evidence that accrue from data gathered using multiple genetic backgrounds. In this case, data from diverse sources add to the human relevance of animal models because humans are genetically and environmentally heterogeneous, whereas laboratory rodents are not. Therefore, using different genetic backgrounds in animals enhances (rather than limits) human relevance, since if a finding stands up across strains, species, and different labo- ratory settings it is a more robust scientific finding than if it is only found in one strain of one species in one laboratory at one point in time without replication. This concept is fundamental to biological research, and therefore, it is difficult to understand why these authors argue against it.

Related to rat strain, McDonnell-Dowling and Kelly (2015) place importance on the body weight of females used at the time of breeding. They propose that females should be mated between 225 and 250 g and that because of age differences at the time of mating across studies, they claim, have occurred up to 40 days, this is an important factor that needs to be standardized. There are problems with this rec- ommendation. Firstly, body weight is dependent on the strain of rat. Secondly, restricting age/weight in animal models limits human relevance, since there is variation in the age (body weight) of women who use methamphetamine during pregnancy (e.g., Della Grotta et al. 2010). Thirdly, there are no data to support the claim that rat maternal body weight has a significant modifying effect on the prenatal effects of amphetamines on offspring neurodevelopment. The females used in all of the reviewed studies were sexually mature, regardless of their exact age (e.g., P60 or P180), and this is the case in the entire field of developmental drug studies. There is no evidence that within a broad range that maternal weight matters. Fourthly, from the literature reviewed herein, especially for those studies that deliver the drug during gestation, the data show that maternal behavior is more important in determining outcome than maternal age/weight. Fifthly, they misrepresent the body weight of animals used in the studies from this laboratory. They state that female Sprague-Dawley rats in our laboratory were bred when they weighed 151–175 g. This is incorrect. The females were significantly heavier than this at the time of breeding. The average female rat's weight at the time of breeding in this laboratory is approximately 276 g. What we stated in those articles was that females were received from Charles River at 151–175 g and given a minimum of 2 weeks before being placed with a male and in most cases much longer than 2 weeks. The first set of females put in breeding weighed over 200 g when cohabitated with males, and the majority of females weighed 250 g or more when mated. This fact was not reported in these studies because the exposure to the drug was postnatal to the pups; therefore, minor variations in maternal body weight were regarded as of no significance to the effects of the drug injected directly into the offspring. Hence, McDonnell-Dowling and Kelly (2015) misread what we wrote, but more importantly emphasize one of the least important aspects of a postnatal drug study.

McDonnell-Dowling and Kelly (2015) state that "the purpose of these preclinical studies should be to investigate the margins of safety of these controlled substances at doses that are typically abused in humans rather than at doses that we know to be neurotoxic." Not only is this not the focus of most developmental drug of abuse studies, but also there is no reason to think it should be. These studies are aimed at the effects that may occur when drug users take abusive or at least recreational levels of these substances, not therapeutic doses. Drug users do not take these drugs "safely" as demonstrated by the number of women seeking drug treatment (Substance Abuse and Mental Health Services Administration 2013); therefore, the safety margin concept is not relevant in this context. If these were studies of therapeutic drugs, this reasoning would apply, but these are studies modeling high-dose use; hence, there is no "clinical" relevance in the usual sense, but there is relevance to how people misuse these drugs. The authors seem to miss the fact that pregnant women addicted to methamphetamine use the drug during pregnancy because they either do not understand the risks, or are unable to quit. These studies are not designed to test the effects of taking Adderall as an ADHD therapeutic drug, for example. While such studies may be worth doing, they are not the intent of the studies these authors cited. Hence, McDonnell-Dowling and Kelly (2015) advocate that low doses be studied as "more relevant." If only low doses are to be used in animal models, then the range of outcomes caused by higher levels will be missed, including those found in chronic, heavy using methamphetamine addicts. Human imaging studies report drug abusers using up to 10 g of methamphetamine per day (Volkow et al. 2001b), not the 30 mg McDonnell-Dowling and Kelly (2015) cite from a study that attempted to characterize recreational use. Further, McDonnell-Dowling and Kelly (2015) only mention the role of drug metabolism as a factor for consideration in dose selection but fail to incorporate the meaning of such information in creating appropriate exposure models. The drug half-life for human and rats is different and allometric scaling of dose approximations that they recommend do not take this into consideration (see Sect. 4). For example, the elimination half-life of methamphetamine in humans is

10–12 h but in rats is 1–2 h. No matter what the allometric scaling generates for a given dose, if that dose is given to a rat the drug is effectively eliminated in 5 half-lives, i.e., in about 10 h, but in a human not for 60 h. To not take this into account in the design of experiments ignores one of the most basic factors that determine drug effects, internal dose at the site of action. In many of our studies, we gave the drug 4 times at 2 h intervals or 2 times at 6 h intervals. We stated that the reason was to partially compensate for the rapid metabolism and clearance of the drug in rats, a fact that McDonnell-Dowling and Kelly (2015) fail to acknowledge in their criticism of this design. Dose has limited meaning in the absence of the absorption, distribution, metabolism, and elimination (ADME) of a drug. It is surprising that this is not acknowledged in the McDonnell-Dowling and Kelly (2015) review. Given that there are consistent physiological effects observed in human studies (i.e., low birth weight), along with replicated neurological and behavioral effects, in animals the focus should be on recapitulating the outcomes of human studies and predicting what human studies may find as the children grow older and mature in their behavioral and cognitive capacities. It is more important that the outcomes align as much as possible even if doses are different in order to account for species differences in ADME and brain ontogeny. It is the effect of the drug that is of primary interest, not how dose, route, and frequency should be copied between rats and humans. Outcome, mechanism of action, and internal dose are the critical factors, not maternal body weight at breeding in a postnatal study or what the authors think about the known limitations of pair-feeding or foster techniques.

Related to dose, McDonnell-Dowling and Kelly (2015) report that the delivery of amphetamines via subcutaneous injection is irrelevant and should not be done. They cite FDA guidelines that the route for a therapeutic drug should, to the extent possible, mimic the route to be used in humans. There are problems with this recommendation. Even though humans do not self-administer methamphetamine subcutaneously, a variety of drug exposure methods are used by methamphetamine abusers (e.g., Volkow et al. 2001a), such that no one route constitutes a gold standard above all others. Furthermore, the pregnant female is not the end point for these studies, but rather the progeny. Consequently, the route of administration is less important than the dose crossing the placenta and reaching the embryo and fetus or directly to the pup. In this sense, it makes little difference whether the mother takes the drug by inhalation, snorting, injection, or orally unless these other routes induce a pattern of ADME that is not approximated in animals by subcutaneous injection. So far there are no data suggesting that the different routes pregnant humans use is radically different from that achieved by subcutaneous injection in rats. The FDA recommendation is that "therapeutic" (not abused) agents in the context of New Drug Applications be given by the route that "most closely mimics the human route." Therapeutic drugs are taken typically orally by humans; therefore, in preclinical studies, gavage is used to collect safety data for submission to the FDA, despite the obvious fact that patients do not take drugs by stomach tube (the human equivalent of gavage), but rather swallow tablets or liquids. If one accepts gavage as the approximate equivalent of taking a pill by a person, which the FDA does, then the FDA guidance is entirely logical. But drugs of abuse are not typically taken orally. Most drugs of abuse, including amphetamines, are snorted, smoked, or i.v. injected. To follow the logic of McDonnell-Dowling and Kelly (2015), rat developmental methamphetamine studies should use these routes as well as gavage. However, rats will not snort a drug, and inhalation in rats is not only technically more difficult, but rats do not inhale vapor the same way people do. Rats do not inhale deeply into their alveolar spaces the way humans do; therefore, even an inhalation model would be subject to a number of caveats. Methamphetamine can be given orally (gavage) or injected i.v. and such models may be worth investigating, as these authors have done using gavage in one experiment (McDonnell-Dowling et al. 2014). This is a welcome addition to the literature, but it does not *ipso facto* undermine studies using other routes as they advocate.

McDonnell-Dowling and Kelly (2015) also construct three arbitrary categories of drug effects, referring to them as neurotoxic, toxic, or pharmacological. They assign doses to these categories and then label studies as irrelevant if they fall into their pejorative categories. These distinctions do not stand up to examination. There are doses in adult rats that are sometimes referred to as neurotoxic when they cause large and lasting neurotransmitter reductions or induce neuropathological changes, but even here the dose, pattern of doses, and environmental conditions (particularly ambient room temperature) can make a "neurotoxic" dose benign if each parameter is not appropriately calibrated. Dose alone, therefore, does not determine the effect of the drug, and such simplistic categories are not helpful. In developing rats, doses that can be neuropathological in adult rats, if conditions are set correctly, induce none of the same CNS changes found in adult rat brain (Cappon et al. 1997). It has been shown multiple times that the dopamine, 5-HT, and TH depleting effects and GFAP, FluoroJade, and silver stain increases seen in adult rats after a "neurotoxic regimen" of methamphetamine does not occur in rats younger than P40. Since McDonnell-Dowling and Kelly (2015) apparently derive their categories based on adult rat effects, they are meaningless in relation to prenatal and neonatal rat studies.

McDonnell-Dowling and Kelly (2015) err when they recommend against fostering–cross-fostering designs. They argue that leaving pups with their mother is preferable because it more closely mimics human families where the mother took amphetamines during pregnancy. Firstly, they miss the point that the purpose of a fostering–cross-fostering study is to isolate maternal effects from offspring effects. It is not intended to be "realist" but rather to control for maternal mediation of effects found in the offspring. Interestingly, McDonnell-Dowling and Kelly (2015) ignore the fact that human studies find that many children of drug-abusing mothers are in fact removed from the home and raised in foster homes (e.g., Sarkola et al. 2007), not unlike a rat being fostered to a control dam. Sorting out different factors is what scientific experiments are about; hence, the fostering–cross-fostering experiments have a legitimate place in prenatal drug studies. Not everything in an experiment is designed to mimic the often chaotic lifestyles of human pregnant drug abusers. Secondly, a fostering–cross-fostering experiment should not be viewed in isolation. Rather, such a study is part of a series of studies with the goal of factoring out as many factors as possible that play a role in the effects of interest in the offspring. McDonnell-Dowling and Kelly (2015) also criticize fostering–cross-fostering studies because they claim there "could possibly" be residual drug in the dam given pups from control dams and this could contaminate the effects by exposing controls to drug. Given the half-life of methamphetamine and other amphetamines in rats, the drug would be gone by the time pups were fostered even if fostered on P1. Secondly, virtually all published prenatal methamphetamine studies stop giving the drug one or more days before parturition, often on E18 or E20. This allows ample time for the drug to be cleared. Thirdly, even if a study gave the drug on E21 to a Sprague-Dawley rat that typically delivers on E22, the amount of drug left in the dam would be trivial (using a 3-h half-life, the percentage of drug remaining after 8 half-lives would be 0.4 % of the starting value). Not only is that level unlikely to have a pharmacological effect even if given at a high dose on E21, but since the amount of any drug that passes to milk during lactation is known to be a fraction of that in maternal circulation, the pups would only receive a fraction of 0.4 % of the drug from its peak concentration.

McDonnell-Dowling and Kelly (2015) acknowledge that rodent brain development is different than in humans and note that the neonatal period in rats more closely resembles third trimester in utero brain development in humans and cite well-known literature to support this (Clancy et al. 2001, 2007a). After having cited this ontological difference, they then criticize studies giving the drug to pups to mimic exposure during this period of brain development stating that in humans, the drug would pass through maternal milk during the postnatal period. But if the neonatal rat does not model a human infant, but rather a human fetus, then this rationale is illogical. The neonatal rat either models human third trimester brain development or it models human infant brain development, but the same period cannot model both simultaneously. It appears that McDonnell-Dowling and Kelly (2015) do not grasp the contradiction in their own writing.

They also criticize the use of pair-feeding in prenatal drug studies because they note that a food-deprived animal does not consume its ration the same as an amphetamine-treated rat with anorexia-induced appetite suppression. This is valid, and the limitations of pair-feeding have been noted before, but it is an established tool that experimentalists use, and use effectively, to isolate nutritional factors as potential confounders when studying a drug that suppresses food consumption and reduces body weight. Pair-feeding is not a perfect technique, but it does serve a purpose, one that McDonnell-Dowling and Kelly (2015) indicate is possible to use but is mostly used incorrectly according to them. They argue that typically the daily ration of pair-fed animals is given and eaten rapidly soon after it is given, and this is what most experimenters observe. But McDonnell-Dowling and Kelly (2015) state that this "problem" can be solved by giving the ration during the dark cycle in the vivarium. However, this would not solve the issue. The pair-fed animal will still consume its ration immediately, whereas the experimental animal will consume its food spread over many hours. There is no evidence that simply giving the ration during the dark cycle solves the rapid food consumption pattern seen in hungry pair-fed animals. If one wanted to match food consumption patterns, one would have to have a device that fed the pair-fed animal whenever the experimental animal ate. While technologically possible, the value of such a method is questionable.

McDonnell-Dowling and Kelly (2015) also recommend that maternal–pup interactions be observed in detail every day in every study throughout lactation to the age of weaning that in most rat studies is P21, but is often P28. The labor-intensive nature of such observations make this prohibitive, except perhaps in a single study done once to document whether the drug under investigation causes changes to maternal–pup interactions. Oddly, McDonnell-Dowling, and Kelly (2015) then state that this kind of observation should begin after cross-fostering which in previous paragraphs they stated should not be done at all.

McDonnell-Dowling and Kelly (2015) inexplicably criticize one and only one behavioral method: rotorod, for no reason they ever explain. They state that testing younger rats using 5-min trials is "an incredibly long time for a young pup" and cite one study (Barenys et al. 2010). Firstly, no very young rats were tested on a rotorod in this study; the test age was P22, 24 h after weaning. This is an age when rats are capable of running on a running wheel. Secondly, the rotorod test began by experimenters placing rats on the rod at low speeds and retesting them at speeds gradually incremented (4, 7, 12, 20, 25, 30, and 40 RPM). Once rats were tested at each of these speeds separately, they were then tested a second time in which the rod was started at 4 RPM and gradually accelerated to 40 RPM. The authors state that the time limit of the test was 5 min, but they did not state that any rats stayed on that long. Rotorod procedures are designed to find the speed at which the rat falls, thereby ending the test. No data are shown by Barenys et al. (2010), they merely state that no effects of developmental MDMA were found. What is it about this single behavioral test from all the 112 papers cited that makes it worth singling out for criticism, when in fact there is nothing apparently wrong with the procedure described by Barenys et al. (2010)? Secondly, how do McDonnell-Dowling and Kelly (2015) know what is "incredibly long" for a P22 rat in the apparatus used in that study? McDonnell-Dowling and Kelly (2015) present no data to support the notion that it is "incredibly long." Experimentalists have to make many judgments in setting up procedures, and not every aspect is fully vetted along every possible dimension; most good investigators use procedures they have tried in their laboratory and found to work sufficiently well to believe they would detect a treatment difference if there was one. This may only be that animals perform the task, but that is often the basis for many behavioral methods.

McDonnell-Dowling and Kelly (2015) state that two developmental methamphetamine studies even failed to mention the strain of rats used. One of the studies they cite is Adams et al. (1982), but that report clearly states on page 64 that Sprague-Dawley rats were used. Another error already mentioned was about maternal body weight in our studies; another is that that giving 25 mg/kg 4 times a day to rat pups is "neurotoxic." We have published data showing that this is not the case (Cappon et al. 1997). McDonnell-Dowling and Kelly (2015) argue that studies should use one dose per day regardless of all other considerations, including whether a once per day dosing regimen provides sufficient drug to reach its phar-macodynamic site(s) of action.

Overall, McDonnell-Dowling and Kelly (2015) miss the point concerning the central theme of their review—that animal models should aim for standardization in order to be relevant. An alternative way of striving for consistency is to define related outcomes, both within and across species (e.g., body weight changes, behavioral alterations, neurofunctional profiles, and mechanisms), rather than trying to restrict an understanding of amphetamine-induced neurotoxicity to using the same rodent strains and other parameters without knowing if they are relevant to outcome or not. As the authors recognize, controlling variables, such as dating conception the same (E0 vs. E1 as the date of inferred conception), can be helpful if it reduces confusion, but different experimental designs are more often helpful in applying animal data to human populations than impairing it. If experimental methodology were to follow the recommendations from McDonnell-Dowling and Kelly (2015), rat models would fail because of species differences alone. Rodents produce multiple births, do not live as long as people, have four rather than two legs, etc., but so what? The point is that rats are not used as miniature humans but to function as models, and like all models are inherently different than what they are modeling but are similar enough in key ways to be useful. The idea of standardizing procedures is appropriate up to a point, but suggesting that procedures that do not have face validity or match the "average" methamphetamine user are too narrow a perspective. Models can have different types of validity based on different parameters chosen by the investigator to answer particular experimental questions. By using different approaches and obtaining converging lines of evidence, broader inferences may be reached, and this provides the strongest approach to understanding a drug's mechanism and effects.

As the present review shows, there have been few laboratories that consistently do research on the effects of developmental methamphetamine. In fact, most of the published papers on this particular drug were by two laboratories. All the others were produced by laboratories that did one or two papers, never replicated any of their findings, nor extended their data to any significant degree. But from the two laboratories that have done the most on developmental methamphetamine, there is extensive replication and extension of findings within the laboratories, and considerable convergence of findings across these two laboratories. Hence, the developmental effects of methamphetamine are more consistent than for any other drug of abuse in a developmental context.

In order to obtain more robust data, the solution that is more effective than standardization of small experimental procedural details is to validate findings across laboratories. In order to get a better understanding of how different procedures in different laboratories effect outcomes, one can design experiments to have one laboratory treat animals in the way they find causes clear effects, then send the animals to another laboratory for testing to see whether the outcomes shown to be affected in the first laboratory are also seen in the second laboratory. While this approach may create some shipping stress, it is a step in the right direction for determining doses, strains, and exposure periods that produce reliable long-term effects that are not specific to one laboratory. An example of this with developmental methamphetamine was done by the present authors (Williams et al. 2002), and it provided convergent evidence that the effect of neonatal methamphetamine on spatial learning and memory in the MWM could be replicated in the hands of a different laboratory. Replication, cross-laboratory validation, accurate model development (based on pharmacokinetics and related factors), relevant dependent variables (rather than early reflexes that have not proven to be valuable in rodents or humans), dose–response, critical period, and neurochemical mechanism consider-ations all matter more than the issues raised by McDonnell-Dowling and Kelly (2015) in their review. In sum, we take issue with a number of points offered by McDonnell-Dowling and Kelly (2015), but it should be understood that these disagreements are scientific and not directed at those authors.

6 Conclusions

Because rat prenatal development only approximates the first and second trimester of human brain development, modeling prenatal methamphetamine in rodents presents challenges. Perhaps, the most difficult of these is how to best model human third trimester brain development given that the related events in rats and mice occur postnatally in these species. Since very little maternal drug in rats gets into milk during lactation, giving a drug to the dam during the preweaning period is not an effective way to model late gestational effects. The only alternative without changing species is to administer the drug directly to the pups during the preweaning period that spans approximately the first three postnatal weeks in rodents. While there is debate about whether the third trimester equivalent to humans in terms of brain develop ends at P10, or as late as P20 depending on the reviewers (Bayer et al. 1993; Clancy et al. 2007a, b; Herlenius and Lagercrantz 2004; Quinn 2005; Rice and Barone 2000; Ross et al. 2015; Semple et al. 2013), all are in agreement that birth in a rodent is not equivalent to birth in humans in terms of brain development. Given this and given that pup drug exposure via maternal milk is ineffective, there is no alternative but to give the drug directly to the pups. Many regulatory agencies now recognize this and for the compounds they regulate have recommended direct treatment of pups (US EPA, US FDA, US NTP, OECD). The obvious problem is that there is no maternal metabolism, no placenta, and different mechanisms of absorption and clearance between a fetus and a neonatal rodent. Other than using a different species, there is no solution to this problem except to model human third trimester based on brain ontogeny if one wants to test the effects of a drug on brain structures that develop postnatally in the rat and prenatally in humans.

Given this unavoidable difference, what are the major effects in rodents from prenatal methamphetamine exposure given during rat gestation or during rat lac-tation directly to the offspring? The most reliable effect found so far from intrauterine exposure to methamphetamine in rodent models is reduced locomotor activity and transient changes in striatal dopamine and 5-HT content, but there is also a lack of clear learning and memory deficits during adulthood and the absence of increased sensitivity of the same drug on these tests during adulthood. The most reliable effects found so far from certain critical periods of neonatal development in rodents are impaired allo- and egocentric learning and memory, reduced open-field activity, increased acoustic startle response, increased locomotor sensitivity to challenge with a D_1 agonist and an NMDA antagonist, transient changes in striatal dopamine and 5-HT content, reductions in D_2 receptor binding and PKA activity, short-term increases in ACTH and corticosterone release, with the latter not associated with the long-term learning and memory deficits. Preweaning methamphetamine exposure does not induce neuropathological changes, i.e., no increase in GFAP, argyrophilia, or FluoroJade staining, nor does it induce hyperthermia as it does in adult animals. In mice, it has been reported that prenatal methamphetamine generates ROS (Wells et al. 2005). In the adult rat, the effects of methamphetamine can be blocked by pretreatment with a spin-trapping drug (PBN; Cappon et al. 1996), but whether such a block of methamphetamine-induced postnatal effects on learning and memory can occur after neonatal exposure is currently being investigated. At present, the leading candidates for how prenatal and neonatal methamphetamine induces long-term effects on the CNS are through generation of ROS, or actions on specific neurotransmitter receptors, with the most evidence pointing toward dopamine D_1, glutamatergic NMDA, and histaminergic H_1 receptors. It should be noted, however, that little attention has been given to other neurotransmitters or receptors, neurotrophic factors other than BDNF, and almost no attention to transcription factor regulation or epigenetic modulation. Like most drugs, methamphetamine has multiple effects and when acting on the rapidly changing substrate of the developing brain, identifying all its effects is challenging, but much progress has occurred and more is within sight.

References

Abar B et al (2013) Examining the relationships between prenatal methamphetamine exposure, early adversity, and child neurobehavioral disinhibition. Psychol Addict Behav 27:662–673

Abar B et al (2014) Cross-national comparison of prenatal methamphetamine exposure on infant and early child physical growth: a natural experiment. Prev Sci 15:767–776

Acevedo SF, de Esch IJ, Raber J (2007) Sex- and histamine-dependent long-term cognitive effects of methamphetamine exposure. Neuropsychopharmacology 32:665–672

Acevedo SF et al (2008) Role of histamine in short- and long-term effects of methamphetamine on the developing mouse brain. J Neurochem 107:976–986

Acuff-Smith KD et al (1996) Stage-specific effects of prenatal d-methamphetamine exposure on behavioral and eye development in rats. Neurotoxicol Teratol 18:199–215

Adams J et al (1982) Behavioral alterations in rats prenatally exposed to low doses of d-amphetamine. Neurobehav Toxicol Teratol 4:63–70

Barenys M et al (2010) MDMA (ecstasy) delays pubertal development and alters sperm quality after developmental exposure in the rat. Toxicol Lett 197:135–142

Bartu A, Dusci LJ, Ilett KF (2009) Transfer of methylamphetamine and amphetamine into breast milk following recreational use of methylamphetamine. Br J Clin Pharmacol 67:455–459

Bayer SA (1982) Changes in the total number of dentate granule cells in juvenile and adult rats: a correlated volumetric and 3H-thymidine autoradiographic study. Exp Brain Res 46:315–323

Bayer SA et al (1993) Timetables of neurogenesis in the human brain based on experimentally determined patterns in the rat. Neurotoxicology 14:83–144

Bernaskova K, Matejovska I, Slamberova R (2011) Postnatal challenge dose of methamphetamine amplifies anticonvulsant effects of prenatal methamphetamine exposure on epileptiform activity induced by electrical stimulation in adult male rats. Exp Neurol 229:282–287

Bowyer JF et al (1998) Long-term effects of amphetamine neurotoxicity on tyrosine hydroxylase mRNA and protein in aged rats. J Pharmacol Exp Ther 286:1074–1085

Braren SH et al (2014) Methamphetamine-induced short-term increase and long-term decrease in spatial working memory affects protein Kinase M zeta (PKMzeta), dopamine, and glutamate receptors. Front Behav Neurosci 8:438

Braun AA et al (2012) Dorsal striatal dopamine depletion impairs both allocentric and egocentric navigation in rats. Neurobiol Learn Mem 97:402–408

Braun A, Amos-Kroohs R, Gutierrez A, Lundgren K, Seroogy K, Skelton M, Vorhees CV, Williams MT (2014) Dopamine depletion in either the dorsomedial or dorsolateral striatum impairs egocentric Cincinnati water maze performance while sparing allocentric Morris water maze learning. Neurobiol Learn Mem 118:55–63

Brecht ML, Herbeck DM (2014) Pregnancy and fetal loss reported by methamphetamine-using women. Subst Abuse 8:25–33

Broening HW, Morford LL, Vorhees CV (2005) Interactions of dopamine D_1 and D_2 receptor antagonists with D-methamphetamine-induced hyperthermia and striatal dopamine and serotonin reductions. Synapse 56:84–93

Bubenikova-Valesova V et al (2009) Prenatal methamphetamine exposure affects the mesolimbic dopaminergic system and behavior in adult offspring. Int J Dev Neurosci 27:525–530

Burchfield DJ et al (1991) Disposition and pharmacodynamics of methamphetamine in pregnant sheep. JAMA 265(15):1968–1973

Cabrera TM et al (1993) Prenatal methamphetamine attenuates serotonin mediated renin secretion in male and female rat progeny: evidence for selective long-term dysfunction of serotonin pathways in brain. Synapse 15:198–208

Campbell LF, Bedi KS (1989) The effects of undernutrition during early life on spatial learning. Physiol Behav 45:883–890

Cappon GD, Vorhees CV (2001) Plasma and brain methamphetamine concentrations in neonatal rats. Neurotoxicol Teratol 23:81–88

Cappon GD et al (1996) alpha-Phenyl-N-tert-butyl nitrone attenuates methamphetamine-induced depletion of striatal dopamine without altering hyperthermia. Synapse 24:173–181

Cappon GD, Morford LL, Vorhees CV (1997) Ontogeny of methamphetamine-induced neurotoxicity and associated hyperthermic response. Brain Res Dev Brain Res 103:155–162

Cappon GD, Pu C, Vorhees CV (2000) Time-course of methamphetamine-induced neurotoxicity in rat caudate-putamen after single-dose treatment. Brain Res 863:106–111

Cass WA, Manning MW (1999) Recovery of presynaptic dopaminergic functioning in rats treated with neurotoxic doses of methamphetamine. J Neurosci 19:7653–7660

Chang L et al (2004) Smaller subcortical volumes and cognitive deficits in children with prenatal methamphetamine exposure. Psychiatry Res 132:95–106

Chang L et al (2007) Structural and metabolic brain changes in the striatum associated with methamphetamine abuse. Addiction 102(Suppl 1):16–32

Chang L et al (2009) Altered neurometabolites and motor integration in children exposed to methamphetamine in utero. Neuroimage 48:391–397

Chapillon P et al (2002) Effects of pre- and postnatal stimulation on developmental, emotional, and cognitive aspects in rodents: a review. Dev Psychobiol 41:373–387

Cho DH et al (1991) Behavioral teratogenicity of methamphetamine. J Toxicol Sci 16(Suppl 1):37–49

Cho AK et al (2001) Relevance of pharmacokinetic parameters in animal models of methamphetamine abuse. Synapse 39:161–166

Chomchai C et al (2004) Methamphetamine abuse during pregnancy and its health impact on neonates born at Siriraj Hospital, Bangkok, Thailand. Southeast Asian J Trop Med Public Health 35:228–231

Clancy B, Darlington RB, Finlay BL (2001) Translating developmental time across mammalian species. Neuroscience 105:7–17

Clancy B et al (2007a) Extrapolating brain development from experimental species to humans. Neurotoxicology 28:931–937

Clancy B et al (2007b) Web-based method for translating neurodevelopment from laboratory species to humans. Neuroinformatics 5:79–94

Cloak CC et al (2009) Lower diffusion in white matter of children with prenatal methamphetamine exposure. Neurology 72:2068–2075

Colby JB et al (2012) White matter microstructural alterations in children with prenatal methamphetamine/polydrug exposure. Psychiatry Res 204:140–148

Crawford CA et al (2003) Methamphetamine exposure during the preweanling period causes prolonged changes in dorsal striatal protein kinase A activity, dopamine D_2-like binding sites, and dopamine content. Synapse 48:131–137

Della Grotta S et al (2010) Patterns of methamphetamine use during pregnancy: results from the infant development, environment, and lifestyle (IDEAL) study. Matern Child Health J 14:519–527

Derauf C et al (2009) Neuroimaging of children following prenatal drug exposure. Semin Cell Dev Biol 20(4):441–454

Derauf C et al (2012a) Prenatal methamphetamine exposure and inhibitory control among young school-age children. J Pediatr 161:452–459

Derauf C et al (2012b) Subcortical and cortical structural central nervous system changes and attention processing deficits in preschool-aged children with prenatal methamphetamine and tobacco exposure. Dev Neurosci 34:327–341

Dixon SD, Bejar R (1989) Echoencephalographic findings in neonates associated with maternal cocaine and methamphetamine use: incidence and clinical correlates. J Pediatr 115:770–778

Eastwood E, Allen CN, Raber J (2012) Effects of neonatal methamphetamine and thioperamide exposure on spatial memory retention and circadian activity later in life. Behav Brain Res 230:229–236

Fenoglio KA, Brunson KL, Baram TZ (2006) Hippocampal neuroplasticity induced by early-life stress: functional and molecular aspects. Front Neuroendocrinol 27:180–192

Fialova M et al (2015) The effect of prenatal methamphetamine exposure on recognition memory in adult rats. Prague Med Rep 116:31–39

Frost DO, Cadet JL (2000) Effects of methamphetamine-induced neurotoxicity on the development of neural circuitry: a hypothesis. Brain Res Rev 34:103–118

Fukumura M et al (1998) A single dose model of methamphetamine-induced neurotoxicity in rats: effects on neostriatal monoamines and glial fibrillary acidic protein. Brain Res 806:1–7

Gil-Mohapel J et al (2010) Hippocampal cell loss and neurogenesis after fetal alcohol exposure: insights from different rodent models. Brain Res Rev 64:283–303

Gomes-da-Silva J et al (2000) Neonatal methamphetamine in the rat: evidence for gender-specific differences upon tyrosine hydroxylase enzyme in the dopaminergic nigrostriatal system. Ann NY Acad Sci 914:431–438

Gomes-da-Silva J et al (2002) Prenatal exposure to methamphetamine in the rat: ontogeny of tyrosine hydroxylase mRNA expression in mesencephalic dopaminergic neurons. Ann NY Acad Sci 965:68–77

Gomes-da-Silva J et al (2004) Effects of neonatal exposure to methamphetamine: catecholamine levels in brain areas of the developing rat. Ann NY Acad Sci 1025:602–611

Goodlett CR et al (1986) Spatial cue utilization in chronically malnourished rats: task-specific learning deficits. Dev Psychobiol 19:1–15

Gould E et al (1991) Adrenal steroids regulate postnatal development of the rat dentate gyrus: II. Effects of glucocorticoids and mineralocorticoids on cell birth. J Comp Neurol 313:486–493

Grace CE et al (2008) (+)-Methamphetamine increases corticosterone in plasma and BDNF in brain more than forced swim or isolation in neonatal rats. Synapse 62:110–121

Grace CE et al (2010a) Effects of inhibiting neonatal methamphetamine-induced corticosterone release in rats by adrenal autotransplantation on later learning, memory, and plasma corticosterone levels. Int J Dev Neurosci 28:331–342

Grace CE et al (2010b) Neonatal methamphetamine-induced corticosterone release in rats is inhibited by adrenal autotransplantation without altering the effect of the drug on hippocampal serotonin. Neurotoxicol Teratol 32:356–361

Graham DL et al (2013) Neonatal (+)-methamphetamine exposure in rats alters adult locomotor responses to dopamine D_1 and D_2 agonists and to a glutamate NMDA receptor antagonist, but not to serotonin agonists. Int J Neuropsychopharmacol 16:377–391

Herlenius E, Lagercrantz H (2004) Development of neurotransmitter systems during critical periods. Exp Neurol 190(Suppl 1):S8–21

Himes SK et al (2014) Risk of neurobehavioral disinhibition in prenatal methamphetamine-exposed young children with positive hair toxicology results. Ther Drug Monit 36:535–543

Hruba L et al (2008) Does cross-fostering modify the impairing effect of methamphetamine on postnatal development of rat pups? Prague Med Rep 109:50–61

Hruba L et al (2009) Does cross-fostering modify the prenatal effect of methamphetamine on learning of adult male rats? Prague Med Rep 110:191–200

Hruba L, Vaculin S, Slamberova R (2010) Effect of prenatal and postnatal methamphetamine exposure on nociception in adult female rats. Dev Psychobiol 52:71–77

Hruba L, Schutova B, Slamberova R (2012) Sex differences in anxiety-like behavior and locomotor activity following prenatal and postnatal methamphetamine exposure in adult rats. Physiol Behav 105:364–370

Jeng W et al (2005) Methamphetamine-enhanced embryonic oxidative DNA damage and neurodevelopmental deficits. Free Radic Biol Med 39:317–326

Kaewsuk S et al (2009) Melatonin attenuates methamphetamine-induced reduction of tyrosine hydroxylase, synaptophysin and growth-associated protein-43 levels in the neonatal rat brain. Neurochem Int 55:397–405

Kiblawi ZN et al (2013) The effect of prenatal methamphetamine exposure on attention as assessed by continuous performance tests: results from the infant development, environment, and lifestyle study. J Dev Behav Pediatr 34:31–37

Kirlic N et al (2013) Cortisol reactivity in two-year-old children prenatally exposed to methamphetamine. J Stud Alcohol Drugs 74:447–451

Kokoshka JM et al (1998) Methamphetamine treatment rapidly inhibits serotonin, but not glutamate, transporters in rat brain. Brain Res 799:78–83

LaGasse LL et al (2011) Prenatal methamphetamine exposure and neonatal neurobehavioral outcome in the USA and New Zealand. Neurotoxicol Teratol 33:166–175

LaGasse LL et al (2012) Prenatal methamphetamine exposure and childhood behavior problems at 3 and 5 years of age. Pediatrics 129:681–688

Lauder JM (1990) Ontogeny of the serotonergic system in the rat: serotonin as a developmental signal. Ann NY Acad Sci 600:297–313 (discussion 314)

Little BB, Snell LM, Gilstrap LC 3rd (1988) Methamphetamine abuse during pregnancy: outcome and fetal effects. Obstet Gynecol 72:541–544

Lu LH et al (2009) Effects of prenatal methamphetamine exposure on verbal memory revealed with functional magnetic resonance imaging. J Dev Behav Pediatr 30:185–192

Lucot JB et al (1982) Decreased sensitivity of rat pups to long-lasting dopamine and serotonin depletions produced by methylamphetamine. Brain Res 247:181–183

Macuchova E, Nohejlova-Deykun K, Slamberova R (2013) Effect of methamphetamine on cognitive functions of adult female rats prenatally exposed to the same drug. Physiol Res 62 (Suppl 1):S89–S98

Macuchova E, Nohejlova K, Slamberova R (2014) Gender differences in the effect of adult amphetamine on cognitive functions of rats prenatally exposed to methamphetamine. Behav Brain Res 270:8–17

Malinova-Sevcikova M et al (2014) Differences in maternal behavior and development of their pups depend on the time of methamphetamine exposure during gestation period. Physiol Res 63(Suppl 4):S559–S572

Mark KA, Soghomonian JJ, Yamamoto BK (2004) High-dose methamphetamine acutely activates the striatonigral pathway to increase striatal glutamate and mediate long-term dopamine toxicity. J Neurosci 24:11449–11456

Matejovska I, Bernaskova K, Slamberova R (2014) Effect of prenatal methamphetamine exposure and challenge dose of the same drug in adulthood on epileptiform activity induced by electrical stimulation in female rats. Neuroscience 257:130–138

Mazer C et al (1997) Serotonin depletion during synaptogenesis leads to decreased synaptic density and learning deficits in the adult rat: a possible model of neurodevelopmental disorders with cognitive deficits. Brain Res 760:68–73

McDonnell-Dowling K, Kelly JP (2015) Sources of variation in the design of preclinical studies assessing the effects of amphetamine-type stimulants in pregnancy and lactation. Behav Brain Res 279:87–99

McDonnell-Dowling K, Donlon M, Kelly JP (2014) Methamphetamine exposure during pregnancy at pharmacological doses produces neurodevelopmental and behavioural effects in rat offspring. Int J Dev Neurosci 35:42–51

McEwen BS et al (1992) Paradoxical effects of adrenal steroids on the brain: protection versus degeneration. Biol Psychiatry 31:177–199

Meaney MJ et al (1988) Stress-induced occupancy and translocation of hippocampal glucocorticoid receptors. Brain Res 445:198–203

Melega WP et al (2007) Methamphetamine blood concentrations in human abusers: application to pharmacokinetic modeling. Synapse 61:216–220

Melo P et al (2006) Myelination changes in the rat optic nerve after prenatal exposure to methamphetamine. Brain Res 1106:21–29

Melo P et al (2008) Correlation of axon size and myelin occupancy in rats prenatally exposed to methamphetamine. Brain Res 1222:61–68

Mirjalili T et al (2013) Congenital abnormality effect of methamphetamine on histological, cellular and chromosomal defects in fetal mice. Iran J Reprod Med 11:39–46

Moriceau S et al (2004) Corticosterone controls the developmental emergence of fear and amygdala function to predator odors in infant rat pups. Int J Dev Neurosci 22:415–422

Motz BA, Alberts JR (2005) The validity and utility of geotaxis in young rodents. Neurotoxicol Teratol 27:529–533

Nguyen D et al (2010) Intrauterine growth of infants exposed to prenatal methamphetamine: results from the infant development, environment, and lifestyle study. J Pediatr 157:337–339

Oro AS, Dixon SD (1987) Perinatal cocaine and methamphetamine exposure: maternal and neonatal correlates. J Pediatr 111:571–578

Pei L et al (2004) Regulation of dopamine D_1 receptor function by physical interaction with the NMDA receptors. J Neurosci 24:1149–1158

Peters DA (1984) Prenatal stress: effect on development of rat brain adrenergic receptors. Pharmacol Biochem Behav 21:417–422

Piper BJ et al (2011) Abnormalities in parentally rated executive function in methamphetamine/polysubstance exposed children. Pharmacol Biochem Behav 98:432–439

Plessinger MA (1998) Prenatal exposure to amphetamines—risks and adverse outcomes in pregnancy. Obstet Gynecol Clin N Am 25:119–138

Pu C, Vorhees CV (1993) Developmental dissociation of methamphetamine-induced depletion of dopaminergic terminals and astrocyte reaction in rat striatum. Brain Res Dev Brain Res 72:325–328

Quinn R (2005) Comparing rat's to human's age: how old is my rat in people years? Nutrition 21:775–777

Rambousek L et al (2014) Sex differences in methamphetamine pharmacokinetics in adult rats and its transfer to pups through the placental membrane and breast milk. Drug Alcohol Depend 139:138–144

Ricaurte GA, Schuster CR, Seiden LS (1980) Long-term effects of repeated methylamphetamine administration on dopamine and serotonin neurons in the rat brain: a regional study. Brain Res 193:153–163

Ricaurte GA et al (1982) Dopamine nerve terminal degeneration produced by high doses of methylamphetamine in the rat brain. Brain Res 235:93–103

Rice D, Barone S Jr (2000) Critical periods of vulnerability for the developing nervous system: evidence from humans and animal models. Environ Health Perspect 108(Suppl 3):511–533

Richards JB et al (1993) A high-dose methamphetamine regimen results in long-lasting deficits on performance of a reaction-time task. Brain Res 627:254–260

Roos A et al (2014) Structural brain changes in prenatal methamphetamine-exposed children. Metab Brain Dis 29:341–349

Roos A et al (2015) White matter integrity and cognitive performance in children with prenatal methamphetamine exposure. Behav Brain Res 279:62–67

Ross EJ et al (2015) Developmental consequences of fetal exposure to drugs: what we know and what we still must learn. Neuropsychopharmacology 40:61–87

Roussotte FF et al (2011) Abnormal brain activation during working memory in children with prenatal exposure to drugs of abuse: the effects of methamphetamine, alcohol, and polydrug exposure. Neuroimage 54:3067–3075

Roussotte FF et al (2012) Frontostriatal connectivity in children during working memory and the effects of prenatal methamphetamine, alcohol, and polydrug exposure. Dev Neurosci 34:43–57

Sabol KE et al (2001) Long-term effects of a high-dose methamphetamine regimen on subsequent methamphetamine-induced dopamine release in vivo. Brain Res 892:122–129

Substance Abuse and Mental Health Services Administration (2013) Results from the 2012 national survey on drug use and health: summary of national findings, NSDUH Series H-46, HHS Publication No. (SMA) 13-4795. Rockville, MD: Substance Abuse and Mental Health Services Administration

Sapolsky RM (1996) Stress, glucocorticoids, and damage to the nervous system: the current state of confusion. Stress 1:1–19

Sapolsky RM, Meaney MJ (1986) Maturation of the adrenocortical stress response—neuroendocrine control mechanisms and the stress hyporesponsive period. Brain Res Rev 11:65–76

Sarkola T et al (2007) Risk factors for out-of-home custody child care among families with alcohol and substance abuse problems. Acta Paediatr 96:1571–1576

Sato M, Fujiwara Y (1986) Behavioral and neurochemical changes in pups prenatally exposed to methamphetamine. Brain Dev 8:390–396

Schaefer TL et al (2006) Comparison of monoamine and corticosterone levels 24 h following (+) methamphetamine, (±)3,4-methylenedioxymethamphetamine, cocaine, (+)fenfluramine or (±) methylphenidate administration in the neonatal rat. J Neurochem 98:1369–1378

Schaefer TL et al (2008) Short- and long-term effects of (+)-methamphetamine and (±)-3,4-methylenedioxymethamphetamine on monoamine and corticosterone levels in the neonatal rat following multiple days of treatment. J Neurochem 104:1674–1685

Schaefer TL et al (2010) Effects on plasma corticosterone levels and brain serotonin from interference with methamphetamine-induced corticosterone release in neonatal rats. Stress 13:469–480

Schutova B et al (2008) Impact of methamphetamine administered prenatally and in adulthood on cognitive functions of male rats tested in Morris water maze. Prague Med Rep 109:62–70

Schutova B et al (2009a) Cognitive functions and drug sensitivity in adult male rats prenatally exposed to methamphetamine. Physiol Res 58:741–750

Schutova B et al (2009b) Impact of prenatal and acute methamphetamine exposure on behaviour of adult male rats. Prague Med Rep 110:67–78

Schutova B et al (2013) Gender differences in behavioral changes elicited by prenatal methamphetamine exposure and application of the same drug in adulthood. Dev Psychobiol 55:232–242

Sekine Y et al (2006) Brain serotonin transporter density and aggression in abstinent methamphetamine abusers. Arch Gen Psychiatry 63:90–100

Semple BD et al (2013) Brain development in rodents and humans: identifying benchmarks of maturation and vulnerability to injury across species. Prog Neurobiol 106–107:1–16

Shah R et al (2012) Prenatal methamphetamine exposure and short-term maternal and infant medical outcomes. Am J Perinatol 29:391–400

Siegel JA, Craytor MJ, Raber J (2010) Long-term effects of methamphetamine exposure on cognitive function and muscarinic acetylcholine receptor levels in mice. Behav Pharmacol 21:602–614

Simon SL et al (2000) Cognitive impairment in individuals currently using methamphetamine. Am J Addict 9:222–231

Skelton MR et al (2007) Neonatal (+)-methamphetamine increases brain derived neurotrophic factor, but not nerve growth factor, during treatment and results in long-term spatial learning deficits. Psychoneuroendocrinology 32(6):734–745

Slamberova R (2005) Flurothyl seizures susceptibility is increased in prenatally methamphetamine-exposed adult male and female rats. Epilepsy Res 65:121–124

Slamberova R, Rokyta R (2005a) Occurrence of bicuculline-, NMDA- and kainic acid-induced seizures in prenatally methamphetamine-exposed adult male rats. Naunyn Schmiedebergs Arch Pharmacol 372:236–241

Slamberova R, Rokyta R (2005b) Seizure susceptibility in prenatally methamphetamine-exposed adult female rats. Brain Res 1060:193–197

Slamberova R, Charousova P, Pometlova M (2005a) Maternal behavior is impaired by methamphetamine administered during pre-mating, gestation and lactation. Reprod Toxicol 20:103–110

Slamberova R, Charousova P, Pometlova M (2005b) Methamphetamine administration during gestation impairs maternal behavior. Dev Psychobiol 46:57–65

Slamberova R et al (2005c) Learning in the Place navigation task, not the new-learning task, is altered by prenatal methamphetamine exposure. Brain Res Dev Brain Res 157:217–219

Slamberova R, Pometlova M, Charousova P (2006) Postnatal development of rat pups is altered by prenatal methamphetamine exposure. Prog Neuropsychopharmacol Biol Psychiatry 30:82–88

Slamberova R, Pometlova M, Rokyta R (2007) Effect of methamphetamine exposure during prenatal and preweaning periods lasts for generations in rats. Dev Psychobiol 49:312–322

Slamberova R et al (2008) Does prenatal methamphetamine exposure affect seizure susceptibility in adult rats with acute administration of the same drug? Epilepsy Res 78:33–39

Slamberova R et al (2009) Effects of a single postnatal methamphetamine administration on NMDA-induced seizures are sex- and prenatal exposure-specific. Naunyn Schmiedebergs Arch Pharmacol 380:109–114

Slamberova R et al (2010a) Effect of cross-fostering on seizures in adult male offspring of methamphetamine-treated rat mothers. Int J Dev Neurosci 28:429–435

Slamberova R et al (2010b) Challenge dose of methamphetamine affects kainic acid-induced seizures differently depending on prenatal methamphetamine exposure, sex, and estrous cycle. Epilepsy Behav 19:26–31

Slamberova R et al (2011) Impact of prenatal methamphetamine exposure on the sensitivity to the same drug in adult male rats. Prague Med Rep 112:102–114

Slamberova R et al (2012a) Do prenatally methamphetamine-exposed adult male rats display general predisposition to drug abuse in the conditioned place preference test? Physiol Res 61 (Suppl 2):S129–S138

Slamberova R et al (2012b) Does prenatal methamphetamine exposure induce cross-sensitization to cocaine and morphine in adult male rats? Prague Med Rep 113:189–205

Slamberova R et al (2013) Gender differences in the effect of prenatal methamphetamine exposure and challenge dose of other drugs on behavior of adult rats. Physiol Res 62(Suppl 1):S99–S108

Slamberova R et al (2014a) Effect of amphetamine on adult male and female rats prenatally exposed to methamphetamine. Prague Med Rep 115:43–59

Slamberova R et al (2014b) Prenatal methamphetamine exposure induces long-lasting alterations in memory and development of NMDA receptors in the hippocampus. Physiol Res 63(Suppl 4):S547–S558

Slamberova R et al (2015) Do the effects of prenatal exposure and acute treatment of methamphetamine on anxiety vary depending on the animal model used? Behav Brain Res 292:361–369

Smith LM et al (2001) Brain proton magnetic resonance spectroscopy and imaging in children exposed to cocaine in utero. Pediatrics 107:227–231

Smith L et al (2003) Effects of prenatal methamphetamine exposure on fetal growth and drug withdrawal symptoms in infants born at term. J Dev Behav Pediatr 24:17–23

Smith LM et al (2006) The infant development, environment, and lifestyle study: effects of prenatal methamphetamine exposure, polydrug exposure, and poverty on intrauterine growth. Pediatrics 118:1149–1156

Smith KJ et al (2007) Methamphetamine exposure antagonizes N-methyl-D-aspartate receptor-mediated neurotoxicity in organotypic hippocampal slice cultures. Brain Res 1157:74–80

Smith LM et al (2008) Prenatal methamphetamine use and neonatal neurobehavioral outcome. Neurotoxicol Teratol 30:20–28

Smith LM et al (2011) Motor and cognitive outcomes through three years of age in children exposed to prenatal methamphetamine. Neurotoxicol Teratol 33:176–184

Smith LM et al (2015) Developmental and behavioral consequences of prenatal methamphetamine exposure: a review of the infant development, environment, and lifestyle (IDEAL) study. Neurotoxicol Teratol 51:35–44

Sonsalla PK, Nicklas WJ, Heikkila RE (1989) Role for excitatory amino acids in methamphetamine-induced nigrostriatal dopaminergic toxicity. Science 243:398–400

Sowell ER et al (2010) Differentiating prenatal exposure to methamphetamine and alcohol versus alcohol and not methamphetamine using tensor-based brain morphometry and discriminant analysis. J Neurosci 30:3876–3885

Sulzer D et al (2005) Mechanisms of neurotransmitter release by amphetamines: a review. Prog Neurobiol 75:406–433

Twomey J et al (2013) Prenatal methamphetamine exposure, home environment, and primary caregiver risk factors predict child behavioral problems at 5 years. Am J Orthopsychiatry 83:64–72

Volkow ND et al (2001a) Loss of dopamine transporters in methamphetamine abusers recovers with protracted abstinence. J Neurosci 21:9414–9418

Volkow ND et al (2001b) Association of dopamine transporter reduction with psychomotor impairment in methamphetamine abusers. Am J Psychiatry 158:377–382

Vorhees CV et al (1994a) Methamphetamine exposure during early postnatal development in rats: I. Acoustic startle augmentation and spatial learning deficits. Psychopharmacology 114:392–401

Vorhees CV et al (1994b) Methamphetamine exposure during early postnatal development in rats: II. Hypoactivity and altered responses to pharmacological challenge. Psychopharmacology 114:402–408

Vorhees CV et al (1996) Neonatal methamphetamine-induced long-term acoustic startle facilitation in rats as a function of prepulse stimulus intensity. Neurotoxicol Teratol 18:135–139

Vorhees CV et al (1998) CYP2D1 polymorphism in methamphetamine-treated rats: genetic differences in neonatal mortality and effects on spatial learning and acoustic startle. Neurotoxicol Teratol 20:265–273

Vorhees CV et al (1999) Genetic differences in spatial learning between Dark Agouti and Sprague-Dawley strains: possible correlation with the CYP2D2 polymorphism in rats treated neonatally with methamphetamine. Pharmacogenetics 9(2):171–181

Vorhees CV et al (2000) Adult learning deficits after neonatal exposure to D-methamphetamine: selective effects on spatial navigation and memory. J Neurosci 20:4732–4739

Vorhees CV et al (2004) Exposure to 3,4-methylenedioxymethamphetamine (MDMA) on postnatal days 11-20 induces reference but not working memory deficits in the Morris water maze in rats: implications of prior learning. Int J Dev Neurosci 22(5–6):247–259

Vorhees CV et al (2005) Periadolescent rats (P41-50) exhibit increased susceptibility to D-methamphetamine-induced long-term spatial and sequential learning deficits compared to juvenile (P21–30 or P31–40) or adult rats (P51–60). Neurotoxicol Teratol 27:117–134

Vorhees CV et al (2008) Effects of neonatal (+)-methamphetamine on path integration and spatial learning in rats: effects of dose and rearing conditions. Int J Dev Neurosci 26:599–610

Vorhees CV et al (2009) Effects of (+)-methamphetamine on path integration and spatial learning, but not locomotor activity or acoustic startle, align with the stress hyporesponsive period in rats. Int J Dev Neurosci 27:289–298

Vorhees CV, Williams MT (2014) Assessing spatial learning and memory in rodents. ILAR J 55 (2):310–332

Vorhees CV, Skelton MR, Williams MT (2007) Age-dependent effects of neonatal methamphetamine exposure on spatial learning. Behav Pharmacol 18(5–6):549–562

Vrajova M et al (2014) Age-related differences in NMDA receptor subunits of prenatally methamphetamine-exposed male rats. Neurochem Res 39:2040–2046

Wagner GC et al (1980) Long-lasting depletions of striatal dopamine and loss of dopamine uptake sites following repeated administration of methamphetamine. Brain Res 181:151–160

Wallace TL, Gudelsky GA, Vorhees CV (1999) Methamphetamine-induced neurotoxicity alters locomotor activity, stereotypic behavior, and stimulated dopamine release in the rat. J Neurosci 19:9141–9148

Weissman AD, Caldecott-Hazard S (1993) In utero methamphetamine effects: I. Behavior and monoamine uptake sites in adult offspring. Synapse 13:241–250

Wells PG et al (2005) Molecular and biochemical mechanisms in teratogenesis involving reactive oxygen species. Toxicol Appl Pharmacol 207:354–366

Wijetunga M et al (2003) Crystal methamphetamine-associated cardiomyopathy: Tip of the iceberg? J Toxicol Clin Toxicol 41:981–986

Williams MT et al (2000) Preweaning treatment with methamphetamine induces increases in both corticosterone and ACTH in rats. Neurotoxicol Teratol 22:751–759

Williams MT et al (2002) Methamphetamine exposure from postnatal day 11 to 20 causes impairments in both behavioral strategies and spatial learning in adult rats. Brain Res 958:312–321

Williams MT et al (2003a) Long-term effects of neonatal methamphetamine exposure in rats on spatial learning in the Barnes maze and on cliff avoidance, corticosterone release, and neurotoxicity in adulthood. Brain Res Dev Brain Res 147:163–175

Williams MT, Moran MS, Vorhees CV (2003b) Refining the critical period for methamphetamine-induced spatial deficits in the Morris water maze. Psychopharmacology 168:329–338

Williams MT et al (2003c) Developmental D-methamphetamine treatment selectively induces spatial navigation impairments in reference memory in the Morris water maze while sparing working memory. Synapse 48:138–148

Williams MT et al (2006) Ontogeny of the adrenal response to (+)-methamphetamine in neonatal rats: the effect of prior drug exposure. Stress 9:153–163

Williams MT, Moran MS, Vorhees CV (2004a) Behavioral and growth effects induced by low dose methamphetamine administration during the neonatal period in rats. Int J Dev Neurosci 22:273–283

Williams MT, Brown RW, Vorhees CV (2004b) Neonatal methamphetamine administration induces region-specific long-term neuronal morphological changes in the rat hippocampus, nucleus accumbens and parietal cortex. Eur J Neurosci 19(12):3165–3170

Won L et al (2001) Methamphetamine concentrations in fetal and maternal brain following prenatal exposure. Neurotoxicol Teratol 23(4):349–354

Won L, Bubula N, Heller A (2002) Fetal exposure to methamphetamine in utero stimulates development of serotonergic neurons in three-dimensional reaggregate tissue culture. Synapse 43:139–144

Wong AW et al (2008) Oxoguanine glycosylase 1 protects against methamphetamine-enhanced fetal brain oxidative DNA damage and neurodevelopmental deficits. J Neurosci 28:9047–9054

Wouldes TA et al (2014) Prenatal methamphetamine exposure and neurodevelopmental outcomes in children from 1 to 3 years. Neurotoxicol Teratol 42:77–84

Yamamoto Y et al (1992) Teratogenic effects of methamphetamine in mice. Nihon Hoigaku Zasshi 46:126–131

Yu S et al (2015) Recent advances in methamphetamine neurotoxicity mechanisms and its molecular pathophysiology. Behav Neurol 2015:103969

Zabaneh R et al (2012) The effects of prenatal methamphetamine exposure on childhood growth patterns from birth to 3 years of age. Am J Perinatol 29:203–210

Early-Life Toxic Insults and Onset of Sporadic Neurodegenerative Diseases—an Overview of Experimental Studies

Anna Maria Tartaglione, Aldina Venerosi and Gemma Calamandrei

Abstract The developmental origin of health and disease hypothesis states that adverse fetal and early childhood exposures can predispose to obesity, cardiovascular, and neurodegenerative diseases (NDDs) in adult life. Early exposure to environmental chemicals interferes with developmental programming and induces subclinical alterations that may hesitate in pathophysiology and behavioral deficits at a later life stage. The mechanisms by which perinatal insults lead to altered programming and to disease later in life are still undefined. The long latency between exposure and onset of disease, the difficulty of reconstructing early exposures, and the wealth of factors which the individual is exposed to during the life course make extremely difficult to prove the developmental origin of NDDs in clinical and epidemiological studies. An overview of animal studies assessing the long-term effects of perinatal exposure to different chemicals (heavy metals and pesticides) supports the link between exposure and hallmarks of neurodegeneration at the adult stage. Furthermore, models of maternal immune activation show that brain inflammation in early life may enhance adult vulnerability to environmental toxins, thus supporting the multiple hit hypothesis for NDDs' etiology. The study of prospective animal cohorts may help to unraveling the complex pathophysiology of sporadic NDDs. In vivo models could be a powerful tool to clarify the mechanisms through which different kinds of insults predispose to cell loss in the adult age, to establish a cause–effect relationship between "omic" signatures and disease/dysfunction later in life,

A.M. Tartaglione · A. Venerosi · G. Calamandrei (✉)
Unit of Neurotoxicology and Neuroendocrinology, Department of Cell Biology
and Neurosciences, Istituto Superiore di Sanità, Viale Regina Elena 299,
00161 Rome, Italy
e-mail: gemma.calamandrei@iss.it

A.M. Tartaglione
e-mail: annamaria.tartaglione@gmail.com

A. Venerosi
e-mail: aldina.venerosi@iss.it

A.M. Tartaglione
Department of Neurology and Psychiatry, University of Rome "Sapienza",
Viale dell'Università 30, 00185 Rome, Italy

© Springer International Publishing Switzerland 2015
Curr Topics Behav Neurosci (2016) 29: 231–264
DOI 10.1007/7854_2015_416
Published Online: 23 December 2015

and to identify peripheral biomarkers of exposure, effects, and susceptibility, for translation to prospective epidemiological studies.

Keywords Neurodevelopment · Heavy metals · Pesticides · Maternal immune activation · Alzheimer's disease · Parkinson's disease · Exposome

Contents

1 General Introduction 232
2 Developmental Neurotoxicity and Neurodegenerative Diseases:
  A Common Playground? 234
  2.1 Introduction 234
  2.2 Brain Plasticity and Vulnerability 235
  2.3 The Multiple Hit Hypothesis for Neurodegenerative Diseases................ 236
  2.4 Environmental Chemicals and Neurodegenerative Diseases: Limits
      and Constraints of Epidemiological Studies and the Need
      of Experimental Models... 237
3 Assessing the Role of Environmental Exposures in Neurodegenerative Diseases'
  Etiology: An Overview of the in Vivo Models.. 238
  3.1 Introduction 238
  3.2 Alzheimer's Disease (AD).. 239
  3.3 Parkinson's Disease (PD) 249
4 Concluding Remarks 255
References... 258

1 General Introduction

Since the publication in 1993 in *The Lancet* of the paper by Barker et al. (1993) indicating that fetal undernutrition was a major risk factor for cardiovascular diseases, a great body of experimental and epidemiological evidence has been accumulating, indicating that exposure to an unfavorable environment in early life is associated with a significantly increased risk of later disease, a phenomenon termed "early life programming." The Barker's hypothesis formed the basis for the developmental origins of health and disease (DOHaD) hypothesis that, further to fetal nutrition, considered a wider spectrum of adverse fetal and childhood exposures that can predispose to chronic diseases later in life. In such expanded view of the original Barker's hypothesis, not only malnutrition, but also the quality of the maternal diet, the intake of micronutrients, maternal smoking, and prenatal maternal stress may account for the relationship of the prenatal environment to adult disorders including cancer, metabolic, neuroendocrine, and neurodegenerative and cognitive diseases in adult life (Gillman 2005; Hanson 2013; Tarantal and Berglund 2014).

Knowledge about the mechanism(s) by which fetal insults lead to altered programming and to disease later in life is still scarce, and as discussed in the following paragraphs, it is mainly derived from experimental studies. The long latency between exposure and disease and the lack of a clear pathological phenotype before disease onset make extremely difficult to assess the underlying mechanisms in clinical studies. Lahiri et al. (2009) suggested that environmental factors play a role in chronic disease etiology by inducing latent epigenetic changes. These authors described the association of early environment with disease onset especially with respect to Alzheimer's disease (AD), proposing a "Latent Early-life Associated Regulation" (LEARn) model that considers environmental exposures as "hits". According to this model, all neurodegenerative disorders come under the category of a "n" hit latent model, where early-life exposure leads to epigenetic perturbations in the genes, which remain silent until a second hit triggers the development of disease. The multiple hit hypothesis appears as particularly suited to explain the etiology of complex neuropsychiatric and neurologic diseases that share a highly sporadic nature characterized by genetic and environmental risk factors. It also explains the relatively long latency between exposure and disease, as subclinical deficits occurring during specific windows of developmental plasticity do not become manifest without a second hit occurring later in life (Heindel et al. 2015).

The DOHaD models, to date, have focused mainly on skewed nutritional states and/or low birthweight (Morley 2006), while the role played by exposures to environmental agents, such as water and air pollution and chemicals in food, either alone or in combination, has only recently come to attention. This is somewhat surprising since there is vast literature on the developmental neurotoxicity that indicates an inverse association between chemical exposure and child neurodevelopment (Bellinger 2014 for an updated review). Extensive evidence also shows that some environmental agents alter developmental programming via changes in gene expression or imprinting that do not result in malformations but in functional deficits that become apparent later in life. However, much less attention has been posed to the role that early exposure to environmental toxins may play in predisposing to neuropsychiatric and neurodegenerative in adult or even old age. This can be explained by several factors: the methodological constraints related to the difficulty of establishing cause–effect relationships between early exposure and disease manifestation (discussed in Sect. 2.4) and the lack of reliable biomarkers that can predict increased risk in early life stages, together with the still diffused idea that neurodegenerative diseases (NDDs) are linked to pathological brain aging.

In the present review, we will describe the general frame of the developmental origin of NDDs and the challenges posed to experimental and epidemiological research investigating the role of environmental pollutants as risk factors. We will then present an overview of animal studies that have assessed the delayed effects of either single or multiple perinatal toxic insults, evidencing neurobehavioral alterations resembling the major hallmarks of human NDDs.

2 Developmental Neurotoxicity and Neurodegenerative Diseases: A Common Playground?

2.1 Introduction

The human nervous system develops over a very long period extending from the embryonic period through puberty. About two weeks after conception, the neural plate folds to form the neural tube, and cells formed during this period of rapid proliferation migrate to a different position where they differentiate into neurons and neuroglia. In the human fetus, cell migration is nearly complete in the neocortex and in most of the brain by the sixth month of gestation (Gupta et al. 2005). During the remainder of intrauterine development, neurons differentiate and connect with each other. The period of rapid synaptogenesis or the so-called brain growth spurt (in humans occurring between the last three months of pregnancy and the 2nd postnatal year) is considered one of the most important processes that take place during brain development (Garner et al. 2002). This process is crucial not only in neurodevelopment but also plays a vital role in synaptic plasticity, learning, memory, and adaptation throughout life. Without this process, no complex brain network can be established, as synapse is the fundamental unit of connectivity and communication between neurons (Tau and Peterson 2010). During the brain growth spurt, substantially more connections are formed than those that will be eventually retained as many of them will be gradually discarded by apoptotic mechanisms. The process of synapse elimination is a normal part of development (Tau and Peterson 2010; Shonkoff 2000), and both endogenous (neurotrophic factors, synthesis, and release of neurotransmitters, hormones) and environmental factors (i.e., sensorial inputs) concur to influence the fine matchup between pre- and postsynaptic neurons. Altogether, brain development is made up of several interactive and temporally overlapping stages. A plethora of mechanisms regulates each of these sequential steps, and some of the major signaling cascades and trophic factors in the developing brain play a role in the aging or neurodegenerating brain. In a recent review, Kovacs et al. (2014) highlighted the commonalities between neuronal loss and neuronal development. These authors performed an overview of NDDs according to major proteins deposited (i.e., β-amyloid and tau protein); they observed that genes that are mutated in NNDs code for protein, which are highly expressed throughout neural development. Characterization of the mechanisms underlying hippocampal neurogenesis in the fetal and adult brain points to the involvement of the same gene products, namely of proteins that seem to drive either synaptogenesis or synaptic dysfunction depending on the age factor. This is the case of reelin, presenilins (PS), Notch-1, and brain-derived neurotrophic factor (BDNF). Furthermore, proteins representing hallmarks of NDDs might also play a crucial physiological role during brain development and adult neurogenesis. Specifically, the amyloid precursor protein (APP) significantly contributes to neurite outgrowth and hippocampal development (Billnitzer et al. 2013), α-synuclein has a key role in synaptic vesicle recycling and is expressed in normal fetal brain, and phosphorylation of the tau protein takes place during fetal development (Goedert et al. 1993). Altogether, though the majority of data originate from animal model research and still await confirmation in humans, there is substantial support to the view that pathological changes observed in NDDs could be in part the result of aberrant neuroplasticity mechanisms settled during the fetal life (Bugiani 2011). In this light, the study of the toxicants' effects on brain development is possibly crucial for unraveling the complex NDDs' etiology.

2.2 Brain Plasticity and Vulnerability

The remarkable plasticity of the brain during the fetal and neonatal periods is a double-edged sword as it renders this organ extremely responsive to environmental factors. On the one hand, early environmental inputs can interact with genetic factors to promote the development of neuronal circuits and functions (Sale et al. 2014); on the other hand, early negative conditions (i.e., stress, toxic insults) can act in the opposite direction, increasing the chances to develop neurodevelopmental disorders or diseases in adulthood. Diverse environmental stressors—chemical pollutants, drugs, nutritional factors, maternal infection, stress, and deprivation—may interfere with typical brain developmental trajectories. In order to cope with insults perturbing prenatal environment, the fetus develops compensatory strategies that could be adaptive or disruptive, resulting in interindividual variability in the total neuronal and glial burden and functional reserve.

Exposure to toxicants at any point during brain development may result in aberrant neural structure or function. The end points affected may vary depending on the timing and duration of exposure. Toxic disruption of early maturational events are more likely to result in neural tube defects or major malformation (i.e., anencephaly, spina bifida), while toxic interference with later developmental events results more commonly in cytoarchitectural and molecular alterations that might be expressed as behavioral dysfunction. Specifically, damage or destruction of neurons by chemical compounds when they are in the process of synapses formation, integration, and formation of neural networks will derange the organization and function of these networks, thereby setting the stage for subsequent impairment of learning and memory (Kovacs et al. 2014).

It was not until the early 1990s that significant evidence emerged showing that low-dose exposures to metals such as methylmercury and lead, though not producing teratogenic effects evident at birth, could result in later behavioral dysfunctions in children (Gilbert and Grant-Webster 1995; Banks et al. 1997). Similar results were later identified for the perinatal exposure to other environmental contaminants such as pesticides (Eskenazi et al. 2007; Perera et al. 2009), polychlorobiphenyls (Winneke 2011), polybrominated diphenyl ethers (Herbstman et al. 2010), and phthalates (Whyatt et al. 2012). These observations were consistent with the emergent DOHaD concept and shifted the emphasis of developmental toxicology from the study of teratogenic effects to functional changes occurring during development and becoming manifest only after a latent period (Bellinger 2009; Miller 1986). Mechanistic studies carried out in animals and in in vitro models point to multiple pathways and targets of toxicity for several established neurotoxicants, depending on the dosage or time window of exposure (Neal et al. 2011; Roy et al. 2004; Slotkin and Seidler 2009). A reconsideration of the targets of toxicity of the major environmental pollutants in the developing brain could possibly reveal significant commonalities between pathophysiology of neurodevelopmental disorders and the cognitive decline associated to NDDs (see the case of developmental lead exposure, Sect. 3.2.1).

2.3 The Multiple Hit Hypothesis for Neurodegenerative Diseases

Because the most of cases of NDDs onset late in life, they have been traditionally considered as linked to aging, but recent evidence suggests that the aging process is less important than was thought in the past. According to the LEARn model, all sporadic NDDs are characterized by 'n' hit latent model. Their onset in adulthood may be the consequence of alteration in gene regulation induced by early-life exposures that, after a silent period, lead to disease. Under this perspective, the old age may be the point in time where any change in developmental programming produced by multiple hits interacts sufficiently to cause overt disease (Logroscino 2005). Under LEARn, early-life stressors modify potential expression levels of disorder-associated genes in a latent manner. Latent changes in these genes are maintained by epigenetic mechanisms such as DNA methylation, DNA oxidation, and chromatin reorganization. Alterations in gene expression lead to altered protein synthesis and eventually to altered numbers and/or location of cells.

The multiple hit hypothesis firstly advanced to explain the etiology of PD (Calne and Langston 1983) might be applied to other major NNDs. In this light, the environmental insult could act via a one hit or even two/three hit scenarios, where the fetal insult may be the first event. Fetal exposure to a toxicant would lead by itself to disease later in life, as the neurotoxicant may induce subclinical injury to neural or glial cells, which leads to progressive loss of function. The developing organism initially compensates for the mild deficit, but the functional reserve and plasticity of the brain would be overcome with time, and loss of function (including behavioral deficits) would appear (Reuhl 1991). A mechanistic hypothesis proposed that early exposures to neurotoxic chemicals reduce the number of neurons in critical areas of the brain, setting in motion a self-perpetuating process of neurodegeneration (Langston et al. 1999). In some cases, the fetal exposure needs to be combined with a neonatal exposure either to the same or different compound or to an adult exposure to cause the pathophysiology. Thus, the initial deficit would only be unmasked by subsequent exogenous influences (stress, disease, additional chemical exposure) or by the natural aging process (Costa et al. 2004).

2.4 Environmental Chemicals and Neurodegenerative Diseases: Limits and Constraints of Epidemiological Studies and the Need of Experimental Models

A role for environmental toxicants in increasing risk of NNDs, such as PD and AD, has been advanced based on case–control retrospective studies (Santibáñez et al. 2007; Van Maele-Fabry et al. 2012). Specifically, AD and PD have been associated to metals and pesticide occupational exposure, but data are far from being conclusive (Hertzman et al. 1990; Liou et al. 1997).

Testing the hypothesis of the neurodevelopmental basis of NDDs in human population presents significant methodological problems. One of the most significant challenges is the issue of silent toxicity, as the developing nervous system is endowed with marked ability to compensate for insults. At variance from neurodevelopmental disorders, exposures may have occurred many decades before the diagnosis of AD or PD; thus, the assessment of chemical exposures is almost impossible to reconstruct in retrospective studies. In addition, the lack of reliable biomarkers makes difficult to obtain information on the fetal environment of an adult individual. Prospective studies linking fetal exposure to adult diseases require decades of follow-up and may be confounded by multiple factors (i.e., coexposure, lifestyle, genetic vulnerability) modulating the risk of the original environmental insult. The methodological constraints widely recognized in neurodevelopmental toxicity studies, namely dose–response estimation, individual dimension of the exposure history, and robustness of the clinical outcome (De Felice et al. 2015), are amplified by the "time" factor in studies on the developmental origin of NDDs. Of the 112 existing EU human cohorts established for studying NDDs, only five of them have followed people continuously from birth into adulthood; though many factors are currently under study (i.e., gene background, nutrition, lifestyle), the issue of early exposure history is scarcely considered (JPND Action Group 2013).

Overall, the epidemiological and clinical data collected so far, while supporting the hypothesis of a contribution of environmental factors to NDDs, highlight the enormous difficulty of establishing causative links between each of these factors and the health outcome. The simultaneous exposure of an individual to multiple risk factors, which may interact in an additive, synergistic, or even antagonistic way along the life course, remains to be explained and defined. Recently, the exposomic approach, considering all the endogenous and environmental factors to which an individual is exposed during the life span, has suggested a procedural framework to understand the fetal origin of adult diseases. In the exposome framework, the complex origin of human chronic diseases can be unfolded by means of several omic signatures possibly describing exposure, effects, individual vulnerability, and their dynamic interplay (Wild 2012). However, peripheral biomarkers of exposure so far available for many environmental chemicals are indeed poor predictors of effects on the brain (Stangle et al. 2004); to complicate the picture further, in the case of late-onset neurologic diseases, biomarkers of effect should be able to capture early biological changes and predict later health effects.

In this framework, studies with laboratory rodents allow evaluation of important parameters, such as dose–response relationships, critical periods of susceptibility, and the relative contribution of genetic, epigenetic, and environmental factors. So far, the use of animal models has been of paramount importance to investigate the mechanisms by which chemicals influence brain development in humans (Vorhees 1986). In particular, animal models are a powerful tool to test what combination of adverse environmental factors produces significant disruption of neurobehavioral development that leads to clinically relevant outcome in later stages. Notably, this kind of information can only be achieved by testing the living organism, as the face validity of an animal model of NDD consists primarily in a robust behavioral phenotype that can be associated to the appearance of neurodegenerative features (Crawley 2012).

3 Assessing the Role of Environmental Exposures in Neurodegenerative Diseases' Etiology: An Overview of the in Vivo Models

3.1 Introduction

The role of environmental chemicals and their mechanisms of action in the etiology of sporadic AD and PD have been extensively discussed in the recent review by Chin-Chan et al. (2015). Based on epidemiological and experimental evidence, the authors identified neurotoxic metals, such as lead (Pb), mercury (Hg), aluminum (Al), cadmium (Cd), manganese (Mn), and arsenic (As), as well as some classes of pesticides as significant environmental risk factors. As mentioned above, most of studies suggesting association between these chemical agents and idiopathic AD or PD refer to occupational exposure (Santibáñez et al. 2007; Firestone et al. 2010). Many are case–control studies (Parrón et al. 2011) and very few population studies considered internal exposure to a specific compound in relation to disease occurrence (Richardson et al. 2014; Kim et al. 2015). The mechanisms by which chemicals with very different modes of action might promote neurodegeneration are still undetermined. For example, the pesticide class includes more than 1000 active ingredients including insecticides, herbicides, and fungicides. Among those possibly implicated in AD or PD etiology, the organophosphorus compounds such as carbamates and organophosphate (OP) are potent inhibitors of acetylcholinesterase (AChE) and could affect APP processing (Giacobini et al. 1996); organochlorines (OC) can impair the functioning of the mitochondrial system and produce free radicals increasing oxidative stress and promoting cell death; bipyridyls such as paraquat (PQ) can generate free radicals that are able to cross the blood–brain barrier and damage neurons (Kumar et al. 2015).

In their extensive review, Chin-Chan et al. (2015) described several experimental studies (in vitro and in vivo) that have focused on metabolic pathways relevant for AD and PD pathogenesis. It is worth noting that in the large majority of the in vivo studies reviewed, rodents or non-human primates were exposed to chemicals at the adult stage. The metabolism of APP, phosphorylation of tau protein, aggregation of α-synuclein, production of proinflammatory cytokines, and reactive oxygen species generation appear as common targets of metals and pesticides' neurotoxicity following adult exposure. Furthermore, transgenic (Tg) models carrying mutation in candidate genes support the hypothesis that both genetic vulnerability and environmental neurotoxicants converge on neuroinflammation as an outcome linked or leading to neurodegeneration. It is worth noting that the very earliest neuronal and pathological changes characteristic of AD suggest oxidative stress as a very early contributor to the disease (Nunomura et al. 2001) and that oxidative stress appears to play a major role in degeneration of DA neurons in PD (Dias et al. 2013).

In the following sections, we will present the more robust experimental studies including fetal and/or neonatal exposure to known environmental neurotoxicants, which support the link between chemical exposure and hallmarks of neurodegeneration at the adult stage. In many of the studies, the significant alteration in one or more of these hallmarks was also paralleled by motor and/or cognitive disturbances, reproducing in the rodent model the clinical behavioral traits of AD and PD affected individuals in the initial stages of the disease. We chose to include in this overview also rodent studies exploring the delayed neurobehavioral effects of maternal infection. The paradigm of maternal infection models a condition of early immune activation and consequent exposure of the fetal brain to proinflammatory cytokines: This can affect the development of brain circuitries leading to neurobehavioral dysfunctions in the progeny (Knuesel et al. 2014). As shown below, studies combining maternal infection with a subsequent toxic insults support the validity of the multiple hit hypothesis, as an unfavorable fetal environment increases later vulnerability to neurotoxicants. Tables 1, 2 and 3 illustrate synthetically the experimental studies discussed in this chapter.

3.2 Alzheimer's Disease (AD)

AD is the most common neurodegenerative disease worldwide. The main pathological hallmarks of AD are beta-amyloid plaques (Aβ) and neurofibrillary tangles (NTFs) composed of hyperphosphorylated tau protein (p-tau), which spread through the cortex in a predictable pattern as Alzheimer's disease progresses causing nerve cell degeneration and cognitive impairment (Selkoe and Schenk 2003). Early onset familial AD is rare and is characterized by autosomal dominant heritability, due to mutations in the genes for APP, presenilin-1 (PSEN1), and presenilin-2 (PSEN2). The prevalent form of AD typically develops after 60 years of age; the etiology of late-onset sporadic AD (LOAD) has not yet been resolved, but it has been suggested that it could result from a multifactorial process involving both genetic and environmental factors (Reitz et al. 2011). Diseases presenting

Table 1 Experimental studies assessing the role of early environmental factors in AD

AD experimental models

	Compound	Species	Exposure window	Outcome in offspring	References
Metals	Pb	Rat	PND 0 to 20	↑APP and Aβ expression	Basha et al. (2005)
	Pb	Rat	PND 1 to 20	↑Oxidative DNA damage	Bolin et al. (2006)
			PND 18 to 20 months	NE	
	Pb	Monkey	PND 0 to 400	↑Oxidative DNA damage, ↓DNA methyltransferase, ↑APP, BACE1, and Sp1 mRNA levels	Wu et al. (2008)
	Pb	Mouse	GD1 to PND 21	↓Learning and memory ability	Li et al. (2010)
				↑p-tau and Aβ expression	
	Pb	Mouse	GD 13 to PND 20	↓Immune response, metal-binding, metabolism, and transcription/transduction coupling genes	Dosunmu et al. (2012)
	Pb	Monkey	PND 0 to 400	↑Tau protein and mRNA levels	Bihaqi and Zawia (2013)
	Pb	Mouse	PND 1 to 20	↑Tau protein and mRNA levels	Bihaqi et al. (2014)
			7 to 9 months	NE	
	As + Cd + Pb	Rat	GD 5 to PND 180	↑Aβ expression, ↓learning and memory abilities, ↑oxidative stress, ↑inflammatory responses	Ashok et al. (2015)
Pesticides	CPF	Tg2576	6 months	↑Aβ expression, ↓spatial retention	Salazar et al. (2011)
	CPF	Tg2576	4 to 6 months	NE	Peris-Sampedro et al. (2014)
	PQ	Tg2576	3 to 4 months	↓Associative learning and memory, ↑Aβ expression,	Chen et al. (2012a)
				↑Oxidative and mitochondrial damage	

(continued)

Table 1 (continued)

AD experimental models

	Compound	Species	Exposure window	Outcome in offspring	References
Inflammation	LPS	Mouse	GD 15 to 17	↓Learning and memory abilities	Chen et al. (2011),
	PolyI:C	Mouse	GD 17	↑Inflammatory cytokines, ↑APP expression, ↑p-tau expression, ↓working memory	Krstic et al. (2012)

NE no effects
Tg transgenic mouse

Table 2 Experimental studies assessing the role of early environmental factors in PD

PD experimental models

	Compound	Species	Exposure window	Outcome in offspring	References
Metals	Fe2+	Mouse	GD 10 to 12	↓Motor behavior, ↓spatial learning ability	Fredriksson et al. (1999)
	Fe2+	Mouse	PND 3 to 5; PND 10 to 12 / PND 19 to 21	↓Motor behavior, ↓spatial learning ability NE	Fredriksson et al. (2000)
	Fe2+	Mouse	PND 10 to 17	↑Oxidative stress, ↓DA in ST, ↓DA neurons in SN, ↑neuronal vulnerability to MPTP	Kaur et al. (2007)
	Fe2+	Rat	PND 10 to 12	↑Oxidative stress, ↓SOD activity	Dal-Pizzol et al. (2001)
	Mn	Rat	PND 8 to 12	Altered intracellular signaling pathways, ↑stress and mitochondrial dysfunction, ↑caspase activity, ↓motor alterations	Cordova et al. (2012)
	Mn	Rat	PND 8 to 27	Altered intracellular signaling pathways, ↑stress and mitochondrial dysfunction, ↑caspase activity, ↓motor alterations	Cordova et al. (2013)
Pesticides	Dieldrin	Mouse	gestation and lactation	↑DAT and VMAT2 levels, ↑neuronal vulnerability to MPTP	Richardson et al. (2006)
	Endosulfan	Mouse	2 weeks before breeding until lactation	↓In DAT and TH in SN, ↑neuronal vulnerability to MPTP	Wilson et al. (2014)
	PQ, MB or PQ + MB	Mouse	6 weeks (1st)	↓DA and its metabolites in ST, ↓DA neurons in the SN*pc*	Thiruchelvam et al. (2000)
	ATZ	Rat	GD 0 to PND 1	↓DA levels in ST, ↓Nurr1, VMAT2, DAT, and TH expression in SN	Li et al. (2014)

(continued)

Table 2 (continued)

PD experimental models

	Compound	Species	Exposure window	Outcome in offspring	References
	ATZ	Rat	GD 5 to PND 22	↓DA levels in ST, ↓Nurr1, VMAT2, DAT, and TH expression in SN	Sun et al. (2014)
Inflammation	LPS	Rat	GD 10.5	Glutathione homeostasis disturbance, ↓DA neurons in the SN, ↑neuronal vulnerability to secondary LPS exposure, ↓neuron in SN and DRN, ↓DA and 5-HT levels	Zhu et al. (2007)
	LPS	Rat	GD 10.5	↓DA and 5-HT neurons in SN and DRN, ↓DA and 5-HT levels	Wang et al. (2009)
	LPS	Rat	PND 5	↓Motor functions, ↓DA neurons in SN and VTA, ↓microglial activation, and inflammatory responses,↑mitochondrial damage, ↑behavioral reaction to METH	Fan et al. (2011b), Tien et al. (2011)

NE no effects

Table 3 Experimental studies assessing the role of early environmental factors in PD: multiple hit studies

PD experimental models

	Compound	Species	Exposure window (multiple hit)	Outcome in offspring after multiple hit	References
Combined chemicals	PQ or MB	Mouse	10–17 GD (1st)	↓DA neurons in the SN*pc*, ↓DA level in ST, ↓Motor behavior only under combined exposure (MB + PQ)	Barlow et al. (2004)
	PQ or MB		6 weeks (2nd)		
	PQ, MB or PQ + MB	Mouse	PND 5 to 19 (1st)	↓Motor behavior, ↓DA level in ST, ↓DA neurons in the SN*pc* only under combined exposure	Thiruchelvam et al. (2002)
	PQ, MB or PQ + MB		6.5–7.5 months (2nd)		
Combined chemicals and Inflammation	LPS rotenone	Rat	PND 5 (1st) PND 70 (2nd)	↓Motor behavior, ↓DA neurons in the SN, ↑Mitochondrial damage, ↑microglial activation, ↑Neurobehavioral vulnerability to rotenone exposure	Fan et al. (2011a)
	LPS	Rat	PND 5 (1st)	↑α-synuclein aggregation, ↑DAT expression, ↑Mitochondrial damage	Tien et al. (2013)
	Rotenone		PND 70 (2nd)		
	LPS	Rat	GD 10.5 (1st)	↓DA neurons in the SN, ↓DA levels in ST, ↑microglial activation, ↑oxidative stress, ↑α-synuclein, and eosin-positive inclusions	Ling et al. (2004a, b)
	Rotenone		PND 21 (2nd)		
	LPS	Rat	GD 10.5 (1st)	↓DA neurons in SN and VTA, ↓DA levels in SN, ↑cytokine levels in SN	Ling et al. (2004a, b)
	Rotenone		PND 99 (2nd)		
	LPS	Rat	PND 5 (1st)	↑Brain inflammation, ↓motor behavior, ↑Neuronal vulnerability to rotenone exposure	Cai et al. (2013)
	Rotenone		PND 70 (2nd)		
	CAPS	Mouse	PND 4–7 s; PND 10–13 (1st)	↓DA neurons in ST and SN*pc*, ↑glutamate levels, ↓motor behavior	Allen et al. (2014)
	PB + MB		PND 60 (2nd)		

NE no effects

increased inflammatory processes such as hypercholesterolemia, hypertension, atherosclerosis, coronary heart disease, obesity, and diabetes are indeed risk factors for AD (Blennow et al. 2006), and thus, inflammation is suggested to have a causal role also in LOAD pathogenesis. Several candidate risk genes have been identified for LOAD: They mainly involve proteins implicated in critical pathogenic pathways, namely Aβ synthesis and processing, oxidative stress, and inflammation/apoptosis (Ballard et al. 2011).

Adverse environmental factors might further increase the risk impose by genetic makeup, by acting on cell pathways implicated in AD. Neurotoxic metals such as Pb, Hg, Al, Cd, and As, as well as some pesticides and metal-based nanoparticles, have been involved in AD due to their ability to increase Aβ peptide and the phosphorylation of tau protein, causing senile/amyloid plaques and NFTs characteristic of AD, and to enhance oxidative stress and inflammatory responses (Casado et al. 2008). In spite of this consistent evidence, studies considering environmental exposure have been underrepresented in the AD literature, possibly due to the challenges of retrospective exposure assessment in older adults (Bakulski et al. 2012).

3.2.1 Heavy Metals and AD

Lead

Among heavy metals, a large body of clinical data supports an association between occupational Pb exposure and cognitive decline with aging as higher levels of Pb in blood and/or bone were accompanied by poor cognitive performance in different cognitive tests (Nordberg et al. 2000; Wright et al. 2003; Weisskopf et al. 2004). Pb is a well-known neurotoxicant in children. Even at relatively low and subclinical levels, epidemiologic studies demonstrate prenatal and childhood Pb exposure affects IQ and behavior up to young adulthood (Lanphear et al. 2005; McFarlane et al. 2013). Recent studies using both in vivo models and primary neuronal cultures indicate that Pb exposure during pregnancy may negatively modify important neuronal pathways implicated in synaptic plasticity, learning, and memory (Neal and Guilarte 2010). In this framework, the potential link between developmental exposure to Pb and increased risk of developing AD later in life has been thoroughly investigated in experimental studies with rodents and primates. Li et al. (2010) found tau hyperphosphorylation and Aβ increase in hippocampus associated with deficits in learning and memory in mice offspring exposed to Pb from conception to weaning.

A line of research investigated the link between perinatal Pb exposure, alteration of APP metabolism, and cognitive impairment at later life stages. Basha et al. (2005) exposed rats to Pb from birth through postnatal day (PND) 20 by systemic administration and monitored the lifetime expression of the APP gene, finding overexpression of APP levels and Aβ aggregation 20 months after exposure to Pb had ceased. These results have been replicated on monkeys, exposed to Pb from birth until the end of the first year of age (Wu et al. 2008). The expression of AD-related genes (APP, β-site APP cleaving enzyme 1 (BACE1)), as well as their transcriptional regulator (Sp1), was elevated in aged (23-year-old) monkeys following early exposure to Pb. Furthermore, levels of amyloid plaques increased in the frontal cortex. These latent effects were paralleled by a decrease in DNA methyltransferase activity and higher levels of oxidative damage to DNA, indicating that epigenetic mechanisms elicited by Pb in early life influenced the expression of AD-related genes and promoted DNA damage and pathogenesis.

Bihaqi et al. (2014) evaluated the impact of Pb exposure at different lifetimes (during either development or adulthood) on cognitive functions and biochemical pathways associated with AD in mice. This study showed that only mice early exposed to Pb (0.2 % Pb acetate), namely during the first 20 days of postnatal life through the mother's milk, showed cognitive impairment at 18 months of age, which further progressed by 24 months. In Pb-exposed rats, cognitive impairment was accompanied by up-regulation of AD biomarkers (APP, Aβ, and BACE1) and overabundance of p-tau. Long-term effects of developmental exposure to Pb has been evaluated also in relation to tau pathology in non-human primates (Bihaqi and Zawia 2013). In this study, female monkeys were exposed to Pb (1.5 mg/kg/day) from birth until the end of the first year of life, and then, they were examined at 23 years of age. Results showed that Pb exposure elevated the mRNA and protein levels of tau as well as its transcriptional regulators and favoured abundant tau phosphorylation in the frontal cortex of aged primates. This study suggests some evidence that Pb during development triggers neurodegenerative processes that have much in common with those implicated in human AD.

Due to the role of developmental exposure of Pb in increase of Aβ, which is known to generate reactive oxygen species in the aging brain, Bolin et al. (2006) investigated the hypothesis that early exposure to Pb, in specific developmental windows, determines the outcome of oxidative damage late in life. Rat pups were exposed to Pb (0.2 % Pb acetate) at different ages (from PND 1 to 20 or from 18 to 20 months), showing that the level of 8-hydroxy-2′-deoxyguanosine (8-oxo-dG), one of the major products of DNA oxidation and a reliable marker of oxidative stress, was elevated 20 months after exposure to Pb had ceased. The effect of Pb on 8-oxo-dG levels did not occur if animals were exposed to Pb only in old age. These increases in DNA damage occurred in the absence of any Pb-induced changes in general antioxidant capacity in the cerebral cortex.

The existence of a critical phase of sensitivity to Pb effects was also supported by other studies. Dosunmu et al. (2012) assessed global gene expression patterns in mice, developmentally exposed to Pb (0.2 % Pb acetate) from gestational day (GD) 13 until PND 20. The selected time points of PND 20 and 700 represented early and late time points of the animal for microarray analysis, where genes that were down- and up-regulated were identified, clustered, and analyzed for their relationship to DNA methylation. Prior exposure to Pb revealed a repression of genes related to the immune response, metal-binding, metabolism, and transcription/transduction coupling. More in detail, the authors identified 150 genes for a transcriptional profile of normal aging and environmentally perturbed aging.

In normal aging, those genes are up-regulated, but prior developmental Pb exposure caused a strong repression of these genes. DNA methylation appears to play an important role in the down-regulation of these genes suggesting that early-life exposure to Pb interferes with the methylation pattern of genes and has an impact on the organism's capacity to respond in old age. These findings would support the LEARn model of NDDs, as the toxic effects of Pb are maintained through the life span in the form of latent epigenetic changes.

Recently, Ashok et al. (2015) treated rats with a mixture of As + Cd + Pb at environmentally relevant concentration, from GD 5 to 6 months of age. They identified dose-dependent increase of $A\beta$ in frontal cortex and hippocampus as early as the post-weaning age. The metal mixture activated the proamyloidogenic pathway, mediated by increase in APP, and subsequent BACE1 and PS-mediated APP processing. Investigating the mechanism of $A\beta$ induction revealed an augmentation in oxidative stress-dependent neuroinflammation that stimulated APP expression. When the authors examined the effects of individual metals and binary mixtures in comparison with the tertiary mixtures, they found that Pb triggered maximum induction of $A\beta$, whereas individual As or Cd had a relatively non significant effect on $A\beta$ despite enhanced APP. Together, these data demonstrate that exposure to As + Cd + Pb induces earlier manifestation of AD-like pathology that is synergistic, and oxidative stress, and inflammation dependent. These data open the way to further experimental research modeling real exposure scenarios to mixtures of chemicals.

Developmental origin of Pb-induced neurodegenerative disease has been investigated also in zebrafish (*Dania rerio*), an aquatic vertebrate model system, which presents a high sequence homology (i.e., 70 %) with the human genome (Howe et al. 2013). Zebrafish are increasingly used to understand the mechanisms of developmental Pb neurotoxicity through evaluation of changes in gene expression and neurobehavioral alterations (Lee and Freeman 2014). Data collected in this model generally supported the hypothesis that developmental exposure to Pb exhibits latent effects on the expression of genes that are involved in neurodegenerative processes during old age.

3.2.2 Pesticides and AD

Several epidemiological studies have suggested the association between pesticides and impairment of cognitive functions and AD-like dementia, but the underlying mechanisms have been poorly explored in experimental models (Baldi et al. 2011; Hayden et al. 2010). Among the class of pesticides with neurotoxic activity, OPs are the subject of increasing concern. OPs make up approximately 50 % of all insecticides used in the world (Colborn 2006) and are intensively investigated for their suspected developmental neurotoxicity (Rosas and Eskenazi 2008). These compounds, largely used in agriculture, as well as in the home and garden, for pest control, exert their acute neurotoxic effects through AChE inhibition and consequent cholinergic stimulation. However, increasing evidence indicates that, as other environmental chemicals, OPs exert developmental toxicity at low doses with mechanisms different from those observed at higher doses, including modulation of several family of genes involved in brain development (Moreira et al. 2010). In vitro studies show that OP compounds impair mitochondrial bioenergetics and induce ROS generation (Wani et al. 2014). In particular, as concern chlorpyrifos (CPF), the most widely applied compound in the OP class in the USA, it interferes with brain and behavior development through a variety of cellular and molecular mechanisms that appear as independent from inhibition of AChE (Androutsopolous et al. 2013; Eaton et al. 2008; Venerosi et al. 2012).

Notwithstanding the recognized effect of this OP insecticide on oxidative stress end points (Cole et al. 2011), developmental exposure to these compounds has never been considered under the perspective of neurodegenerative disease etiology. However, three recent studies using a genetic model of vulnerability to AD, the APP Tg mice (Tg2576 mice) evidenced a potential increased susceptibility to both OP insecticides and bipyridyl herbicide PQ. Specifically, CPF caused a significant increase in Aβ levels in the cortex and hippocampus, as well as increased memory loss and reduced motor activity in Tg2576 mice 6 months after a single administration (Salazar et al. 2011), an effect that has not been confirmed by Peris-Sampedro et al. (2014). Another study showed that treatment of wild type and Tg2576 mice with PQ (10 mg/kg/twice a week/3 weeks) produced a significant increase in Aβ levels in Tg mice, associated with mitochondrial oxidative damage in cerebral cortex and impairment of learning and memory. Interestingly, the overexpression of peroxiredoxin 3, a mitochondrial antioxidant defense enzyme, produced an improvement in cognitive functions and a significant reduction in Aβ levels in Tg2576 exposed to PQ (Chen et al. 2012), suggesting that pro-oxidant xenobiotics like PQ can contribute to AD.

3.2.3 Maternal Infection and AD

As illustrated previously, experimental and clinical data support the inflammation hypothesis of AD etiology. It suggests that neuroinflammatory response triggers and follows the Aβ increase in the AD brain and significantly interferes with memory processes (Heneka et al. 2015; Heppner et al. 2015). The administration of the bacterial endotoxin lipopolysaccharide (LPS) or polyinosinic:polycytidylic acid (PolyI:C) to pregnant rodents was among the first animal models examining the long-term consequences of prenatal immune challenge on adult brain functioning (Meyer 2014). In experimental studies, LPS is used to increase proinflammatory cytokines in the pregnant female's organism, mimicking an infective state that can induce abnormalities in the fetus neurodevelopment (Knuesel et al. 2014).

Chen et al. (2011) injected female mice with LPS (50 mg/kg) daily during late pregnancy (from GD 15 to GD 17). The offspring exposed prenatally to LPS, although showing normal development of sensorimotor and cognitive functions, had an accelerated age-related impairment of both spatial and non-spatial learning and memory in middle age. In a subsequent study by Krstic et al. (2012), female mice received a single injection of PolyI:C (5 mg/kg) at GD 17. Adult and aged offspring showed cognitive decline and AD-like neuropathology such as increase of APP protein and its proteolytic fragments and altered tau phosphorylation, in parallel with impairments in working memory. A second immune challenge in adulthood exacerbated this pathological phenotype, with appearance of Aβ-like plaque deposition. In this same study, Tg AD mice (3xTg-AD), overexpressing the human variants of AD-relevant genes, were challenged with PolyI:C at the pre-plaque stage of 4 months and showed accumulation of APP and increase of Aβ deposits resembling the morphology of Aβ plaques in human patients with AD. Taken together, these results indicated that systemic immune challenges are able to promote AD-like neuropathology in both wild-type (PolyI:C exposure at GD17) and genetically (3xTg-AD) predisposed animals, in agreement with both causative and exacerbating role of systemic immune challenges on AD development.

3.3   Parkinson's Disease (PD)

Parkinson's disease (PD) is the second most common neurodegenerative disease after AD, characterized by depletion of dopaminergic cell bodies in the substantia nigra *pars compacta* (SN*pc*) with subsequent loss of dopamine (DA) in the nigrostriatal system, and intracytoplasmic inclusions termed as Lewy bodies, containing aggregates of the protein α-synuclein as well as other substances (Barlow et al. 2007). Studies on familial PD show an autosomal pattern, either dominant or recessive, identifying some genetic mutations and chromosal loci responsible for this form. Several loci (PARK1-15) and genes have been linked to familial forms of the disease. The majority of PD cases are, however, sporadic "idiopathic" forms, and the recent application of genome-wide screening revealed almost 20 genes implicated in mitochondrial function, in detoxification, and/or protection against oxidative stress that might contribute to disease risk. It is striking that none of the genes implicated to date is specifically expressed in DA neurons. Instead, many of the implicated genes have rather general neuronal or cellular functions in and outside of the brain, suggesting that DA neurons are more susceptible to stress than other neurons (Lees et al. 2009; Westerlund et al. 2010).

The cause of sporadic PD is unknown, with uncertainty about the role of environmental toxins and genetic factors. The environmental hypothesis posits that PD-related neurodegeneration results from exposure to a dopaminergic neurotoxin. The finding that people intoxicated with the chemical 1-methyl-4pheny 1-1, 2, 3, 6-tetrahydropyridine (MPTP) developed a syndrome nearly identical to PD (Langston et al. 1983) prompted the interest in environmental causes of PD (Tanner et al. 2014). The herbicide PQ is structurally similar to 1-methyl-4-phenylpyridinium (MPP+), the active metabolite of MPTP. Similar to MPP+, the insecticide rotenone is also a mitochondrial poison present in the environment. Yet, there are no convincing data to implicate any specific toxin as a cause of sporadic PD, and chronic environmental exposure to MPP+ or rotenone is unlikely to cause PD per se. Another possibility is that an endogenous toxin may be responsible for PD neurodegeneration.

Distortions of normal DA metabolism might create toxic substances (i.e., reactive oxygen species) because of environmental exposures or inherited differences in metabolic pathways (Dauer and Przedborski 2003; Sandy et al. 1996).

Several environmental contaminants, known to cause nigrostriatal damage at toxic concentrations, could contribute to PD by interference with mitochondrial respiratory functions and ROS generation (Di Monte 2003; Jenner 2003). The insecticides PQ, rotenone, and the fungicide MB are agricultural chemicals that have each been linked to nigrostriatal damage and the emergence of parkinsonian symptoms, via epidemiological surveys, clinical case reports, and/or animal models. Exposure to various metals has long been suggested to increase risk of PD, based on experimental studies, but human evidence remains inconclusive (Lai et al. 2002; Tanner et al. 2014).

3.3.1 Heavy Metals and PD

Iron

Iron (Fe2+), the most abundant metal present in the human brain, has been proposed to play a role in the pathogenesis of PD for its pro-oxidants characteristics that may lead to ROS generation (Sian-Hülsmann et al. 2011). Several studies have focused on the ability of Fe2+, as well as other metals to facilitate the aggregation of the PD-related protein α-synuclein, but very few studies investigated the impact of early exposure to Fe2+ on development of PD hallmarks in adult life. In the study by Fredriksson et al. (1999) at the age of 3 months, offspring exposed in utero to Fe2+ (3.7 or 37.0 mg/kg) on GD 10–12 showed marked hypokinesia and lack of habituation in spontaneous activity. These effects were more pronounced in mice treated with the higher dose that showed also a dose-dependent impaired performance in the radial arm maze. Brain Fe2+ content was significantly increased in the basal ganglia, but not in the frontal cortex at the higher dose group. In a second study from the same group (Fredriksson et al. 2000), mice were treated in the neonatal stage with Fe2+ (7.5 mg/kg) on either PND 3–5, 10–12, or 19–21. Adult mice treated with Fe2+ on PND 3–5 and 10–12 showed effects similar to those found after gestational exposure on spontaneous motor behavior and learning performance, more marked in animals treated at PND 10–12.

The role of Fe2+ in oxidative stress and superoxide dismutase (SOD) activity may be implicated in the adverse delayed effects of this metal (Dal-Pizzol et al. 2001). Male rats were treated with Fe2+ (7.5 or 15 mg/kg) from PND 10 to 12, measuring at 3 months different indexes of oxidative stress. Notably and according to previous clinical and experimental studies, this study demonstrated that Fe2 + exposure during a critical neonatal period induced oxidative stress and modulated SOD activity restricted to SN in adult rats. Finally, Kaur et al. (2007) investigated the role of early exposure of Fe2+ at a dose (120 µg/g of carbonyl iron) equivalent to that found in iron-fortified human infant formula as a risk factor for PD-like neurodegeneration. Neonatally exposed mice were challenged at adulthood with the toxin MPTP that induces depletion of neostriatal dopamine. Neonatal Fe2+ feeding alone did not induce decrease in striatum (ST) dopamine levels in 2-month-old mice, also when they were challenged with MPTP. However, at 12, 16, and 24 months of age, mice treated with Fe2+ showed more marked DA depletion following acute administration of MPTP than vehicle-treated mice. It is worth noting that Fe2+ is important for many biological reactions such as the synthesis and release of neurotransmitters. In particular, the neonatal period is critical for the establishment of normal Fe2+ content in the adult brain and its regional distribution; deficiency such as excess of this metal can perturb the typical neurological development. These results highlight the need of epidemiological research in humans, especially on the potential long-term effects of Fe2+ supplementation during infancy (i.e., milk) on neurological functions in adulthood.

Manganese

Manganese (Mn) was related to PD since 1837, when it was noted that high Mn exposures caused an extrapyramidal syndrome that resembles the dystonic movements associated with parkinsonian symptoms (Couper 1837; Jankovic 2005). Several evidence on the influence of Mn on neurodegenerative processes has been collected in in vitro models. A single study comparing in vitro and in vivo experiments showed that oxidative stress, mitochondrial dysfunction, and neuroinflammation are implicated in Mn-induced neurodegeneration (Milatovic et al. 2009).

Two experimental studies suggest that early Mn exposure might have profound effects on later-life susceptibility to neurotoxins and subsequent degeneration (Cordova et al. 2012, 2013). Cordova et al. (2012) show for the first time in vivo the link between early Mn exposure (from PND 8 to 12) and the modulation of intracellular signaling pathways. Oxidative stress, DA cell death, and later-life impairment in motor function were also observed. In a subsequent study, the same authors investigated the effects of longer exposure to Mn in developing rats. They demonstrated that exposure to Mn (5, 10, and 20 mg/kg) from PND 8 to 27 caused significant deficits in motor coordination and increased signs of oxidative stress in the ST in rats at 3, 4, or 5 weeks of age (Cordova et al. 2013).

3.3.2 Pesticides and PD

One-hit studies

Experimental models assessing the causative link between exposure to pesticides and PD-like neurodegeneration have focused on the effects of environmental toxins on the DA nigrostriatal system. This system develops both pre- and postnatally, with receptor development occurring predominantly in the postnatal period (Giorgi et al. 1987; Voorn et al. 1988). As described above, the chemical structure of the insecticide PQ resembles that of MPP+, the toxic metabolite that mediates the effects of the parkinsonism-inducing agent MPTP. Similarly, administration of PQ to mice produces several neurotoxic effects, including damage of the DA nigrostriatal system (McCormack et al. 2002). Dithiocarbamate fungicides also possess DA activity: Maneb (MB) administered acutely exacerbates the behavioral symptoms induced by MPTP in mice (hypokinesia and catalepsy). Similar effects have been described with the OC herbicide dieldrin, an inhibitor of mitochondrial respiration.

Richardson et al. (2006) showed that perinatal exposure to low doses of dieldrin (0.3, 1, or 3 mg/kg every 3 days) during gestation and lactation altered the levels of the dopamine transporter (DAT) and vesicular monoamine transporter 2 (VMAT2) at 12 weeks of age. The alterations of both the DA markers were exacerbated by MPTP injections at 12 weeks. Wilson et al. (2014) investigated the effects of in utero exposure to endosulfan, an OC insecticide, in male offspring at 3 months of age. Female mice received endosulfan (1 mg/kg) every 2 days from 2 weeks before breeding until end of lactation. The authors found a reduced expression of DAT and tyrosine hydroxylase (TH) in the ST of treated mice, exacerbated by exposure to additional insults such as MPTP. Endosulfan failed to cause the same effects when administered at the same dose and for the same period to adult mice. These findings indicate that gestation and lactation are critical windows of sensitivity to endosulfan exposure and development of nigrostriatal DA system.

Two studies evaluated the effects of developmental exposure to the herbicide atrazine (ATR) during gestation and lactation on DA system development (Li et al. 2014; Sun et al. 2014). The in utero exposure to ATR (25 or 50 mg/kg) from GD 0 to PND 1 reduced the level of DA and expression of orphan nuclear hormone (Nurr1), VMAT2, DAT, and TH genes in ST and SN, respectively, at 6 months of age (Li et al. 2014). In a second study, the authors assessed the effects of a more prolonged exposure to ATR given to pregnant mice from GD 5 to PND 22 at the same doses of the previous study: DA concentrations and mRNA levels of Nurr1 were decreased in the offspring at 12 months of age. Decreased Nurr1 levels paralleled changes in the mRNA levels of VMAT2, which controls the transport and reuptake of DA (Sun et al. 2014).

Two-hit studies

In line with the double hit hypothesis, it has been suggested that developmental exposure to either PQ or MB alone, or in combination, would result in permanent nigrostriatal DA system neurotoxicity, and secondly, it would render this system more susceptible to a second chemical challenge later in life. In the study of Thiruchelvam et al. (2002), mice were exposed to PQ, MB, and PQ + MB from PND 5 to 19 and then rechallenged with the single agents or with the combination at 6.5 months of age. Developmental and adult exposure to PQ + MB combination produced the greatest damage of the nigrostriatal DA system and a significant reduction in motor activity. Furthermore, increased sensitivity to the rechallenge with PQ and MB either alone or in combination was observed in only those animals treated neonatally with the combination. Vice versa, exposure to PQ or MB alone produced minimal changes, but after adult rechallenge, a significant decrease in DA levels and nigral cell count was found, confirming that adult re-exposure might unmask a condition of silent toxicity. Thus, developmental exposures of mice to PQ + MB combination led not only to a permanent and selective loss of DA neurons in the SN*pc* but also enhanced the impact of these pesticides administered during adulthood relative to developmental only or adult only treatment. Exposure to MB alone during gestation resulted in enhanced response to PQ in adulthood, including notable reductions in levels of DA and its metabolites and loss of nigral DA neurons, despite the structural dissimilarity and the different mode of action of PQ and MB (Cory-Slechta et al. 2005).

Another line of research considered the role of in utero exposure to MB, combined with subsequent exposure to PQ, in disrupting the nigrostriatal system (Barlow et al. 2004). Pregnant mice were exposed to MB or PQ on GD 10–17, at doses ten times lower than those used in young adult studies; 2-month-old offspring received either PQ or MB at the doses used for adult mice for 8 days. One week after the termination of the second treatment, mice were assessed in a motor activity task and then sacrificed to measure brain DA markers. The authors observed only in males exposed prenatally to MB and then in the adulthood to PQ, a 95 % decrease of locomotor activity, associated to decreased levels of striatal DA, increased striatal DA turnover, and selective reduction in DA neurons of the SN*pc*. The authors concluded that the sequence of exposures is critical, since prenatal exposure to PQ followed by adult challenges with MB failed to produce marked alteration of locomotor activity or any DA neuron loss (Barlow et al. 2004).

3.3.3 Maternal Infection and PD

Inflammation and reduced antioxidant defenses (i.e., reduced glutathione activity) have been implicated in PD pathogenesis (McGeer et al. 2001). Intrauterine infection might lead to nigral cell loss in the fetal brain and subsequent permanent decrease in DA neurons, secondary to inflammation. In such framework, Zhu et al. (2007) assessed the effects of prenatal LPS exposure on glutathione (GSH) metabolism in rat brain. They observed a disturbance of GSH homeostasis in offspring of dams administered with LPS at GD 10.5, which possibly makes DA neurons more susceptible to the secondary chemical insult. Wang et al. (2009) have observed significant reduction of DA and serotonin (5-HT) levels accompanied by loss of DA and 5-HT neurons in SN and in the dorsal raphe nucleus (DRN) in adult rats offspring of pregnant dams exposed to a single injection of LPS (33 µg/kg) at GD 10.5. Fan et al. (2011) assessed the effects of intracerebral injection of LPS (1 mg/kg) in male rat pups at PND 5. Although on PND 70 rats spontaneously recovered the neurobehavioral dysfunction induced by LPS, the authors observed persistent injury to DA system and strong inflammatory response including increased activation of microglia and inflammatory cytokines in the brain. When challenged with METH (0.5 mg/kg), LPS-exposed rats showed increased METH-induced locomotion and stereotyped behaviors as compared to control rats (Tien et al. 2011).

3.3.4 Multiple Hits in PD: Maternal Infection and Chemical Insult Combination

Other their life span, human beings are exposed to mixtures of chemicals as well as to several non-chemical risk factors. One of the main questions in the assessment of the environmental hypothesis of PD etiology is whether sequential exposures to different kinds of risk factors across the lifetime would result in cumulative neurotoxicity to the nigrostriatal DA system. Various models based on "multiple hits" have been developed, assuming that prenatal inflammation makes the brain more susceptible to subsequent exposure to DA neurotoxins.

The first studies combining maternal infection with exposure to an environmental neurotoxicant at a later life stage were carried out by Ling et al. (2004a, b). Rats were exposed to LPS prenatally and then administered with a subtoxic dose of the DA neurotoxin rotenone (1.25 mg/kg per day for 14 days) when 16 months old. The combined effects of prenatal LPS and postnatal rotenone exposure produced higher DA cell loss compared with the effects of single exposure to either LPS or rotenone. Prenatal LPS exposure also led to increased levels of oxidized proteins and the formation of α-synuclein and eosin-positive inclusions resembling Lewy bodies. In the second study, the same authors (Ling et al. 2004a, b) injected on PND 99 the neurotoxin 6-hydroxydopamine (6-OHDA) or saline into animals exposed to either LPS or saline prenatally. The results showed that animals exposed to prenatal LPS or postnatal 6-OHDA alone had fewer DA neurons than controls, while the two toxins combined produced a greater loss. They conclude that prenatal exposure to LPS produces permanent cell loss accompanied by an inflammatory state that leads to further DA neuron loss once in the presence of subsequent neurotoxin exposure.

Fan et al. (2011a) found that neonatal intracerebral injection of LPS following administration of rotenone at low dose (1.25 mg/kg) for 14 days from PND 70 resulted in PD-like neurobehavioral deficits later in life. In agreement with this hypothesis, Tien et al. (2013) demonstrated that neonatal LPS exposure enhanced the rotenone-induced accumulation of α-synuclein aggregation and DAT protein expression in the SN. Overall, these studies show that brain inflammation in early life may enhance adult susceptibility to develop neurodegenerative hallmarks triggered by environmental toxins.

To compare the effects of gestational LPS exposure with neonatal systemic LPS exposure on adult DA neuron susceptibility to rotenone neurotoxicity, rats were administered with LPS at PND 5 (Cai et al. 2013) and, at adulthood, were challenged with rotenone at a dose of 1.25 mg/kg per day for 14 days. Briefly, rotenone induced loss of DA neurons and PD-like motor impairment in 3-month-old rats that had experienced neonatal LPS exposure, but not in those without the LPS exposure. These results support the view that although neonatal systemic LPS exposure may not necessarily lead to loss of DA neurons in the SN, it could cause persistent functional alterations that predispose the nigrostriatal system to be damaged by environmental toxins.

Another recent line of evidence concerns air pollution, which has been associated with adverse neurological and behavioral effects in children and adults. Increases in neuroinflammation, oxidative stress, and glial activation have been identified as putative mechanisms by which air pollution exposures may impair central nervous system function in adults. Particulate matter, one of many components of ambient outdoor air pollution, causes elevation in cytokines and oxidative stress in the brain (Campbell et al. 2005). Given the evidence suggesting susceptibility of both the SN and ST to components of air pollution, two studies assessed whether exposure of mice to concentrated ambient ultrafine particles (CAPS; <100 nm diameter) during the first two weeks of life would alter susceptibility to induction of the PD phenotype by the combined administration of PQ + MB, utilizing 10 mg/kg PQ and 30 mg/kg MB 2 × per week for 6 weeks (Thiruchelvam et al. 2000, 2003). Animals treated with CAPS in the early postnatal period show enhanced response to PQ + MB combination. Both CAPS and PQ + MB elevated glutamate levels in ST, consistent with potential excitotoxicity. These findings demonstrate the ability of postnatal CAPS to produce locomotor dysfunction and dopaminergic and glutamatergic changes, independent of PQ + MB, in brain regions involved in PD (Allen et al. 2014).

4 Concluding Remarks

Increasing evidence indicates that events occurring in the earliest stages of human development may influence both resilience and vulnerability to several adult diseases, including neuropsychiatric and neurologic disorders. As stated by Olden et al. (2011), the timing of environmental exposure is a crucial factor that influences the size of the gene–environment interaction: Whereas genes may contain the potential for adverse health outcomes, subsequent exposure to environmental triggers is required to initiate physiological or pathological pathways responsible for health and disease.

Experimental models confirm this general view. Research in animals has clearly shown that diverse environmental stressors may interfere with typical brain developmental trajectories in critical time windows, inducing silent subclinical alterations that may hesitate in pathophysiology and overt functional (e.g., behavioral) deficits at a later life stage. The bulk of data produced by developmental neurotoxicologists since the 1980s has started a series of epidemiological studies to assess the possible link between subtoxic exposure to chemical pollutants and neurobehavioral effects in children. Thanks to such combined effort, recent advances in research offer important clues into the complex etiology of autism and other neurodevelopmental disorders, indicating that developmental exposure to chemicals cannot be ruled out (Rossignol et al. 2014). Specifically, variations in candidate genes may confer higher individual vulnerability to environmental toxicants. Of note, the convergence of risk factors on the same cell/molecular pathways in critical developmental windows, combined with a vulnerable gene makeup, would trigger the development of the disorders or predispose to adverse health outcome later on. As an example, children at risk for autism may present constitutionally reduced antioxidant defenses: This may render the individual more vulnerable to oxidative stress and inflammatory processes produced by both xenobiotic exposure and maternal infection during pregnancy (Rose et al. 2012).

A similar research effort has not been taking place as concerns the developmental origin of NDDs and the role of environmental factors in increasing risk. As discussed in Chap. 1, the long latency between exposure and disease occurrence, the difficulty of reconstructing exposures that occurred several decades before disease manifestation, and the wealth of factors which the individual is exposed to during the life course make the developmental origin of NDD a promising but difficulty provable hypothesis. Nonetheless, given the huge public health impact of NDDs, identification of early risk factors would be crucial to design prevention and/or intervention strategies: In this perspective, the comprehension of the environmental etiology of major NDDs requires innovative study paradigms with multifaceted and multidisciplinary approaches.

The experimental findings presented in Chap. 2 prove the utility of in vivo models and support a role of environmental chemicals in increasing risk of NDDs. These experiments showed not only that an environmental insult in early life has long-term irreversible consequences, but also that the insult must occur during a critical period in development to have its maximal effect. Gestational and neonatal exposure to metals and pesticides as well as experimentally induced maternal infection both lead to alterations in developmental programming that are expressed as an irreversibly altered function. The functional effects often include significant motor and/or cognitive impairments resembling the pathological phenotypes of major human NDDs. Most of these models present strong construct validity, as the more robust hallmarks of AD and PD are associated with the functional impairment. Furthermore, epigenetic processes seem to play a key role in the mechanisms underlying these phenomena, as shown by the finding that perinatal exposure to Pb modulates the expression of AD-related genes, influencing the course of amyloidogenesis and oxidative DNA damage via DNA methylation (Dosunmu et al. 2012). Among environmental pollutants, Pb is a paradigmatic example of a developmental neurotoxicant able to interfere with developmental programming through different and possibly interrelated mechanisms. These include the antagonism of NMDA receptors in the early developmental phases, which is the key event at the basis of Pb interference with synaptogenesis, neural network formation, and behavioral plasticity (Toscano and Guilarte 2005), but Pb also induces latent changes in antioxidant defenses and mitochondrial redox dysfunction that may trigger neurodegeneration-related pathways at later life stages (Caito and Aschner 2015).

The experiments assessing the double or multiple hit hypothesis, either by rechallenging the animal with the same or a different chemical compounds at different life stages or by presenting a chemical insult in the adulthood following fetal exposure to maternal inflammation, support the LEARn hypothesis advanced by Lahiri et al. (2007). The end result of the fetal insult in fact is an animal that is sensitized such that it will be more susceptible to diseases later in life, provided the occurrence of a second hit. In such framework, the exposure of the fetal brain to proinflammatory cytokines secondary to maternal infection appears as a major causative factor of altered developmental programming and enhanced vulnerability. This supports the view that neurodevelopmental and neurodegenerative disorders do share pathogenic pathways: Abnormal immune activation in the early stage of brain development may constitute the common event predisposing to disease at some point in the life span of the individual.

The question arising at this point is as to whether experimental findings may help to fill the huge gaps in knowledge that lay beneath the developmental hypothesis for NDDs. Whereas some of the studies here reviewed offer interesting clues for the mechanistic underpinnings of chemical-related effects, experimental models on this topic are still at their beginning. The very nature of the problem requires the establishment of prospective animal cohorts: This implies the use of developmental exposures with multilevel analysis of the embryos, fetus, and pups, as well as of later life stages to trace dysfunction/disease. Research must use environmentally relevant doses, dose–response curves, and the examination of the relationship between the molecular mechanism proposed and the disease/dysfunction studied. This would also require the use of the new technologies of gene expression profiling and epigenetics: A critical component is the development of a direct correlation and eventually a cause–effect relationship between the alterations in gene expression during the development (either increased, decreased, or inappropriate timing) to alterations in "omic" signatures that may predict a specific disease or dysfunction later in life. To date, no available biomarker is a clear and validated indicator of typical brain development. Recent studies suggest that placental miRNA expression profiles and DNA methylation of specific genes are associated with measures of neurobehavioral outcome in the infants as well as with increased risks of neurological and NDDs (Sheinerman and Umansky 2013); altered Aβ protein in plasma is related with neurodegenerative risk after prenatal Pb exposure (Mazumdar et al. 2012). In vivo models could be instrumental to identify robust peripheral biomarkers of exposure, effects, and susceptibility, to be subsequently verified in large prospective human cohort studies to establish biologically plausible links between early chemical exposure and later health effects.

Acknowledgments Supported by FP7 HEALS Grant No. 603946 "Health and Environment-wide Associations based on Large Population Surveys," by CROME Grant No. LIFE12 ENV/GR/001040 "Cross-Mediterranean Environment and Health Network," and by Ministry of Health-ISS Project 13CAL "Identification of early markers of pathology in rodent models of autism: role of endogenous endoretroviruses."

References

Allen JL, Liu X, Weston D et al (2014) Consequences of developmental exposure to concentrated ambient ultrafine particle air pollution combined with the adult paraquat and maneb model of the Parkinson's disease phenotype in male mice. Neurotoxicology 41:80–88

Androutsopoulos VP, Hernandez AF, Liesivuori J et al (2013) A mechanistic overview of health associated effects of low levels of organochlorine and organophosphorous pesticides. Toxicology 307:89–94

Ashok A, Rai NK, Tripathi S et al (2015) Exposure to As-, Cd-, and Pb-mixture induces Aβ, amyloidogenic APP processing and cognitive impairments via oxidative stress-dependent neuroinflammation in young rats. Toxicol Sci 143(1):64–80

Bakulski KM, Rozek LS, Dolinoy DC et al (2012) Alzheimer's disease and environmental exposure to lead: the epidemiologic evidence and potential role of epigenetics. Curr Alzheimer Res 9(5):563–573

Baldi I, Gruber A, Rondeau V et al (2011) Neurobehavioral effects of long-term exposure to pesticides: results from the 4-year follow-up of the PHYTONER study. Occup Environ Med 68 (2):108–115

Ballard C, Gauthier S, Corbett A et al (2011) Alzheimer's disease. Lancet 377(9770):1019–1031

Banks EC, Ferretti LE, Shucard DW (1997) Effects of low level lead exposure on cognitive function in children: a review of behavioral, neuropsychological and biological evidence. Neurotoxicology 18(1):237–281

Barker DJ, Gluckman PD, Godfrey KM et al (1993) Fetal nutrition and cardiovascular disease in adult life. Lancet 341(8850):938–941

Barlow BK, Richfield EK, Cory-Slechta DA et al (2004) A fetal risk factor for Parkinson's disease. Dev Neurosci 26(1):11–23

Barlow BK, Cory-Slechta DA, Richfield EK et al (2007) The gestational environment and Parkinson's disease: evidence for neurodevelopmental origins of a neurodegenerative disorder. Reprod Toxicol 23(3):457–470

Basha MR, Wei W, Bakheet SA et al (2005) The fetal basis of amyloidogenesis: exposure to lead and latent overexpression of amyloid precursor protein and beta-amyloid in the aging brain. J Neurosci 25(4):823–829

Bellinger DC (2009) Interpreting epidemiologic studies of developmental neurotoxicity: conceptual and analytic issues. Neurotoxicol Teratol 31:267–274

Bellinger DC (2014) Mercury and pregnancy. Birth Defects Res A Clin Mol Teratol 100:1–3

Bihaqi SW, Zawia NH (2013) Enhanced taupathy and AD-like pathology in aged primate brains decades after infantile exposure to lead (Pb). Neurotoxicology 39:95–101

Bihaqi SW, Bahmani A, Subaiea GM et al (2014) Infantile exposure to lead and late-age cognitive decline: relevance to AD. Alzheimers Dement 10(2):187–195

Billnitzer AJ, Barskaya I, Yin C et al (2013) APP independent and dependent effects on neurite outgrowth are modulated by the receptor associated protein (RAP). J Neurochem 124(1):123–132

Blennow K, de Leon MJ, Zetterberg H (2006) Alzheimer's disease. Lancet 368(9533):387–403

Bolin CM, Basha R, Cox D et al (2006) Exposure to lead and the developmental origin of oxidative DNA damage in the aging brain. FASEB J 20(6):788–790

Bugiani O (2011) Alzheimer's disease: ageing-related or age-related? New hypotheses from an old debate. Neurol Sci 32:1241–1247

Cai Z, Fan LW, Kaizaki A et al (2013) Neonatal systemic exposure to lipopolysaccharide enhances susceptibility of nigrostriatal dopaminergic neurons to rotenone neurotoxicity in later life. Dev Neurosci 35(2–3):155–171

Caito SW, Aschner M (2015) Mitochondrial redox dysfunction and environmental exposures. Antioxid Redox Signal 23(6):578–595

Calne DB, Langston JW (1983) Aetiology of Parkinson's disease. Lancet 2(8365–66):1457–1459

Campbell A, Oldham M, Becaria A et al (2005) Particulate matter in polluted air may increase biomarkers of inflammation in mouse brain. Neurotoxicology 26:133–140

Casado A, Encarnación López-Fernández M, Concepción Casado M et al (2008) Lipid peroxidation and antioxidant enzyme activities in vascular and Alzheimer dementias. Neurochem Res 33(3):450–458

Chen GH, Wang H, Yang QG et al (2011) Acceleration of age-related learning and memory decline in middle-aged CD-1 mice due to maternal exposure to lipopolysaccharide during late pregnancy. Behav Brain Res 218(2):267–279

Chen L, Yoo SE, Na R et al (2012) Cognitive impairment and increased Aβ levels induced by paraquat exposure are attenuated by enhanced removal of mitochondrial H(2)O(2). Neurobiol Aging 33(2):432.e15–26)

Chin-Chan M, Navarro-Yepes J, Quintanilla-Vega B (2015) Environmental pollutants as risk factors for neurodegenerative disorders: Alzheimer and Parkinson diseases. Front Cell Neurosci 9:124

Colborn T (2006) A case for revisiting the safety of pesticides: a closer look at neurodevelopment. Environ Health Perspect 114(1):10–17

Cole TB, Beyer RP, Bammler TK et al (2011) Repeated developmental exposure of mice to chlorpyrifos oxon is associated with paraoxonase 1 (PON1)-modulated effects on cerebellar gene expression. Toxicol Sci 123(1):155–169

Cordova FM, Aguiar AS Jr, Peres TV et al (2012) In vivo manganese exposure modulates Erk, Akt and Darpp-32 in the striatum of developing rats, and impairs their motor function. PLoS ONE 7(3):e33057

Cordova FM, Aguiar AS Jr, Peres TV et al (2013) Manganese-exposed developing rats display motor deficits and striatal oxidative stress that are reversed by Trolox. Arch Toxicol 87 (7):1231–1244

Cory-Slechta DA, Thiruchelvam M, Barlow BK et al (2005) Developmental pesticide models of the Parkinson disease phenotype. Environ Health Perspect 113(9):1263–1270

Costa LG, Aschner M, Vitalone A et al (2004) Developmental neuropathology of environmental agents. Annu Rev Pharmacol Toxicol 44:87–110

Couper J (1837) On the effects of black oxide of manganese when inhaled into the lungs. Br Ann Med Pharm Vital Stat Gen Sci 1:41–42

Crawley JN (2012) Translational animal models of autism and neurodevelopmental disorders. Dialogues Clin Neurosci 14:293–305

Dal-Pizzol F, Klamt F, Frota ML Jr et al (2001) Neonatal iron exposure induces oxidative stress in adult Wistar rat. Brain Res Dev Brain Res 130(1):109–114

Dauer W, Przedborski S (2003) Parkinson's disease: mechanisms and models. Neuron 39 (6):889–909

De Felice A, Ricceri L, Venerosi A, Chiarotti F, Calamandrei G (2015) Multifactorial origin of neurodevelopmental disorders: approaches to understanding complex etiologies. Toxics 3 (1):89–129

Di Monte DA (2003) The environment and Parkinson's disease: is the nigrostriatal system preferentially targeted by neurotoxins. 2(9):531–538

Dias V, Junn E, Mouradian MM (2013) The role of oxidative stress in Parkinson's disease. J Parkinsons Dis 3(4):461–491

Dosunmu R, Alashwal H, Zawia NH (2012) Genome-wide expression and methylation profiling in the aged rodent brain due to early-life Pb exposure and its relevance to aging. Mech Ageing Dev 133(6):435–443

Eaton DL, Daroff RB, Autrup H et al (2008) Review of the toxicology of chlorpyrifos with an emphasis on human exposure and neurodevelopment. Crit Rev Toxicol 38(Suppl 2):1–125

Eskenazi B, Marks AR, Bradman A et al (2007) Organophosphate pesticide exposure and neurodevelopment in young Mexican-American children. Environ Health Perspect 115:792–798

Fan LW, Tien LT, Lin RC et al (2011a) Neonatal exposure to lipopolysaccharide enhances vulnerability of nigrostriatal dopaminergic neurons to rotenone neurotoxicity in later life. Neurobiol Dis 44(3):304–316

Fan LW, Tien LT, Zheng B et al (2011b) Dopaminergic neuronal injury in the adult rat brain following neonatal exposure to lipopolysaccharide and the silent neurotoxicity. Brain Behav Immun 25(2):286–297

Firestone JA, Lundin JI, Powers KM et al (2010) Occupational factors and risk of Parkinson's disease: a population-based case-control study. Am J Ind Med 53(3):217–223

Fredriksson A, Schröder N, Eriksson P et al (1999) Neonatal iron exposure induces neurobehavioural dysfunctions in adult mice. Toxicol Appl Pharmacol 159(1):25–30

Fredriksson A, Schröder N, Eriksson P et al (2000) Maze learning and motor activity deficits in adult mice induced by iron exposure during a critical postnatal period. Brain Res Dev Brain Res 119(1):65–74

Garner CC, Zhai RG, Gundelfinger ED et al (2002) Molecular mechanisms of CNS synaptogenesis. Trends Neurosci 25(5):243–251

Giacobini E, Mori F, Lai CC (1996) The effect of cholinesterase inhibitors on the secretion of APPS from rat brain cortex. Ann N Y Acad Sci 777:393–398

Gilbert SG, Grant-Webster KS (1995) Neurobehavioral effects of developmental methylmercury exposure. Environ Health Perspect 103(Suppl 6):135–142

Gillman MW (2005) Developmental origins of health and disease. N Engl J Med 353(17):1848–1850

Giorgi O, De Montis G, Porceddu ML et al (1987) Developmental and age-related changes in D1-dopamine receptors and dopamine content in the rat striatum. Brain Res 432(2):283–290

Goedert M, Jakes R, Crowther RA et al (1993) The abnormal phosphorylation of tau protein at Ser-202 in Alzheimer disease recapitulates phosphorylation during development. Proc Natl Acad Sci USA 90(11):5066–5070

JPND Action Group (2013) Report of the JPND action group. www.jpnd.eu

Gupta RK, Hasan KM, Trivedi R et al (2005) Diffusion tensor imaging of the developing human cerebrum. J Neurosci Res 81:172–178

Hanson MA (2013) Developmental origins of obesity and non-communicable disease. Endocrinol Nutr 60(Suppl 1):10–11

Hayden KM, Norton MC, Darcey D et al (2010) Occupational exposure to pesticides increases the risk of incident AD: the cache county study. Neurology 74(19):1524–1530

Heindel JJ, Balbus J, Birnbaum L et al (2015) developmental origins of health and disease: integrating environmental influences. Endocrinology 156(10):3416–3421

Heneka MT, Carson MJ, El Khoury J et al (2015) Neuroinflammation in Alzheimer's disease. Lancet Neurol 14(4):388–405

Heppner FL, Ransohoff RM, Becher B (2015) Immune attack: the role of inflammation in Alzheimer disease. Nat Rev Neurosci 16(6):358–372

Herbstman JB, Sjodin A, Kurzon M et al (2010) Prenatal exposure to PBDEs and neurodevelopment. Environ Health Perspect 118:712–719

Hertzman C, Wiens M, Bowering D et al (1990) Parkinson's disease: a case-control study of occupational and environmental risk factors. Am J Ind Med 17(3):349–355

Howe K, Clark MD, Torroja CF et al (2013) The zebrafish reference genome sequence and its relationship to the human genome. Nature 496(7446):498–503

Jankovic J (2005) Searching for a relationship between manganese and welding and Parkinson's disease. Neurology 64(12):2021–2028

Jenner P (2003) Oxidative stress in Parkinson's disease. Ann Neurol 53(Suppl 3):S26–S36

Kaur D, Peng J, Chinta SJ et al (2007) Increased murine neonatal iron intake results in Parkinson-like neurodegeneration with age. Neurobiol Aging 28(6):907–913

Kim JT, Son MH, Lee DH et al (2015) Partitioning behavior of heavy metals and persistent organic pollutants among feto-maternal bloods and tissues. Environ Sci Technol 49(12):7411–7422

Knuesel I, Chicha L, Britschgi M et al (2014) Maternal immune activation and abnormal brain development across CNS disorders. Nat Rev Neurol 10(11):643–660

Kovacs GG, Adle-Biassette H, Milenkovic I et al (2014) Linking pathways in the developing and aging brain with neurodegeneration. Neuroscience 269:152–172

Krstic D, Madhusudan A, Doehner J et al (2012) Systemic immune challenges trigger and drive Alzheimer-like neuropathology in mice. J Neuroinflammation 9:151

Kumar A, Leinisch F, Kadiiska MB et al (2015) Formation and implications of alpha-synuclein radical in Maneb- and Paraquat-induced models of Parkinson's disease. Mol Neurobiol (Epub ahead of print)

Lahiri DK, Maloney B, Basha MR et al (2007) How and when environmental agents and dietary factors affect the course of Alzheimer's disease: the "LEARn" model (latent early-life associated regulation) may explain the triggering of AD. Curr Alzheimer Res 4(2):219–228

Lahiri DK, Maloney B, Zawia NH (2009) The LEARn model: an epigenetic explanation for idiopathic neurobiological diseases. Mol Psychiatry 14(11):992–1003

Lai BC, Marion SA, Teschke K et al (2002) Occupational and environmental risk factors for Parkinson's disease. Parkinsonism Relat Disord 8(5):297–309

Langston J, Ballard P, Tetrud J et al (1983) Chronic parkinsonism in humans due to a product of meperidine-analog synthesis. Science 219:979–980

Langston JW, Forno LS, Tetrud J et al (1999) Evidence of active nerve cell degeneration in the substantia nigra of humans years after 1-methyl-4-phenyl-1,2,3,6-tetrahydropyridine exposure. Ann Neurol 46(4):598–605

Lanphear BP, Hornung R, Khoury J et al (2005) Low level environmental lead exposure and children's intellectual function: an international pooled analysis. Environ Health Perspect 113:894–899

Lee J, Freeman JL (2014) Zebrafish as a model for investigating developmental lead (Pb) neurotoxicity as a risk factor in adult neurodegenerative disease: a mini-review. Neurotoxicology 43:57–64

Lees AJ, Hardy J, Revesz T (2009) Parkinson's disease. Lancet 373:2055–2066

Li N, Yu ZL, Wang L et al (2010) Increased tau phosphorylation and betaamyloid in the hippocampus of mouse pups by early life lead exposure. Acta Biologica Hungarica 61(2): 123–134

Li Y, Sun Y, Yang J et al (2014) Age-dependent dopaminergic dysfunction following fetal exposure to atrazine in SD rats. Environ Toxicol Pharmacol 37(3):1275–1282

Ling ZD, Chang Q, Lipton JW et al (2004a) Combined toxicity of prenatal bacterial endotoxin exposure and postnatal 6-hydroxydopamine in the adult rat midbrain. Neuroscience 124 (3):619–628

Ling Z, Chang QA, Tong CW et al (2004b) Rotenone potentiates dopamine neuron loss in animals exposed to lipopolysaccharide prenatally. Exp Neurol 190(2):373–383

Liou HH, Tsai MC, Chen CJ et al (1997) Environmental risk factors and Parkinson's disease: a case-control study in Taiwan. Neurology 48(6):1583–1588

Logroscino G (2005) The role of early life environmental risk factors in Parkinson disease: what is the evidence? Environ Health Perspect 113(9):1234–1238

Mazumdar M, Xia W, Hofmann O et al (2012) Prenatal lead levels, plasma amyloid beta levels, and gene expression in young adulthood. Environ Health Perspect 120:702–707

McCormack AL, Thiruchelvam M, Manning-Bog AB et al (2002) Environmental risk factors and Parkinson's disease: selective degeneration of nigral dopaminergic neurons caused by the herbicide paraquat. Neurobiol Dis 10(2):119–127

McFarlane AC, Searle AK, Van Hooff M et al (2013) Prospective associations between childhood low-level lead exposure and adult mental health problems: the Port Pirie cohort study. Neurotoxicology 39:11–17

McGeer PL, Yasojima K, McGeer EG (2001) Inflammation in Parkinson's disease. Adv Neurol 86:83–89

Meyer U (2014) Prenatal poly(i:C) exposure and other developmental immune activation models in rodent systems. Biol Psychiatry 75(4):307–315

Milatovic D, Zaja-Milatovic S, Gupta RC et al (2009) Oxidative damage and neurodegeneration in manganese-induced neurotoxicity. Toxicol Appl Pharmacol 240(2):219–225

Miller MW (1986) Effects of alcohol on the generation and migration of cerebral cortical neurons. Science 233:1308–1311

Moreira EG, Yu X, Robinson JF et al (2010) Toxicogenomic profiling in maternal and fetal rodent brains following gestational exposure to chlorpyrifos. Toxicol Appl Pharmacol 245(3):310–325

Morley R (2006) Fetal origins of adult disease. Semin Fetal Neonatal Med 11(2):73–78

Neal AP, Guilarte TR (2010) Molecular neurobiology of lead (Pb(2+)): effects on synaptic function. Mol Neurobiol 42(3):151–160

Neal AP, Worley PF, Guilarte TR (2011) Lead exposure during synaptogenesis alters NMDA receptor targeting via NMDA receptor inhibition. Neurotoxicology 32:281–289

Nordberg M, Winblad B, Fratiglioni L et al (2000) Lead concentrations in elderly urban people related to blood pressure and mental performance: results from a population-based study. Am J Ind Med 38(3):290–294

Nunomura A, Perry G, Aliev G et al (2001) Oxidative damage is the earliest event in Alzheimer disease. J Neuropathol Exp Neurol 60(8):759–767

Olden K, Isaac L, Roberts L (2011) Neighborhood-specific epigenome analysis: the pathway forward to understanding gene-environment interactions. N C Med J 72(2):125–127

Parrón T, Requena M, Hernández AF et al (2011) Association between environmental exposure to pesticides and neurodegenerative diseases. Toxicol Appl Pharmacol 256(3):379–385

Perera FP, Li Z, Whyatt R et al (2009) Prenatal airborne polycyclic aromatic hydrocarbon exposure and child IQ at age 5 years. Pediatrics 124:e195–e202

Peris-Sampedro F, Salazar JG, Cabré M et al (2014) Impaired retention in AβPP Swedish mice six months after oral exposure to chlorpyrifos. Food Chem Toxicol 72:289–294

Reitz C, Brayne C, Mayeux R (2011) Epidemiology of Alzheimer disease. Nat Rev Neurol 7 (3):137–152

Reuhl KR (1991) Delayed expression of neurotoxicity: the problem of silent damage. Neurotoxicology 12(3):341–346

Richardson JR, Caudle WM, Wang M et al (2006) Developmental exposure to the pesticide dieldrin alters the dopamine system and increases neurotoxicity in an animal model of Parkinson's disease. FASEB J 20(10):1695–1697

Richardson JR, German D, Levey A (2014) Alzheimer disease risk factors. JAMA Neurol 71 (8):1051

Rosas LG, Eskenazi B (2008) Pesticides and child neurodevelopment. Curr Opin Pediatr 20 (2):191–197

Rose S, Melnyk S, Pavliv O et al (2012) Evidence of oxidative damage and inflammation associated with low glutathione redox status in the autism brain Transl. Psychiatry 2(7):e134

Rossignol DA, Genuis SJ, Frye RE (2014) Environmental toxicants and autism spectrum disorders: a systematic review. Trans Psychiatry 4:e360

Roy TS, Seidler FJ, Slotkin TA (2004) Morphologic effects of subtoxic neonatal chlorpyrifos exposure in developing rat brain: regionally selective alterations in neurons and glia. Brain Res Dev Brain Res 148:197–206

Salazar JG, Ribes D, Cabré M et al (2011) Amyloid β peptide levels increase in brain of AβPP Swedish mice after exposure to chlorpyrifos. Curr Alzheimer Res 8(7):732–740

Sale A, Berardi N, Maffei L (2014) Environment and brain plasticity: towards an endogenous pharmacotherapy. Physiol Rev 94(1):189–234

Sandy MS, Armstrong M, Tanner CM et al (1996) CYP2D6 allelic frequencies in young-onset Parkinson's disease. Neurology 47(1):225–230

Santibáñez M, Bolumar F, García AM (2007) Occupational risk factors in Alzheimer's disease: a review assessing the quality of published epidemiological studies. Occup Environ Med 64 (11):723–732

Selkoe DJ, Schenk D (2003) Alzheimer's disease: molecular understanding predicts amyloid-based therapeutics. Annu Rev Pharmacol Toxicol 43:545–584

Sheinerman KS, Umansky SR (2013) Circulating cell-free microRNA as biomarkers for screening, diagnosis and monitoring of neurodegenerative diseases and other neurologic pathologies. Front Cell Neurosci 7:150

Shonkoff JP (2000) From neurons to neighborhoods: the science of early childhood development. National Academy Press, Washington

Sian-Hülsmann J, Mandel S, Youdim MB et al (2011) The relevance of iron in the pathogenesis of Parkinson's disease. J Neurochem 118(6):939–957

Slotkin TA, Seidler FJ (2009) Oxidative and excitatory mechanisms of developmental neurotoxicity: transcriptional profiles for chlorpyrifos, diazinon, dieldrin, and divalent nickel in PC12 cells. Environ Health Perspect 117:587–596

Stangle DE, Strawderman MS, Smith D et al (2004) Reductions in blood lead overestimate reductions in brain lead following repeated succimer regimens in a rodent model of childhood lead exposure. Environ Health Perspect 112:302–308

Sun Y, Li YS, Yang JW et al (2014) Exposure to atrazine during gestation and lactation periods: toxicity effects on dopaminergic neurons in offspring by downregulation of Nurr1 and VMAT2. Int J Mol Sci 15(2):2811–2825

Tanner CM, Goldman SM, Ross GW et al (2014) The disease intersection of susceptibility and exposure: chemical exposures and neurodegenerative disease risk. Alzheimers Dement 10(3 Suppl):S213–S225

Tarantal AF, Berglund L (2014) Obesity and lifespan health–importance of the fetal environment. Nutrients 6(4):1725–1736

Tau GZ, Peterson BS (2010) Normal development of brain circuits. Neuropsychopharmacology 35 (1):147–168

Thiruchelvam M, Brockel BJ, Richfield EK et al (2000) Potentiated and preferential effects of combined paraquat and maneb on nigrostriatal dopamine systems: environmental risk factors for Parkinson's disease? Brain Res 873(2):225–234

Thiruchelvam M, Richfield EK, Goodman BM et al (2002) Developmental exposure to the pesticides paraquat and maneb and the Parkinson's disease phenotype. Neurotoxicology 23(4–5):621–633

Thiruchelvam M, McCormack A, Richfield EK et al (2003) Age-related irreversible progressive nigrostriatal dopaminergic neurotoxicity in the paraquat and maneb model of the Parkinson's disease phenotype. Eur J Neurosci 18(3):589–600

Tien LT, Cai Z, Rhodes PG et al (2011) Neonatal exposure to lipopolysaccharide enhances methamphetamine-induced reinstated behavioral sensitization in adult rats. Behav Brain Res 224(1):166–173

Tien LT, Kaizaki A, Pang Y et al (2013) Neonatal exposure to lipopolysaccharide enhances accumulation of α-synuclein aggregation and dopamine transporter protein expression in the substantia nigra in responses to rotenone challenge in later life. Toxicology 308:96–103

Toscano CD, Guilarte TR (2005) Lead neurotoxicity: from exposure to molecular effects. Brain Res Brain Res Rev 49(3):529–554

Van Maele-Fabry G, Hoet P, Vilain F et al (2012) Occupational exposure to pesticides and Parkinson's disease: a systematic review and meta-analysis of cohort studies. Environ Int 1 (46):30–43

Venerosi A, Ricceri L, Tait S et al (2012) Sex dimorphic behaviors as markers of neuroendocrine disruption by environmental chemicals: the case of chlorpyrifos. Neurotoxicology 33(6):1420–1426

Voorn P, Kalsbeek A, Jorritsma-Byham B et al (1988) The pre- and postnatal development of the dopaminergic cell groups in the ventral mesencephalon and the dopaminergic innervation of the striatum of the rat. Neuroscience 25(3):857–887

Vorhees RA (1986) Handbook of behavioral teratology. Plenum Press, New York

Wang S, Yan JY, Lo YK et al (2009) Dopaminergic and serotoninergic deficiencies in young adult rats prenatally exposed to the bacterial lipopolysaccharide. Brain Res 1265:196–204

Wani WY, Sunkaria A, Sharma DR et al (2014) Caspase inhibition augments Dichlorvos-induced dopaminergic neuronal cell death by increasing ROS production and PARP1 activation. Neuroscience 258:1–15

Weisskopf MG, Wright RO, Schwartz J et al (2004) Cumulative lead exposure and prospective change in cognition among elderly men: the VA normative aging study. Am J Epidemiol 160 (12):1184–1193

Westerlund M, Hoffer B, Olson L (2010) Parkinson's disease: exit toxins, enter genetics. Prog Neurobiol 90:146–156

Whyatt RM, Liu X, Rauh VA et al (2012) Maternal prenatal urinary phthalate metabolite concentrations and child mental, psychomotor, and behavioral development at 3 years of age. Environ Health Perspect 120:290–295

Wild CP (2012) The exposome: from concept to utility. Int J Epidemiol 41:24–32

Wilson WW, Shapiro LP, Bradner JM et al (2014) Developmental exposure to the organochlorine insecticide endosulfan damages the nigrostriatal dopamine system in male offspring. Neurotoxicology 44:279–287

Winneke G (2011) Developmental aspects of environmental neurotoxicology: Lessons from lead and polychlorinated biphenyls. J Neurol Sci 308:9–15

Wright RO, Tsaih SW, Schwartz J et al (2003) Lead exposure biomarkers and mini-mental status exam scores in older men. Epidemiology 14(6):713–718

Wu J, Basha MR, Brock B et al (2008) Alzheimer's disease (AD)-like pathology in aged monkeys after infantile exposure to environmental metal lead (Pb): evidence for a developmental origin and environmental link for AD. J Neurosci 28(1):3–9

Zhu Y, Carvey PM, Ling Z (2007) Altered glutathione homeostasis in animals prenatally exposed to lipopolysaccharide. Neurochem Int 50(4):671–680

The Use of Perinatal 6-Hydroxydopamine to Produce a Rodent Model of Lesch–Nyhan Disease

Darin J. Knapp and George R. Breese

Abstract Lesch–Nyhan disease is a neurologically, metabolically, and behaviorally devastating condition that has eluded complete characterization and adequate treatment. While it is known that the disease is intimately associated with dysfunction of the hypoxanthine phosphoribosyltransferase 1 (HPRT1) gene that codes for an enzyme of purine metabolism (hypoxanthine-guanine phosphoribosyltransferase) and is associated with neurological, behavioral, as well as metabolic dysfunction, the mechanisms of the neurobehavioral manifestations are as yet unclear. However, discoveries over the past few decades not only have created useful novel animal models (e.g., the HPRT-deficient mouse and the serendipitously discovered perinatal 6-hydroxydopamine (6-OHDA lesion model), but also have expanded into epigenetic, genomic, and proteomic approaches to better understand the mechanisms underlying this disease. The perinatal 6-OHDA model, in addition to modeling self-injury and dopamine depletion in the clinical condition, also underscores the profound importance of development in the differential course of maladaptive progression in the face of a common/single neurotoxic insult at different ages. Recent developments from clinical and basic science efforts attest to the fact that while the disease would seem to have a simple single gene defect at its core, the manifestations of this defect are profound and unexpectedly diverse. Future efforts employing the 6-OHDA model and others in the context of the novel technologies of genome editing, chemo- and opto-genetics, epigenetics, and further studies on the mechanisms of stress-induced maladaptations in brain all hold promise in taking our understanding of this disease to the next level.

Keywords Lesch–Nyhan disease · Rats · 6-hydroxydopamine 6-OHDA · Perinatal · Self-injurious behavior · L-DOPA · Animal models

D.J. Knapp (✉) · G.R. Breese
School of Medicine, Department of Psychiatry and Bowles
Center for Alcohol Studies, University of North Carolina, Chapel Hill,
NC 27599, USA
e-mail: darin_knapp@med.unc.edu

G.R. Breese
e-mail: George_breese@med.unc.edu

© Springer International Publishing Switzerland 2016
Curr Topics Behav Neurosci (2016) 29: 265–278
DOI 10.1007/7854_2016_444
Published Online: 01 April 2016

Contents

1 Introduction... 266
2 Overview of Experimental Models 267
3 Perinatal 6-OHDA Model of Lesch–Nyhan Disease.. 268
4 Recent Developments 270
   4.1 Clinical Science... 270
   4.2 Basic Science 271
5 Future Directions 272
References... 274

1 Introduction

Lesch–Nyhan disease is a rare X-linked recessive disorder that affects approximately 3 in 1 million people, almost all male (Jinnah and Friedmann 2001). A cardinal feature is excessive production of uric acid, which leads to elevated serum uric acid concentrations and its increased urinary excretion. Uric acid is near its limit of solubility in the body, and its overproduction leads to calcium precipitates in the urogenital system to cause kidney stones and renal failure; in subcutaneous tissue to cause tophi; and in the joints to cause gouty arthritis. Other clinical manifestations of Lesch–Nyhan disease include motor dysfunction (generalized dystonia), intellectual disability, and self-injurious behavior including biting, hitting, and eye poking.

Lesch–Nyhan disease appears to manifest as a spectrum of phenotypes (Fu et al. 2014). Isolated overproduction of uric acid without neurobehavioral abnormalities constitutes the mildest Lesch–Nyhan variant. The disease is caused by mutations in the hypoxanthine phosphoribosyltransferase 1 (HPRT1) gene, which encodes hypoxanthine-guanine phosphoribosyltransferase (HPRT), an enzyme involved in purine metabolism. HPRT catalyzes the conversion of the purine bases hypoxanthine and guanine into the purine nucleotide pools (Jinnah 2009). Without HPRT, hypoxanthine and guanine are not converted to nucleotides but are instead degraded to uric acid, a by-product of purine metabolism. These effects are accompanied by enhanced purine synthesis. Failed conversion of hypoxanthine and guanine and consequent enhanced purine synthesis cause excess uric acid production in Lesch–Nyhan disease.

Overproduction of uric acid in Lesch–Nyhan disease can be treated with allopurinol, a xanthine oxidase inhibitor that inhibits conversion of hypoxanthine and guanine to uric acid. Allopurinol and hydration reduce the occurrence of gouty arthritis, kidney stones, and tophi. Efficacious treatments for other aspects of Lesch–Nyhan disease have proven to be elusive.

The pathophysiologic basis of the neurobehavioral manifestations of Lesch–Nyhan disease appears to differ from somatic manifestations. Mitigation of uric acid overproduction in Lesch–Nyhan disease does not affect its neurobehavioral manifestations. While no generally effective treatment for the neurobehavioral manifestations of Lesch–Nyhan disease has been identified to date, muscle relaxants have been used as palliative treatment for generalized dystonia (http://www. lesch–nyhan.org/en/treatment/neurological-disability/). Some neurobehavioral manifestations have been managed with behavioral therapy including therapy for short attention spans and associated learning disabilities. Overall, behavioral manifestations such as self-injury are mitigated by physical restraint which includes exclusion of sharp objects from the patient and limiting arm movement with ties or splints to control finger biting or hitting. In some cases, self-biting is managed by removing the teeth. Behavioral therapy, also employed in some cases, focuses attention on good behaviors to the exclusion of bad behaviors which are further minimized by avoiding situations in which they arise. Finally, other medications including gabapentin, carbamazepine, and diazepam have been tried in the management of self-injury with mixed or limited results. Risperidone has shown some evidence of efficacy against self-injury (Allen and Rice 1996); however, it is unclear whether any apparent effects on self-injury are non-specifically related to sedation. Given the limitations of the therapeutic approaches tried to date, new-generation treatments for behavioral symptoms remain a high priority.

The physiologic basis of the neurobehavioral abnormalities in Lesch–Nyhan disease is not fully elucidated. Dopamine pathways of the basal ganglia are affected, as demonstrated by marked (60–80 %) dopamine loss in postmortem neurochemical studies (Lloyd et al. 1981; Saito et al. 1999) and reductions in dopamine neuronal markers (Ernst et al. 1996; Wong et al. 1996) in this region. Detailed neurobiological descriptions of many other neurobiological features have been provided in extensive reviews (e.g., Papadeas and Breese 2014). However, circuit functions and definitive descriptions of maladaptive pathways are poorly understood.

2 Overview of Experimental Models

The rareness of Lesch–Nyhan disease renders studies in patients challenging and elevates the importance of experimental models in elucidating its biological aspects (Jinnah 2009). Several tissue culture and animal models that elucidate various aspects of Lesch–Nyhan disease have been developed. They have been reviewed previously by Jinnah (2009) and are summarized briefly herein.

Tissue culture models with non-neuronal cells including erythrocytes, lymphocytes, and fibroblasts derived from Lesch–Nyhan-affected patients have shed light on the metabolic and cellular impacts of HPRT deficiency. Cells deficient in HPRT accumulate purine waste products because of the failure to recycle hypoxanthine and guanine. Unexpectedly, however, purine deficiency has not been consistently associated with the failure of purine recycling (Shirley et al. 2007).

Tissue culture models with neuronal and glial cells developed as HPRT-deficient sublines of established neuroblastoma or glioma cell lines have been used to study the central nervous system dysfunction in Lesch–Nyhan disease. These models reveal neuronal structural and chemical abnormalities including reduction in dopamine in HPRT-deficient dopamine neuron-like lines (Bitler and Howard 1986; Yeh et al. 1998; Shirley et al. 2007; Lewers et al. 2008). This dopamine deficiency is consistent with the evidence of dopamine pathway damage in patients with Lesch–Nyhan disease (Lloyd et al. 1981; Ernst et al. 1996; Wong et al. 1996; Saito et al. 1999).

Several animal models of Lesch–Nyhan disease or specific manifestations such as self-injurious behavior have also been developed. For example, self-injurious behavior is observed after high doses of chronic amphetamine, caffeine, or opiates and single-dose clonidine (Jinnah and Breese 1997; Jinnah et al. 1999a).

Introduction of the HPRT-deficient mouse in 1987 marked the first time that a genetically engineered mouse modeled a specific human disease (Hooper et al. 1987; Kuehn et al. 1987; Jinnah 2009). Like people with Lesch–Nyhan disease, the mutant mice cannot recycle hypoxanthine or guanine and exhibit increased purine synthesis (Jinnah et al. 1992, 1993). In humans, these biochemical abnormalities lead to increased uric acid with clinical consequences of gout or kidney stones (Jinnah and Friedmann 2001). In the mutant mice, in contrast, the biochemical abnormalities do not result in increased uric acid with clinical consequences because of the presence of uricase, which degrades uric acid into allantoin. Furthermore, while mutant mice, like people with Lesch–Nyhan disease, show a loss of dopamine in the basal ganglia (Jinnah et al. 1999b), they do not exhibit the motor or neurobehavioral abnormalities that people do (Engle et al. 1996; Jinnah 2009). The reason that the dopamine loss is not associated with neurobehavioral consequences in the mutant mice is unknown. Further subsequent attempts to create a mouse with comparable behavioral phenotype as Lesch–Nyhan disease has proven difficult, in part because of failure to obtain viable mice after lesioning dopaminergic neurons during development (Breese et al. unpublished observation). However, the accidental discovery of a perinatal-6-hydroxydopamine (6-OHDA)-induced Lesch–Nyhan-type behavioral profile of self-injury in adult rats provided a new avenue for research.

3 Perinatal 6-OHDA Model of Lesch–Nyhan Disease

The relevance of the perinatal 6-OHDA model to self-injurious behavior in Lesch–Nyhan disease was discovered serendipitously. Breese and colleagues initially administered 6-OHDA intracisternally to young rats in investigations designed to examine the specificity of 6-OHDA for dopamine reductions in brain. Young animals were chosen in an attempt to develop models to complement a focus in Breese's center on childhood diseases at that time. In young rats, 6-OHDA caused reductions in tyrosine hydroxylase and dopamine (Breese and Taylor 1972; Jinnah et al. 1999b)—observations that mirrored results previously found in animals administered 6-OHDA during adulthood (Breese and Taylor 1970).

From these early studies (Breese and Traylor 1972), several of the original young rats that had been perinatally treated with 6-OHDA were maintained until adulthood and challenged with the dopamine precursor L-3,4-dihydroxyphenylalanine (L-DOPA)—an exposure which had previously been observed to enhance loco-motor activity when given to rats treated as adults with 6-OHDA. Unexpectedly, the non-specific enhancement of locomotion was not observed in rats treated perinatally with 6-OHDA. Rather, the surprising and disturbing observation was made that all of these animals exhibited self-mutilation of the front legs and abdominal area during a time frame when hyperlocomotion was the only expected phenotype. Such an outcome had never been observed in L-DOPA-treated rats treated as adults with 6-OHDA (Breese et al. 1984).

The consistent observation of self-mutilation among rats perinatally treated with 6-OHDA and challenged as adults with L-DOPA led Breese and colleagues to abandon this model for several years. However, based on emerging biochemical evidence for dysfunction of brain neurotransmitters in Lesch–Nyhan patients (Lloyd et al. 1981), the animal model was revived. These observations in patients with Lesch–Nyhan disease were consistent with the Breese et al. (1984) publication on self-mutilation associated with perinatal depletion of dopamine with 6-OHDA. Later studies went on to show that patients with Lesch–Nyhan disease have abnormally few dopaminergic nerve terminals and cell bodies as assessed by positron emission tomography (Ernst et al. 1996). Compared with healthy control patients (n = 15), patients with Lesch–Nyhan disease (n = 12) had 31 % lower dopaminergic activity in the putamen, 39 % lower in the caudate nucleus, and 57 % lower in the substantia nigra and ventral tegmentum. These abnormalities were observed even in the youngest patients (range 10–20 years), a finding that led the investigators to conclude that the dopaminergic deficits in Lesch–Nyhan disease are developmental rather than degenerative in origin.

Together, the data from patients and the perinatal 6-OHDA model suggest that injury to dopaminergic systems early in life might underlie the self-mutilation that characterizes Lesch–Nyhan disease. Since its original development, this model has been used extensively to probe the neurobiological basis of the self-injury with a modified behavioral endpoint (first nick of skin due to biting, with immediate follow-on treatment with a dopamine D_1 receptor antagonist to prevent progression to abject self-injury (reviewed by Papadeas and Breese 2014).

Jinnah notes that no "ideal" model that replicates all features of Lesch–Nyhan disease has been developed (Jinnah 2009). Rather, the development of multiple models, each elucidating a specific aspect of this multifaceted disease and directed at answering specific research questions, has proven to be a more useful approach. Tissue culture models and the HPRT-deficient mouse model have shed light on the effects of HPRT deficiency on purine metabolism but have not helped to elucidate the basis of the neurobehavioral manifestations of the disease. The neonatal 6-OHDA model does not directly address purine metabolic effects but does help to elucidate the relationship between dopamine neuronal and basal ganglia dysfunc-tion and self-injurious behavior and provides a useful system for probing circuits, neurotransmission, interactions with stress, and preclinical pharmacology. While evidence from pharmacological, postmortem neurochemical, and imaging studies in humans and from animal models suggests that dopamine deficiency might underlie the self-injurious behavior across disorders (Devine 2012), a diversity of contributing mechanisms might better represent the disease. Thus, having multiple models, including the perinatal 6-OHDA model, is essential.

While Lesch–Nyhan disease is an orphan disease affecting just a few per million people, models of Lesch–Nyhan disease potentially shed light on the pathology and management of numerous disorders involving self-injury such as mental retardation and autism. In fact, the behavioral profile of Lesch–Nyhan disease and its animal models suggest that this condition is not only generalizable to the spectrum of neurobehavioral pathology in neurology and psychiatry, but may also constitute one of the most profound neuroadaptations in neuroscience. Thus, rare genetic disorders such as Lesch–Nyhan disease can have significant relevance to overall human health. The insights of these past decades of research into this disease have had general implications for neuroscience (e.g., Breese et al. 2005; Fu et al. 2014) all the while providing hope for elusive therapeutic options for patients with Lesch–Nyhan disease themselves.

4 Recent Developments

4.1 Clinical Science

Recent clinical research is consistent with the prenatal 6-OHDA model in continuing to support dopaminergic involvement in the neurobehavioral manifestations of Lesch–Nyhan disease. New clinical data also suggest an important role of epigenetics (e.g., Nguyen 2015; Trigueros Genao and Torres 2014).

4.1.1 Additional Support Consistent with Dopaminergic Involvement in Lesch–Nyhan Disease

Evidence consistent with the presence of basal ganglia dysfunction in Lesch–Nyhan disease comes from a case study in which a 29-year-old patient with refractory generalized severe dystonia and self-injurious behavior was treated with chronic bilateral globus pallidus internus deep brain stimulation (Piedimonte et al. 2015). During a 5-year follow-up period, self-injurious behavior ceased and dystonia as assessed by the Burke–Fahn–Marsden Dystonia Rating Scale and the Mean Disability Scale significantly improved. In another case report involving chronic bilateral globus pallidus internus deep brain stimulation, dystonia and self-mutilation in a 15-year-old boy ceased over several weeks after initiation of stimulation (Abel et al. 2014). After a right lead fracture, the dystonia and self-mutilation returned on the left side of the body only. These results suggest that these neurobehavioral manifestations of Lesch–Nyhan disease are lateralized. Assessment of potential laterality in the perinatal 6-OHDA model is warranted.

In a study investigating the basis of the dopamine deficiency in Lesch–Nyhan disease, substantia nigral neurons in 5 Lesch–Nyhan patients compared with 6 controls at autopsy were found to have reduced melanization and reduced tyrosine hydroxylase immunoreactivity but no signs of a degenerative process (Göttle et al. 2014). The authors suggested that the neurochemical phenotype of Lesch–Nyhan disease is not associated with neurodegeneration.

Reductions in gray matter and white matter in basal ganglia and other brain regions were observed in Lesch–Nyhan disease as revealed by observational studies using voxel-based morphometry in 21 patients with classic Lesch–Nyhan disease; 17 with Lesch–Nyhan variant disease; and 33 age-, sex-, and race-matched healthy controls (Schretlen et al. 2013, 2015). Reductions in volume versus controls were greater in patients with classic Lesch–Nyhan than in patients with Lesch–Nyhan variant disease.

Intrathecal baclofen was associated with reductions in dystonia, cessation of self-injurious behavior, and improvement in sleep that persisted through 5–16 months of follow-up in 3 patients with Lesch–Nyhan disease (Pozzi et al. 2014). The authors speculated that the results might be explained by an interaction between baclofen and dopamine complemented by baclofen-associated anxiolysis.

4.1.2 Role of Epigenetics

Lesch–Nyhan disease has been diagnosed in patients lacking an HPRT mutation, and family members with the same genetic defect can manifest with differing severities of Lesch–Nyhan disease (Trigueros Genao and Torres 2014; Ceballos-Picot et al. 2013). Epigenetic processes (that affect gene expression without affecting DNA sequence) might account for this clinical variability (Trigueros Genao and Torres 2014).

Consistent with this possibility, Nguyen (2015) recently found a wide range of amyloid precursor protein mRNA isoforms in Lesch–Nyhan patients, the implications of which are unknown, yet suggest further complexity in the full neurobiological manifestations underlying the disease. Relatedly, Kang and Friedmann (2015) recently reported genetic dysregulations common to both Alzheimer's and Lesch–Nyhan disease so the intriguing question arises as to the potential epigenetic mechanisms underlying both.

4.2 Basic Science

In recent studies, neuronal perturbations in the HPRT-deficient mouse extended outside the dopamine system to the histamine system where 1-methylhistamine and 1-methylimidazole-4-acetic acid were found to be reduced in multiple regions of brain (Tschirner et al. 2015). Whether histamine-based manipulations into the perinatal 6-OHDA self-injury might be effective or whether therapeutic interventions could result from these new observations is unknown. In gene screening research, Torres and Puig (2015) demonstrated that HPRT actions on transcription factor genes critical to neuronal differentiation led to deregulated WNT4 from the WNT/β-caterin pathway, engrailed homeobox 2 gene, and increased tyrosine hydroxylase, DRD1, and adenosine and HTR7 serotonin receptors. Relatedly, Kang et al. (2013) demonstrated that the HPRT gene regulates a host of developmental and metabolic pathways in embryonic stem cell neuronal differentiation. The implications of these findings have yet to be determined but again suggest broad impacts of HPRT beyond its expected role. These broad impacts have been suggested by Dammer et al. (2015) who examined the proteome profiles in an HPRT deficiency model and found extensive changes in protein expression depending on whether cells were differentiated or not. They further report that not all of the identified were related to purine recycling, a finding consistent with heretofore underappreciated mechanisms that may underlie Lesch–Nyhan disease. Comparable efforts to scan the transcriptome have led to similar findings in genes and gene clusters (Dauphinot et al. 2014). One could speculate that such unexpected changes could provide links to common behavioral phenotypes across models such as the perinatal 6-OHDA model and others.

5 Future Directions

In clinical care, palliative care around minimization of injury and continued use of allopurinol to limit the consequences of the HPRT deficiency will likely remain primary and important foci. In both the clinic and research, it has been observed that stress complicates the effective management of behavior in Lesch–Nyhan patients (e.g., Anderson and Ernst 1994). Thus, a role for stress-related circuitry is apparent. In this regard, understanding the neurobiological mechanisms of stress induction of behavioral pathology (e.g., in drug addiction or induction of clinical depression) could prove useful in targeting circuits as well. It is known that select circuits engaging the amygdala and its complex circuitry along with select neurochemical systems such as corticotropin-releasing factor, serotonin, GABA, and norepinephrine are integrally involved in stress effects in other neurobehavioral pathologies (e.g., Knapp et al. 2007, 2011a, b; Huang et al. 2010). While a full understanding of the contributions of these and other systems to behavioral pathology generally is not available, it is clear that Lesch–Nyhan disease is a chronic stressful condition for patients and should be considered further. The underappreciated stress component in the animal models could also be influencing the course of the neurobiological adaptations (e.g., those mediated by corticosterone, corticotropin-releasing hormone, or others) that in turn elicit differential behavioral sensitivity. In a pemoline-induced self-injury model, for example, stress worsened the severity of injury (Muehlmann et al. 2012). Further, Stodgell et al. (1998) showed that footshock stress exacerbated self-injurious behavior in the perinatal 6-OHDA model of Lesch–Nyhan disease.

On other fronts, new technologies such as CRISPR9, while in early phases, have captured the attention of neuroscience and biology in general. It is hoped that such technologies will help hasten the elucidation of the role of genes in pathology through insertion/deletion of genes in whole animal or tissue systems. MicroRNAs, too, may help to advance Lesch–Nyhan research. Across the neurobiological spectrum, research is accelerating toward the discovery of functional microRNAs as players in the course of neurobehavioral adaptation and behavioral responsivity. With regard to Lesch–Nyhan disease, Guibinga (2015) has recently proposed a model that accommodates the idea that mRNA transcripts of the HPRT gene can exert pleiotropic effects across a variety of genes and signaling pathways that expand the suspected actions of this gene (or gene system) beyond its well-known basic metabolic housekeeping functions. Specifically, Guibinga (2015) proposes that competitive endogenous RNAs (ceRNAs) are engaged and may regulate cross talk between neural transcripts and miRNAs. A focus on genes and gene screening more generally may provide new clues to the molecular and cellular processing underlying Lesch–Nyhan disease. Moreover, differential profiles of gene expression over developmental windows post-lesion are in all likelihood critical in the course of neuroadaptations in animal models. While select genes have been examined (e.g., Torres and Puig 2015), the most prominent being the purine metabolic gene encoding the HPRT enzyme itself, comprehensive gene expression profiling across brain regions thought to be engaged in the neurobehavioral pathology of this disease has not been widely reported. Given the critical narrow window of time in which inducing the lesion will ultimately come to elicit the behavioral sensitivity in rats that models Lesch–Nyhan disease, gene screening techniques could identify gene clusters and gene expression profiles that are prominent in the underlying adaptations. Profiling soon after the perinatal 6-OHDA-induced lesion would reveal one pattern of gene expression, while profiling after pharmacological induction of hyperactivity and/or self-injurious behavior in older lesioned animals could reveal another. Both approaches could focus attention on new targets not only for therapeutics, but also for research on relevant mechanistic etiologies in this developmentally dependent pathology.

Another avenue of research that could impact both self-injury in Lesch–Nyhan disease and in other conditions characterized by self-injury is the application of contemporary technologies pertaining to chemogenetic and/or optogenetic manipulations of circuits as well as functional imaging of the consequences of these interventions (Deisseroth 2015). It is clear that circuit dysfunction is paramount in self-injury, but the nature of this dysfunction is not known. Classical thinking about dopamine D_1 or D_2 receptor pathways from the striatum has long been instructive (e.g., Surmeier et al. 2007). Assessing them directly via tracing studies, or via exciting or silencing select neuronal populations in neurobehavioral circuits suspected of regulating locomotor behavior or self-injury, constitutes an important opportunity. Combined with fMRI in rats, the optogenetic or chemogenetic approach can help to isolate and control maladaptive circuits upon which self-injurious behavior depends, and lead to discoveries of other regions, and differential connectivities among regions, that may not have been suspected.

Adeno-associated viruses (AAVs) manufactured to deliver and functionally express inhibitory halorhodopsin or excitatory channel rhodopsin to select neuronal phenotypes (e.g., subtypes of GABA output neurons of the striatum that project differentially throughout the motor integration and output circuitry) would be central players in this effort. Myriad variants of these viruses provide neuroscientists with an incredible array of tools to probe the necessity or sufficiency of select neuronal groups in specific brain regions in mediating behavioral phenotypes (Murlidharan et al. 2014; Weinberg et al. 2013). These tools could readily be employed to probe the nature of the circuit dysfunction that regulates self-injury in the perinatal 6-OHDA model. Moreover, parallel employment of these tools (or related ones described below) to the HPRT model may well help isolate common pathways and mechanisms that lead to self-injury.

Relatedly, with the advent of DREADD (Designer Receptors Exclusively Activated by Designer Drugs) (e.g., Lee et al. 2014; Urban and Roth 2015), it is now possible to modify existing brain circuits through custom-designed viral systems to respond to either peripherally or centrally administered compounds such as the designer drug clozapine-N-Oxide (CNO), which on its own has no effects in brain. New-generation DREADD-type tools are being designed to increase the sophistication and control that experimenters have in circuit discovery. Further, it is important to remember that the loss of dopamine during the development in the animal model results in a maladaptive neurobiological condition that must differ from that seen in Parkinson's disease and models thereof as the behavioral profiles of the two conditions are very different. Thus, a persistent maladaptive circuit function arising over the course of development post-6-OHDA lesion must be present, at least somewhat unique to Lesch–Nyhan disease, and should be approachable with employment of these novel tools in the model. Employing these tools in the perinatal 6-OHDA model of Lesch–Nyhan disease and associated disorders could lead to a more refined, heuristically valuable, and therapeutically relevant understanding of the behavioral components of this devastating disease.

References

Abel TJ, Dalm BD, Grossbach AJ, Jackson AE, Thomsen T, Greenlee JD (2014) Lateralized effect of pallidal stimulation on self-mutilation in Lesch-Nyhan disease. J Neurosurg Pediatr 14:594–597

Allen SM, Rice SN (1996) Risperidone antagonism of self-mutilation in a Lesch-Nyhan patient. Prog Neuropsychopharmacol Biol Psychiatry 20:793–800

Anderson LT, Ernst M (1994) Self-injury in Lesch-Nyhan disease. J Autism Dev Disord 24(1):67–81

Bitler CM, Howard BD (1986) Dopamine metabolism in hypoxanthine-guanine phosphoribosyltransferase-deficient variants of PC12 cells. J Neurochem 47:107–112

Breese GR, Baumeister AA, McCown TJ, Emerick SG, Frye GD, Mueller RA (1984) Neonatal-6-hydroxydopamine treatment: model of susceptibility for self-mutilation in the Lesch-Nyhan syndrome. Pharmacol Biochem Behav 21(3):459–461

Breese GR, Knapp DJ, Criswell HE, Moy SS, Papadeas ST, Blake BL (2005) The neonate-6-hydroxydopamine-lesioned rat: a model for clinical neuroscience and neurobiological principles. Brain Res Rev 48(1):57–73

Breese GR, Traylor TD (1970) Effect of 6-hydroxydopamine on brain norepinephrine and dopamine: evidence for selective degeneration of catecholamine neurons. J Pharmacol Exp Ther 174:413–420

Breese GR, Traylor TD (1972) Developmental characteristics of brain catecholamines and tyrosine hydroxylase in the rat: effects of 6-hydroxydopamine. Br J Pharmcol 44:210–222

Ceballos-Picot I, Auge F, Fu R, Olivier-Bandini A, Cahu J, Chabrol B, Aral B, de Martinville B, Lecain JP, Jinnah HA (2013) Phenotypic variation among seven members of one family with deficiency of hypoxanthine-guanine phosphoribosyltransferase. Mol Genet Metab 110:268–274

Dammer EB, Göttle M, Duong DM, Hanfelt J, Seyfried NT, Jinnah HA (2015) Consequences of impaired purine recycling on the proteome in a cellular model of Lesch-Nyhan disease. Mol Genet Metab 114(4):570–579

Dauphinot L, Mockel L, Cahu J, Jinnah HA, Ledroit M, Potier MC, Ceballos-Picot I (2014) Transcriptomic approach to Lesch-Nyhan disease. Nucleosides, Nucleotides Nucleic Acids 33(4–6):208–217

Deisseroth K (2015) Optogenetics: 10 years of microbial opsins in neuroscience. Nat Neurosci 18(9):1213–1225

Devine DP (2012) Animal models of self-injurious behavior: an overview. Methods Mol Biol 829:65–84

Engle SJ, Womer DE, Davies PM, Boivin G, Sahota A, Simmonds HA, Stambrook PJ, Tischfield JA (1996) HPRT-APRT-deficient mice are not a model for Lesch-Nyhan syndrome. Hum Mol Genet 5:1607–1610

Ernst M, Zametkin AJ, Matochik JA, Pascualvaca D, Jons PH, Hardy K, Hankerson JG, Doudet DJ, Cohen RM (1996) Presynaptic dopaminergic deficits in Lesch-Nyhan disease. N Engl J Med 334(24):1568–1572

Fu R, Ceballos-Picot I, Torres RJ, Larovere LE, Yamada Y, Nguyen KV, Hegde M, Visser JE, Schretlen DJ, Nyhan WL, Puig JG, O'Neill PJ, Jinnah HA, Lesch-Nyhan Disease International Study Group (2014) Genotype-phenotype correlations in neurogenetics: Lesch-Nyhan disease as a model disorder. Brain 137:1282–1303

Göttle M, Prudente CN, Fu R, Sutcliffe D, Pang H, Cooper D, Veledar E, Glass JD (2014) Loss of dopamine phenotype among midbrain neurons in Lesch-Nyhan disease. Ann Neurol 76(1):95–107

Guibinga GH (2015) MicroRNAs: tools of mechanistic insights and biological therapeutics discovery for the rare neurogenetic syndrome Lesch-Nyhan disease (LND). Adv Genet 90:103–131

Hooper M, Hardy K, Handyside A, Hunter S, Monk M (1987) HPRT-deficient (Lesch-Nyhan) mouse embryos derived from germline colonization by cultured cells. Nature 326:292–295

Huang MM, Overstreet DH, Knapp DJ, Angel R, Wills TA, Navarro M, Rivier J, Vale W, Breese GR (2010) Corticotropin-releasing factor (CRF) sensitization of ethanol withdrawal-induced anxiety-like behavior is brain site specific and mediated by CRF-1 receptors: relation to stress-induced sensitization. J Pharmacol Exp Ther 332(1):298–307

Jinnah HA (2009) Lesch-Nyhan disease: from mechanism to model and back again. Dis Model Mech 2:116–121

Jinnah HA, Breese GR (1997) Anmal models for Lesch-Nyhan disease. In: Iannocconne PM, Scarpelli DG (eds) Biological aspects of disease: contributions from animcal models. Harwood Academic Publishers, Amersterdam, pp 94–143

Jinnah HA, Friedmann T (2001) Lesch-Nyhan disease and its variants. In: Scriver CR, Beaudet AL, Sly WS, Valle D (eds) The metabolic and molecular bases of inherited disease. McGraw Hill, New York, pp 2537–2570

Jinnah HA, Hess EJ, Wilson MC, Gage FH, Friedmann T (1992) Localization of hypoxanthine-guanine phosphoribosyltransferase mRNA in the mouse brain by in situ hybridization. Mol Cell Neurosci 3:64–78

Jinnah HA, Page T, Friedmann T (1993) Brain purines in a genetic mouse model of Lesch-Nyhan disease. J Neurochem 60:2036–2045

Jinnah HA, Yitta S, Drew T, Kim BS, Visser JE, Rothstein JD (1999a) Calcium channel activation and self-biting in mice. Proc Nat Acad Sci USA 96:15228–15232

Jinnah HA, Jones MD, Wojcik BE, Rothstein JD, Hess EJ, Friedmann T, Breese GR (1999b) Influence of age and strain on striatal dopamine loss in a genetic mouse model of Lesch-Nyhan disease. J Neurochem 72:225–229

Kang TH, Friedmann T (2015) Alzheimer's disease shares gene expression aberrations with purinergic dysregulation of HPRT deficiency (Lesch-Nyhan disease). Neurosci Lett 590:35–39

Kang TH, Park Y, Bader JS, Friedmann T (2013) The housekeeping gene hypoxanthine guanine phosphoribosyltransferase (HPRT) regulates multiple developmental and metabolic pathways of murine embryonic stem cell neuronal differentiation. PLoS ONE 8(10):e74967

Knapp DJ, Overstreet DH, Angel RA, Navarro M, Breese GR (2007) The amygdala regulates the antianxiety sensitization effect of flumazenil during repeated chronic ethanol or repeated stress. Alcohol Clin Exp Res 31(11):1872–1882

Knapp DJ, Whitman BA, Wills TA, Angel RA, Overstreet DH, Criswell HE, Ming Z, Breese GR (2011a) Cytokine involvement in stress may depend on corticotrophin releasing factor to sensitize ethanol withdrawal anxiety. Brain Behav Immun 25(Suppl 1):S146–S154

Knapp DJ, Overstreet DH, Huang M, Wills TA, Whitman BA, Angel RA, Sinnett SE, Breese GR (2011b) Effects of a stressor and corticotrophin releasing factor on ethanol deprivation-induced ethanol intake and anxiety-like behavior in alcohol-preferring P rats. Psychopharmacology 218 (1):179–189

Kuehn MR, Bradley A, Robertson EJ, Evans MJ (1987) A potential animal model for Lesch-Nyhan syndrome through introduction of HPRT mutations into mice. Nature 326:295–298

Lee HM, Giguere PM, Roth BL (2014) DREADDs: novel tools for drug discovery and development. Drug Discov Today 19(4):469–473

Lewers JC, Ceballos-Picot I, Shirley TL, Mockel L, Egami K, Jinnah HA (2008) Consequences of impaired purine recycling in dopaminergic neurons. Neuroscience 152:761–772

Lloyd KG, Hornykiewicz O, Davidson L, Shannak K, Farley I, Goldstein M, Shibuya M, Kelley WN, Fox IH (1981) Biochemical evidence of dysfunction of brain neurotransmitters in the Lesch-Nyhan syndrome. N Engl J Med 305(19):1106–1111

Muehlmann AM, Kies SD, Turner CA, Wolfman S, Lewis MH, Devine DPM (2012) Self-injurious behaviour: limbic dysregulation and stress effects in an animal model. J Intellect Disabil Res 56(5):490–500

Murlidharan G, Samulski RJ, Asokan A (2014) Biology of adeno-associated viral vectors in the central nervous system. Front Mol Neurosci 7:76

Nguyen KV (2015) Epigenetic regulation in amyloid precursor protein with genomic rearrangements and the Lesch-Nyhan syndrome. Nucleosides, Nucleotides Nucleic Acids 34(10):674–690

Papadeas ST, Breese GR (2014) 6-Hydroxydopamine lesioning of dopamine neurons in neonatal and adult rats induces age-dependent consequences. In: Kostrzewa RM (ed) Handbook of neurotoxicity. Springer, New York, pp 133–198

Piedimonte F, Andreani JC, Piedimonte L, Micheli F, Graff P, Bacaro V (2015) Remarkable clinical improvement with bilateral globus pallidus internus deep brain stimulation in a case of Lesch-Nyhan disease: five-year follow-up. Neuromodulation 18:118–122

Pozzi M, Piccinini L, Gallo M, Motta F, Radice S, Clementi E (2014) Treatment of motor and behavioural symptoms in three Lesch-Nyhan patients with intrathecal baclofen. Orphanet J Rare Dis 9:208

Saito Y, Ito M, Hanaoka S, Ohama E, Akaboshi S, Takashima S (1999) Dopamine receptor upregulation in Lesch-Nyhan syndrome: a postmortem study. Neuropediatrics 30:66–71

Schretlen FJ, Varvaria M, Te Ho, Vannorsdall TD, Gordon B, Harris JC, Jinnah HA (2013) Regional brain volume abnormalities in Lesch-Nyhan disease and its variants: a cross-sectional study. Lancet Neurol 12:1151–1158

Schretlen DJ, Varvaris M, Vannorsdall TD, Gordon B, Harris JC, Jinnah HA (2015) Brain white matter volume abnormalities in Lesch-Nyhan disease and its variants. Neurology 84:190–196

Shirley TL, Lewers JC, Egami K, Majumdar A, Kelly M, Ceballos-Picot I, Seidman MM, Jinnah HA (2007) A human neuronal tissue culture model for Lesch-Nyhan disease. J Neurochem 101:841–853

Stodgell CJ, Loupe PS, Schroeder SR, Tessel RE (1998) Cross-sensitization between footshock stress and apomorphine on self-injurious behavior and neostriatal catecholamines in a rat model of Lesch-Nyhan syndrome. Brain Res 783(1):10–18

Surmeier DJ, Ding J, Day M, Wang Z, Shen W (2007) D_1 and D_2 dopamine-receptor modulation of striatal glutamatergic signaling in striatal medium spiny neurons. Trends Neurosci 30(5):228–235

Torres RJ, Puig JG (2015) Hypoxanthine deregulates genes involved in early neuronal development: implications in Lesch-Nyhan disease pathogenesis. J Inherit Metab Dis 38:1109–1118

Trigueros Genao M, Torres RJ (2014) From genotype to phenotype; clinical variability in Lesch-Nyhan disease. The role of epigenetics. Rev Clin Esp 214:461–465

Tschirner SK, Gutzki F, Kaever V, Seifert R, Schneider EH (2015) Altered histamine neurotransmission in HPRT-deficient mice. Neurosci Lett 609:74–80

Urban DJ, Roth B (2015) DREADDs (designer receptors exclusively activated by designer drugs): chemogenetic tools with therapeutic utility. Annu Rev Pharmacol Toxicol 55:399–417

Weinberg MS, Samulski RJ, McCown TJ (2013) Adeno-associated virus (AAV) gene therapy for neurological disease. Neuropharmacology 69:82–88

Wong DF, Harris JC, Naidu S, Yokoi F, Marenco S, Dannals RF, Ravert HT, Yaster M, Evans A, Rousset O et al (1996) Dopamine transporters are markedly reduced in Lesch-Nyhan disease in vivo. Proc Nat Acad Sci USA 93:5539–5543

Yeh J, Zheng S, Howard BD (1998) Impaired differentiation of HPRT-deficient dopaminergic neurons: a possible mechanism underlying neuronal dysfunction in Lesch-Nyhan syndrome. J Neurosci Res 53:78–85

Perinatal 6-Hydroxydopamine Modeling of ADHD

John P. Kostrzewa, Rose Anna Kostrzewa, Richard M. Kostrzewa, Ryszard Brus and Przemysław Nowak

Abstract The neonatally 6-hydroxydopamine (n6-OHDA)-lesioned rat has been the standard for 40 years, as an animal model of attention-deficit hyperactivity disorder (ADHD). Rats so lesioned during postnatal ontogeny are characterized by ~99 % destruction of dopaminergic nerves in pars compacta substantia nigra, with comparable destruction of the nigrostriatal tract and lifelong ~99 % dopaminergic denervation of striatum, with lesser destructive effect on the ventral tegmental nucleus and associated lesser dopaminergic denervation of nucleus accumbens and prefrontal cortex. As a consequence of striatal dopaminergic denervation, reactive serotoninergic hyperinnervation of striatum ensues. The striatal extraneuronal milieu of DA and serotonin is markedly altered. Also, a variety of sensitization changes occur for dopaminergic D_1 and D_2 receptors, and for serotoninergic receptors. Behaviorally, these rats in adulthood display spontaneous hyperlocomotor activity, attentional deficits, and cognitive impairment—all of which are acutely attenuated by the psychostimulants amphetamine (AMPH) and methylphenidate (MPH) (i.e., opposite to the acute effects of AMPH and MPH in intact control rats). The acute behavioral effects of AMPH and MPH in intact and lesioned rats are analogous to their respective acute effects in non-ADHD and in ADHD humans. The neurochemical template of brain, and behavioral series of

J.P. Kostrzewa
North Alabama ENT Associates, Huntsville, AL 35801, USA

R.A. Kostrzewa
Walgreen's Pharmacy, Colonial Heights, TN 37663, USA

R.M. Kostrzewa (✉)
Department of Biomedical Sciences, Quillen College of Medicine,
East Tennessee State University, 70577, Johnson City, TN 37614, USA
e-mail: kostrzew@etsu.edu

R. Brus
Department of Nurse, High School of Strategic Planning, Koscielna 6,
41-303 Dabrowa Gornicza, Poland

P. Nowak
Public Health Faculty, Department of Toxicology and Occupational Health Protection,
Medical University of Silesia, Medykow 18, 40-752 Katowice Ligota, Poland

© Springer International Publishing Switzerland 2015
Curr Topics Behav Neurosci (2016) 29: 279–294
DOI 10.1007/7854_2015_397
Published Online: 17 October 2015

changes in n6-OHDA-lesioned rats, is described in the review. Despite the fact that nigrostriatal damage is not an underlying pathophysiological process of human ADHD (i.e., lacking construct validity), the described animal model has face validity (behavioral profile) and predictive validity (mirror of ADHD/MPH effects, as well as putative and new ADHD treatment effects). Also described in this review is a modification of the n6-OHDA rat, produced by adulthood partial lesioning of the serotoninergic fiber overgrowth. This ADHD model has even more accentuated hyperlocomotor and attentional deficits, counteracted by AMPH—thus providing a more robust means of animal modeling of ADHD. The n6-OHDA rat as a model of ADHD continues to be important in the search for new ADHD treatments.

Keywords Attention-deficit hyperactivity disorder · ADHD · Hyperlocomotor activity · 6-hydroxydopamine · 6-OHDA · Amphetamine · Methylphenidate

Contents

1 Introduction 280
2 Features of the n6-OHDA-Lesioned Rat Model of ADHD 281
   2.1 Dopaminergic Striatal Denervation in n6-OHDA Rats 281
   2.2 Dopamine D_1-R and D_2-R Status and Sensitization in n6-OHDA Rats 282
   2.3 Serotoninergic Striatal Hyperinnervation in n6-OHDA Rats 282
   2.4 Serotonin Receptor Status and Sensitization in n6-OHDA Rats 283
   2.5 Behavioral Status of n6-OHDA Rats 283
   2.6 Hyperactivity in n6-OHDA Rats: Effects of AMPH and MPH 284
3 Accentuated Hyperactivity in n6-OHDA Rats by Adulthood 5,7-DHT
   (a5,7-DHT) Lesioning 284
4 Comparison of n6-OHDA Rats with Other Animal Models of ADHD 285
5 Conclusions 286
References 287

1 Introduction

The neurotoxin 6-hydroxydopamine (6-OHDA), having high affinity to dopamine (DA) transporters (DATs) and norepinephrine (NE) transporters (NETs) (Thoenen and Tranzer 1968; Ljungdahl et al. 1971; Saner and Thoenen 1971), is accumulated with relative selectivity by dopaminergic and noradrenergic neuronal phenotypes. Inside the nerve 6-OHDA auto-oxidizes to a series of catechol analogs including self-perpetuating 6-OHDA-semiquinone, all of which evoke cellular reactive oxygen species (ROS) formation. Additionally, 6-OHDA and auto-oxidized analogs inhibit mitochondrial complex I, to further generate hydrogen peroxide, superoxide, and hydroxyl radical (HO·) (Heikkila and Cohen 1971, 1972a, b, 1973; Heikkila et al. 1973; Cohen and Heikkila 1974) and impair ATP production (Senoh and Witkop 1959a, b; Senoh et al. 1959; Graham et al. 1978; Bindoli et al. 1999; Segura-Aguilar 2001; Segura-Aguilar and Paris 2014). This cascade of events overwhelms neuronal antioxidant capacity and depletes energy stores, resulting in overt destruction of dopaminergic and noradrenergic nerves. Selectivity for dopaminergic nerves can be attained by coadministering 6-OHDA with the NET inhibitor, desipramine (Smith et al. 1973).

When high-dose 6-OHDA is administered centrally to adult rats, 6-OHDA bilateral destruction of dopaminergic nerves produces a series of behavioral deficits including akinesia. In marked contrast, when 6-OHDA is administered centrally to neonatal rats (n6-OHDA), there is likewise profound destruction of dopaminergic nerves, but with few observable behavioral deficits (see Papadeas and Breese 2014; Kostrzewa et al. 2015). However, n6-OHDA-lesioned rats in adulthood display spontaneous hyperlocomotor activity (Breese et al. 1975; Shaywitz et al. 1976a, b) with cognitive deficits (Cooper et al. 1973; Archer et al. 1988), mirroring to some extent, human attention-deficit hyperactivity disorder (ADHD). Significantly, amphetamine (AMPH) and methylphenidate (MPH) attenuate the behavioral abnormalities, analogous to their respective effects in human ADHD (Shaywitz et al. 1976a, b). Over the past 40 years, n6-OHDA-lesioned rats have become the most commonly used animal model of ADHD.

2 Features of the n6-OHDA-Lesioned Rat Model of ADHD

2.1 Dopaminergic Striatal Denervation in n6-OHDA Rats

Behavioral hyperactivity occurs in adult rats that were n6-OHDA-lesioned, if DA depletion is greater than 55 % (Shaywitz et al. 1976b). Typically, high-dose 6-OHDA is used, resulting in permanent and near-complete destruction of the ~7000 pars compacta substantia nigra neurons (Berger et al. 1985; Descarries et al. 1992), resulting in near-total dopaminergic denervation of striatum (Doucet et al. 1986; Snyder et al. 1986). Dopaminergic perikarya in the ventral tegmental nucleus are largely spared from n6-OHDA destruction (Snyder et al. 1986; Fernandes Xavier et al. 1994), although there is partial dopaminergic denervation of septum, nucleus accumbens, and frontal cortex (Luthman et al. 1990a). The n6-OHDA-lesioned rat has been extensively characterized neurochemically and behaviorally and in terms of dopaminergic sensitization status (Breese and Traylor 1972; Breese et al. 1973, 1984a, b, 1985a, b, 1987).

2.2   Dopamine D_1-R and D_2-R Status and Sensitization in n6-OHDA Rats

DA D_1-R number is little affected in adulthood in n6-OHDA-lesioned rats (Breese et al. 1987a, b; Broaddus and Bennett 1990; Luthman et al. 1990b), and there is no change in the numbers of high- and low-affinity D_1-R (Gong et al. 1994). DA D_2-R number is reported to be unchanged when assessed in vitro by [^{3}H]spiperone binding (Breese et al. 1987; Duncan et al. 1987; Kostrzewa and Brus 1991; Kostrzewa and Hamdi 1991), but is found to be increased when assessed in vitro by [^{3}H]raclopride binding (Dewar et al. 1990; Radja et al. 1993b). Also, there is an increase in the D_4-R subclass of D_2-R in the striatum of rats in which spontaneous hyperlocomotor activity is observed, and D_4-R antagonists attenuate the hyperlocomotor activity (Zhang et al. 2002a, b).

The initial treatment with a D_1-R agonist produces an elevation in the number of vacuous chewing movements (i.e., oral activity) in n6-OHDA rats versus that of intact control rats, indicating overt D_1-R supersensitization in n6-OHDA rats (Kostrzewa and Gong 1991; Gong et al. 1993a). However, for most stereotypic and locomotor effects, repeated D_1-R agonist treatments are required before an enhanced effect is observed, indicating that there is "latent" sensitization of this pool of D_1-R which must be "primed" in order for sensitization to occur (Breese et al. 1985a, 1987; Hamdi and Kostrzewa 1991; Gong et al. 1993a). D_1-R supersensitization is not associated with an change in the overall D_1-R number (Hamdi and Kostrzewa 1991), nor in numbers of D_1-R high- or low-affinity receptors (Gong et al. 1994).

While repeated D_2-R agonist treatments do not sensitize for classic stereotypic and locomotor effects in adult n6-OHDA rats (Criswell et al. 1989; Brus et al. 2003), repeated pre-weanling D_2-R agonist treatments evoke profound enhancement of vertical jumping in n6-OHDA rats during the third and fourth weeks post-birth (Kostrzewa and Kostrzewa 2012). The status of D_1-R and D_2-R supersensitization in n6-OHDA rats has been reviewed (Kostrzewa 1995).

2.3   Serotoninergic Striatal Hyperinnervation in n6-OHDA Rats

Perinatal destruction of dopaminergic neurons by 6-OHDA is the initiating factor for reactive sprouting of serotoninergic innervation of the striatum (Breese et al. 1984a; Stachowiak et al. 1984; Snyder et al. 1986; Luthman et al. 1987; Towle et al. 1989). In adult n6-OHDA rats, there is a doubling of the numbers of serotoninergic fibers innervating rostral striatum (Descarries et al. 1992, 1995; Molina-Holgado et al. 1994; Soucy et al. 1994; Mrini et al. 1995) and a 20 % increase in caudal striatum versus intact controls (Mrini et al. 1995).

Reactive serotoninergic fiber sprouting, arising from dorsal and ventral raphe nuclei (Snyder et al. 1986; Descarries et al. 1992), is initiated when striatal DA is depleted by >80 % (Towle et al. 1989; Gong et al. 1993b) within the first ten days after birth (Kostrzewa et al. 1993).

2.4   Serotonin Receptor Status and Sensitization in n6-OHDA Rats

In adult n6-OHDA rats, quantitative ligand-binding autoradiography has established that there is an approximate 30–40 % increase in the striatal number of $5\text{-HT}_{1B/2C/1D/1E/2A}$-R, with no change in 5-HT_{1A}-R. The change in 5-HT-R numbers extends in part to other brain regions, as well (Radja et al. 1993a). Also, in adult n6-OHDA rats, 5-HT_2-R, likely 5-HT_{2C}-R, is overtly supersensitized, as demonstrated by enhanced behavioral responses, versus controls to 5-HT_2-R agonist treatments (Gong and Kostrzewa 1992; Gong et al. 1992; Plech et al. 1995). 5-HT_2-R supersensitization, which occurs in n6-OHDA rats when striatal DA is reduced by ≥88 % and within the first 3 days post-birth (Gong et al. 1993b; Kostrzewa et al. 1993), is eliminated by partial serotoninergic fiber destruction with 5,7-dihydroxytryptamine (5,7-DHT) treatment, either perinatally (Brus et al. 1994) or in adulthood (Kostrzewa et al. 1994). Moreover, 5-HT-R supersensitization persists in n6-OHDA rats even when experimental conditions have been altered to prevent or eliminate DA-R supersensitization (Gong et al. 1993b; Kostrzewa et al. 1993).

2.5   Behavioral Status of n6-OHDA Rats

The overall behavioral status of n6-OHDA-lesioned rats is largely unaltered, so that in many ways, they are indistinguishable from unlesioned control rats (Kostrzewa et al. 2015). With specific laboratory testing, however, there are several definitive deficits, including lack of avoidance in aversive learning (Smith et al. 1973; Shaywitz et al. 1976b; Pappas et al. 1980; Raskin et al. 1983; Whishaw et al. 1987), which could relate to ADHD modeling in these rats. Spontaneous hyperlocomotor activity in adult n6-OHDA-lesioned rats is the outcome most related to ADHD modeling (Shaywitz et al. 1976a, b; see Kostrzewa et al. 2008). High-dose L-3,4-dihydroxyphenylalanine (L-DOPA) has the potential to produce self-injurious behavior in n6-OHDA rats (Breese et al. 1984b; 1985a, 1989, 1990a, b; Criswell et al. 1992), and n6-OHDA rats have been used to model Lesch-Nyhan syndrome (Breese et al. 1986, 1987b, 1990b, 2005; Breese and Breese 1998; Papadeas and Breese 2014; Knapp and Breese 2015). However, L-DOPA and other such test agents are not likely to be used in testing of n6-OHDA rats as animal models of ADHD. Consequently, this bizarre behavior is not likely to be a confound in n6-OHDA modeling of ADHD.

2.6   Hyperactivity in n6-OHDA Rats: Effects of AMPH and MPH

Breese et al. (1975) and Shaywitz et al. (1976a, b) first demonstrated that n6-OHDA-lesioned rats displayed adulthood spontaneous hyperlocomotor activity and attentional deficits. In n6-OHDA rats, it is now known that this hyperlocomotor activity is related to the extent of dopaminergic denervation (Miller et al. 1981) and perinatal age at the time of 6-OHDA lesioning (Erinoff et al. 1979). DA D_4-R number is transiently elevated in striatum and nucleus accumbens at postnatal day 25, but absents a change at postnatal days 37 and 60 (Zhang et al. 2001b, 2002a). Both AMPH, which evokes DA release, and MPH, a DAT ligand (i.e., inhibitor), suppress hyperlocomotor activity and allay attentional deficits (Shaywitz et al. 1976a, b; Miller et al. 1981; Heffner and Seiden 1982; Luthman et al. 1989). Also, atomoxetine, a NET ligand, alters hyperlocomotor activity and attentional deficits in n6-OHDA rats in an analogous manner (Moran-Gates et al. 2005). There are several reviews on n6-OHDA rats as animal models of ADHD (Davids et al. 2003; Breese et al. 2005; van der Kooij and Glennon 2007).

3   Accentuated Hyperactivity in n6-OHDA Rats by Adulthood 5,7-DHT (a5,7-DHT) Lesioning

In an attempt to negate the influence of serotoninergic hyperinnervation in n6-OHDA rats used to model ADHD, 5,7-DHT (pretreatment with desipramine and pargyline) was bilaterally administered intraventricularly at 10 weeks. This treatment reduced striatal 5-HT content to 2.7 nmol/g, similar to the intact control level of 2.32 nmol/g, but much less than the 3.84 nmol/g in the striatum of rats lesioned solely with n6-OHDA. In the group of rats lesioned with n6-OHDA+a5,7-DHT spontaneous locomotor activity at 16 weeks increased to ~ 550 s in a 600-s session versus ~ 125 s in n6-OHDA rats, and ~ 20 s in control rats. Moreover, acute AMPH treatment attenuated hyperlocomotor activity in the n6-OHDA group and in the n6-OHDA+a5,7-DHT group (Kostrzewa et al. 1994). This finding indicates that (1) serotoninergic hyperinnervation acts to suppress spontaneous locomotor activity in n6-OHDA rats, and (2) partial destruction of hyperinnervating serotoninergic fibers enhances spontaneous hyperlocomotor activity in n6-OHDA rats. Stereotypic effects among the treatment groups were variable. However, D_1-R agonist-induced VCM number was reduced after a5,7-DHT lesioning, while 5-HT$_2$-R agonist-induced VCM number was increased 2-fold after a5,7-DHT lesioning (Kostrzewa et al. 1994).

These findings thus indicate that n6-OHDA+a5,7-DHT rats represent a more robust animal model of ADHD than n6-OHDA rats (Kostrzewa et al. 1994). To determine the possible potency of 5-HT-R antagonists in abating m-chlorophenylpiperazine (mCPP) induction of locomotor activity in n6-OHDA+5,7-DHT rats, several

5-HT-R antagonists were tested. While the largely 5-HT$_3$-R antagonist phenylbiguanide had no effect on mCPP-induced locomotor activity, the largely 5-HT$_2$-R antagonist mianserin fully blocked the mCPP increase in locomotor activity—analogous to the AMPH effect in the n6-OHDA+5,7-DHT rats (Brus et al. 2004). This finding further supports the contention that striatal serotoninergic fibers influence (i.e., increase) locomotor activity in rats with near-complete striatal dopaminergic denervation, via an action at 5-HT$_2$-R (Brus et al. 2004).

In intact freely moving rats, the striatal in vivo microdialysate baseline level of DA was elevated nearly 2-fold in a5,7-DHT rats, but reduced in n6-OHDA rats and in n6-OHDA+5,7-DHT rats. In the latter group, the in vivo microdialysate (i.e., extraneuronal) level was reduced $\sim$99 %. The AMPH-induced increase in extraneuronal DA was observable in intact controls and markedly increased in a5,7-DHT rats. However, extraneuronal DOPAC was most increased in n6-OHDA+a5,7-DHT rats (Nowak et al. 2007). These findings support the association between AMPH-induced DA release (i.e., DA turnover = ↑ DOPAC) and AMPH suppression of hyperlocomotor activity. Enhancement of hyperlocomotor activity arising by a5,7-DHT lesioning in n6-OHDA rats indicates the superior advantage of such rats, versus n6-OHDA alone, for modeling ADHD (Kostrzewa et al. 1994).

4　Comparison of n6-OHDA Rats with Other Animal Models of ADHD

The n6-OHDA rat has suitable face validity, demonstrating hyperactivity and attentional deficits as is found in humans with ADHD. Also, DA D$_4$-R number is increased in the striatum of n6-OHDA rat, similar to what has been found in human ADHD (LaHoste et al. 1996; Smalley et al. 1999; Zhang et al. 2001a, b, 2002a, b), and abnormalities of the D$_4$-R are considered to be a risk factor for ADHD (LaHoste et al. 1996; Smalley et al. 1999; Eisenberg et al. 2000).

The n6-OHDA rat has excellent predictive validity, since AMPH, MPH, atomoxetine, and other agents that suppress hyperactivity and attentional deficits in human ADHD have like effects in the n6-OHDA rat. However, the n6-OHDA rat lacks construct validity, since perinatal dopaminergic denervation—a feature of the n6-OHDA rat—is not observed in human ADHD (Davids et al. 2003; Breese et al. 2005; Moran-Gates et al. 2005; Kostrzewa et al. 2008).

There are several genetic animal models of ADHD. ***Spontaneously hypertensive rat (SHR)***. The SHR displays hyperlocomotor activity in a novel environment and exhibits attentional and learning deficits (Sagvolden et al. 1992, 1998; Sagvolden 2000). These behavioral alterations are abated by AMPH and METH (Myers et al. 1982; Sagvolden et al. 1992). The SHR is second only to the n6-OHDA rats in frequency of use in modeling ADHD (see Davids et al. 2003). ***DAT knockout (DAT-KO) mouse.*** Hyperactivity and cognitive deficits are inherent in the DAT-KO mouse (Gainetdinov et al. 2001), with these effects being attenuated by AMPH and

MPH (Gainetdinov et al. 1998, 2001; Jones et al. 1998). There is a characteristic elevation in the striatal DA level in the DAT-KO mouse (Jones et al. 1998). *Coloboma mutant mouse.* This mouse model of ADHD with a dysfunctional abnormality of SNAP-25, a synaptically associated protein necessary for DA and NE exocytosis in dorsal striatum, displays spontaneous hyperlocomotor activity (Hess et al. 1992, 1996; Wilson 2000), with behavioral abnormalities being moderated by AMPH but not MPH (Hess et al. 1996). *Hypoxia in neonatal rats.* In rats rendered hypoxic at birth, hyperactivity and learning deficits occur in adulthood (Gramatté and Schmidt 1986)—effects attenuated by acute AMPH treatment (Speiser et al. 1983). *Ethanol in neonatal rats.* Rats exposed to ethanol during gestation display hyperactivity (Fahlke and Hansen 1999) and attentional deficits (Hausknecht et al. 2005), and acute suppression of these behavioral deficits by AMPH (Highfield et al. 1999). *Heavy metal exposure in neonatal rats.* In rats neonatally exposed to lead (Silbergeld and Goldberg 1975) or cadmium (Ruppert et al. 1985), both hyperactivity and attentional deficits are observed in adulthood, with AMPH and/or MPH acutely attenuating these behaviors. Additional neonatal exposure to ethanol exacerbates the behavioral abnormalities (Nowak et al. 2006). *Methylazoxymethanol in neonatal rats* results in adulthood hyperactivity (Archer et al. 1988). *Cerebellar injury in neonatal rats* also produces adulthood hyperactivity (Ferguson 1996). Additional animal models, infrequently used to model ADHD, have been enumerated (Davids et al. 2003).

5 Conclusions

The most commonly engaged animal model of ADHD is derived by 6-OHDA lesioning of rat pups, produced within the first three days of birth, and with approximately 90–99 % destruction of nigrostriatal dopaminergic neurons. As a consequence of dopaminergic denervation, there is a reactive serotoninergic hyperinnervation of striatum, an event that likely contributes to the signature behavioral abnormalities, notably adulthood hyperlocomotor activity with attentional and learning deficits which are acutely abated by AMPH or MPH—drugs typically used to treat human ADHD. Adulthood partial lesioning of serotoninergic fiber overgrowth of striatum in these rats does not attenuate the hyperactivity nor the responsiveness to AMPH. In fact, rats become even more hyperactive, thus representing a variation of the neonatally 6-OHDA-lesioned (n6-OHDA) rat, and another model of ADHD. This latter model has demonstrated usefulness of a serotoningergic 5-HT$_2$ receptor agonist in attenuating hyperactivity. The classic n6-OHDA rat is valuable in screening potential agents and drugs that are or can be useful in treating human ADHD, including the norepinephrine transporter inhibitor, atomoxetine. There are alternate animal models of ADHD, some being genetic variants such as the SHR which displays AMPH-reversible hyperactivity and attentional deficits. Other alternate animal models of ADHD are produced by perinatal lesioning of rodents with a variety of substances. However, because of the convenience of scheduling, low cost, and ease of producing perinatal 6-OHDA lesioning of rats—with no lethality and with little behavioral modification during perinatal development—and reliable production of adulthood hyperactivity and attentional deficits, the n6-OHDA-lesioned rat remains the most studied animal model of ADHD.

References

Archer T, Danysz W, Fredriksson A, Jonsson G, Luthman J, Sundström E, Teiling A (1988) Neonatal 6-hydroxydopamine-induced dopamine depletions: motor activity and performance in maze learning. Pharmacol Biochem Behav 31(2):357–364 PMID: 3149743

Berger TW, Kaul S, Stricker EM, Zigmond MJ (1985) Hyperinnervation of the striatum by dorsal raphe afferents after dopamine-depleting brain lesions in neonatal rats. Brain Res 336(2):354–358 PMID: 3924346

Bindoli A, Scutari G, Rigobello MP (1999) The role of adrenochrome in stimulating the oxidation of catecholamines. Neurotox Res 1(2):71–80 PMID: 12835103

Breese GR, Traylor TD (1972) Developmental characteristics of brain catecholamines and tyrosine hydroxylase in the rat: effects of 6-hydroxydopamine. Br J Pharmacol 44(2):210–222 PMID: 4148915

Breese GR, Smith RD, Cooper BR, Grant LD (1973) Alterations in consummatory behavior following intracisternal injection of 6-hydroxydopamine. Pharmacol Biochem Behav 1(3):319–328 PMID: 4775860

Breese GR, Cooper BR, Hollister AS (1975) Involvement of brain monoamines in the stimulant and paradoxical inhibitory effects of methylphenidate. Psychopharmacologia 44(1):5–10 PMID: 128026

Breese GR, Baumeister AA, McCown TJ, Emerick SG, Frye GD, Crotty K, Mueller RA (1984a) Behavioral differences between neonatal and adult 6-hydroxydopamine-treated rats to dopamine agonists: relevance to neurological symptoms in clinical syndromes with reduced brain dopamine. J Pharmacol Exp Ther 231(2):343–354 PMID: 6149306

Breese GR, Baumeister AA, McCown TJ, Emerick SG, Frye GD, Mueller RA (1984b) Neonatal-6-hydroxydopamine treatment: model of susceptibility for self-mutilation in the Lesch-Nyhan syndrome. Pharmacol Biochem Behav 21(3):459–461 PMID: 6436840

Breese GR, Baumeister A, Napier TC, Frye GD, Mueller RA (1985a) Evidence that D-1 dopamine receptors contribute to the supersensitive behavioral responses induced by L-dihydroxyphenylalanine in rats treated neonatally with 6-hydroxydopamine. J Pharmacol Exp Ther 235(2):287–295 PMID: 3932640

Breese GR, Napier TC, Mueller RA (1985b) Dopamine agonist-induced locomotor activity in rats treated with 6-hydroxydopamine at differing ages: functional supersensitivity of D-1 dopamine receptors in neonatally lesioned rats. J Pharmacol Exp Ther 234(2):447–455 PMID: 3926987

Breese GR, Mueller RA, Napier TC, Duncan GE (1986) Neurobiology of D_1 dopamine receptors after neonatal-6-OHDA treatment: relevance to Lesch-Nyhan disease. Adv Exp Med Biol 204:197–215 (Review. PMID: 2947429)

Breese GR, Duncan GE, Napier TC, Bondy SC, Iorio LC, Mueller RA (1987a) 6-hydroxydopamine treatments enhance behavioral responses to intracerebral microinjection of D_1- and D_2-dopamine agonists into nucleus accumbens and striatum without changing dopamine antagonist binding. J Pharmacol Exp Ther 240(1):167–176 PMID: 3100767

Breese GR, Duncan GE, Napier TC, Bondy SC, Iorio LC, Mueller RA (1987b) 6-hydroxydopamine treatments enhance behavioral responses to intracerebral microinjection of D1- and D2- dopamine agonists into nucleus accumbens and striatum without changing dopamine antagonist binding. J Pharmacol Exp Ther 240(1):167-176 PMID: 3100767

Breese GR, Criswell HE, Duncan GE, Mueller RA (1989) Dopamine deficiency in self-injurious behavior. Psychopharmacol Bull 25(3):353–357. (Review. Erratum in: Psychopharmacol Bull 1990;26(3):296. PMID: 2697009)

Breese GR, Criswell HE, Duncan GE, Mueller RA (1990a) A dopamine deficiency model of Lesch-Nyhan disease–the neonatal-6-OHDA-lesioned rat. Brain Res Bull 25(3):477–484 (Review. PMID: 2127238)

Breese GR, Criswell HE, Mueller RA (1990b) Evidence that lack of brain dopamine during development can increase the susceptibility for aggression and self-injurious behavior by influencing D_1-dopamine receptor function. Prog Neuropsychopharmacol Biol Psychiatry 14 Suppl:S65-S80 (Review. PMID: 1982973)

Breese CR, Breese GR (1998) The use of neurotoxins to lesion catecholamine-containing neurons tomodel clinical disorders: approach for defining adaptive neural mechanisms and role of neurotrophic factors in brain. In: Kostrzewa RM (ed) Highly selective neurotoxins: basic and clinical applications. Humana Press, Totowa NJ, pp 19–73

Breese GR, Knapp DJ, Criswell HE, Moy SS, Papadeas ST, Blake BL (2005) The neonate-6-hydroxydopamine-lesioned rat: a model for clinical neuroscience and neurobiological principles. Brain Res Brain Res Rev 48(1):57–73 (Review. PMID: 15708628)

Broaddus WC, Bennett JP Jr (1990) Postnatal development of striatal dopamine function. II. Effects of neonatal 6-hydroxydopamine treatments on D_1 and D_2 receptors, adenylate cyclase activity and presynaptic dopamine function. Brain Res Dev Brain Res 52(1–2):273–277 PMID: 2110042

Brus R, Kostrzewa RM, Perry KW, Fuller RW (1994) Supersensitization of the oral response to SKF 38393 in neonatal 6-hydroxydopamine-lesioned rats is eliminated by neonatal 5,7-dihydroxytryptamine treatment. J Pharmacol Exp Ther 268:231–237

Brus R, Kostrzewa RM, Nowak P, Perry KW, Kostrzewa JP (2003) Ontogenetic quinpirole treatments fail to prime for D_2 agonist-enhancement of locomotor activity in 6-hydroxydopamine-lesioned rats. Neurotoxicity Res 5(5):329–338

Brus R, Nowak P, Szkilnik R, Mikolajun U, Kostrzewa RM (2004) Serotoninergics attenuate hyperlocomotor activity in rats. Potential new therapeutic strategy for hyperactivity. Neurotox Res 6(4):317–325 PMID: 15545015

Cohen G, Heikkila RE (1974) The generation of hydrogen peroxide, superoxide radical, and hydroxyl radical by 6-hydroxydopamine, dialuric acid, and related cytotoxic agents. J Biol Chem 249(8):2447–2452 PMID: 4362682

Cooper BR, Breese GR, Grant LD, Howard JL (1973) Effects of 6-hydroxydopamine treatments on active avoidance responding: evidence for involvement of brain dopamine. J Pharmacol Exp Ther 185(2):358–370 PMID: 4703827

Criswell H, Mueller RA, Breese GR (1989) Priming of D_1-dopamine receptor responses: long-lasting behavioral supersensitivity to a D_1-dopamine agonist following repeated administration to neonatal 6-OHDA-lesioned rats. J Neurosci 9(1):125–133 PMID: 2521511

Criswell HE, Mueller RA, Breese GR (1992) Pharmacologic evaluation of SCH-39166, A-69024, NO-0756, and SCH-23390 in neonatal-6-OHDA-lesioned rats. Further evidence that self-mutilatory behavior induced by L-dopa is related to D_1 dopamine receptors. Neuropsychopharmacology 7(2):95–103 PMID: 1418306

Davids E, Zhang K, Tarazi FI, Baldessarini RJ (2003) Animal models of attention-deficit hyperactivity disorder. Brain Res Brain Res Rev 42(1):1–21 (Review. PMID: 12668288)

Descarries L, Soghomonian JJ, Garcia S, Doucet G, Bruno JP (1992) Ultrastructural analysis of the serotonin hyperinnervation in adult rat neostriatum following neonatal dopamine denervation with 6-hydroxydopamine. Brain Res 569(1):1–13 PMID: 1611468

Descarries L, Soucy JP, Lafaille F, Mrini A, Tanguay R (1995) Evaluation of three transporter ligands as quantitative markers of serotonin innervation density in rat brain. Synapse 21 (2):131–139 PMID: 8584974

Dewar KM, Soghomonian JJ, Bruno JP, Descarries L, Reader TA (1990) Elevation of dopamine D_2 but not D_1 receptors in adult rat neostriatum after neonatal 6-hydroxydopamine denervation. Brain Res 536(1–2):287–296 PMID: 2150772

Doucet G, Descarries L, Garcia S (1986) Quantification of the dopamine innervation in adult rat neostriatum. Neuroscience 19(2):427–445 PMID: 3095678

Duncan GE, Criswell HE, McCown TJ, Paul IA, Mueller RA, Breese GR (1987) Behavioral and neurochemical responses to haloperidol and SCH-23390 in rats treated neonatally or as adults with 6-hydroxydopamine. J Pharmacol Exp Ther 243(3):1027–1034 PMID: 3121842

Eisenberg J, Zohar A, Mei-Tal G, Steinberg A, Tartakovsky E, Gritsenko I, Nemanov L, Ebstein RP (2000) A haplotype relative risk study of the dopamine D_4 receptor (DRD4) exon III repeat polymorphism and attention deficit hyperactivity disorder (ADHD). Am J Med Genet 96(3):258–261 (Review. PMID: 10898895)

Erinoff L, MacPhail RC, Heller A, Seiden LS (1979) Age-dependent effects of 6-hydroxydopamine on locomotor activity in the rat. Brain Res 164:195–205 PMID: 427556

Fahlke C, Hansen S (1999) Alcohol responsiveness, hyperreactivity, and motor restlessness in an animal model for attention-deficit hyperactivity disorder. Psychopharmacology 146(1):1–9 PMID: 10485958

Ferguson SA (1996) Neuroanatomical and functional alterations resulting from early postnatal cerebellar insults in rodents. Pharmacol Biochem Behav 55(4):663–671 (Review. PMID: 8981598)

Fernandes Xavier FG, Doucet G, Geffard M, Descarries L (1994) Dopamine neoinnervation in the substantia nigra and hyperinnervation in the interpeduncular nucleus of adult rat following neonatal cerebroventricular administration of 6-hydroxydopamine. Neuroscience 59(1):77–87 PMID: 8190274

Gainetdinov RR, Jones SR, Fumagalli F, Wightman RM, Caron MG (1998) Re-evaluation of the role of the dopamine transporter in dopamine system homeostasis. Brain Res Brain Res Rev 26 (2–3):148-153 (Review. PMID: 9651511)

Gainetdinov RR, Mohn AR, Bohn LM, Caron MG (2001) Glutamatergic modulation of hyperactivity in mice lacking the dopamine transporter. Proc Natl Acad Sci USA 98 (20):11047–11054 PMID: 11572967

Gong L, Kostrzewa RM (1992) Supersensitized oral response to a serotonin agonist in neonatal 6-OHDA treated rats. Pharmacol Biochem Behav 41:621–623

Gong L, Kostrzewa RM, Fuller RW, Perry KW (1992) Supersensitization of the oral response to SKF 38393 in neonatal 6-OHDA-lesioned rats is mediated through a serotonin system. J Pharmacol Exp Ther 261:1000–1007

Gong L, Kostrzewa RM, Brus R, Fuller RW, Perry KW (1993a) Ontogenetic SKF 38393 treatments sensitize dopamine D_1 receptors in neonatal 6-OHDA-lesioned rats. Brain Res Dev Brain Res 76(1):59–65 PMID: 8306431

Gong L, Kostrzewa RM, Perry KW, Fuller RW (1993b) Dose-related effects of a neonatal 6-OHDA lesion on SKF 38393- and m-chlorophenylpiperazine-induced oral activity responses of rats. Brain Res Dev Brain Res 76(2):233–238 PMID: 8149589

Gong L, Kostrzewa RM, Li C (1994) Neonatal 6-hydroxydopamine and adult SKF 38393 treatments alter dopamine D_1 receptor mRNA levels: absence of other neurochemical associations with the enhanced behavioral responses of lesioned rats. J Neurochem 63 (4):1282–1290 PMID: 7931280

Graham DG, Tiffany SM, Bell WR Jr, Gutknecht WF (1978) Autoxidation versus covalent binding of quinones as the mechanism of toxicity of dopamine, 6-hydroxydopamine, and related compounds toward C1300 neuroblastoma cells in vitro. Mol Pharmacol 14(4):644–653

Gramatté T, Schmidt J (1986) The effect of early postnatal hypoxia on the effectiveness of drugs influencing motor behaviour in adult rats. Biomed Biochim Acta 45(8):1069–1074 PMID: 3778434

Hamdi A, Kostrzewa RM (1991) Ontogenic homologous supersensitization of dopamine D_1 receptors. Eur J Pharmacol 203:115–120

Hausknecht KA, Acheson A, Farrar AM, Kieres AK, Shen RY, Richards JB, Sabol KE (2005) Prenatal alcohol exposure causes attention deficits in male rats. Behav Neurosci 119(1):302–310 PMID: 15727534

Heffner TG, Seiden LS (1982) Possible involvement of serotonergic neurons in the reduction of locomotor hyperactivity caused by amphetamine in neonatal rats depleted of brain dopamine. Brain Res 244(1):81–90 PMID: 6288184

Heikkila R, Cohen G (1971) Inhibition of biogenic amine uptake by hydrogen peroxide: a mechanism for toxic effects of 6-hydroxydopamine. Science 172(3989):1257–1258 PMID: 5576164

Heikkila R, Cohen G (1972a) Further studies on the generation of hydrogen peroxide by 6-hydroxydopamine. Potentiation by ascorbic acid. Mol Pharmacol 8(2):241–248 PMID: 5025202

Heikkila R, Cohen G (1972b) In vivo generation of hydrogen peroxide from 6-hydroxydopamine. Experientia 28(10):1197–1198 PMID: 5087037

Heikkila RE, Cohen G (1973) 6-Hydroxydopamine: evidence for superoxide radical as an oxidative intermediate. Science 181(4098):456–457 PMID: 4718113

Heikkila RE, Mytilineou C, Côté L, Cohen G (1973) Evidence for degeneration of sympathetic nerve terminals caused by the ortho- and para-quinones of 6-hydroxydopamine. J Neurochem 20(5):1345–1350 PMID: 4716829

Hess EJ, Jinnah HA, Kozak CA, Wilson MC (1992) Spontaneous locomotor hyperactivity in a mouse mutant with a deletion including the Snap gene on chromosome 2. J Neurosci 12 (7):2865–2874 PMID: 1613559

Hess EJ, Collins KA, Wilson MC (1996) Mouse model of hyperkinesis implicates SNAP-25 in behavioral regulation. J Neurosci 16(9):3104–3111 PMID: 8622140

Highfield DA, Lilliquist MW, Amsel A (1999) Reversal of a postnatal alcohol-induced deficit in learned persistence in the rat by d-amphetamine. Alcohol Clin Exp Res 23(6):1094–1101 PMID: 10397296

Jones SR, Gainetdinov RR, Wightman RM, Caron MG (1998) Mechanisms of amphetamine action revealed in mice lacking the dopamine transporter. J Neurosci 18(6):1979–1986 (Review. PMID: 9482784)

Knapp DJ, Breese GR (2015) In Kostrzewa RM (ed) Neurotoxin modeling of brain disorders— life-long outcomes in behavioral teratology. Springer New York (this volume)

Kostrzewa JP, Kostrzewa RA, Kostrzewa RM, Brus R, Nowak P (2015) Perinatal 6-hydroxydopamine to produce a life-long model of severe Parkinson's disease, In Kostrzewa RM (ed) Neurotoxin modeling of brain disorders—life-long outcomes in behavioral teratology. Springer New York (this volume)

Kostrzewa RM (1995) Dopamine receptor supersensitivity. Neurosci Biobehav Rev 19(1):1–17. Review. PMID: 7770190

Kostrzewa RM, Brus R (1991) Ontogenic homologous supersensitization of quinpirole-induced yawning in rats. Pharmacol Biochem Behav 39(2):517–519 PMID: 1682952

Kostrzewa RM, Gong L (1991) Supersensitized D_1 receptors mediate enhanced oral activity after neonatal 6-OHDA. Pharmacol Biochem Behav 39(3):677–682 PMID: 1686103

Kostrzewa RM, Hamdi A (1991) Potentiation of spiperone-induced oral activity in rats after neonatal 6-hydroxydopamine. Pharmacol Biochem Behav 38(1):215–218 PMID: 1901995

Kostrzewa RM, Brus R, Perry KW, Fuller RW (1993) Age-dependence of a 6-hydroxydopamine lesion on SKF 38393- and m-chlorophenylpiperazine-induced oral activity responses of rats. Brain Res Dev Brain Res 76(1):87–93 PMID: 8306435

Kostrzewa RM, Brus R, Kalbfleisch JH, Perry KW, Fuller RW (1994) Proposed animal model of attention deficit hyperactivity disorder. Brain Res Bull 34(2):161–167 PMID: 7913871

Kostrzewa RM, Kostrzewa JP, Kostrzewa RA, Nowak P, Brus R (2008) Pharmacological models of ADHD. J Neural Transm 115(2):287–298. Epub 2007 Nov 12. Review. PMID: 17994186

Kostrzewa RM, Kostrzewa FP (2012) Neonatal 6-hydroxydopamine lesioning enhances quinpirole-induced vertical jumping in rats that were quinpirole-primed during postnatal ontogeny. Neurotoxicity Res. 21(2):231–235. Epub 2011 Aug 19. doi:10.1007/s12640-011-9268-5 PMID: 21853388

LaHoste GJ, Swanson JM, Wigal SB, Glabe C, Wigal T, King N, Kennedy JL (1996) Dopamine D_4 receptor gene polymorphism is associated with attention deficit hyperactivity disorder. Mol Psychiatry 1(2):121–124 PMID: 9118321

Ljungdahl A, Hökfelt T, Jonsson G, Sachs C (1971) Autoradiographic demonstration of uptake and accumulation of ^{3}H-6-hydroxydopamine in adrenergic nerves. Experientia 27(3):297–299 PMID: 5546649

Luthman J, Bolioli B, Tsutsumi T, Verhofstad A, Jonsson G (1987) Sprouting of striatal serotonin nerve terminals following selective lesions of nigro-striatal dopamine neurons in neonatal rat. Brain Res Bull 19(2):269–274 PMID: 3664282

Luthman J, Fredriksson A, Lewander T, Jonsson G, Archer T (1989) Effects of d-amphetamine and methylphenidate on hyperactivity produced by neonatal 6-hydroxydopamine treatment. Psychopharmacology 99(4):550–557 PMID: 2594922

Luthman J, Brodin E, Sundström E, Wiehager B (1990a) Studies on brain monoamine and neuropeptide systems after neonatal intracerebroventricular 6-hydroxydopamine treatment. Int J Dev Neurosci 8(5):549–560 PMID: 1704172

Luthman J, Lindqvist E, Young D, Cowburn R (1990b) Neonatal dopamine lesion in the rat results in enhanced adenylate cyclase activity without altering dopamine receptor binding or dopamine- and adenosine 3':5'-monophosphate-regulated phosphoprotein (DARPP-32) immunoreactivity. Exp Brain Res 83(1):85–95 PMID: 1981564

Miller FE, Heffner TG, Kotake C, Seiden LS (1981) Magnitude and duration of hyperactivity following neonatal 6-hydroxydopamine is related to the extent of brain dopamine depletion. Brain Res 229(1):123–132 PMID: 6796194

Molina-Holgado E, Dewar KM, Descarries L, Reader TA (1994) Altered dopamine and serotonin metabolism in the dopamine-denervated and serotonin-hyperinnervated neostriatum of adult rat after neonatal 6-hydroxydopamine. J Pharmacol Exp Ther 270(2):713–721 PMID: 8071864

Moran-Gates T1, Zhang K, Baldessarini RJ, Tarazi FI (2005) Atomoxetine blocks motor hyperactivity in neonatal 6-hydroxydopamine-lesioned rats: implications for treatment of attention-deficit hyperactivity disorder. Int J Neuropsychopharmacol. 2005 Sep;8(3):439–44. Epub 2005 Apr 7. PMID: 15817135

Mrini A, Soucy JP, Lafaille F, Lemoine P, Descarries L (1995) Quantification of the serotonin hyperinnervation in adult rat neostriatum after neonatal 6-hydroxydopamine lesion of nigral dopamine neurons. Brain Res 669(2):303–308 PMID: 7712186

Myers MM, Musty RE, Hendley ED (1982) Attenuation of hyperactivity in the spontaneously hypertensive rat by amphetamine. Behav Neural Biol 34(1):42–54 PMID: 7073635

Nowak P, Szczerbak G, Biedka I, Drosik M, Kostrzewa RM, Brus R (2006) Effect of ketanserin and amphetamine on nigrostriatal neurotransmission and reactive oxygen species in Parkinsonian rats. In vivo microdialysis study. J Physiol Pharmacol 57(4):583–597 PMID: 17229983

Nowak P, Bortel A, Dabrowska J, Oswiecimska J, Drosik M, Kwiecinski A, Opara J, Kostrzewa RM, Brus R (2007) Amphetamine and mCPP effects on dopamine and serotonin striatal in vivo microdialysates in an animal model of hyperactivity. Neurotox Res 11(2):131–144 PMID: 1744945

Papadeas ST, Breese GR (2014) 6-Hydroxydopamine lesioning of dopamine neurons in neonatal and adult rats induces age-dependent consequences, in Section on Selective Neurotoxins, In Handbook of Neurotoxicity (Kostrzewa RM, Editor), Springer New York, Heidelberg, Dordrecht, London, pp. 133–198. ISBN 978-1-4614-5835-7 (print); ISBN 978-1-4614-5836-4 (eBook); ISBN 978-1-4614-7458-6 (print and electronic bundle). doi:10.1007/978-1-4614-5836-4_59

Pappas BA, Gallivan JV, Dugas T, Saari M, Ings R (1980) Intraventricular 6-hydroxydopamine in the newborn rat and locomotor responses to drugs in infancy: no support for the dopamine depletion model of minimal brain dysfunction. Psychopharmacology 70(1):41–46 PMID: 677533

Plech A, Brus R, Kalbfleisch JH, Kostrzewa RM (1995) Enhanced oral activity responses to intrastriatal SKF 38393 and m-CPP are attenuated by intrastriatal mianserin in neonatal 6-OHDA-lesioned rats. Psychopharmacol. 119:466–473

Radja F, Descarries L, Dewar KM, Reader TA (1993a) Serotonin 5-HT$_1$ and 5-HT$_2$ receptors in adult rat brain after neonatal destruction of nigrostriatal dopamine neurons: a quantitative autoradiographic study. Brain Res 606(2):273–285 PMID: 8490720

Radja F, el Mansari M, Soghomonian JJ, Dewar KM, Ferron A, Reader TA, Descarries L (1993b) Changes of D$_1$ and D$_2$ receptors in adult rat neostriatum after neonatal dopamine denervation: quantitative data from ligand binding, in situ hybridization and iontophoresis. Neuroscience 57 (3):635–648 PMID: 7906013

Raskin LA, Shaywitz BA, Anderson GM, Cohen DJ, Teicher MH, Linakis J (1983) Differential effects of selective dopamine, norepinephrine or catecholamine depletion on activity and learning in the developing rat. Pharmacol Biochem Behav 19(5):743–749 PMID: 6647508

Ruppert PH, Dean KF, Reiter LW (1985) Development of locomotor activity of rat pups exposed to heavy metals. Toxicol Appl Pharmacol 78(1):69–77 PMID: 4035674

Sagvolden T (2000) Behavioral validation of the spontaneously hypertensive rat (SHR) as an animal model of attention-deficit/hyperactivity disorder (AD/HD). Neurosci Biobehav Rev 24 (1):31–39. (Review. PMID: 10654658)

Sagvolden T, Metzger MA, Schiørbeck HK, Rugland AL, Spinnangr I, Sagvolden G (1992) The spontaneously hypertensive rat (SHR) as an animal model of childhood hyperactivity (ADHD): changed reactivity to reinforcers and to psychomotor stimulants. Behav Neural Biol 58(2):103–112 PMID: 1360797

Sagvolden T, Aase H, Zeiner P, Berger D (1998) Altered reinforcement mechanisms in attention-deficit/hyperactivity disorder. Behav Brain Res 94(1):61–71 PMID: 9708840

Saner A, Thoenen H (1971) Model experiments on the molecular mechanism of action of 6-hydroxydopamine. Mol Pharmacol 7(2):147–154 PMID: 5125851

Segura-Aguilar J (2001) DT-Diaphorase: a protective enzyme of the dopaminergic system, In Mechanisms of Degeneration and Protection of the Dopaminergic System (Segura-Aguilar, Editor), F.P. Graham Publishing Co. 2001, pp. 289–299. ISBN 1-929675-03-8

Segura-Aguilar J, Paris I (2014) Mechanisms of dopamine oxidation and Parkinson's Disease, in Handbook of Neurotoxicity (Kostrzewa RM, Editor), Springer New York, Heidelberg, Dordrecht, London, 2014, pp. 3–67. ISBN 978-1-4614-5835-7 (print); ISBN 978-1-4614-5836-4 (eBook); ISBN 978-1-4614-7458-6 (print and electronic bundle). doi:10. 1007/978-1-4614-5836-4_53

Senoh S, Witkop B (1959a) Formation and rearrangements of aminochromes from a new metabolite of dopamine and some of its derivatives. J Am Chem Soc 81:6231–6235

Senoh S, Witkop B (1959b) Non-enzymatic conversions of dopamine to norepinephrine and trihydroxyphenethylamine. J Am Chem Soc 81:6222–6231

Senoh S, Witkop B, Creveling CR, Udenfriend S (1959) 2,4,5-Tri-hydroxyphenethylamine, a new metabolite of 3,4-dihydroxyphenethylamine. J Am Chem Soc 81:1768–1769

Shaywitz BA, Klopper JH, Yager RD, Gordon JW (1976a) Paradoxical response to amphetamine in developing rats treated with 6-hydroxydopamine. Nature 261(5556):153–155 PMID: 944861

Shaywitz BA, Yager RD, Klopper JH (1976b) Selective brain dopamine depletion in developing rats: an experimental model of minimal brain dysfunction. Science 191(4224):305–308 PMID: 942800

Silbergeld EK, Goldberg AM (1975) Pharmacological and neurochemical investigations of lead-induced hyperactivity. Neuropharmacology 14(5–6):431–444 PMID: 1171389

Smalley SL, Bailey JN, Palmer CG, Cantwell DP, McGough JJ, Del'Homme MA, Asarnow JR, Woodward JA, Ramsey C, Nelson SF (1999) Evidence that the dopamine D$_4$ receptor is a susceptibility gene in attention deficit hyperactivity disorder. Mol Psychiatry. 1998 Sep;3 (5):427–30. Erratum in: Mol Psychiatry 1999 Jan;4(1):100 PMID:9774776

Smith RD, Cooper BR, Breese GR (1973) Growth and behavioral changes in developing rats treated intracisternally with 6-hydroxydopamine: evidence for involvement of brain dopamine. J Pharmacol Exp Ther 185(3):609–619 PMID: 4145870

Snyder AM, Zigmond MJ, Lund RD (1986) Sprouting of serotoninergic afferents into striatum after dopamine-depleting lesions in infant rats: a retrograde transport and immunocytochemical study. J Comp Neurol 245(2):274–281 PMID: 2420842

Soucy JP, Lafaille F, Lemoine P, Mrini A, Descarries L (1994) Validation of the transporter ligand cyanoimipramine as a marker of serotonin innervation density in brain. J Nucl Med 35 (11):1822–1830 PMID: 7965165

Speiser Z, Korczyn AD, Teplitzky I, Gitter S (1983) Hyperactivity in rats following postnatal anoxia. Behav Brain Res 7(3):379–382 PMID: 6682332

Stachowiak MK, Bruno JP, Snyder AM, Stricker EM, Zigmond MJ (1984) Apparent sprouting of striatal serotonergic terminals after dopamine-depleting brain lesions in neonatal rats. Brain Res 291(1):164–167 PMID: 6199080

Thoenen H, Tranzer JP (1968) Chemical sympathectomy by selective destruction of adrenergic nerve endings with 6-Hydroxydopamine. Naunyn Schmiedebergs Arch Exp Pathol Pharmakol 261(3):271–288 PMID: 4387076

Towle AC, Criswell HE, Maynard EH, Lauder JM, Joh TH, Mueller RA, Breese GR (1989) Serotonergic innervation of the rat caudate following a neonatal 6-hydroxydopamine lesion: an anatomical, biochemical and pharmacological study. Pharmacol Biochem Behav 34(2):367–374 PMID: 2576138

van der Kooij MA, Glennon JC (2007) Animal models concerning the role of dopamine in attention-deficit hyperactivity disorder. Neurosci Biobehav Rev 31(4):597–618. Epub 2007 Jan 23. (Review. PMID: 17316796)

Whishaw IQ, Funk DR, Hawryluk SJ, Karbashewski ED (1987) Absence of sparing of spatial navigation, skilled forelimb and tongue use and limb posture in the rat after neonatal dopamine depletion. Physiol Behav 40(2):247–253 PMID: 3114777

Wilson MC (2000) Coloboma mouse mutant as an animal model of hyperkinesis and attention deficit hyperactivity disorder. Neurosci Biobehav Rev 24(1):51–57. (Review. PMID: 10654661)

Zhang K, Tarazi FI, Baldessarini RJ (2001a) Nigrostriatal dopaminergic denervation enhances dopamine D(4) receptor binding in rat caudate-putamen. Pharmacol Biochem Behav 69(1–2):111–116 PMID: 11420075

Zhang K, Tarazi FI, Baldessarini RJ (2001b) Role of dopamine D(4) receptors in motor hyperactivity induced by neonatal 6-hydroxydopamine lesions in rats. Neuropsychopharmacology 25(5):624–632 PMID: 11682245

Zhang K, Davids E, Tarazi FI, Baldessarini RJ (2002a) Effects of dopamine D_4 receptor-selective antagonists on motor hyperactivity in rats with neonatal 6-hydroxydopamine lesions. Psychopharmacology 161(1):100–106 Epub 2002 Mar 6. PMID: 11967637

Zhang K, Tarazi FI, Davids E, Baldessarini RJ (2002b) Plasticity of dopamine D_4 receptors in rat forebrain: temporal association with motor hyperactivity following neonatal 6-hydroxydopamine lesioning. Neuropsychopharmacology 26(5):625–633 PMID: 11927187

Attention-Deficit/Hyperactivity Disorder: Focus upon Aberrant N-Methyl-D-Aspartate Receptors Systems

Trevor Archer and Danilo Garcia

Abstract Attention-deficit/hyperactivity disorder (ADHD) pathophysiology persists in an obscure manner with complex interactions between symptoms, staging, interventions, genes, and environments. Only on the basis of increasing incidence of the disorder, the need for understanding is greater than ever. The notion of an imbalance between central inhibitory/excitatory neurotransmitters is considered to exert an essential role. In this chapter, we first review how the default mode network functions and dysfunction in individuals diagnosed with ADHD. We also present and briefly review some of the animal models used to examine the neurobiological aspects of ADHD. There is much evidence indicating that compounds/interventions that antagonize/block glutamic acid receptors and/or block the glutamate signal during the "brain growth spurt" or in the adult animal may induce functional and biomarker deficits. Additionally, we present evidence suggesting that animals treated with glutamate blockers at the period of the "brain growth spurt" fail to perform the exploratory activity, observed invariably with control mice, that is associated with introduction to a novel environment (the test cages). Later, when the control animals show less locomotor and rearing activity, i.e., interest in the test cages, the MK-801, ketamine and ethanol treated mice showed successively greater levels of locomotion and rearing (interest), i.e., they fail to "habituate" effectively, implying a cognitive dysfunction. These disturbances of glutamate signaling during a critical period of brain development may contribute

T. Archer (✉) · D. Garcia
Department of Psychology, University of Gothenburg, Box 500,
SE 40530 Gothenburg, Sweden
e-mail: trevor.archer@psy.gu.se

T. Archer · D. Garcia
Network for Empowerment and Well-Being, University of Gothenburg,
Gothenburg, Sweden

D. Garcia
Blekinge Center of Competence, Blekinge County Council, Karlskrona, Sweden

D. Garcia
Centre for Ethics, Law and Mental Health (CELAM), University of Gothenburg,
Gothenburg, Sweden

© Springer International Publishing Switzerland 2015
Curr Topics Behav Neurosci (2016) 29: 295–312
DOI 10.1007/7854_2015_415
Published Online: 31 December 2015

to the ADHD pathophysiology. As a final addition, we have briefly identified new research venues in the interaction between ADHD, molecular studies, and personality research.

Keywords Hyperactivity · Attention-deficit · Glutamate antagonists · Motor activity · Deficits · Brain regions · Mice

Contents

1   N-Methyl-D-Aspartate Receptors (NMDA-R) Linked ADHD Models 298
2   Aberrant Glutamate in Spontaneously Hypertensive (SHR) Rats 301
3   Conclusions and Final Remarks 303
References 304

The incidence of attention-deficit/hyperactivity disorder (ADHD) is, very likely, increasingly diagnosed, from about 5 % of all children in the USA (Kalat 2001), to 5.1 million or 8.8 % of all children in the age-group 4–17 years (CDC 2011). The incidence over developmental trajectory breaks down as follows: (i) 6.8 % of all children aged 4–10 years (1 in every 14), (ii) 11.4 % of all children aged 11–14 years (1 in every 9), and (iii) 10.2% of all children aged 16–17 years (1 in every 10), with average age at current diagnosis 6.2, with "mild" ADHD diagnosed at 7 years, "moderate" at 6.1 years and "severe" at 4.4 years 3.5 million children taking medication. Boys (12.1 %) remain as more likely to be diagnosed than girls (5.5 %) to be diagnosed. High rates of comorbidity with of oppositional defiant disorder (ODD), anxiety, and depression in children with ADHD have been reported (Anckarsäter et al. 2006; Mitchison and Njardvik 2015; Garcia et al. 2013). The primary symptoms manifested in children with ADHD include: incessant talking in the classroom, restlessness, inattention, impulsiveness, lack of concentration, and hyperactivity (Antonini et al. 2015; Barkley et al. 2002; Bussing et al. 2015; Swanson et al. 1998). In adults with ADHD, symptom profiles are defined by poor attention with excessive distractibility, over-impulsivity, i.e., thoughtless utterances/actions, restlessness/hyperactivity, chronic procrastination, difficulty initiating and completing tasks, frequently losing objects, poor organization, planning and time management, and excessive forgetfulness (Froehlich et al. 2007, 2009, 2010, 2011; Jaber et al. 2015; Lin and Gau 2015; Micoulaud-Franchi et al. 2015). Individuals presenting ADHD exhibit excessive levels of default mode network[1] activity during goal-directed tasks, which are associated with attentional disturbances and performance decrements. However, the process of downregulating the default mode network activity when preparing to switch from rest to task is

---

[1] A network of brain regions that are active when the individual is not focused on the outside world and the brain is at wakeful rest.

unimpaired in adults with ADHD adults and these adults also lack switch-specific deficit in right anterior insula modulation (Sidlauskaite et al. 2015). In addition, individuals presenting ADHD show difficulties in upregulating the default mode network activity when switching from task phase to rest phase (Sidlauskaite et al. 2015). Kucyi et al. (2015) showed evidence of impaired cerebellar areas of the default mode network coupling with cortical networks in adult patients with ADHD and highlights a role of cerebro–cerebellar interactions in cognitive function. ADHD is associated with significantly increased mortality rates and individuals diagnosed with ADHD during adulthood show higher mortality rates than did those diagnosed in childhood and adolescence (Dalsgaard et al. 2015).

The vast number of studies examining neurobiological aspects of ADHD attests to variety of laboratory animal models available including (i) genetically based models, (ii) neurotoxin-induced models, (iii) Neonatal NMDA-R antagonist administration models, (iv) environmentally based models, and (v) sleep disorder problems. (i) Genetically based models present strains of rats/mice with particular, measureable ADHD phenotypes with phenotypic behaviors and biomarkers such as spontaneously hypertensive (SHR) rats, Naples high-excitability (NHE) rat, rats giving poor performers in the 5-choice serial reaction time task, dopamine transporter (DAT) knock-out mice, SNAP-25-deficient mutant coloboma mice, mice expressing a human mutant thyroid hormone receptor, nicotinic receptor knock-out mice, 22q11.2 deletion syndrome (Meechan et al. 2015), and tachykinin-1 receptor knock-out mice. (ii) Neurotoxin-induced models of ADHD applying N-(2-chloroethyl)-N-ethyl-2-bromobenzylamine (DSP-4) or 6-hydroxydopamine (6-OHDA) as catecholamine neurotoxins or 5,7-dihydroxytryptamine (5,7-DHT) to induce deficits. (iii) NMDA-R antagonist administration models administer MK-801, ketamine, ethanol, polychlorinated biphenyls, or phencyclidine (PCP). (iv) Environmentally based models such as pups reared in deprived environments or isolated housing or neonatal anoxia or variations of environmental stress (Ishii and Hashimoto-Torii 2015). (v) Complex regulatory circuits involving clock genes themselves and their influence on circadian rhythms of diverse body functions and behavioral domains form an important aspect of the gene-environment interaction (Dueck et al. 2015).

As indicated by Russell (2011), one major insight provided by animal models was the consistency of findings regarding the involvement of dopaminergic, noradrenergic, and sometimes also serotonergic systems, as well as glutamateric and GABAergic pathways (see also Russell 2007). Blockade of the N-methyl-D-aspartate receptors (NMDA-Rs) during the neonatal period has been to induce long-term behavioral and neurochemical alterations that are applied as laboratory models for ADHD, schizophrenia, borderline personality disorder and depression. During development, GABA exerts a depolarizing action on immature neurons. Ende et al. (2015) studied glutamate and GABA influences in relation to impulsiveness and aggressive behavior associated with the anterior cingulate cortex in the groups of female patients presenting borderline personality disorder and ADHD, respectively. The links between glutamate and GABA levels and further borderline personality disorder (symptom severity) and ADHD aspects (hyperactivity and inattention) were evaluated in an explorative manner. They acquired 1H MR spectra at 3T to determine the glutamate to total creatine ratios (Glu/tCr) and GABA levels from the anterior cingulate cortex in a borderline personality disorder group ($n = 26$), an ADHD group ($n = 22$), as well as a healthy control (HC) group ($n = 30$); all the participants were females. Both patient groups, i.e., borderline personality disorder and ADHD, presented higher scores on self-reported impulsiveness, anger, and aggression compared with the healthy controls. Anterior cingulate cortex GABA levels were significantly lower in ADHD than HC. Although measures of impulsiveness were related positively to glutamate and negatively to GABA, in the case of aggression only a negative correlation with GABA was obtained. This pattern of results may provide human in vivo evidence for the role of anterior cingulate cortex Glu/tCr and GABA in impulsiveness and aggression.

1 N-Methyl-D-Aspartate Receptors (NMDA-R) Linked ADHD Models

ADHD pathophysiology persists in an obscure manner with complex interactions between symptoms, staging, interventions, genes, and environments (Archer and Bright 2012; Archer and Kostrzewa 2012; Archer et al. 2011; Kyeong et al. 2015; Rommel et al. 2013; Schuch et al. 2015). Functional magnetic resonance imaging (fMRI) has shown that almost all test–retest reliability of resting state fMRI metrics presented significantly higher intra-class correlation coefficient in typically developing children than in with ADHD children for one or more brain regions studied (Somandepalli et al. 2015). Several physiological biomarkers besides brain neurochemical show deficits in the disorder: for example, serum levels of oxytocin in total subjects presenting ADHD were reduced significantly compared with those levels of neurotypical control individuals, and serum levels of oxytocin in drug naïve ADHD patients were significantly lower than those in medicated ADHD patients. Interestingly, there was a significant negative correlation between serum oxytocin levels and ADHD-RS total scores, as well as ADHD-RS inattentive scores in all ADHD patients Sasaki et al. 2015). The genetic factor in ADHD is undeniable (Fang et al. 2015; Salatino-Oliveira et al. 2015; Thapar and Cooper 2015; Van Rooij et al. 2015a; Pettersson et al. 2013), connected with sibling associations (Richards et al. 2015; Thissen et al. 2015; Van Rooij et al. 2015b), parents (Costa Dde et al. 2015; Grizenko et al. 2015; Van der Kolk et al. 2015), and with extremely high twin concordance (Arcos-Burgos et al. 2004; Ehli et al. 2012; Langner et al. 2013; McLoughlin et al. 2014; Garcia et al. 2014) and comorbidity (Volh et al. 2005; Garcia et al. 2013). The notion of an imbalance between central inhibitory/excitatory neurotransmitters is considered to exert an essential role in the pathophysiology of ADHD (Purkayastha et al. 2015; Sadile et al. 1996), although this is not always the case (Endres et al. 2015). For instance, in contrast to children, adult patients presenting ADHD display altered cerebral levels of GABA+ in a subcortical voxel (Bollmann et al. 2015).

Additionally, there are also increased cerebral glutamine levels in children with ADHD, but this difference is normalized among adults with ADHD (Bollmann et al. 2015). In other words, suggesting that this alteration might change through development.

There is a plethora of results showing that compounds/interventions that antagonize glutamic acid receptors and/or block the glutamate signal during the "brain growth spurt" or in the adult animal may induce functional and biomarker deficits (Davison and Dobbing 1968; Di Miceli and Gonier 2015; Fredriksson and Archer 2002, 2003; Fredriksson et al. 2004; Pozzi et al. 2011; Zimmermann et al. 2015; Zhou et al. 2011). Normal regional brain development follows an inherited, preprogrammed route that differentiates the specific structural characteristics and functional domains that are expressed in the adult human and animal (Dobbing and Sands 1970, 1979). Any interference with the course of normal brain development threatens the regional structural and functional integrity with more or less perma-nent consequences for the individual (Chen et al. 1999; Dobbing 1970a, b, c, 1971; Huebner et al. 2015). The period of the "brain growth spurt" starts with the final trimester of pregnancy in humans and continues until about three years after birth (Ikonomidou et al. 2001). The corresponding period in rodents is encompassed by a rapid increase in brain weight, proliferation of astroglial and oligodendrocyte cells, axonal elongation, arborization, and synaptogenesis (Byrnes et al. 2001; Davison and Dobbing 1966). Agents affecting glutamate system are implicated highly in the vulnerability of brain development since chronic prenatal exposure to an ethanol regimen throughout gestation induced suppression of the hippocampal glutamate-NMDA receptor-NOS signaling system, decreased number of hip-pocampal CA1 pyramidal cells, increased spontaneous locomotor activity, and impaired performance in the Morris water maze (Byrnes et al. 2001; but see also Byrnes et al. 2003, 2004). Anticonvulsant drugs can initiate neuron and oligo-dendroglia apoptosis, suppress neurogenesis, and inhibit normal synapse develop-ment and regional-sculpting (Turski and Ikonomidou 2012). The behavioral correlates in rodents and non-human primates consist of long-lasting cognitive impairment and motor deficits. Physiological apoptosis and that caused by other agents, a naturally occurring process of the developing brain, modulates regional progressions at cellular and circuitry levels periodically (Dikranian et al. 2001; Ishimaru et al. 1999; Olney et al. 2000).

Several studies have demonstrated marked deficits in behavioral domains in the adult animal following disruptions in glutamate signaling and GABAergic activity (Cohen Kadosh et al. 2015; Kim et al. 2015; Tzanoulinou et al. 2015; Zhang et al. 2015) in the prenatal or neonatal human and animal (Higuera-Matas et al. 2015; Jantzie et al. 2015; Keimpema et al. 2014; Kleteckova et al. 2014; Simões et al. 2015). It was shown more than twenty years ago that chronic neonatal treatment with the glutamate antagonist, MK-801 (postnatal days 8 to 19), induced marked alterations of monoamines (Gorter et al. 1992a) whereby dihydroxyphenylacetic acid (DOPAC) concentrations were elevated (greater than 40 %) in both regions (cortex and striatum) tested, while 5-hydroxyindoleacetic acid (5-HIAA) concentration was significantly elevated only in the cortex (19 %), and 3-methoxy-4-hydroxyphenylglycol (MHPG) only in the striatum (47 %). When tested spatial learning and memory using a water maze, the neonatal MK-801-treated rats were shown to be capable of learning the spatial task as well as control rats but did so at a significantly slower rate. Their performance in a visual cue task was not affected by the neonatal treatment, suggesting that the slower spatial learning is not caused by locomotor or sensory deficits (Gorter and de Bruin 1992). The authors interpreted their findings to imply that chronic NMDA receptor blockade during the neonatal period leads to long-lasting disturbances of hippocampal function (but see also Gorter et al. 1991, 1992b). The glutamate antagonists, MK-801 (3 × 0.5 mg/kg), ketamine (1 × 50 mg/kg), and ethanol (2 × 2.5 mg/kg) were administered postnatally to mouse pups on days 10 or 11 postpartum (cf. Fredriksson and Archer 2004), and behavioral testing was performed at adult ages over and above 65 days-of-age. At testing, it was found that over 60-min periods of motor activity testing, the mice administered the glutamate antagonists showed a somewhat bizarre pattern of activity that was completely different to that shown by untreated, saline- or vehicle-treated, or sham-operated mice: The former presented markedly lower levels of motor activity than the latter during the initial period of activity testing (1st 20 min) and then successively greater levels of motor activity than the latter during the middle and final periods of activity testing (2nd and 3rd 20 min periods). Table 1 presents the motor activity, locomotion, and rearing of

Table 1 Locomotion and rearing behavior expressed as percent of control values (0.9 % saline-vehicle) over successive 20-min periods in the motor activity test cages by adult mice administered either MK-801, ketamine or ethanol, glutamate antagonists at the doses used, or diazepam, GABA agonist at the doses used. MK-801 (0.5 mg/kg, s.c.) was administered to male mouse pups on postnatal day 11 on three occasions over that day. Ketamine (1 × 50 mg/kg, s.c.) and ethanol (2 × 25 mg/kg, s.c., with a 2-h interval between injections) were administered on postnatal day 10

Neonatal treatment	20-min period	Locomotion % of control values	Rearing % of control values
MK-801	20	38	32
	40	199	247
	60	3439	8405
Ketamine	20	36	34
	40	207	225
	60	1967	3812
Diazepam	20	99	86
	40	112	101
	60	189	544
Ketamine + Diazepam	20	35	27
	40	188	261
	60	2617	7422
Ethanol	20	38	29
	40	224	239
	60	2128	4122

mice injected postnatally with either MK-801, ketamine, or ethanol expressed as a percentage of each respective vehicle control group. It will be noted that the mice administered glutamate antagonists evidenced massively greater levels of apoptosis (as measured by fluoro-jade positive staining) in several brain regions, including frontal cortex, hippocampus, cerebellum, parietal cortex, and laterodorsal thalamus, compared to controls, 24 h after administration of the drugs, MK-801, ketamine, or ethanol. The results on Table 1 indicate that these animals fail to perform the exploratory activity, observed invariably with control mice, that is associated with introduction to a novel environment (the test cages). Later, when the control animals show less locomotor and rearing activity, i.e., interest in the test cages, the MK-801-, ketamine-, and ethanol-treated mice showed successively greater levels of locomotion and rearing (interest), i.e., they fail to "habituate" effectively, implying a cognitive dysfunction. The cognitive dysfunctionality of the MK-801-, ketamine-, and ethanol-treated mice was demonstrated in both the radial arm maze and the circular swimming maze (Fredriksson and Archer 2004). Spontaneously hypertensive rats (SHRs) present several of the characteristic behavioral anomalies observed in ADHD children and adults: hyperactivity, impulsiveness and poorly sustained attention, restlessness, and comorbid drug self-administration (Grünblatt et al. 2015; Jordan et al. 2015; Womersley et al. 2015).

2 Aberrant Glutamate in Spontaneously Hypertensive (SHR) Rats

There is emerging evidence that spontaneously hypertensive (SHR) rats possess disruptions in glutamate systems or in glutamate signaling or in region (e.g., nucleus accumbens) characteristics (Russell 2003). For example, it was observed that the glutamatergic system in the prefrontal cortex of the SHR rats was hyper-functional (Miller et al. 2014). Sterley et al. (2015) have provided evidence for a disturbed glutamatergic and GABAergic transmission in the hippocampus of SHRs and that maternal separation induced effects on glutamate uptake in these rats and Wister-Kyoto and Sprague-Dawley rats as well. Furthermore, compared to control animals, SHRs displayed a lower expression of both NMDA (Grin1) and AMPA (Gria1) gene receptors in the nucleus accumbens. It has been observed also that SHRs express decreased levels of several proteins involved in energy metabolism, cytoskeletal structure, myelination, and neurotransmitter function when compared to Wistar-Kyoto rats[2] (Dimatelis et al. 2015). Liso Navarro et al. (2014) found significant correlations between brain metabolites and the behavior registered in the open field and elevated plus maze: SHR rats expressing higher levels of brain total creatine levels and glutamate levels exhibited higher levels of hyperactivity in a familiar environment, but conversely, risk-taking exploratory behavior, an

---

[2]The Wistar rat is an outbred albino rat.

indication of impulsivity, of the elevated plus maze's open arms correlated negatively with forebrain total N-acetylaspartate and lactate levels. It has been shown also that there is a reduction in extracellular concentrations of GABA in the hippocampus of SHR rats, in vivo, by comparison with Wistar-Kyoto and Sprague-Dawley rats (Sterley et al. 2013). The authors suggest that an underlying defect in GABA function may be the underlying cause of the dysfunction in catecholamine transmission noted in SHR and may underlie their ADHD-like behaviors (see also Mc Fie et al. 2012; Miller et al. 2012). Ye et al. (2013) observed that there were increases in presynaptic group II metabotropic glutamate receptor activity at the glutamatergic terminals at hypothalamic paraventricular nucleus sites in SHR rats. The activation of group II metabotropic glutamate receptors in the hypothalamic paraventricular nucleus inhibits sympathetic vasomotor tone through attenuation of increased glutamatergic input and neuronal hyperactivity in SHR rats, thereby affecting sympathetic outflow in hypertension and related conditions. In this context, physical exercise was shown to ameliorate the enhancement in the tonically acting glutamatergic input to the rostral ventrolateral medulla of SHR rats, thereby reducing the sympathetic hyperactivity and blood pressure (Zha et al. 2013). Following exposure to the NMDAR antagonist, MK-801, during postnatal days 5–14, to male Sprague-Dawley rat pups (Li et al. 2015), the animals were tested for object and object-in-context recognition memory during adolescence (PND 35) and adulthood (PND 63). They examined also parvalbumin-positive GABA-ergic interneurons and presynaptic markers for excitatory and inhibitory neurons, vesicular glutamate transporter-1, and vesicular GABA transporter in the hippocampus to reflect the excitatory/Inhibitory balance. They observed that rats that had received MK-801 treatment displayed deficits of recognition memory, reduction in parvalbumin-positive cell counts, and upregulation of the vesicular glutamate transporter-1/vesicular GABA transporter ratio in both adolescence and adulthood. It would appear that the changes of the vesicular glutamate transporter-1/vesicular GABA transporter ratio at the two time points exhibited distinct mechanisms. Furthermore, prenatal alcohol exposure affected cortical angiogenesis negatively both in mice and in fetal alcohol syndrome patients, implying that vascular defects contributed to alcohol-induced brain abnormalities (Jegou et al. 2012). Postnatal treatment with domoic acid, which disturbs glutamate signaling, induced deficits in latent inhibition and sensory gating through prepulse inhibition impairments (Marriott et al. 2012).

Finally, several genetic linkage and association studies point to candidate genes relating to ADHD (Franke et al. 2009). Associations between ADHD and a handful of NMDA-R gene variants [GRM1, GRM5, GRM7, and GRM8: encoding G-protein coupled receptor family] (Diana et al. 2015; Akutagava-Martins et al. 2014). Santoro et al. (2015) compared the gene expression profile of neurotransmitter receptors and regulators in the prefrontal cortex and nucleus accumbens of SHR and control Wistar rats, as well as the DNA methylation pattern of promoter region of the genes differentially expressed. They found that four genes were downregulated significantly in the prefrontal cortex of the SHRs in comparison with Wistar rats (Gad2, Chrnb4, Slc5a7, and Qrfpr) and none of those in nucleus accumbens. Gad2 and Qrfpr showed CpG islands in their promoter region. For both of these genes, the promoter region was hypomethylated in SHR rats and may be linked to the abnormalities displayed by these animals. Since adverse life events, dysfunctional families, pregnancy and birth complications, etc, all increase risk for ADHD (Class et al. 2014; Lindström et al. 2011; Pires et al. 2013; Webb 2013), the epigenetic influences upon glutamatergic integrity seems immeasurable (Grissom and Reyes 2013; Schuch et al. 2015). van Mil et al. (2014) examined the association between DNA methylation levels at different regions and ADHD symptoms. They observed that DNA methylation levels were linked negatively with ADHD symptoms scores in the analysis of eleven brain regions.

3 Conclusions and Final Remarks

The notion of a disruption of the normal brain developmental trajectory, due an over-stimulation of GABAergic systems and/or an understimulation (antagonism) of the glutamate systems, i.e., excitatory-inhibitory imbalance, in the underlying pathophysiology of ADHD, particularly regarding motor and cognitive domains, is appealing. The pattern of behavioral deficits when tested as adult animals and "accelerated apoptosis" 24 h after administrations of the glutamate antagonists offers a useful laboratory model of the disorder. Babenko et al. (2015) have offered a plausible account that describes the complex gene–environment interactions between prenatal stress exposure, whether chemical intervention or social-behavioral, associated changes in miRNA expression and DNA methylation in placenta and brain regions together with the possible links to greater risks of schizophrenia, ADHD, autism, anxiety-, or depression-related disorders that may be expressed later in life.

Finally, ADHD is associated with an increased risk of personality disorders and deficits and specific temperament configurations: high novelty seeking and high harm avoidance (Anckarsäter et al. 2006). Individuals high in novelty seeking tend to be highly active or to direct their attention/behaviors in response to novel stimuli, potential rewards, and punishments. This is expressed as frequent exploration of new unfamiliar places or situations, quick loss of temper, impulsive decision-making, and active avoidance of monotony. High levels of harm avoidance are expressed as the tendency to avoid or cease behaviors due to intense response to aversive stimuli expressed as fear of uncertainty, shyness of strangers, quick fatigability, and pessimistic worry of future problems (Cloninger et al. 1993). This "explosive" temperament profile (high novelty seeking and high harm avoidance) does fit in the ADHD pathophysiology outlined in this chapter. What is more, molecular genetics studies have found an association between novelty seeking and the dopamine-4 receptor (Benjamin et al. 1996; Ebstein et al. 1996; Noble et al. 1998; Ono et al. 1997) and between harm avoidance and the serotonin transporter 5HTTLPR (Rybakowski et al. 2006; Samochowiec et al. 2001).

Fortunately, recent advances using person-centered interventions (i.e., well-being coaching) suggest that the expression of genes as personality traits can be changed (Cloninger 2004; Wong and Cloninger 2010; see also Fahlgren et al. 2015), In particular when the intervention focuses on the development of character traits, such as self-directedness (e.g., sense of control, self-efficacy, self-acceptance), cooperativeness (e.g., tolerance, helpfulness, empathy), and self-transcendence (e.g., spirituality, meaningfulness, ability to experience flow). Relatedly, changes in mean levels of character traits are much greater between 20 and 45 years of age than for temperament traits (Josefsson et al. 2013). Hence, upcoming studies using person-centered interventions among individuals with ADHD are most welcome.

References

Akutagava-Martins GC, Salatino-Oliveira A, Genro JP, Contini V, Polanczyk G, Zeni C, Chazan R, Kieling C, Anselmi L, Menezes AM, Grevet EH, Bau CH, Rohde LA, Hutz MH (2014) Glutamatergic copy number variants and their role in attention-deficit/hyperactivity disorder. Am J Med Genet B Neuropsychiatr Genet 165B(6):502–509. doi:10.1002/ajmg.b.32253

Anckarsäter H, Ståhlberg O, Larson T, Håkansson C, Jutblad SB, Niklasson L, Nydén A, Wentz E, Westergren S, Cloninger CR, Gillberg C, Rastam M (2006) The impact of ADHD and autism spectrum disorders on temperament, character, and personality development. Am J Psychiatry 163(7):1239–1244

Antonini TN, Becker SP, Tamm L, Epstein JN (2015) Hot and cool executive functions in children with attention-deficit/hyperactivity disorder and comorbid oppositional defiant disorder. J Int Neuropsychol Soc 21(8):584–595. doi:10.1017/S1355617715000752

Archer T, Bright P (2012) Functional and structural MRI studies on impulsiveness: attention deficit/hyperactive disorder and borderline personality disorders. In: Bright P (ed) Neuroimaging—cognitive and clinical neuroscience. InTech, London, pp 205–228. ISBN 978-953-51-0606-7

Archer T, Kostrzewa RM (2012) Physical exercise alleviates ADHD symptoms: regional deficits and development trajectory. Neurotox Res 21(2):195–209. doi:10.1007/s12640-011-9260-0

Archer T, Oscar-Berman M, Blum K (2011) Epigenetics in developmental disorder: ADHD and endophenotypes. J Genet Syndr Gene Ther 2(104). pii: 1000104

Arcos-Burgos M, Castellanos FX, Pineda D, Lopera F, Palacio JD, Palacio LG, Rapoport JL, Berg K, Bailey-Wilson JE, Muenke M (2004) Attention-deficit/hyperactivity disorder in a population isolate: linkage to loci at 4q13.2, 5q33.3, 11q22, and 17p11. Am J Hum Genet 75(6):998–1014

Babenko O, Kovalchuk I, Metz GA (2015) Stress-induced perinatal and transgenerational epigenetic programming of brain development and mental health. Neurosci Biobehav Rev 48:70–91. doi:10.1016/j.neubiorev.2014.11.013

Barkley RA, Fisher M, Smallish L, Fletcher K (2002) The persistence of attention-deficit/hyperactivity disorder into young adulthood as a function of reporting source and definition of disorder. J Abnorm Psychol 111:279–289

Benjamin J, Li L, Patterson C, Greenberg BD, Murphy DL, Hamer DH (1996) Population and familial association between the D4 dopamine receptor gene and measures of novelty seeking. Nat Genet 12:81–84

Bollmann S, Ghisleni C, Poil SS, Martin E, Ball J, Eich-Höchli D, Edden RA, Klaver P, Michels L, Brandeis D, O'Gorman RL (2015) Developmental changes in gamma-aminobutyric acid levels in attention-deficit/hyperactivity disorder. Transl Psychiatry 5:e589. doi:10.1038/tp. 2015.79

Bussing R, Meyer J, Zima BT, Mason DM, Gary FA, Garvan CW (2015) Childhood ADHD symptoms: association with parental social networks and mental health service use during adolescence. Int J Environ Res Public Health. 12(9):11893–11909. doi:10.3390/ijerph120911893

Byrnes ML, Reynolds JN, Brien JF (2001) Effect of prenatal ethanol exposure during the brain growth spurt of the guinea pig. Neurotoxicol Teratol 23(4):355–364

Byrnes ML, Reynolds JN, Brien JF (2003) Brain growth spurt-prenatal ethanol exposure and the guinea pig hippocampal glutamate signaling system. Neurotoxicol Teratol 25(3):303–310

Byrnes ML, Richardson DP, Brien JF, Reynolds JN, Dringenberg HC (2004) Spatial acquisition in the Morris water maze and hippocampal long-term potentiation in the adult guinea pig following brain growth spurt–prenatal ethanol exposure. Neurotoxicol Teratol 26(4):543–551

Centers for Disease Control and Prevention (CDC) (2011) Key findings: trends in the parent-report of health care provider-diagnosis and medication treatment for ADHD: United States, 2003–2011

Chen WA, Parnell SE, West JR (1999) Early postnatal alcohol exposure produced long-term deficits in brain weight, but not the number of neurons in the locus coeruleus. Brain Res Dev Brain Res 118(1–2):33–38

Class QA, Abel KM, Khashan AS, Rickert ME, Dalman C, Larsson H, Hultman CM, Långström N, Lichtenstein P, D'Onofrio BM (2014) Offspring psychopathology following preconception, prenatal and postnatal maternal bereavement stress. Psychol Med 44(1):71–84. doi:10.1017/S0033291713000780

Cloninger CR (2004) Feeling good: the science of well-being. Oxford University Press, New York

Cloninger CR, Svrakic DM, Przybeck TR (1993) A psychobiological model of temperament and character. Arch Gen Psychiatry 50:975–989

Cohen Kadosh K, Krause B, King AJ, Near J, Cohen Kadosh R (2015) Linking GABA and glutamate levels to cognitive skill acquisition during development. Hum Brain Mapp. doi:10.1002/hbm.22921

Costa Dde S, Rosa DV, Barros AG, Romano-Silva MA, Malloy-Diniz LF, Mattos P, de Miranda DM (2015) Telomere length is highly inherited and associated with hyperactivity-impulsivity in children with attention deficit/hyperactivity disorder. Front Mol Neurosci. 8:28. doi:10.3389/fnmol.2015.00028

Dalsgaard S, Østergaard SD, Leckman JF, Mortensen PB, Pedersen MG (2015) Mortality in children, adolescents and adults with attention deficit hyperactivity disorder: a nationwide cohort study. Lancet 385(9983):2190–2196. doi:10.1016/S0140-6736(14)61684-6

Davison AN, Dobbing J (1966) Myelination as a vulnerable period in brain development. Br Med Bull 22(1):40–44

Davison AN, Dobbing J (1968) Applied neurochemistry. Blackwell, Oxford, pp 253–316

Di Miceli M, Gronier B (2015) Psychostimulants and atomoxetine alter the electrophysiological activity of prefrontal cortex neurons, interaction with catecholamine and glutamate NMDA receptors. Psychopharmacology 232(12):2191–2205. doi:10.1007/s00213-014-3849-y

Diana MC, Santoro ML, Xavier G, Santos CM, Spindola LN, Moretti PN, Ota VK, Bressan RA, Abilio VC, Belangero SI (2015) Low expression of Gria1 and Grin1 glutamate receptors in the nucleus accumbens of spontaneously hypertensive rats (SHR). Psychiatry Res 229(3):690–694. doi:10.1016/j.psychres.2015.08.021

Dikranian K, Ishimaru MJ, Tenkova T, Labruyere J, Qin YQ, Ikonomidou C, Olney JW (2001) Apoptosis in the in vivo mammalian forebrain. Neurobiol Dis 8(3):359–379

Dimatelis JJ, Hsieh JH, Sterley TL, Marais L, Womersley JS, Vlok M, Russell VA (2015) Impaired energy metabolism and disturbed dopamine and glutamate signalling in the striatum and prefrontal cortex of the spontaneously hypertensive rat model of attention-deficit hyperactivity disorder. J Mol Neurosci 56(3):696–707. doi:10.1007/s12031-015-0491-z

Dobbing J (1970a) Undernutrition and the developing brain. The relevance of animal models to the human problem. Am J Dis Child 120(5):411–415

Dobbing J (1970b) The kinetics of growth. Lancet 2(7687):1358

Dobbing J (1970c) Undernutrition and the developing brain. The relevance of animal models to the human problem. Am J Dis Child 120(5):411–415

Dobbing J, Sands J (1970) Growth and development of the brain and spinal cord of the guinea pig. Brain Res 17(1):115–123

Dobbing J (1971) Undernutrition and the developing brain: the use of animal models ot elucidate the human problem. Psychiatr Neurol Neurochir 74(6):433–442

Dobbing J, Sands J (1979) Comparative aspects of the brain growth spurt. Early Hum Dev. 3 (1):79–83

Dueck A, Berger C, Wunsch K, Thome J, Cohrs S, Reis O, Haessler F (2015) The role of sleep problems and circadian clock genes in attention-deficithyperactivity disorder and mood disorders during childhood and adolescence: an update. J Neural Transm (15 Oct 2015)

Ebstein RP, Novick O, Umansky R, Priel B, Osher Y, Blaine D, Bennett ER, Nemanov L, Katz M, Belmaker RH (1996) Dopamine D4 receptor (D4DR) exon III polymorphism associated with the human personality trait of novelty seeking. Nat Genet 12:78–80

Ehli EA, Abdellaoui A, Hu Y, Hottenga JJ, Kattenberg M, van Beijsterveldt T, Bartels M, Althoff RR, Xiao X, Scheet P, de Geus EJ, Hudziak JJ, Boomsma DI, Davies GE (2012) De novo and inherited CNVs in MZ twin pairs selected for discordance and concordance on attention problems. Eur J Hum Genet 20(10):1037–1043. doi:10.1038/ejhg.2012.49

Ende G, Cackowski S, Van Eijk J, Sack M, Demirakca T, Kleindienst N, Bohus M, Sobanski E, Krause-Utz A, Schmahl C (2015) Impulsivity and aggression in female BPD and ADHD patients: associations with ACC glutamate and GABA concentrations. Neuropsychopharmacology. doi:10.1038/npp.2015.153

Endres D, Perlov E, Maier S, Feige B, Nickel K, Goll P, Bubl E, Lange T, Glauche V, Graf E, Ebert D, Sobanski E, Philipsen A, Tebartz van Elst L (2015) Normal neurochemistry in the prefrontal and cerebellar brain of adults with attention-deficit hyperactivity disorder. Front Behav Neurosci 9:242. doi:10.3389/fnbeh.2015.00242

Falhgren E, Nima AA, Archer T, Garcia D (2015) Person-centered osteopathic practice: patients' personality (body, mind, and soul) and health (ill-being and well-being). PeerJ 3:e1349. doi:10. 7717/peerj.1349

Fang Y, Ji N, Cao Q, Su Y, Chen M, Wang Y, Yang L (2015) Variants of dopamine beta hydroxylase gene moderate atomoxetine response in children with attention-deficit/hyperactivity disorder. J Child Adolesc Psychopharmacol (8 Oct 2015)

Franke B, Neale BM, Faraone SV (2009) Genome wide association studies in ADHD. Hum Genet 126:13–50. doi:10.1007/s00439-009-0663-4

Fredriksson A, Archer T (2002) Functional alteration by NMDA antagonist: effects of l-Dopa, neuroleptic drugs and postnatal administration. Amino Acids 25:111–132

Fredriksson A, Archer T (2003) Hyperactivity following postnatal NMDA antagonist treatment: reversal by d-amphetamine. Neurotox Res 5:549–564

Fredriksson A, Archer T (2004) Neurobehavioral deficits associated with apoptotic neurodegeneration and vulnerability for ADHD. Neurotox Res 6:435–456

Fredriksson A, Archer T, Alm H, Gordh T, Eriksson P (2004) Neurofunctional deficits and potentiated apoptosis by neonatal NMDA antagonist administration. Behav Brain Res 153:367–376

Froehlich TE, Lanphear BP, Epstein JN, Barbaresi WJ, Katusic SK, Kahn RS (2007) Prevalence, recognition, and treatment of attention-deficit/hyperactivity disorder in a national sample of US children. Arch Pediatr Adolesc Med 161(9):857–864

Froehlich TE, Lanphear BP, Auinger P, Hornung R, Epstein JN, Braun J, Kahn RS (2009) Association of tobacco and lead exposures with attention-deficit/hyperactivity disorder. Pediatrics 124(6):e1054–e1063. doi:10.1542/peds.2009-0738

Froehlich TE, McGough JJ, Stein MA (2010) Progress and promise of attention-deficit hyperactivity disorder pharmacogenetics. CNS Drugs. 24(2):99–117. doi:10.2165/11530290-000000000-00000

Froehlich TE, Anixt JS, Loe IM, Chirdkiatgumchai V, Kuan L, Gilman RC (2011) Update on environmental risk factors for attention-deficit/hyperactivity disorder. Curr Psychiatry Rep 13 (5):333–344. doi:10.1007/s11920-011-0221-3

Garcia D, Anckarsäter H, Lundström S (2013) Self-directedness and cooperativeness, psychosocial dysfunction and suffering in ESSENCE. Sci World J 2013, Article ID 416981, 1–10. doi:10.1155/2013/416981

Garcia D, Stråge A, Lundström S, Radovic S, Brändström S, Råstam M, Nilsson T, Cloninger CR, Kerekes N, Anckarsäter H (2014) Responsibility and cooperativeness are constrained not determined. Front Psychol 5:308. doi:10.3389/fpsyg.2014.00308

Gorter JA, de Bruin JP (1992) Chronic neonatal MK-801 treatment results in an impairment of spatial learning in the adult rat. Brain Res 580(1–2):12–17

Gorter JA, Botterblom MH, Feenstra MG, Boer GJ (1992a) Chronic neonatal NMDA receptor blockade with MK-801 alters monoamine metabolism in the adult rat. Neurosci Lett 137 (1):97–100

Gorter JA, Veerman M, Mirmiran M (1992b) Hippocampal neuronal responsiveness to NMDA agonists and antagonists in the adult rat neonatally treated with MK-801. Brain Res 572(1–2):176–181

Gorter JA, Veerman M, Mirmiran M, Bos NP, Corner MA (1991) Spectral analysis of the electroencephalogram in neonatal rats chronically treated with the NMDA antagonist MK-801. Brain Res Dev Brain Res 64(1–2):37–41

Grissom NM, Reyes TM (2013) Gestational overgrowth and undergrowth affect neurodevelopment: similarities and differences from behavior to epigenetics. Int J Dev Neurosci 31(6):406–414. doi:10.1016/j.ijdevneu.2012.11.006

Grizenko N, Fortier MÈ, Gaudreau-Simard M, Jolicoeur C, Joober R (2015) The effect of maternal stress during pregnancy on IQ and ADHD symptomatology. J Can Acad Child Adolesc Psychiatry 24(2):92–99

Grünblatt E, Bartl J, Iuhos DI, Knezovic A, Trkulja V, Riederer P, Walitza S, Salkovic-Petrisic M (2015) Characterization of cognitive deficits in spontaneously hypertensive rats, accompanied by brain insulin receptor dysfunction. J Mol Psychiatry 3(1):6. doi:10.1186/s40303-015-0012-6

Higuera-Matas A, Ucha M, Ambrosio E (2015) Long-term consequences of perinatal and adolescent cannabinoid exposure on neural and psychological processes. Neurosci Biobehav Rev 55:119–146. doi:10.1016/j.neubiorev.2015.04.020

Huebner SM, Tran TD, Rufer ES, Crump PM, Smith SM (2015) Maternal iron deficiency worsens the associative learning deficits and hippocampal and cerebellar losses in a rat model of fetal alcohol spectrum disorders. Alcohol Clin Exp Res. doi:10.1111/acer.12876

Ikonomidou C, Bittigau P, Koch C, Genz K, Hoerster F, Felderhoff-Mueser U, Tenkova T, Dikranian K, Olney JW (2001) Neurotransmitters and apoptosis in the developing brain. Biochem Pharmacol 62(4):401–405

Ishii S, Hashimoto-Torii K (2015) Impact of prenatal environmental stress on cortical development. Front Cell Neurosci 9:207. doi:10.3389/fncel.2015.00207

Ishimaru MJ, Ikonomidou C, Tenkova TI, Der TC, Dikranian K, Sesma MA, Olney JW (1999) Distinguishing excitotoxic from apoptotic neurodegeneration in the developing rat brain. J Comp Neurol 408(4):461–476

Jaber L, Kirsh D, Diamond G, Shuper A (2015) Long-term functional outcomes in Israeli adults diagnosed in childhood with attention-deficit hyperactivity disorder. Isr Med Assoc J 17(8):481–485

Jantzie LL, Talos DM, Jackson MC, Park HK, Graham DA, Lechpammer M, Folkerth RD, Volpe JJ, Jensen FE (2015) Developmental expression of N-methyl-D-aspartate (NMDA) receptor subunits in human white and gray matter: potential mechanism of increased vulnerability in the immature brain. Cereb Cortex 25(2):482–495. doi:10.1093/cercor/bht246

Jégou S, El Ghazi F, de Lendeu PK, Marret S, Laudenbach V, Uguen A, Marcorelles P, Roy V, Laquerrière A, Gonzalez BJ (2012) Prenatal alcohol exposure affects vasculature development in the neonatal brain. Ann Neurol 72(6):952–960. doi:10.1002/ana.23699

Jordan CJ, Taylor DM, Dwoskin LP, Kantak KM (2015) Adolescent d-amphetamine treatment in a rodent model of ADHD: Pro-cognitive effects in adolescence without an impact on cocaine cue reactivity in adulthood. Behav Brain Res. pii: S0166-4328(15)30233-3doi: 10.1016/j.bbr.2015.10.017

Josefsson K, Jokela M, Cloninger CR, Hintsanen M, Salo J, Hintsa T, Pulkki-Råback L, Keltikangas-Järvinen L (2013) Maturity and Change in Personality: De-velopmental Trends of Temperament and Character in Adulthood. Dev Psychopathol 25:713–727

Kalat JW (2001) Biological psychology. Wadsworth/Thomson Learning, Belmont, CA

Keimpema E, Zheng K, Barde SS, Berghuis P, Dobszay MB, Schnell R, Mulder J, Luiten PG, Xu ZD, Runesson J, Langel Ü, Lu B, Hökfelt T, Harkany T (2014) GABAergic terminals are a source of galanin to modulate cholinergic neuron development in the neonatal forebrain. Cereb Cortex 24(12):3277–3288. doi:10.1093/cercor/bht192

Kim J, Son Y, Kim J, Lee S, Kang S, Park K, Kim SH, Kim JC, Kim J, Takayama C, Im HI, Yang M, Shin T, Moon C (2015) Developmental and degenerative modulation of GABAergic transmission in the mouse hippocampus. Int J Dev Neurosci. pii: S0736-5748(15)30048-4. doi: 10.1016/j.ijdevneu.2015.08.009

Kleteckova L, Tsenov G, Kubova H, Stuchlik A, Vales K (2014) Neuroprotective effect of the 3α5β-pregnanolone glutamate treatment in the model of focal cerebral ischemia in immature rats. Neurosci Lett 564:11–15. doi:10.1016/j.neulet.2014.01.057

Kucyi A, Hove MJ, Biederman J, Van Dijk KR, Valera EM (2015) Disrupted functional connectivity of cerebellar default network areas in attention-deficit/hyperactivity disorder. Hum Brain Mapp 36(9):3373–3386. doi:10.1002/hbm.22850

Kyeong S, Park S, Cheon KA, Kim JJ, Song DH, Kim E (2015) A new approach to investigate the association between brain functional connectivity and disease characteristics of attention-deficit/hyperactivitydisorder: topological neuroimaging data analysis. PLoS One 10(9): e0137296. doi:10.1371/journal.pone.0137296

Langner I, Garbe E, Banaschewski T, Mikolajczyk RT (2013) Twin and sibling studies using health insurance data: the example of attention-deficit/hyperactivity disorder (ADHD). PLoS One 8(4):e62177. doi:10.1371/journal.pone.0062177

Li JT, Zhao YY, Wang HL, Wang XD, Su YA, Si TM (2015) Long-term effects of neonatal exposure to MK-801 on recognition memory and excitatory-inhibitory balance in rat hippocampus. Neuroscience 308:134–143. doi:10.1016/j.neuroscience.2015.09.003

Lin HY, Gau SS (2015) Atomoxetine treatment strengthens an anti-correlated relationship between functional brain networks in medication-naïve adults with attention-deficit hyperactivity disorder: a randomized double-blind placebo-controlled clinical trial. Int J Neuropsychopharmacol. pii: pyv094. doi:10.1093/ijnp/pyv094

Lindström K, Lindblad F, Hjern A (2011) Preterm birth and attention-deficit/hyperactivity disorder in schoolchildren. Pediatrics 127:858–865. doi:10.1542/peds.2010-1279

Liso Navarro AA, Sikoglu EM, Heinze CR, Rogan RC, Russell VA, King JA, Moore CM (2014) Effect of diet on brain metabolites and behavior in spontaneously hypertensive rats. Behav Brain Res 270:240–247. doi:10.1016/j.bbr.2014.05.013

Marriott AL, Ryan CL, Doucette TA (2012) Neonatal domoic acid treatment produces alterations to prepulse inhibition and latent inhibition in adult rats. Pharmacol Biochem Behav 103 (2):338–344. doi:10.1016/j.pbb.2012.08.022

Mc Fie S, Sterley TL, Howells FM, Russell VA (2012) Clozapine decreases exploratory activity and increases anxiety-like behaviour in the Wistar-Kyoto rat but not the spontaneously hypertensive rat model of attention-deficit/hyperactivity disorder. Brain Res 1467:91–103. doi:10.1016/j.brainres.2012.05.047

McLoughlin G, Palmer JA, Rijsdijk F, Makeig S (2014) Genetic overlap between evoked frontocentral theta-band phase variability, reaction time variability, and attention-deficit/hyper-activity disorder symptoms in a twinstudy. Biol Psychiatry 75(3):238–247. doi:10.1016/j.biopsych.2013.07.020

Meechan DW, Maynard TM, Tucker ES, Fernandez A, Karpinski BA, Rothblat LA, LaMantia AS (2015) Modeling a model: Mouse genetics, 22q11.2 Deletion Syndrome, and disorders of cortical circuit development. Prog Neurobiol 130:1–28. doi:10.1016/j.pneurobio.2015.03.004

Micoulaud-Franchi JA, Lopez R, Vaillant F, Richieri R, El-Kaim A, Bioulac S, Philip P, Boyer L, Lancon C (2015) Perceptual abnormalities related to sensory gating deficit are core symptoms inadults with ADHD. Psychiatry Res. pii: S0165-1781(15)30319-X. doi:10.1016/j.psychres.2015.09.016

Miller EM, Pomerleau F, Huettl P, Russell VA, Gerhardt GA, Glaser PE (2012) The spontaneously hypertensive and Wistar Kyoto rat models of ADHD exhibit sub-regional differences in dopamine release and uptake in the striatum and nucleus accumbens. Neuropharmacology 63(8):1327–1334. doi:10.1016/j.neuropharm.2012.08.020

Miller EM, Pomerleau F, Huettl P, Gerhardt GA, Glaser PE (2014) Aberrant glutamate signaling in the prefrontal cortex and striatum of the spontaneously hypertensive rat model of attention-deficit/hyperactivity disorder. Psychopharmacology (Berl) 231(15):3019–3029. doi:10.1007/s00213-014-3479-4

Mitchison GM, Njardvik U (2015) Prevalence and Gender Differences of ODD, Anxiety, and Depression in a Sample of Children With ADHD. J Atten Disord. pii:1087054715608442

Noble EP, Ozkaragoz TZ, Ritchie TL, Zhang X, Belin TR, Sparkes RS (1998) D2 and D4 dopamine receptor polymorphisms and personality. Am J Med Genet 81:257–267

Ono Y, Manki H, Yoshimura K, Muramatsu T, Mizushima H, Higuchi S, Yagi G, Kanba S, Asai M (1997) Association between dopamine D4 receptor (D4DR) exon III polymorphism and novelty seeking in Japanese subjects. Am J Med Genet 74:501–503

Olney JW, Ishimaru MJ, Bittigau P, Ikonomidou C (2000) Ethanol-induced apoptotic neurodegeneration in the developing brain. Apoptosis 5(6):515–521

Pettersson E, Anckarsäter H, Gillberg C, Lichtenstein P (2013) Different neurodevelopmental symptoms have a common genetic etiology. J Child Psychol Psychiatr Allied Discip 54 (12):1356–1365

Pires T de O, de Silva CM, de Assis SG (2013) Association between family environment and attention deficit hyperactivity disorder in children—mothers' and teachers' views. BMC Psychiatr 13:215–230. doi:10.1186/1471-244X-13-215

Pozzi L, Baviera M, Sacchetti G, Calcagno E, Balducci C, Invernizzi RW, Carli M (2011) Attention deficit induced by blockade of N-methyl D-aspartate receptors in the prefrontal cortex is associated with enhanced glutamate release and cAMP response element binding protein phosphorylation: role of metabotropic glutamate receptors 2/3. Neuroscience 176:336–348. doi:10.1016/j.neuroscience.2010.11.060

Purkayastha P, Malapati A, Yogeeswari P, Sriram D (2015) A review on GABA/glutamate pathway for therapeutic intervention of ASD and ADHD. Curr Med Chem (9 Feb 2015)

Richards JS, Hartman CA, Franke B, Hoekstra PJ, Heslenfeld DJ, Oosterlaan J, Arias Vásquez A, Buitelaar JK (2015) Differential susceptibility to maternal expressed emotion in children with ADHD and their siblings? Investigating plasticity genes, prosocial and antisocial behaviour. Eur Child Adolesc Psychiatry 24(2):209–217. doi:10.1007/s00787-014-0567-2

Rommel AS, Halperin JM, Mill J, Asherson P, Kuntsi J (2013) Protection from genetic diathesis in attention-deficit/hyperactivity disorder: possible complementary roles of exercise. J Am Acad Child Adolesc Psychiatry 52(9):900–910. doi:10.1016/j.jaac.2013.05.018

Russell VA (2003) In vitro glutamate-stimulated release of dopamine from nucleus accumbens core and shell of spontaneous hypertensive rats. Metab Brain Dis 18:161–168

Russell VA (2007) Neurobiology of animal models of attention-deficit hyperactivity disorder. J Neurosci Methods 161(2):185–198

Russell VA (2011) Overview of animal models of attention deficit hyperactivity disorder (ADHD). Curr Protoc Neurosci Chapter 9:Unit9.35. doi:10.1002/0471142301.ns0935s54

Rybakowski F, Slopien A, Dmitrzak-Weglarz M, Czerski P, Rajewski A, Hauser J (2006) The 5-HT2A-1438 A/G and 5-HTTLPR polymorphisms and personality dimensions in adolescent anorexia nervosa: association study. Neuropsychobiology 53:33–39

Sadile AG, Pellicano MP, Sagvolden T, Sergeant JA (1996) NMDA and non-NMDA sensitive [L-3H]glutamate receptor binding in the brain of the Naples high- and low-excitability rats: an autoradiographic study. Behav Brain Res 78(2):163–174

Salatino-Oliveira A, Wagner F, Akutagava-Martins GC, Bruxel EM, Genro JP, Zeni C, Kieling C, Polanczyk GV, Rohde LA, Hutz MH (2015) MAP1B and NOS1 genes are associated with working memory in youths withattention-deficit/hyperactivity disorder. Eur Arch Psychiatry Clin Neurosci (2 Aug 2015)

Samochowiec J, Rybakowski F, Czerski P, Zakrzewska M, Stepien G, Pelka-Wysiecka J, Horodnicki J, Rybakowski JK, Hauser J (2001) Polymorphisms in the dopamine, serotonin, and norepinephrine transporter genes and their relationship to temperamental dimensions measured by the Temperament and Character Inventory in healthy volunteers. Neuropsychobiology 43:248–253

Santoro ML, Santos CM, Ota VK, Gadelha A, Stilhano RS, Diana MC, Silva PN, Spíndola LM, Melaragno MI, Bressan RA, Han SW, Abílio VC, Belangero SI (2015) Expression profile of neurotransmitter receptor and regulatory genes in the prefrontal cortex of spontaneously hypertensive rats: relevance to neuropsychiatric disorders. Psychiatry Res 219(3):674–679. doi:10.1016/j.psychres.2014.05.034

Sasaki T, Hashimoto K, Oda Y, Ishima T, Kurata T, Takahashi J, Kamata Y, Kimura H, Niitsu T, Komatsu H, Ishikawa M, Hasegawa T, Shiina A, Hashimoto T, Kanahara N, Shiraishi T, Iyo M (2015) Decreased levels of serum oxytocin in pediatric patients with AttentionDeficit/Hyperactivity Disorder. Psychiatry Res 228(3):746–751. doi:10.1016/j.psychres.2015.05.029

Schuch V, Utsumi DA, Costa TV, Kulikowski LD, Muszkat M (2015) Attention deficit hyperactivity disorder in the light of the epigenetic paradigm. Front Psychiatry 6:126. doi:10.3389/fpsyt.2015.00126

Sidlauskaite J, Sonuga-Barke E, Roeyers H, Wiersema JR (2015) Default mode network abnormalities during state switching in attention deficit hyperactivity disorder. Psychol Med 1–10 (12 Oct 2015)

Simões RV, Cruz-Lemini M, Bargalló N, Gratacós E, Sanz-Cortés M (2015) Brain metabolite differences in one-year-old infants born small at term and association with neurodevelopmental outcome. Am J Obstet Gynecol. 213(2):210.e1–210.e11. doi:10.1016/j.ajog.2015.04.011

Somandepalli K, Kelly C, Reiss PT, Zuo XN, Cameron Craddock R, Yan CG, Petkova E, Xavier Castellanos F, Milham MP, Di Martino A (2015) Short-term test-retest reliability of resting state fMRI metrics in children with and without attention-deficit/hyperactivity disorder. Dev Cogn Neurosci. pii: S1878-9293(15)00082-1. doi:10.1016/j.dcn.2015.08.003

Sterley TL, Howells FM, Dimatelis JJ, Russell VA (2015) Genetic predisposition and early life experience interact to determine glutamate transporter (GLT1) and solute carrier family 12 member 5 (KCC2) levels in rat hippocampus. Metab Brain Dis (14 Oct 2015)

Sterley TL, Howells FM, Russell VA (2013) Evidence for reduced tonic levels of GABA in the hippocampus of an animal model of ADHD, the spontaneously hypertensive rat. Brain Res 1541:52–60. doi:10.1016/j.brainres.2013.10.023

Swanson JM, Sergeant JA, Taylor E, Sonuga-Barke EJS, Jensen PS, Cantwell DP (1998) Attention deficit hyperactivity disorder and hyperkynetic disorder. Lancet 351:329–343

Thapar A, Cooper M (2015) Attention deficit hyperactivity disorder. Lancet. pii: S0140-6736(15)00238-X. doi: 10.1016/S0140-6736(15)00238-X

Thissen AJ, Bralten J, Rommelse NN, Arias-Vasquez A, Greven CU, Heslenfeld D, Luman M, Oosterlaan J, Hoekstra PJ, Hartman C, Franke B, Buitelaar JK (2015) The role of age in association analyses of ADHD and related neurocognitive functioning: A proof of concept for dopaminergic and serotonergic genes. Am J Med Genet B Neuropsychiatr Genet. doi:10.1002/ajmg.b.32290

Tzanoulinou S, García-Mompó C, Riccio O, Grosse J, Zanoletti O, Dedousis P, Nacher J, Sandi C (2015) Neuroligin-2 expression in the prefrontal cortex is involved in attention deficits induced by peripubertal stress. Neuropsychopharmacology. doi:10.1038/npp.2015.200

Turski CA, Ikonomidou C (2012) Neuropathological sequelae of developmental exposure to antiepileptic and anesthetic drugs. Front Neurol 3:120. eCollection 2012

van der Kolk A, Bouwmans CA, Schawo SJ, Buitelaar JK, van Agthoven M, Hakkaart-van Roijen L (2015) Association between societal costs and treatment response in children and adolescents with ADHD and their parents. A cross-sectional study in the Netherlands. Springerplus. 4:224. doi:10.1186/s40064-015-0978-7

van Mil NH, Steegers-Theunissen RP, Bouwland-Both MI, Verbiest MM, Rijlaarsdam J, Hofman A, Steegers EA, Heijmans BT, Jaddoe VW, Verhulst FC, Stolk L, Eilers PH, Uitterlinden AG, Tiemeier H (2014) DNA methylation profiles at birth and child ADHD symptoms. J Psychiatry Res 49:51–59. doi:10.1016/j.jpsychires.2013.10.017

van Rooij D, Hartman CA, van Donkelaar MM, Bralten J, von Rhein D, Hakobjan M, Franke B, Heslenfeld DJ, Oosterlaan J, Rommelse N, Buitelaar JK, Hoekstra PJ (2015a) Variation in serotonin neurotransmission genes affects neural activation during response inhibition in adolescents and young adults with ADHD and healthy controls. World J Biol Psychiatry 1–10 (Oct 1 2015)

van Rooij D, Hoekstra PJ, Bralten J, Hakobjan M, Oosterlaan J, Franke B, Rommelse N, Buitelaar JK, Hartman CA (2015b) Influence of DAT1 and COMT variants on neural activation during response inhibition in adolescents with attention-deficit/hyperactivity disorder and healthy controls. Psychol Med 45(15):3159–3170. doi: 10.1017/S0033291715001130

Volk HE, Neuman RJ, Todd RD (2005) A systematic evaluation of ADHD and comorbid psychopathology in a population-based twin sample. J Am Acad Child Adolesc Psychiatry 44 (8):768–775

Webb E (2013) Poverty, maltreatment and attention deficit hyperactivity disorder. Arch Dis Child 98:397–400. doi:10.1136/archdischild-2012-303578

Wong KM, Cloninger CR (2010) A person-centered approach to clinical practice. Focus 8 (2):199–215

Womersley JS, Kellaway LA, Stein DJ, Gerhardt GA, Russell VA (2015) Effect of cocaine on striatal dopamine clearance in a rat model of developmental stress and attention-deficit/hyperactivity disorder. Stress 1–17 (23 Sep 2015)

Ye ZY, Li DP, Pan HL (2013) Regulation of hypothalamic presympathetic neurons and sympathetic outflow by group II metabotropic glutamate receptors in spontaneously hypertensive rats. Hypertension 62(2):255–262. doi:10.1161/HYPERTENSIONAHA.113.01466

Zimmermann AM, Jene T, Wolf M, Görlich A, Gurniak CB, Sassoè-Pognetto M, Witke W, Friauf E, Rust MB (2015) Attention-deficit/hyperactivity disorder-like phenotype in a mouse model with impaired actin dynamics. Biol Psychiatry 78(2):95–106. doi:10.1016/j.biopsych.2014.03.011

Zha YP, Wang YK, Deng Y, Zhang RW, Tan X, Yuan WJ, Deng XM, Wang WZ (2013) Exercise training lowers the enhanced tonically active glutamatergic input to the rostral ventrolateral medulla in hypertensive rats. CNS Neurosci Ther. 19(4):244–251. doi:10.1111/cns.12065

Zhang K, Chammas C, Soghomonian JJ (2015) Loss of glutamic acid decarboxylase (Gad67) in striatal neurons expressing the Drdr1a dopamine receptor prevents l-DOPA-induced dyskinesia in 6-hydroxydopamine-lesioned mice. Neuroscience. 303:586–594. doi:10.1016/j.neuroscience.2015.07.032

Zhou R, Bai Y, Yang R, Zhu Y, Chi X, Li L, Chen L, Sokabe M, Chen L (2011) Abnormal synaptic plasticity in basolateral amygdala may account for hyperactivity and attention-deficit in male rat exposed perinatally to low-dose bisphenol-A. Neuropharmacology. 60(5):789–798. doi:10.1016/j.neuropharm.2011.01.031

Perinatal 6-Hydroxydopamine to Produce a Lifelong Model of Severe Parkinson's Disease

John P. Kostrzewa, Rose Anna Kostrzewa, Richard M. Kostrzewa, Ryszard Brus and Przemysław Nowak

Abstract The classic rodent model of Parkinson's disease (PD) is produced by unilateral lesioning of pars compacta substantia nigra (SNpc) in adult rats, producing unilateral motor deficits which can be assessed by dopamine (DA) D_2 receptor (D_2-R) agonist induction of measurable unilateral rotations. Bilateral SNpc lesions in adult rats produce life-threatening aphagia, adipsia, and severe motor disability resembling paralysis—a PD model that is so compromised that it is seldom used. Described in this paper is a PD rodent model in which there is bilateral 99 % loss of striatal dopaminergic innervation, produced by bilateral intracerebroventricular or intracisternal 6-hydroxydopamine (6-OHDA) administration to perinatal rats. This procedure produces no lethality and does not shorten the life span, while rat pups continue to suckle through the pre-weaning period; and eat without impairment post-weaning. There is no obvious motor deficit during or after weaning, except with special testing, so that parkinsonian rats are indistinguishable from control and thus allow for behavioral assessments to be conducted in a blinded manner. L-DOPA (L-3,4-dihydroxyphenylalanine) treatment increases DA content in striatal tissue, also evokes a rise in extraneuronal (i.e., in vivo microdialysate) DA, and is able to evoke dyskinesias. D_2-R agonists produce effects similar to those of L-DOPA. In addition, effects of both D_1- and D_2-R agonist effects

J.P. Kostrzewa
ENT Associates, North Alabama, Huntsville, AL 35801, USA

R.A. Kostrzewa
Walgreen's Pharmacy, Colonial Heights, TN 37663, USA

R.M. Kostrzewa (✉)
Department of Biomedical Sciences, Quillen College of Medicine, East Tennessee State University, PO Box 70577, Johnson City, TN 37614, USA
e-mail: kostrzew@etsu.edu

R. Brus
Department of Nurse, High School of Strategic Planning, Koscielna 6, 41-303 Dabrowa Gornicza, Poland

P. Nowak
Department of Toxicology and Occupational Health Protection, Public Health Faculty, Medical University of Silesia, Medykow 18, 40-752 Katowice Ligota, Poland

© Springer International Publishing Switzerland 2015
Curr Topics Behav Neurosci (2016) 29: 313–332
DOI 10.1007/7854_2015_396
Published Online: 17 October 2015

on overt or latent receptor supersensitization are amenable to study. Elevated basal levels of reactive oxygen species (ROS), namely hydroxyl radical, occurring in dopaminergic denervated striatum are suppressed by L-DOPA treatment. Striatal serotoninergic hyperinnervation ensuing after perinatal dopaminergic denervation does not appear to interfere with assessments of the dopaminergic system by L-DOPA or D_1- or D_2-R agonist challenge. Partial lesioning of serotonin fibers with a selective neurotoxin either at birth or in adulthood is able to eliminate serotoninergic hyperinnervation and restore the normal level of serotoninergic innervation. Of all the animal models of PD, that produced by perinatal 6-OHDA lesioning provides the most pronounced destruction of nigrostriatal neurons, thus representing a model of severe PD, as the neurochemical outcome resembles the status of severe PD in humans but without obvious motor deficits.

Keywords Parkinson's disease · 6-hydroxydopamine · 6-OHDA · Animal model · Nigrostriatal tract · Dopamine · Serotonin · Receptor supersensitivity

Contents

1 Introduction 315
2 Nigrostriatal Dopaminergic Nerve Damage by Neonatal 6-OHDA 316
3 Striatal DA-R After n6-OHDA 317
   3.1 DA D_1-R Number 317
   3.2 DA D_2-R Number 318
4 DA-R Sensitization Status After n6-OHDA 318
   4.1 Latent DA D_1-R Sensitization 318
   4.2 Overt DA D_1-R Sensitization 319
   4.3 Reliance of D_1-RSS on the Serotoninergic System 319
   4.4 DA D_2-R Sensitization 320
5 Serotonin Neural Adaptations to n6-OHDA Lesioning of Nigrostriatal Neurons 320
6 L-DOPA Effects on DA Neurochemistry and Reactive Oxygen Species
in n6-OHDA-Lesioned Rats 321
   6.1 L-DOPA Effects in Striatal Tissue of n6-OHDA-Lesioned Rats 321
   6.2 Striatal DA and DOPAC 322
   6.3 Striatal 5-HT and 5-HIAA 322
   6.4 L-DOPA Effects on Striatal DA and DOPAC 322
   6.5 L-DOPA Effects on Striatal 5-HT and 5-HIAA 322
   6.6 Striatal HO· 323
   6.7 L-DOPA Effects on Striatal Tissue of n6-OHDA-Lesioned Rats 323
   6.8 Summary on Striatal Tissue 323
7 Striatal Microdialysates of n6-OHDA-Lesioned Rats 323
8 Summary on n6-OHDA-Lesioned Rats as a Model for Severe PD 324
References 325

1 Introduction

Parkinson's disease (PD), a neurodegenerative disorder characterized by age-related spontaneous degeneration of pars compacta substantia nigra (SNpc) dopaminergic neurons, has become readily amenable to animal modeling because of the discovery of relatively selective neurotoxins for dopaminergic neurons. The most commonly used of such neurotoxins are (1) the mitochondrial complex I inhibitor 1-methyl-4-phenyl-1,2,3,6-tetrahydropyridinium ion (MPP^+) (Langston et al. 1984b), also (2) its metabolic precursor 1-methyl-4-phenyl-1,2,3,6-tetrahydropyridine (MPTP) (Langston and Ballard 1984; Langston et al. 1984a; Manning-Bog and Langston 2007; Pasquali et al. 2014), (3) daily administration of rotenone, another complex I inhibitor (Ernster et al. 1963; Gutman et al. 1970) which destroys nigrostriatal neurons (Betarbet et al. 2000; Sherer et al. 2002, 2003), and (4) 6-hydroxydopamine (6-OHDA) which can be rendered dopaminergic selective when co-administered with desipramine, a ligand inhibitor of the norepinephrine transporter (Smith et al. 1973). Unilateral 6-OHDA lesioning of rodent SNpc is commonly used to model PD (Ungerstedt 1968), since motor behavior is maintained and vital behaviors persist after lesioning (e.g., eating, drinking, and grooming) (Ungerstedt 1971a). Putative anti-parkinsonian agents, typically dopamine (DA) D_2 receptor (D_2-R) agonists tested in this model, evoke unilateral rotational activity contralateral to the lesioned side (Ungerstedt 1971c)—an event related to the development of D_2-R supersensitivity (D_2-RSS) on the lesioned side (Ungerstedt 1971b, c). In black mice, MPTP, administered systemically, crosses the blood–brain barrier and is metabolized to MPP^+ which selectively destroys SNpc dopaminergic neurons bilaterally. Nevertheless, a reasonable number of SNpc neurons survive MPTP treatment, as well as dopaminergic innervation of striatum. White rats are virtually unaffected by MPTP and MPP^+, largely limiting these neurotoxins to black mice or primate modeling of PD.

Severe SNpc bilateral lesioning would represent an ideal rodent model for many experimental studies (Zigmond and Stricker 1984). Unfortunately such a lesion produced by intra-nigral 6-OHDA treatment of adult rodents produces aphagia, adipsia, cessation of grooming, and virtual motor paralysis (Ungerstedt 1971a; Zigmond and Stricker 1972). Rodents die within a few days except with extraordinary measures to maintain hydration, body heat, and cleanliness. In time, behavioral recovery from partial nigral lesions can occur (Neve et al. 1982; Dravid et al. 1984; Altar et al. 1987). These liabilities of bilateral 6-OHDA lesioning of rodents are overcome by administering 6-OHDA intracerebroventricularly or intracisternally to rodents—rats in particular—within a few days of birth. This treatment is non-lethal, virtually all rats survive and continue to eat, drink, and ambulate. In adulthood, except for a slight diminishment in body mass, 6-OHDA-lesioned rats are indistinguishable from control non-lesioned rats except by special testing (Breese et al. 1984a, b; Hamdi and Kostrzewa 1991).

The neontally 6-OHDA (n6-OHDA)-lesioned rat is described in detail in this paper as a valuable model of severe PD, characterized by ~99 % loss of SNpc and associated striatal dopaminergic denervation but with good motor activity (Kostrzewa et al. 1998). The elements of DA D_1-R sensitization are considered, as well as the slight influence on D_2-R sensitization and the involvement of serotonin (5-HT) receptor sensitization, namely 5-HT$_{2C}$-R, on dopaminergic systems. Effects of L-DOPA (L-3,4-dihydroxyphenylalanine) are described in terms of partial restoration of striatal tissue content of DA and influence on extraneuronal (i.e., in vivo microdialysate) levels of DA and L-3,4-dihydroxyphenylacetic acid (DOPAC) levels. The influence of DA-lesioning and L-DOPA treatment on intraneuronal and extraneuronal ROS, namely hydroxyl radical (HO·), is likewise described. The overall composite of neurochemical and behavioral effects of 6-OHDA, and its partial negation by L-DOPA, provide convincing evidence for the use of n6-OHDA rats as a near-ideal model of severe PD.

2 Nigrostriatal Dopaminergic Nerve Damage by Neonatal 6-OHDA

Neonatal 6-OHDA (n6-OHDA), either by a non-central route (sc, ip) (Jonsson and Sachs 1976; Jonsson et al. 1974; Sachs and Jonsson 1972, 1975) or central route (intracisternal) (Breese and Traylor 1972), produces marked destruction of both noradrenergic and dopaminergic nerves in brain. By pretreating with desipramine, protection is conferred on noradrenergic nerves and thereby 6-OHDA becomes relatively selective, producing predominately dopaminergic nerve destruction (Smith et al. 1973). While there is partial preservation of dopaminergic perikarya in the ventral tegmental nucleus (VTA) after 6-OHDA (Snyder et al. 1986; Fernandes Xavier et al. 1994), destruction of dopaminergic perikarya in substantia nigra (SN) is near total (Berger et al. 1985; Fernandes Xavier et al. 1994). As a consequence, dopaminergic innervation of VTA targets—namely septum, nucleus accumbens, and frontal cortex—is reduced (Luthman et al. 1990a), while dopaminergic innervation of the SN target, namely striatum, is near total (Snyder et al. 1986; Descarries et al. 1992). The extent of dopamine (DA) depletion in the respective brain regions is reflective of the degree of dopaminergic denervation, and the described effects of n6-OHDA are lifelong (Breese and Traylor 1972; Breese et al. 1984a, 1994; Breese and Breese 1998; Stachowiak et al. 1984).

Despite the drastic loss of dopaminergic innervation of brain by n6-OHDA treatment, survival is not altered. Also, overt behavioral effects are not obvious, although 6-OHDA-lesioned rats have slightly reduced body weight (Bruno et al. 1984). However, the aphasia and adipsia observed in rats lesioned as adults with 6-OHDA do not occur in rats lesioned neonatally with 6-OHDA (Breese and Traylor 1972; Smith et al. 1973). Sensorimotor function is maintained (Weihmuller and Bruno 1989a, b; Potter and Bruno 1989).

Nevertheless, with special testing there are "behavioral" deficits that can be revealed in adulthood in n6-OHDA-lesioned rats, analogous to deficits observed in adulthood 6-OHDA-lesioned rats (Papadeas and Breese 2014). n6-OHDA-lesioned rats, in adulthood, do not drink a sucrose solution (Smith et al. 1973); are unable to acquire an avoidance response during aversive learning (Smith et al. 1973; Shaywitz et al. 1976b; Pappas et al. 1980; Raskin et al. 1983; Whishaw et al. 1987); display skilled motor deficits (Whishaw et al. 1987); lack the ability for high-rate operant responding (Takeichi et al. 1986; Stellar et al. 1988); and have acquisition deficits in operant responding (Heffner and Seiden 1983).

The overall preservation of function in n6-OHDA-lesioned rats is considered to be related to increased DA synthesis by surviving dopaminergic nerves (Molina-Holgado et al. 1994), also increased storage of DA (Reader and Dewar 1999), and release of greater amounts of DA (Castañeda et al. 1990a, b). Notably, further adulthood reduction of DA in the n6-OHDA-lesioned rats, either by impairing tyrosine hydroxylase (TOH) activity or by adulthood 6-OHDA treatment, produces deficits resembling that observed by single adulthood 6-OHDA treatment (Rogers and Dunnett 1989b).

The most obvious change from controls is the absence of catalepsy following adulthood treatment with DA receptor (DA-R) antagonists (Bruno et al. 1984; Duncan et al. 1987). Also, n6-OHDA-lesioned rats are hyperactive in adulthood and have been used as an animal model of attention-deficit hyperactivity disorder (ADHD), since amphetamine or methylphenidate suppresses hyperactivity in these rats versus their promotion of hyperactivity in intact control rats (Shaywitz et al. 1976a, b; see Kostrzewa et al. 2008; Kostrzewa et al. 2016).

One negative aspect of n6-OHDA-lesioned rats is the potential for L-DOPA induction of self-injurious behavior (SIB). Haloperidol partially attenuates L-DOPA-induced SIB (Breese et al. 1984a, b, 1985b), while D_1-R antagonists fully block the effect (Breese et al. 1985a, 1989, 1990a, b; Criswell et al. 1992). Notably, a D_1-R agonist, alone, does not evoke SIB, nor does a D_2-R agonist, alone, but the combination of a D_1-R agonist with a D_2-R agonist can evoke SIB (Breese et al. 1985a).

It is noteworthy that in rats treated with haloperidol in drinking water for 11 months, to model tardive dyskinesia (Huang et al. 1997), there is a much higher level of haloperidol-evoked VCMs in n6-OHDA-lesioned rats versus intact control rats. Also, when haloperidol treatment is terminated for 9 months, 5-HT_2-R antagonists abate the high level of VCMs in n6-OHDA-lesioned rats, while D_1-R antagonists do not do so (Huang et al. 1997; see Kostrzewa and Brus 2016).

3   Striatal DA-R After n6-OHDA

3.1   DA D_1-R Number

The relative number of striatal DA D_1 receptors (D_1-R) following n6-OHDA treatment is reported to be slightly increased (Broaddus and Bennett 1990), unchanged (Breese et al. 1987; Luthman et al. 1990b; Duncan et al. 1993; Gong et al. 1994), or slightly decreased (Dewar et al. 1990, 1997; Molina-Holgado et al. 1995). Regardless, there is no change in the relative number of both high-affinity D_1-R and low-affinity D_1-R (Gong et al. 1994).

3.2 DA D_2-R Number

When striatal D_2-R number was assessed by [^{3}H]spiperone binding, there was no change from control in the n6-OHDA-lesioned rats (Breese et al. 1987; Duncan et al. 1987; Kostrzewa and Hamdi 1991). However, with [^{3}H]raclopride binding striatal, D_2-R was found to be increased at 1–3 months in rostral striatum of n6-OHDA-lesioned rats (Dewar et al. 1990); and increased at a later time, throughout the striatum (Radja et al. 1993b).

Because the D_1-R class includes both D_1-R and D_5-R subclasses, and because the D_2-R class includes D_2-R, D_3-R and D_4-R subclasses (see Strange 1993), it is possible that different subclasses of these receptors are altered, in different ways in each subregion of brain, and according to 6-OHDA dosage. Accordingly, it is difficult to ascribe a particular behavior alteration with a specific receptor subclass in any defined brain region. With this caveat, it appears that alterations in the D_4-R subclass in striatum and nucleus accumbens are more closely associated with the hyperactivity of n6-OHDA-lesioned rats; and D_4-R antagonists attenuate the hyperactivity (Zhang et al. 2002a, b).

4 DA-R Sensitization Status After n6-OHDA

4.1 Latent DA D_1-R Sensitization

Regardless of whether DA D_1-R number is altered in n6-OHDA-lesioned rats, there is no obvious increase in stereotypic and locomotor effects of an initial D_1-R agonist treatment. However, with repeated D_1-R agonist treatments, there is ultimate development of D_1-RSS (Breese et al. 1985a, b), as evidenced by the fact that repeated D_1-R agonist treatments either during postnatal ontogeny (Hamdi and Kostrzewa 1991; Gong et al. 1993a) or in adulthood (Breese et al. 1987) produce an abnormal increase in the stereotypic (nucleus accumbens/striatum sites) and loco-motor (nucleus accumbens site) response. The D_1-RSS produced by ontogenetic D_1-R agonist treatments is incomplete, as additional D_1-R agonist treatments pro-duce further D_1-R agonist supersensitization (Gong et al. 1993a). The development of RSS is designated as a "priming" phenomenon (Breese et al. 1987). D_1-RSS is not accompanied by an increase in the D_1-R number (i.e., B_{max}) or affinity (K_d) (Hamdi and Kostrzewa 1991); nor in the percentage of high-affinity or low-affinity receptors, nor by a change in DA-stimulated adenylate cyclase activity (Gong et al. 1994). However, D_1-RSS is permanent (Criswell et al. 1989; Kostrzewa and Gong 1991).

Repeated pre-weanling D_2-R agonist treatments also prime (i.e., supersensitize) D_1-R in adulthood for locomotor and stereotyped effects (Criswell et al. 1989).

4.2 Overt DA D_1-R Sensitization

In n6-OHDA-lesioned rats, the first D_1-R agonist dose produces a marked increase in the numbers of vacuous chewing movements (VCMs) (i.e., oral activity = oral dyskinesia) versus the D_1-R agonist response in non-lesioned control rats (Kostrzewa and Gong 1991). Pre-weaning D_1-R agonist treatments, moreover, promotes an even further increase in the adulthood D_1-R induction of VCMs, indicating further sensitization, priming, of D_1-R associated with VCMs (Gong et al. 1993a).

4.3 Reliance of D_1-RSS on the Serotoninergic System

This overt D_1-RSS is accompanied by simultaneous serotonin (5-HT) receptor supersensitization (RSS) (Gong and Kostrzewa 1992; el Mansari et al. 1994), which appears to reside primarily with the 5-HT_{2C}-R subtype (Gong et al. 1992). This 5-HT_{2C}-RSS and its inductive effect on VCMs are not abated by D_1-R antagonists. In contrast, D_1-R induction of VCMs is abated by 5-HT_2-R antagonists, indicating that D_1-R-mediated effects are reliant of 5-HT_2 receptor effects (Gong and Kostrzewa 1992; Gong et al. 1992, 1993b), residing at least in part in ventral striatum (Plech et al. 1995). In fact, the development of overt D_1-RSS after n6-OHDA is suppressed if 5-HT innervation is largely destroyed by the serotoninergic neurotoxin 5,7-dihydroxytryptamine (5,7-DHT), administered simultaneously with n6-OHDA administration (Brus et al. 1994), or when the serotoninergic hyperinnervation is eliminated in adulthood by 5,7-DHT treatment (unpublished).

Development of 5-HT_2-RSS occurs when n6-OHDA treatment produces a loss of striatal DA $\geq$ 97 %, but not when the loss of striatal DA is $\leq$88 % (Gong et al. 1993b) or when n6-OHDA treatment is later than three days of post-birth (Kostrzewa et al. 1993a). In fact, 5-HT_2-RSS can persist when D_1-RSS does not exist in n6-OHDA-lesioned rats (Gong et al. 1993b; Kostrzewa et al. 1993a). Therefore, serotoninergic adaptations to n6-OHDA treatment may account for multiple behavioral interactions, as well as responsiveness to DA-R and 5-HT-R agonists (Luthman et al. 1991; Kostrzewa et al. 1993b).

4.4 DA D₂-R Sensitization

In contrast to the noted priming of D_1-R in rats lesioned neonatally with 6-OHDA, there is no obvious sensitization, overt or latent, of D_2-R (Criswell et al. 1989). However, when the D_2-R antagonist spiperone is administered in adult n6-OHDA-lesioned rats, there is enhanced oral activity (VCMs) compared to unlesioned controls (Hamdi and Kostrzewa 1991). Repeated pre-weanling D_2-R agonist treatments in n6-OHDA-lesioned rats sensitize for D_2-R agonist-induced vertical jumping at 3 weeks of age or later (Kostrzewa and Kostrzewa 2012). However, ontogenetic quinpirole treatments fail to prime for D_2 agonist-enhancement of locomotor activity in adult n6-OHDA-lesioned rats (Brus et al. 2003). Descriptions of DA D_1-R and D_2-R sensitization in n6-OHDA-lesioned rats are discussed in detail in an earlier report (Kostrzewa 1995).

5 Serotonin Neural Adaptations to n6-OHDA Lesioning of Nigrostriatal Neurons

The relative dopaminergic denervation of striatum attending n6-OHDA treatment is accompanied by an elevation of striatal serotonin (5-hydroxytryptamine, 5-HT) levels in adulthood (Breese et al. 1984a; Stachowiak et al. 1984)—an outcome occurring only if striatal DA is depleted by >80 % (Towle et al. 1989; Gong et al. 1993b) and when 6-OHDA is administered within the first ten days of post-birth (Kostrzewa et al. 1993a). The increase in striatal 5-HT is associated with increased striatal synaptosomal uptake of [^{3}H]5-HT (Stachowiak et al. 1984) or other ligands of the 5-HT transporter (Molina-Holgado et al. 1994; Soucy et al. 1994; Descarries et al. 1995).

Elevated striatal 5-HT content is attributable to serotoninergic terminal hyper-innervation of striatum (Snyder et al. 1986; Luthman et al. 1987; Towle et al. 1989)—mainly in rostral striatum (Stachowiak et al. 1984; Snyder et al. 1986; Luthman et al. 1987; Descarries et al. 1992), a near doubling of serotoninergic fiber number in this region in n6-OHDA-lesioned rats (Mrini et al. 1995). In caudal striatum, serotoninergic innervation is increased only by $\sim$20 % (Mrini et al. 1995), thereby inverting the normal gradient of increasing striatal rostro-caudal innervation (Ternaux et al. 1977; Soghomonian et al. 1989). Serotoninergic hyperinnervation develops fully by 2–3 months of post-birth (Dewar et al. 1990).

By retrograde tracing of horseradish peroxidase, it appears that medial and dorsal raphe perikarya, which normally provide serotoninergic innervation to caudal striatum, largely account for serotoninergic hyperinnervation of rostral striatum (Snyder et al. 1986), and still retaining the typical 10 % ratio of synaptic to non-synaptic striatal serotoninergic terminations (Descarries et al. 1992).

While serotoninergic hyperinnervation does not produce an elevation in extra-neuronal 5-HT levels (Jackson and Abercrombie 1992), serotoninergic hyperinnervation is associated with reduced acetylcholine release (Jackson et al. 1988). Serotoninergic hyperinnervation does not preserve functions that are impaired by dopaminergic denervation, since perinatal treatment with the serotoninergic neurotoxin 5,7-DHT does not produce greater functional impairment (Breese et al. 1978; Bruno et al. 1987).

As determined by quantitative ligand-binding autoradiography of serotoninergic hyperinnervated striatum, there is an approximate increase in numbers of 5-HT_{1B} (+30 %), 5-HT_{1nonAB} (i.e., 5-HT_{2C}, 5-HT_{1D}, 5-HT_{1E}) (+40 %), and 5-HT_{2A} receptor number, but no change in $5\text{-HT}_{1A}\text{-R}$ number (Radja et al. 1993a). Moreover, elevations in $5\text{-HT}_{1B}\text{-R}$ occurred also in globus pallidum and substantia nigra (i.e., striatopallidal and striatonigral pathways); elevations in $5\text{-HT}_{1nonAB}\text{-R}$ were observed in substantia nigra (i.e., striatonigral pathway) (Radja et al. 1993a). GABA (gamma-aminobutyric acid), cholinergic (Jackson et al. 1993; Kostrzewa and Neely 1993), and other systems are altered in adulthood in n6-OHDA-lesioned rats (see Papadeas and Breese 2014).

6 L-DOPA Effects on DA Neurochemistry and Reactive Oxygen Species in n6-OHDA-Lesioned Rats

In vitro DA in high concentration is overtly toxic to cells in culture, possibly by virtue of its auto-oxidation to DA-hydroquinone, DA-o-quinone, DA-p-quinone, adrenochrome, and other analogs including DA-semiquinone which recycles in a schema to generate additional ROS including hydrogen peroxide (H_2O_2), superoxide anion (O_2^{-}), and hydroxyl radical (HO·) (Senoh and Witkop 1959a, b; Senoh et al. 1959; Graham et al. 1978; Kaur and Halliwell 1996; Bindoli et al. 1999; Segura-Aguilar 2001; Segura-Aguilar and Paris 2014).

There is thus the quandary as to whether L-DOPA, the most efficacious drug for treatment of PD, might pose the risk of promoting dopaminergic cell death by virtue of greater intraneuronal formation of neurotoxic ROS, while acutely and simultaneously alleviating neuromuscular symptoms. Accordingly, a series of studies was conducted in the n6-OHDA-lesioned rats to explore this possible outcome.

6.1 L-DOPA Effects in Striatal Tissue of n6-OHDA-Lesioned Rats

Intact and n6-OHDA-lesioned rats at ten weeks of post-birth were acutely treated with L-DOPA (60 mg/kg i.p.) subsequent to carbidopa (12.5 mg/kg i.p., 30 min) pretreatment. Following ketamine–xylazine anesthetization and surgical implantation of a cannula guide, salicylic acid (8 micromoles) was injected intracerebroventricularly so that HO· could be detected in striatal tissue as the salicylate spin trap products, 2,3- and 2,5-dihydroxybenzoic acid (2,3-DHBA; 2,5-DHBA), 45 min after the prior L-DOPA treatment (Kostrzewa et al. 2000). 2,5-DHBA is reflective of cytochrome P450 metabolism (Giovanni et al. 1995; Dajas-Bailador et al. 1998). The following was determined.

6.2 Striatal DA and DOPAC

Endogenous DA and DOPAC in striatal tissue of n6-OHDA-lesioned rats at 10 weeks of post-birth were reduced by ~ 99 % (Kostrzewa et al. 2000), which was shown in related studies to represent an equivalent reduction in TOH immunoreactive fibers and DA transporters (DATs) in striatum, along with a similar loss of SNpc perikarya (Berger et al. 1985; Snyder et al. 1986; Descarries et al. 1992; Fernandez Xavier et al. 1994).

6.3 Striatal 5-HT and 5-HIAA

Endogenous 5-HT and 5-hydroxyindoleacetic acid (5-HIAA) in neostriatal tissue of n6-OHDA-lesioned at 10 weeks of post-birth were elevated by ~ 75 and ~ 50 %, respectively (Kostrzewa et al. 2000)—shown in related studies to represent serotoninergic hyperinnervation of striatum (Stachowiak et al. 1984; Berger et al. 1985; Snyder et al. 1986; Descarries et al. 1992).

6.4 L-DOPA Effects on Striatal DA and DOPAC

In intact control rats, L-DOPA treatment elevated striatal DOPAC content by 100 % but had no effect on striatal DA content—indicating rapid metabolism of newly formed DA to DOPAC. In n6-OHDA-lesioned rats, L-DOPA produced (1) ~ 30-fold elevation in striatal DA, to a level equivalent to ~ 25 % of that of untreated intact striatum, and (2) ~ 30-fold elevation in striatal DOPAC, to a level equivalent to ~ 100 % of that of untreated intact striatum (Kostrzewa et al. 2000).

6.5 L-DOPA Effects on Striatal 5-HT and 5-HIAA

In intact control rats, L-DOPA treatment reduced endogenous striatal 5-HT content by ~ 30 %, while elevating 5-HIAA by ~ 25 %. This is likely reflective of L-DOPA uptake by serotoninergic fibers followed by DA formation and consequent displacement of 5-HT by newly formed DA, which would be a false-transmitter in serotoninergic nerve terminals (Kannari et al. 2000). 5-HIAA elevation would reflect rapid metabolism of displaced 5-HT (Kostrzewa et al. 2000).

6.6   Striatal HO·

Tissue content of HO·, as indicated by 2,3-DHBA and 2,5-DHBA contents, was elevated more than 3-fold in the striatum of n6-OHDA-lesioned rats at ten weeks of post-birth, versus intact control striatum (Kostrzewa et al. 2000). This finding implies that DA would normally maintain low tissue levels of HO·; dopaminergic denervation is accompanied by greater HO· formation/retention.

6.7   L-DOPA Effects on Striatal Tissue of n6-OHDA-Lesioned Rats

L-DOPA treatment reduced HO· content in intact control striatum, as indicated by $\sim$60 % reduction in 2,3-DHBA and $\sim$95 % reduction in 2,5-DHBA levels. In n6-OHDA-lesioned rats, L-DOPA similarly reduced 2,5-DHBA by $\sim$50 % but failed to alter 2,3-DHBA content. These findings similarly indicate that L-DOPA-derived DA was slightly neuroprotective (Kostrzewa et al. 2000, 2002).

6.8   Summary on Striatal Tissue

In the n6-OHDA-lesioned rat model of PD, dopaminergic denervation of striatum is associated with a marked elevation of tissue HO·, while L-DOPA treatment is acutely associated with a slight decrease in enzymatically formed HO· (i.e., 2,5-DHBA). This implies that DA is reflective of a neuroprotective species, while its relative absence leads to greater ROS formation.

7   Striatal Microdialysates of n6-OHDA-Lesioned Rats

L-DOPA effects were also assessed by in vivo microdialysis in the striatum of awake and freely moving intact and n6-OHDA-lesioned rats. In these rats in adulthood, basal levels of striatal extraneuronal levels of DA and DOPAC were much lower, as expected since there was far less striatal dopaminergic innervation in the lesioned rats (Kostrzewa et al. 2005; Nowak et al. 2010). However, basal extraneuronal levels of 2,3- and 2,5-DHBA in the striatum of n6-OHDA-lesioned rats were elevated from that of intact control rats by 2-fold and 3-fold, respectively (Nowak et al. 2010). This change is comparable to that observed for striatal tissue levels of 2,3- and 2,5-DHBA in n6-OHDA-lesioned rats.

L-DOPA administration (100 mg/kg i.p.; carbidopa, 12.5 mg/kg i.p., 30 min pretreatment) increased the striatal microdialysate level of DA in intact control rats by ~5-fold and in n6-OHDA-lesioned rats by ~25-fold. Consequently, the attained extraneuronal level of DA following L-DOPA was actually ~3 times higher in lesioned versus intact controls (Abercrombie et al. 1990; Kostrzewa et al. 2005). In the striatum of n6-OHDA-lesioned rats, there is a relative dopaminergic denervation and near-total absence of DATs. Also in striatum, DA exocytosis is predominately volume transmission versus synaptic transmission. Therefore, once DA is released from the few remaining dopaminergic terminals in n6-OHDA-lesioned rats, there is virtually no recapture of release DA—accounting for the high extraneuronal levels after L-DOPA (Fuxe et al. 1988). In contrast, because the total amount of DOPAC formed from DA in the striatum of intact rats is so much greater than that for n6-OHDA-lesioned rats, L-DOPA-induced extraneuronal levels of DOPAC are much higher in intact control than in n6-OHDA-lesioned rats (Nowak et al. 2010).

In n6-OHDA-lesioned rats, striatal extraneuronal levels of HO·, reflected by measures of 2,3- and 2,5-DHBA, were elevated ~2- to 3-fold—similar to the higher striatal tissue levels of HO· in these rats. However, acute L-DOPA treatment had no influence on extraneuronal HO· levels (Nowak et al. 2010).

8 Summary on n6-OHDA-Lesioned Rats as a Model for Severe PD

In rats lesioned shortly after birth with 6-OHDA, there is near-total striatal dopaminergic denervation which is maintained throughout the life span. Rat pups do not die from the 6-OHDA treatment and are able to eat, drink, and maintain motor control. Growth rate is only slightly less than that of controls, from which n6-OHDA-lesioned rats are otherwise nearly indistinguishable except by special testing. The destructive effect of n6-OHDA treatment on nigrostriatal dopaminergic nerves is near complete, is reproducible and has a variability of only ~1 %.

The n6-OHDA model of PD fulfills neurochemical criteria of other rodent models of PD, but does not suffer from the debilitating (aphagia, adipsia, paralysis) and oft fatal effects of adulthood bilateral 6-OHDA injections (Ungerstedt 1971a; Kostrzewa et al. 2006). In adulthood, n6-OHDA-lesioned rats have a basal increase in HO·/ROS levels in tissue and in microdialysates (i.e., intracellularly and extracellularly). Also, acute L-DOPA treatment of adulthood n6-OHDA-lesioned rats produces a greater increase in extraneuronal levels of DA, an increase in striatal tissue levels of both DA and DOPAC, and a reduction in striatal tissue HO·. Effects are analogous to findings in other rodent models of PD. One negative aspect of n6-OHDA-lesioned rats is a risk for L-DOPA induction of self-injurious behavior.

The striatal serotoninergic hyperinnervation in n6-OHDA-lesioned rats represents an effect that is not repeatable with other rodent models, but this element does not appear to alter the expected basal levels relating to DA nor the expected effects of L-DOPA. Also, it would be possible to (1) attenuate the serotoninergic hyperinnervation by administering a fixed dose of the serotoninergic neurotoxin 5,7-DHT simultaneous with n6-OHDA administration (see Brus et al. 1994) or (2) eliminate the serotoninergic hyperinnervation in adulthood by 5,7-DHT treatment (Kostrzewa et al. 1994).

The n6-OHDA-lesioned rat is considered to be a good model of severe PD. Advantages relating to behavioral elements are the maintained ambulation, maintained nutrition, and maintained grooming—such that adult n6-OHDA-lesioned rats are healthy and able to live the full life span, as per intact control rats. Advantages relating to neurochemical status are reflected in the near-total destruction of nigrostriatal fibers and near-total dopaminergic denervation of striatum—akin to the neurochemical status approached in severe PD in humans. The n6-OHDA-lesioned rat displays behavioral activation with acute L-DOPA treatment, and there is a corresponding increase in striatal tissue DA content and increase in extraneuronal DA content. The n6-OHDA-lesioned rat reliably fulfills the criteria for an animal model of severe PD.

References

Abercrombie ED, Bonatz AE, Zigmond MJ (1990) Effects of L-dopa on extracellular dopamine in striatum of normal and 6-hydroxydopamine-treated rats. Brain Res 525(1):36–44

Altar CA, Marien MR, Marshall JF (1987) Time course of adaptations in dopamine biosynthesis, metabolism, and release following nigrostriatal lesions: implications for behavioral recovery from brain injury. J Neurochem 48(2):390–399

Berger TW, Kaul S, Stricker EM, Zigmond MJ (1985) Hyperinnervation of the striatum by dorsal raphe afferents after dopamine-depleting brain lesions in neonatal rats. Brain Res 336 (2):354–358

Betarbet R, Sherer TB, MacKenzie G, Garcia-Osuna M, Panov AV, Greenamyre JT (2000) Chronic systemic pesticide exposure reproduces features of Parkinson's disease. Nat Neurosci 3(12):1301–1306

Bindoli A, Scutari G, Rigobello MP (1999) The role of adrenochrome in stimulating the oxidation of catecholamines. Neurotox Res 1(2):71–80

Breese CR, Breese GR (1998) The use of neurotoxins to lesion catecholamine-containing neurons to model clinical disorders: approach for defining adaptive neural mechanisms and role of neurotrophic factors in brain. In: Kostrzewa RM (ed) Highly selective neurotoxins: basic and clinical applications. Humana Press, Totowa, NJ, pp 19–73

Breese GR, Traylor TD (1972) Developmental characteristics of brain catecholamines and tyrosine hydroxylase in the rat: effects of 6-hydroxydopamine. Br J Pharmacol 44(2):210–222

Breese GR, Vogel RA, Mueller RA (1978) Biochemical and behavioral alterations in developing rats treated with 5,7-dihydroxytryptamine. J Pharmacol Exp Ther 205(3):587–595

Breese GR, Baumeister AA, McCown TJ, Emerick SG, Frye GD, Crotty K, Mueller RA (1984a) Behavioral differences between neonatal and adult 6-hydroxydopamine-treated rats to dopamine agonists: relevance to neurological symptoms in clinical syndromes with reduced brain dopamine. J Pharmacol Exp Ther 231(2):343–354

Breese GR, Baumeister AA, McCown TJ, Emerick SG, Frye GD, Mueller RA (1984b) Neonatal-6-hydroxydopamine treatment: model of susceptibility for self-mutilation in the Lesch–Nyhan syndrome. Pharmacol Biochem Behav 21(3):459–461

Breese GR, Baumeister A, Napier TC, Frye GD, Mueller RA (1985a) Evidence that D-1 dopamine receptors contribute to the supersensitive behavioral responses induced by L-dihydroxyphenylalanine in rats treated neonatally with 6-hydroxydopamine. J Pharmacol Exp Ther 235 (2):287–295

Breese GR, Napier TC, Mueller RA (1985b) Dopamine agonist-induced locomotor activity in rats treated with 6-hydroxydopamine at differing ages: functional supersensitivity of D-1 dopamine receptors in neonatally lesioned rats. J Pharmacol Exp Ther 234(2):447–455

Breese GR, Duncan GE, Napier TC, Bondy SC, Iorio LC, Mueller RA (1987) 6-hydroxydopamine treatments enhance behavioral responses to intracerebral microinjection of D_1- and D_2-dopamine agonists into nucleus accumbens and striatum without changing dopamine antagonist binding. J Pharmacol Exp Ther 240(1):167–176

Breese GR, Criswell HE, Duncan GE, Mueller RA (1989) Dopamine deficiency in self-injurious behavior. Psychopharmacol Bull 25(3):353–357. Review. Erratum in: Psychopharmacol Bull 1990; 26(3):296. PMID: 2697009

Breese GR, Criswell HE, Duncan GE, Mueller RA (1990a) A dopamine deficiency model of Lesch–Nyhan disease–the neonatal-6-OHDA-lesioned rat. Brain Res Bull 25(3):477–484. Review. PMID: 2127238

Breese GR, Criswell HE, Mueller RA (1990b) Evidence that lack of brain dopamine during development can increase the susceptibility for aggression and self-injurious behavior by influencing D_1-dopamine receptor function. Prog Neuropsychopharmacol Biol Psychiatry 14 (Suppl):S65–S80. Review. PMID: 1982973

Breese GR, Criswell HE, Johnson KB, O'Callaghan JP, Duncan GE, Jensen KF, Simson PE, Mueller RA (1994) Neonatal destruction of dopaminergic neurons. Neurotoxicology 15 (1):149–159. Review. PMID: 8090354

Broaddus WC, Bennett JP Jr (1990) Postnatal development of striatal dopamine function. II. Effects of neonatal 6-hydroxydopamine treatments on D_1 and D_2 receptors, adenylate cyclase activity and presynaptic dopamine function. Brain Res Dev Brain Res 52(1–2):273–277

Bruno JP, Snyder AM, Stricker EM (1984) Effect of dopamine-depleting brain lesions on suckling and weaning in rats. Behav Neurosci 98(1):156–161

Bruno JP, Stricker EM, Zigmond MJ (1985) Rats given dopamine-depleting brain lesions as neonates are subsensitive to dopaminergic antagonists as adults. Behav Neurosci 99 (4):771–775

Bruno JP, Jackson D, Zigmond MJ, Stricker EM (1987) Effect of dopamine-depleting brain lesions in rat pups: role of striatal serotonergic neurons in behavior. Behav Neurosci 101(6):806–811

Brus R, Kostrzewa RM, Perry KW, Fuller RW (1994) Supersensitization of the oral response to SKF 38393 in neonatal 6-hydroxydopamine-lesioned rats is eliminated by neonatal 5,7-dihydroxytryptamine treatment. J Pharmacol Exp Ther 268:231–237

Brus R, Kostrzewa RM, Nowak P, Perry KW, Kostrzewa JP (2003) Ontogenetic quinpirole treatments fail to prime for D2 agonist-enhancement of locomotor activity in 6-hydroxydopamine-lesioned rats. Neurotoxicity Res 5(5):329–338

Castañeda E, Whishaw IQ, Robinson TE (1990a) Changes in striatal dopamine neurotransmission assessed with microdialysis following recovery from a bilateral 6-OHDA lesion: variation as a function of lesion size. J Neurosci 10(6):1847–1854

Castañeda E, Whishaw IQ, Lermer L, Robinson TE (1990b) Dopamine depletion in neonatal rats: effects on behavior and striatal dopamine release assessed by intracerebral microdialysis during adulthood. Brain Res 508(1):30–39

Criswell H, Mueller RA, Breese GR (1989) Priming of D_1-dopamine receptor responses: long-lasting behavioral supersensitivity to a D_1-dopamine agonist following repeated administration to neonatal 6-OHDA-lesioned rats. J Neurosci 9(1):125–133

Criswell HE, Mueller RA, Breese GR (1992) Pharmacologic evaluation of SCH-39166, A-69024, NO-0756, and SCH-23390 in neonatal-6-OHDA-lesioned rats. Further evidence that self-mutilatory behavior induced by L-dopa is related to D_1 dopamine receptors. Neuropsychopharmacology 7(2):95–103

Dajas-Bailador FA, Martinez-Borges A, Costa G, Abin JA, Martignoni E, Nappi G, Dajas F (1998) Hydroxyl radical production in the substantia nigra after 6-hydroxydopamine and hypoxia-reoxygenation. Brain Res 813(1):18–25

Descarries L, Soghomonian JJ, Garcia S, Doucet G, Bruno JP (1992) Ultrastructural analysis of the serotonin hyperinnervation in adult rat neostriatum following neonatal dopamine denervation with 6-hydroxydopamine. Brain Res 569(1):1–13

Descarries L, Soucy JP, Lafaille F, Mrini A, Tanguay R (1995) Evaluation of three transporter ligands as quantitative markers of serotonin innervation density in rat brain. Synapse 21 (2):131–139

Dewar KM, Soghomonian JJ, Bruno JP, Descarries L, Reader TA (1990) Elevation of dopamine D_2 but not D_1 receptors in adult rat neostriatum after neonatal 6-hydroxydopamine denervation. Brain Res 536(1–2):287–296

Dewar KM, Paquet M, Reader TA (1997) Alterations in the turnover rate of dopamine D_1 but not D_2 receptors in the adult rat neostriatum after a neonatal dopamine denervation. Neurochem Int 30(6):613–621

Dravid A, Jaton AL, Enz A, Frei P (1984) Spontaneous recovery from motor asymmetry in adult rats with 6-hydroxydopamine-induced partial lesions of the substantia nigra. Brain Res 311 (2):361–365

Duncan GE, Criswell HE, McCown TJ, Paul IA, Mueller RA, Breese GR (1987) Behavioral and neurochemical responses to haloperidol and SCH-23390 in rats treated neonatally or as adults with 6-hydroxydopamine. J Pharmacol Exp Ther 243(3):1027–1034

Duncan GE, Breese GR, Criswell HE, Johnson KB, Schambra UB, Mueller RA, Caron MG, Fremeau RT Jr (1993) D_1 dopamine receptor binding and mRNA levels are not altered after neonatal 6-hydroxydopamine treatment: evidence against dopamine-mediated induction of D_1 dopamine receptors during postnatal development. J Neurochem 61(4):1255–1262

el Mansari M, Radja F, Ferron A, Reader TA, Molina-Holgado E, Descarries L (1994) Hypersensitivity to serotonin and its agonists in serotonin-hyperinnervated neostriatum after neonatal dopamine denervation. Eur J Pharmacol 261(1–2):171–178

Ernster L, Dallner G, Azzone GF (1963) Differential effects of rotenone and amytal on mitochondrial electron and energy transfer. J Biol Chem 238:1124–1131

Fernandes Xavier FG, Doucet G, Geffard M, Descarries L (1994) Dopamine neoinnervation in the substantia nigra and hyperinnervation in the interpeduncular nucleus of adult rat following neonatal cerebroventricular administration of 6-hydroxydopamine. Neuroscience 59(1):77–87

Fuxe K, Cintra A, Agnati LF, Härfstrand A, Goldstein M (1988) Studies on the relationship of tyrosine hydroxylase, dopamine and cyclic amp-regulated phosphoprotein-32 immunoreactive neuronal structures and D1 receptor antagonist binding sites in various brain regions of the male rat-mismatches indicate a role of D1 receptors in volume transmission. Neurochem Int 13 (2):179–197

Giovanni A, Liang LP, Hastings TG, Zigmond MJ (1995) Estimating hydroxyl radical content in rat brain using systemic and intraventricular salicylate: impact of methamphetamine. J Neurochem 64(4):1819–1825

Gong L, Kostrzewa RM (1992) Supersensitized oral response to a serotonin agonist in neonatal 6-OHDA treated rats. Pharmacol Biochem Behav 41:621–623

Gong L, Kostrzewa RM, Fuller RW, Perry KW (1992) Supersensitization of the oral response to SKF 38393 in neonatal 6-OHDA-lesioned rats is mediated through a serotonin system. J Pharmacol Exp Ther 261:1000–1007

Gong L, Kostrzewa RM, Brus R, Fuller RW, Perry KW (1993a) Ontogenetic SKF 38393 treatments sensitize dopamine D_1 receptors in neonatal 6-OHDA-lesioned rats. Brain Res Dev Brain Res 76(1):59–65

Gong L, Kostrzewa RM, Perry KW, Fuller RW (1993b) Dose-related effects of a neonatal 6-OHDA lesion on SKF 38393- and m-chlorophenylpiperazine-induced oral activity responses of rats. Brain Res Dev Brain Res 76(2):233–238

Gong L, Kostrzewa RM, Li C (1994) Neonatal 6-hydroxydopamine and adult SKF 38393 treatments alter dopamine D_1 receptor mRNA levels: absence of other neurochemical associations with the enhanced behavioral responses of lesioned rats. J Neurochem 63 (4):1282–1290

Graham DG, Tiffany SM, Bell WR Jr, Gutknecht WF (1978) Autoxidation versus covalent binding of quinones as the mechanism of toxicity of dopamine, 6-hydroxydopamine, and related compounds toward C1300 neuroblastoma cells in vitro. Mol Pharmacol 14(4):644–653

Gutman M, Singer TP, Casida JE (1970) Studies on the respiratory chain-linked reduced nicotinamide adenine dinucleotide dehydrogenase: XVII. Reaction sites of piericidin A and rotenone. J Biol Chem 245:1992–1997

Hamdi A, Kostrzewa RM (1991) Ontogenic homologous supersensitization of dopamine D_1 receptors. Eur J Pharmacol 203:115–120

Heffner TG, Seiden LS (1983) Impaired acquisition of an operant response in young rats depleted of brain dopamine in neonatal life. Psychopharmacology 79(2–3):115–119

Huang NY, Kostrzewa RM, Li C, Perry KW, Fuller RW (1997) Persistent spontaneous oral dyskinesias in haloperidol-withdrawn rats neonatally lesioned with 6-hydroxydopamine: absence of an association with the Bmax for [^{3}H]raclopride binding to neostriatal homogenates. J Pharmacol Exp Ther 280(1):268–276

Jackson D, Abercrombie ED (1992) In vivo neurochemical evaluation of striatal serotonergic hyperinnervation in rats depleted of dopamine at infancy. J Neurochem 58(3):890–897

Jackson D, Bruno JP, Stachowiak MK, Zigmond MJ (1988) Inhibition of striatal acetylcholine release by serotonin and dopamine after the intracerebral administration of 6-hydroxydopamine to neonatal rats. Brain Res 457(2):267–273

Jackson D, Abercrombie ED, Zigmond MJ (1993) Impact of L-dopa on striatal acetylcholine release: effects of 6-hydroxydopamine. J Pharmacol Exp Ther 267(2):912–918

Jonsson G, Sachs C (1976) Regional changes in [^{3}H]-noradrenaline uptake, catecholamines and catecholamine synthetic and catabolic enzymes in rat brain following neonatal 6-hydroxydopamine treatment. Med Biol 54(4):286–297

Jonsson G, Pycock C, Fuxe K, Sachs C (1974) Changes in the development of central noradrenaline neurones following neonatal administration of 6-hydroxydopamine. J Neurochem 22(3):419–426

Kannari K, Tanaka H, Maeda T, Tomiyama M, Suda T, Matsunaga M (2000) Reserpine pretreatment prevents increases in extracellular striatal dopamine following L-DOPA administration in rats with nigrostriatal denervation. J Neurochem 74(1):263–269

Kaur H, Halliwell B (1996) Measurement of oxidized and methylated DNA bases by HPLC with electrochemical detection. Biochem J 318(1):21–23. PMID: 8761446

Kostrzewa RM (1995) Dopamine receptor supersensitivity. Neurosci Biobehav Rev 19(1):1–17. Review. PMID: 7770190

Kostrzewa RM, Brus R (2016?) Life-long rodent model of tardive dyskinesia—persistence after antipsychotic drug withdrawal, In: Kostrzewa RM, Archer T (eds) Neurotoxin modeling of brain disorders—life-long outcomes in behavioral teratology. Springer, New York (this volume)

Kostrzewa RM, Gong L (1991) Supersensitized D_1 receptors mediate enhanced oral activity after neonatal 6-OHDA. Pharmacol Biochem Behav 39(3):677–682

Kostrzewa RM, Hamdi A (1991) Potentiation of spiperone-induced oral activity in rats after neonatal 6-hydroxydopamine. Pharmacol Biochem Behav 38(1):215–218

Kostrzewa RM, Kostrzewa FP (2012) Neonatal 6-hydroxydopamine lesioning enhances quinpirole-induced vertical jumping in rats that were quinpirole-primed during postnatal ontogeny. Neurotoxicity Res 21(2):231–235. Epub 19 Aug 2011. doi:10.1007/s12640-011-9268-5. PMID: 21853388

Kostrzewa RM, Neely D (1993) Enhanced pilocarpine-induced oral activity responses in neonatal 6-OHDA-treated rats. Pharmacol Biochem Behav 45:737–740

Kostrzewa RM, Brus R, Perry KW, Fuller RW (1993a) Age-dependence of a 6-hydroxydopamine lesion on SKF 38393- and m-chlorophenylpiperazine-induced oral activity responses of rats. Brain Res Dev Brain Res 76(1):87–93. PMID: 8306435

Kostrzewa RM, Gong L, Brus R (1993b) Serotonin (5-HT) systems mediate dopamine (DA) receptor supersensitivity. Acta Neurobiol Exp (Wars) 53(1):31–41

Kostrzewa RM, Brus R, Kalbfleisch JH, Perry KW, Fuller RW (1994) Proposed animal model of attention deficit hyperactivity disorder. Brain Res Bull 34(2):161–167

Kostrzewa RM, Reader TA, Descarries L (1998) Serotonin neural adaptations to ontogenetic loss of dopamine neurons in rat brain. J Neurochem 70(3):889–898. Review. PMID: 9489707

Kostrzewa RM, Kostrzewa JP, Brus R (2000) Dopaminergic denervation enhances susceptibility to hydroxyl radicals in rat neostriatum. Amino Acids 19(1):183–199

Kostrzewa RM, Kostrzewa JP, Brus R (2002) Neuroprotective and neurotoxic roles of levodopa (L-DOPA) in neurodegenerative disorders relating to Parkinson's disease. Amino Acids 23(1–3): 57–63. Review. PMID: 12373519

Kostrzewa RM, Nowak P, Kostrzewa JP, Kostrzewa RA, Brus R (2005) Peculiarities of L-DOPA treatment of Parkinson's disease. Amino Acids 28(2):157–164. Epub 9 Mar 2005. Review. PMID: 15750845

Kostrzewa RM, Kostrzewa JP, Brus R, Kostrzewa RA, Nowak P (2006) Proposed animal model of severe Parkinson's disease: neonatal 6-hydroxydopamine lesion of dopaminergic innervation of striatum. J Neural Transm Suppl 70:277–279

Kostrzewa RM, Kostrzewa JP, Kostrzewa RA, Nowak P, Brus R (2008) Pharmacological models of ADHD. J Neural Transm 115(2):287–298. Epub 12 Nov 2007. Review. PMID: 17994186

Kostrzewa JP, Kostrzewa RA, Kostrzewa RM, Brus R, Nowak P (2016) Perinatal 6-Hydroxydopamine Modeling of ADHD. In: Kostrzewa RM, Archer T (eds) Neurotoxin modeling of brain disorders—life-long outcomes in behavioral teratology. Springer, New York (this volume)

Langston JW, Ballard P (1984) Parkinsonism induced by 1-methyl-4-phenyl-1,2,3,6-tetrahydropyridine (MPTP): implications for treatment and the pathogenesis of Parkinson'sdisease. Can J Neurol Sci 11(1 Suppl):160–165. PMID: 6608979

Langston JW, Forno LS, Rebert CS, Irwin I (1984a) Selective nigral toxicity after systemic administration of 1-methyl-4-phenyl-1,2,5,6-tetrahydropyrine (MPTP) in the squirrel monkey. Brain Res 292(2):390–394

Langston JW, Irwin I, Langston EB, Forno LS (1984b) 1-Methyl-4-phenylpyridinium ion (MPP$^+$): identification of a metabolite of MPTP, a toxin selective to the substantia nigra. Neurosci Lett 48(1):87–92

Luthman J, Bolioli B, Tsutsumi T, Verhofstad A, Jonsson G (1987) Sprouting of striatal serotonin nerve terminals following selective lesions of nigro-striatal dopamine neurons in neonatal rat. Brain Res Bull 19(2):269–274

Luthman J, Brodin E, Sundström E, Wiehager B (1990a) Studies on brain monoamine and neuropeptide systems after neonatal intracerebroventricular 6-hydroxydopamine treatment. Int J Dev Neurosci 8(5):549–560

Luthman J, Lindqvist E, Young D, Cowburn R (1990b) Neonatal dopamine lesion in the rat results in enhanced adenylate cyclase activity without altering dopamine receptor binding or dopamine- and adenosine 3′:5′-monophosphate-regulated phosphoprotein (DARPP-32) immunoreactivity. Exp Brain Res 83(1):85–95

Luthman J, Fredriksson A, Plaznik A, Archer T (1991) Ketanserin and mianserin treatment reverses hyperactivity in neonatally dopamine-lesioned rats. J Psychopharmacol 5(4):418–425

Manning-Bog AB, Langston JW (2007) Model fusion, the next phase in developing animal models for Parkinson's disease. Neurotox Res 11(3–4):219–240. Review. PMID: 17449461

Molina-Holgado E, Dewar KM, Descarries L, Reader TA (1994) Altered dopamine and serotonin metabolism in the dopamine-denervated and serotonin-hyperinnervated neostriatum of adult rat after neonatal 6-hydroxydopamine. J Pharmacol Exp Ther 270(2):713–721

Molina-Holgado E, Van Gelder NM, Dewar KM, Reader TA (1995) Dopamine receptor alterations correlate with increased GABA levels in adult rat neostriatum: effects of a neonatal 6-hydroxydopamine denervation. Neurochem Int 27(4–5):443–451

Mrini A, Soucy JP, Lafaille F, Lemoine P, Descarries L (1995) Quantification of the serotonin hyperinnervation in adult rat neostriatum after neonatal 6-hydroxydopamine lesion of nigral dopamine neurons. Brain Res 669(2):303–308

Neve KA, Kozlowski MR, Marshall JF (1982) Plasticity of neostriatal dopamine receptors after nigrostriatal injury: relationship to recovery of sensorimotor functions and behavioral supersensitivity. Brain Res 244(1):33–44

Nowak P, Kostrzewa RA, Skaba D, Kostrzewa RM (2010) Acute L-DOPA effect on hydroxyl radical- and DOPAC-levels in striatal microdialysates of parkinsonian rats. Neurotox Res 17 (3):299–304. doi:10.1007/s12640-009-9105-2 Epub 4 Sep 2009

Papadeas ST, Breese GR (2014) 6-hydroxydopamine lesioning of dopamine neurons in neonatal and adult rats induces age-dependent consequences, in Section on Selective Neurotoxins. In: Kostrzewa RM (ed) Handbook of neurotoxicity. Springer, New York, pp. 133–198. ISBN 978-1-4614-5835-7 (print); ISBN 978-1-4614-5836-4 (eBook); ISBN 978-1-4614-7458-6 (print and electronic bundle). doi:10.1007/978-1-4614-5836-4_59

Pappas BA, Gallivan JV, Dugas T, Saari M, Ings R (1980) Intraventricular 6-hydroxydopamine in the newborn rat and locomotor responses to drugs in infancy: no support for the dopamine depletion model of minimal brain dysfunction. Psychopharmacology 70(1):41–46

Pasquali L, Caldarazzo-Ienco C, Fornai F (2014) MPTP neurotoxicity: actions, mechanisms, and animal modeling of Parkinson's disease. In: Kostrzewa RM (ed) Handbook of neurotoxicity. Springer, New York, pp. 133–198. ISBN 978-1-4614-5835-7 (print); ISBN 978-1-4614-5836-4 (eBook); ISBN 978-1-4614-7458-6 (print and electronic bundle). doi:10.1007/978-1-4614-5836-4_170

Plech A, Brus R, Kalbfleisch JH, Kostrzewa RM (1995) Enhanced oral activity responses to intrastriatal SKF 38393 and m-CPP are attenuated by intrastriatal mianserin in neonatal 6-OHDA-lesioned rats. Psychopharmacol 119:466–473

Potter BM, Bruno JP (1989) Food intake of rats depleted of dopamine as neonates is impaired by inhibition of catecholamine biosynthesis. Neurosci Lett 107(1–3):295–300

Radja F, Descarries L, Dewar KM, Reader TA (1993a) Serotonin 5-HT$_1$ and 5-HT$_2$ receptors in adult rat brain after neonatal destruction of nigrostriatal dopamine neurons: a quantitative autoradiographic study. Brain Res 606(2):273–285

Radja F, el Mansari M, Soghomonian JJ, Dewar KM, Ferron A, Reader TA, Descarries L (1993b) Changes of D$_1$ and D$_2$ receptors in adult rat neostriatum after neonatal dopamine denervation: quantitative data from ligand binding, in situ hybridization and iontophoresis. Neuroscience 57 (3):635–648

Raskin LA, Shaywitz BA, Anderson GM, Cohen DJ, Teicher MH, Linakis J (1983) Differential effects of selective dopamine, norepinephrine or catecholamine depletion on activity and learning in the developing rat. Pharmacol Biochem Behav 19(5):743–749

Reader TA, Dewar KM (1999) Effects of denervation and hyperinnervation on dopamine and serotonin systems in the rat neostriatum: implications for human Parkinson's disease. Neurochem Int 34(1):1–21. Review. PMID: 10100192

Rogers DC, Dunnett SB (1989b) Neonatal dopamine-rich grafts and 6-OHDA lesions independently provide partial protection from the adult nigrostriatal lesion syndrome. Behav Brain Res 34(1–2):131–146. PMID: 2504223

Sachs C, Jonsson G (1972) Degeneration of central noradrenaline neurons after 6-hydroxydopamine in newborn animals. Res Commun Chem Pathol Pharmacol 4(1):203–220

Sachs C, Jonsson G (1975) Effects of 6-hydroxydopamine on central noradrenaline neurons during ontogeny. Brain Res 99(2):277–291

Segura-Aguilar J (2001) DT-diaphorase: a protective enzyme of the dopaminergic system. In: Segura-Aguilar (ed) Mechanisms of degeneration and protection of the dopaminergic system. F.P. Graham Publishing Co., USA, pp 289–299. ISBN 1-929675-03-8

Segura-Aguilar J, Paris I (2014) Mechanisms of dopamine oxidation and Parkinson's disease. In: Kostrzewa RM (ed) Handbook of neurotoxicity. Springer, New York, pp 3–67. ISBN 978-1-4614-5835-7 (print); ISBN 978-1-4614-5836-4 (eBook); ISBN 978-1-4614-7458-6 (print and electronic bundle). doi:10.1007/978-1-4614-5836-4_53

Senoh S, Witkop B (1959a) Formation and rearrangements of aminochromes from a new metabolite of dopamine and some of its derivatives. J Amer Chem Soc 81:6231–6235

Senoh S, Witkop B (1959b) Non-enzymatic conversions of dopamine to norepinephrine and trihydroxyphenethylamine. J Amer Chem Soc 81:6222–6231

Senoh S, Witkop B, Creveling CR, Udenfriend S (1959) 2,4,5-tri-hydroxyphenethylamine, a new metabolite of 3,4-dihydroxyphenethylamine. J Amer Chem Soc 81:1768–1769

Shaywitz BA, Klopper JH, Yager RD, Gordon JW (1976a) Paradoxical response to amphetamine in developing rats treated with 6-hydroxydopamine. Nature 261(5556):153–155

Shaywitz BA, Yager RD, Klopper JH (1976b) Selective brain dopamine depletion in developing rats: an experimental model of minimal brain dysfunction. Science 191(4224):305–308

Sherer TB, Betarbet R, Stout AK, Lund S, Baptista M, Panov AV, Cookson MR, Greenamyre JT (2002) An in vitro model of Parkinson's disease: linking mitochondrial impairment to altered alpha-synuclein metabolism and oxidative damage. J Neurosci 22(16):7006–7015

Sherer TB, Kim JH, Betarbet R, Greenamyre JT (2003) Subcutaneous rotenone exposure causes highly selective dopaminergic degeneration and alpha-synuclein aggregation. Exp Neurol 179 (1):9–16

Smith RD, Cooper BR, Breese GR (1973) Growth and behavioral changes in developing rats treated intracisternally with 6-hydroxydopamine: evidence for involvement of brain dopamine. J Pharmacol Exp Ther 185(3):609–619

Snyder AM, Zigmond MJ, Lund RD (1986) Sprouting of serotoninergic afferents into striatum after dopamine-depleting lesions in infant rats: a retrograde transport and immunocytochemical study. J Comp Neurol 245(2):274–281

Soghomonian JJ, Descarries L, Watkins KC (1989) Serotonin innervation in adult rat neostriatum. II. Ultrastructural features: a radioautographic and immunocytochemical study. Brain Res 481 (1):67–86

Soucy JP, Lafaille F, Lemoine P, Mrini A, Descarries L (1994) Validation of the transporter ligand cyanoimipramine as a marker of serotonin innervation density in brain. J Nucl Med 35 (11):1822–1830

Stachowiak MK, Bruno JP, Snyder AM, Stricker EM, Zigmond MJ (1984) Apparent sprouting of striatal serotonergic terminals after dopamine-depleting brain lesions in neonatal rats. Brain Res 291(1):164–167

Stellar JR, Waraczynski M, Bruno JP (1988) Neonatal dopamine depletions spare lateral hypothalamic stimulation reward in adult rats. Pharmacol Biochem Behav 30(2):365–370

Strange PG (1993) New insights into dopamine receptors in the central nervous system. Neurochem Int 22(3):223–236. Review. PMID: 8095172

Takeichi T, Kurumiya S, Umemoto M, Olds ME (1986) Roles of catecholamine terminals and intrinsic neurons of the ventral tegmentum in self-stimulation investigated in neonatally dopamine-depleted rats. Pharmacol Biochem Behav 24(4):1101–1109

Ternaux JP, Héry F, Bourgoin S, Adrien J, Glowinski J, Hamon M (1977) The topographical distribution of serotoninergic terminals in the neostriatum of the rat and the caudate nucleus of the cat. Brain Res 121(2):3113–3126

Towle AC, Criswell HE, Maynard EH, Lauder JM, Joh TH, Mueller RA, Breese GR (1989) Serotonergic innervation of the rat caudate following a neonatal 6-hydroxydopamine lesion: an anatomical, biochemical and pharmacological study. Pharmacol Biochem Behav 34(2):367–374

Ungerstedt U (1968) 6-hydroxy-dopamine induced degeneration of central monoamine neurons. Eur J Pharmacol 5(1):107–110. PMID: 5718510

Ungerstedt U (1971a) Adipsia and aphagia after 6-hydroxydopamine induced degeneration of the nigro-striatal dopamine system. Acta Physiol Scand Suppl 367:95–122

Ungerstedt U (1971b) Postsynaptic supersensitivity after 6-hydroxy-dopamine induced degeneration of the nigro-striatal dopamine system. Acta Physiol Scand Suppl 367:69–93

Ungerstedt U (1971c) Striatal dopamine release after amphetamine or nerve degeneration revealed by rotational behaviour. Acta Physiol Scand Suppl 367:49–68

Weihmuller FB, Bruno JP (1989a) Age-dependent plasticity in the dopaminergic control of sensorimotor development. Behav Brain Res 35(2):95–109

Weihmuller FB, Bruno JP (1989b) Drinking behavior and motor function in rat pups depleted of brain dopamine during development. Dev Psychobiol 22(2):101–113

Whishaw IQ, Funk DR, Hawryluk SJ, Karbashewski ED (1987) Absence of sparing of spatial navigation, skilled forelimb and tongue use and limb posture in the rat after neonatal dopamine depletion. Physiol Behav 40(2):247–253

Zhang K, Davids E, Tarazi FI, Baldessarini RJ (2002a) Effects of dopamine D_4 receptor-selective antagonists on motor hyperactivity in rats with neonatal 6-hydroxydopamine lesions. Psychopharmacology 161(1):100–106 Epub 6 Mar 2002

Zhang K, Tarazi FI, Davids E, Baldessarini RJ (2002b) Plasticity of dopamine D_4 receptors in rat forebrain: temporal association with motor hyperactivity following neonatal 6-hydroxydopamine lesioning. Neuropsychopharmacology 26(5):625–633

Zigmond MJ, Stricker EM (1972) Deficits in feeding behavior after intraventricular injection of 6-hydroxydopamine in rats. Science 177(4055):1211–1214

Zigmond MJ, Stricker EM (1984) Parkinson's disease: studies with an animal model. Life Sci 35(1):5–18

Exercise and Nutritional Benefits in PD: Rodent Models and Clinical Settings

Trevor Archer and Richard M. Kostrzewa

Abstract Physical exercise offers a highly effective health-endowering activity as has been evidence using rodent models of Parkinson's disease (PD). It is a particularly useful intervention in individuals employed in sedentary occupations or afflicted by a neurodegenerative disorder, such as PD. The several links between exercise and quality-of-life, disorder progression and staging, risk factors and symptoms-biomarkers in PD all endower a promise for improved prognosis. Nutrition provides a strong determinant for disorder vulnerability and prognosis with fish oils and vegetables with a mediterranean diet offering both protection and resistance. Three factors determining the effects of exercise on disorder severity of patients may be presented: (i) Exercise effects upon motor impairment, gait, posture and balance, (ii) Exercise reduction of oxidative stress, stimulation of mitochondrial biogenesis and up-regulation of autophagy, and (iii) Exercise stimulation of dopamine (DA) neurochemistry and trophic factors. Running-wheel performance, as measured by distance run by individual mice from different treatment groups, was related to DA-integrity, indexed by striatal DA levels. Finally, both nutrition and exercise may facilitate positive epigenetic outcomes, such as lowering the dosage of L-Dopa required for a therapeutic effect.

Keywords Exercise · Nutrition · Parkinson's disease · Dopamine · Impairment · Amelioration · Epigenetics

T. Archer (✉)
Department of Psychology, University of Gothenburg, Gothenburg, Sweden
e-mail: trevor.archer@psy.gu.se

R.M. Kostrzewa
Department of Biomedical Sciences, Quillen College of Medicine,
East Tennessee State University, Johnson City, TN 37604, USA

© Springer International Publishing Switzerland 2015
Curr Topics Behav Neurosci (2016) 29: 333–352
DOI 10.1007/7854_2015_409
Published Online: 05 January 2016

Contents

1 Nutrition and PD 335
2 PD and Exercise 338
3 Exercise and DA-Integrity Relationship 339
4 Epigenetic Influences... 341
References... 342

Parkinson's disease (PD) is a relatively common, idiopathic neurodegenerative movement disorder characterized by impaired motor function, including resting tremors, rigidity, akinesia/bradykinesia, and postural instability as the cardinal symptoms (Gaggelli et al. 2006; Jankovic 2008; Lees et al. 2009). It is a progressive neurodegenerative disorder, and compared with familial forms, is associated most often with advanced age (>55 years-of-age). The pathophysiology of PD involves dopaminergic neuron death and accumulation of Lewy bodies associated with mutations in α-synuclein, a 14-kDa protein predominantly expressed in the brain and CNS (Rasia et al. 2005). PD patients show decreased levels of presynaptic dopamine (DA) neuron terminal markers in the basal ganglia (Felicio et al. 2009), consistent with loss of dopaminergic terminals due to degeneration of neuronal cell bodies in the substantia nigra pars compacta. PD patients exhibit decreased levels of DA transporters (DATs) and vesicular monoamine transporter type 2 (VMAT2), as well as reduced activity of dopa decarboxylase, assessed by striatal conversion of L-Dopa to DA, according to PET and SPECT analyses. Wu et al. (2012) using MRI showed that the substantia nigra pars compacta expressed a decreased connectivity with several regions, including the striatum, globus pallidus, subthalamic nucleus, thalamus, supplementary motor area, dorsolateral prefrontal cortex, insula, default mode network, temporal lobe, cerebellum, and pons in patients compared to controls. They found that L-Dopa administration partially normalized the pattern of connectivity to a similarity such as that expressed by the healthy volunteers involving causal connectivity of basal ganglia networks from the substantia nigra pars compacta. Postsynaptic D_2 DA receptors (D_2Rs) are either unaffected or increased in the striatum of untreated PD patients (Antonini et al. 1994). Oxidative injury appears to be one effect of α-synuclein (α-Syn) aggregates and could ultimately produce neuronal cell death. α-Syn, a 140 residue, intrinsically disordered protein is localized in presynaptic terminals of DA neurons (Yang et al. 2010). Autonomic nervous system involvement occurs at early stages in both PD and incidental Lewy body disease, and affects the sympathetic, parasympathetic, and enteric nervous systems. It has been proposed that α-Syn pathology in PD has a distal to proximal progression along autonomic pathways. According to Braakian notions, the enteric nervous systems is affected before the dorsal motor nucleus of the vagus, and distal axons of cardiac sympathetic nerves degenerate before there is loss of paravertebral sympathetic ganglion neurons. Cersosimo and Benarroch (2012a) have shown that consistent with neuropathological findings, some autonomic manifestations such as constipation or impaired cardiac uptake of

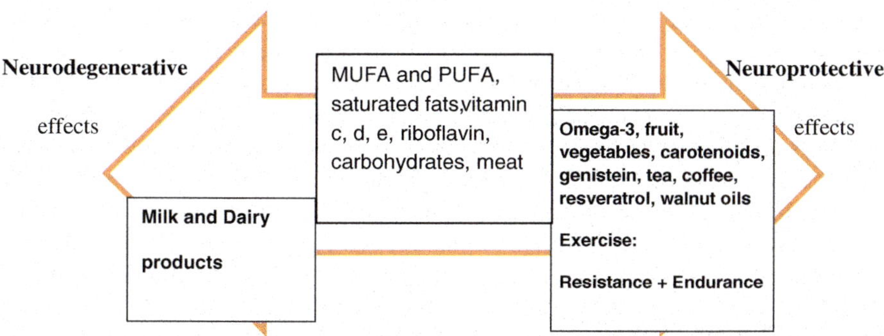

Fig. 1 Dietary ingredients contributing to both neurodegenerative and neuroprotective effects in Parkinsonism; Exercise contributes to both protection and recovery whereas diet may contribute to either neurodegenerative or neuroprotective aspects. Modified after Seidl et al. (2014)

norepinephrine precursors, occur at early stages of the disease even before the onset of motor symptoms (cf. Braak et al. 2007; Cersosimo and Benarroch 2012b; Hawkes et al. 2007) (Fig. 1).

1 Nutrition and PD

Several lines of investigation, including both epidemiological and neurochemical analyses have indicated that certain agents in foodstuffs/diets may provide neuro-protection in PD (Searles Nielsen et al. 2013; Shaltiel-Karyo et al. 2013), and vice versa (Chen et al. 2007; Hellenbrand et al. 1996). Dairy product consumption, e.g. drinking milk, may elevate PD risk, independent of calcium intake (Kyrozis et al. 2013; Park et al. 2005a, b). Serum urate, increased by dairy product consumption, is linked to PD (Andreadou et al. 2009; Schlesinger and Schlesinger 2008; Shen et al. 2013). Additionally, certain dopaminergic neurotoxic elements, both organic and inorganic, may be present in milk (Chen et al. 2002, 2003, 2004; Corrigan et al. 1998). Diet may play an important role in the etiology of Parkinson's disease, either by altering the oxidative balance in the brain or by serving as a vehicle for envi-ronmental neurotoxins (Barichella et al. 2009). In a comprehensive review, Seidl et al. (2014) have outlined the nutrients that are linked to a reduced risk for or progression of the disorder, these include: fruits and vegetables, and what has been referred to as a 'mediterranean diet' (Alcalay et al. 2012), although this relationship is not straightforward since pomegranate juice exacerbated oxidative stress in a rotenone model of PD (Tapias et al. 2013). In a similar PD model, the carotenoid compound, lycopene, reduced oxidative stress (Kaur et al. 2011) and hesperidine, a plant flavanone, alleviated oxidative stress and apoptosis (Tamilselvam et al. 2013).

A mediteranean diet (not necessarily a Mediteranean cuisine) includes generally: proportionally high consumption of olive oil, legumes, unrefined cereals, fruits, and vegetables, moderate to high consumption of fish, moderate consumption of dairy products (mostly as cheese and yogurt), moderate wine consumption, and low consumption of non-fish meat and non-fish meat products. Several other dietary nutrients may, or may not. contribute to an antiparkinsonian effect, e.g. riboflavin (Coimbra and Junqueira 2003), solanaceae (like potatoes, tomatoes and peppers) or the cabbages (Morroni et al. 2013; Searles Nielsen et al. 2013). Other components of diet that may render a putative neuroprotective effect against PD include Omega-3 (Cansev et al. 2008; Ozsoy et al. 2011), soy protein, isoflavone genistein, (Qian et al. 2011, 2012), caffeine (Lee et al. 2013; Sonsalla et al. 2012, 2013; Yadav et al. 2012) which acts against DA neurotoxicity (Morelli et al. 2010; Xiao et al. 2011), and both black and green tea (Gosslau et al. 2011; Ye et al. 2012). Similarly, the presence of neuroprotective factors in walnut oils is emerging (Iwamoto et al. 2000; Jiang et al. 2002; Muthaiyah et al. 2014; Ros et al. 2004; Spaccarotella et al. 2008). The levels of omega 3 and 6, vitamin E, melatonin, folate, phenolic acid and γ-tocopherol content in walnut oils (Braidy et al. 2013; Crews et al. 2005; Maraldi et al. 2014; Muthaiyah et al. 2011; Patl et al. 2014) seem indicative of mitochondrial protective and antioxidant actions. Essa et al. (2015) have showed that the dietary supplementation of walnut oils alleviated several types of MPTP-induced deficits, including diminished TH Western Blot Densitometry, TH Immunoreactivity in the substantia nigra, nigral GSH and GPx, as well as MPTP-enhanced nigral TBARS, SOD, and CAT, implying the mitochondrial protective and antioxidant actions.

Fish oils have to shown to offer neuroprotective effect in animal models of neurodegeneration (Bousquet et al. 2008, 2009. 2011a, b; Flores-Mancilla et al. 2014; Denny Joseph and Muralidhara 2012, 2013, 2015), through actions such as increasing BDNF expression (Vines et al. 2012). Combinations of fish oils with antioxidants may increase neuroprotective propensities thereby alleviating parkinsonism in the laboratory (Luchtman et al. 2012; Ozsoy et al. 2011; Yakunin et al. 2012). For example, the resolvins, converted to the omega-3 fatty acids, eicosapentaenoic acid and docosahexaenoic acid, in the human body, suppressed the inflammatory mediators expression of the liposaccharide-induced inflammation in substantia nigra pars compacta involving increased expression of NO, iNOS, TNF-α, IL-1, IL-18, IL-6, IL-1β, ROS production, the translocation of NF-κB p65, IκBα, and IKKβ expression in glial cells in a rat model of PD (Tian et al. 2015). In a chronic rotenone model of PD, Denny Joseph and Muralidhara (2015) observed that fish oils provided anti-parkinsonian protection against neurotoxic effects in rats. Antioxidants, such as selenium with known role in the maintenance of genomic stability (Ferguson et al. 2012), present in diets have been shown to be useful in alleviating parkinsonian symptoms and biomarkers which includes mitochondrial DNA damage in laboratory models (Hoang et al. 2009; Pickrell et al. 2011). Using paraquat in a rat model of PD, Ellwanger et al. (2015) found that selenium prevented the damaging effects of the toxin upon bradykinesia, and DNA damage in leukocytes.

Several aspects of nutrition that contribute to either a neurodegenerative or a neuroprotective trajectory underlying PD pathogenesis are warranted: (i) the putative occurrence of mutations in genes encoding epigenetic factors may be associated with either disease vulnerability or resistance, (2) putative genetic variations in genes encoding epigenetic factors modify disease risk and/or protection against, (3) biomarker abnormalities in epigenetic factor expression, localization, or function are involved in disease pathophysiology or in alleviation of disorder, (4) epigenetic mechanisms regulate disease-associated genomic loci, gene products, and cellular pathways, and (5) differential epigenetic profiles are present in patient-derived or animal model central and peripheral tissues (Qureshi and Mehler 2012, 2013a, b, 2014, 2015). Early life environments may be crucial to whether or not individuals show susceptibility since motor systems and cognitive processes, responses to stress in adult life, behavior, and the eventual manifestation of neurodegenerative conditions may all be imprinted in the organism by epigenetic modifications that contribute to shape the brain during prenatal or early postnatal life (Desplats 2015). Epigenetic mechanisms contributing to PD involve several related genes (cf. Landgrave-Gomez et al. 2015): (a) *SNCA*, the gene coding for α-synuclein, is implicated in the reduced SNCA substantia nigral methylation in PD patients and animal models (Ammal Kaidey et al. 2013; Desplats et al. 2011, 2012a, b), and *SNCA* gene silencing mediated by histone methylation (Nalls et al. 2015a, b), (b) *LRKK2* (Dardarin), a member of the leucine-rich repeat kinase family, mutant *LRKK2* antagonises miR-184, a short non-coding RNA molecule functioning as posttranscriptional regulators of expression levels of other genes by several mechanisms, in Drosophila melanogaster in PD models (IPDGC 2011), (c) Parkin, protein encoded by the *PARK2* gene, let-7 (lethal-7) family miRNAs (microRNA precursors) are under-expressed in parkin transgenic C. elegans (Asikainen et al. 2010), and (d) *PARK16/Iq32,GPNMB*, Parkinson susceptibility genes, imply aberrant gene methylation in post-mortem PD brains (IPDGC 2011). In this context, there is evidence that a healthy lifestyle consisting of nutrition, neutraceutics and antioxidants both protect against cell death and delay disease progression (Bega et al. 2014).

The role of diet, and in particular dairy products, as a risk factor for PD remains uncertain (Anderson et al. 1999; Park et al. 2005a, b). Nevertheless, exposure to industrial toxicants that may contaminate dairy products, through dietary habits provide a more serious factor (Brouwer et al. 2015). Several avenues imply that pesticides may exacerbate PD risk (Fleming et al. 1994; Priyadarshi et al. 2000). In this regard, the application of serum metabolimics to develop noninvasive biomarkers of PD and/or follow the consequences of nutritional factors requires identification (James and Parkinson 2015). Furthermore, greater consumption of dairy products is linked to lower circulating levels of uric acid (Choi 2005) which may exert neuroprotective effects against oxidative stress (Davis et al. 1996; de Lau et al. 2005). In a meta-analysis study (Chen et al. 2007) it was postulated that dairy consumption did in fact increase the risk of PD. Whereas both diet and exercise provide epigenetic mechanisms that affect the programming of neurodegenerative disorders (Babenko et al. 2012; Park et al. 2012) in positive and negative direction, only exercise the wholly positive aspect. In the study by Honglei Chen et al. (2007), the authors prospectively investigated the association between dairy intake and risk of Parkinson's disease among 57,689 men and 73,175 women from the Cancer Prevention Study II Nutrition Cohort from the American Cancer Society. A total of 250 men and 138 women with Parkinson's disease were identified during the follow-up (1992–2001). Dairy consumption was positively associated with the risk of Parkinson's disease: compared with the lowest intake quintile, the corresponding relative risks (RRs) for quintiles 2–5 were 1.4, 1.4, 1.4, and 1.6. A higher risk among dairy consumers was found in both men and women, although the association in women appeared non-linear. The meta-analysis of all prospective studies confirmed a moderately elevated risk of Parkinson's disease among individuals with high dairy consumption: the RRs between extreme intake categories were 1.6 for men and women combined, 1.8 for men. These data suggest that dairy consumption may increase the risk of Parkinson's disease, particularly in men. More studies are needed to further examine these findings and to explore the underlying mechanisms.

2 PD and Exercise

Any bodily activity that enhances or maintains physical fitness implies the engagement of regular and frequent exercise thereby maintaining physical fitness and the reduction of agents associated with health problems, e.g. cortisol. With regard to the large proportion of individuals with more-or-less sedentary occupations, physical exercise offers probably the most effective health-promoting lifestyle available with positive outcomes for both neurologic and psychiatric conditions (Archer 2012; Archer et al. 2011; Archer and Kostrzewa 2012, 2015; Archer et al. 2014, 2012), as well as in domains defined by cognitive capacity and affective status (Archer 2014; Archer and Garcia 2014, 2015; Archer and Kostrzewa 2015; Garcia and Archer 2014; Garcia et al. 2015; Moradi et al. 2014). Exercise has been defined as a planned, structured physical activity with the purpose of improving one or more aspects of physical fitness and functional capacity (Morris and Schoo 2004). Regular physical exercise/exertion promotes neuroimmune functioning and facilitates the prevention of chronic heart disease, cardiovascular problems, type II diabetes, obesity and psychological illbeing, e.g. depressiveness and apati, all of which may exacerbate the severity of PD (Cugusi et al. 2014; Lamotte et al. 2015; Low et al. 2014; Mattson 2014; Nicolucci et al. 2012; Scalzo et al. 2010; Shumway-Cook et al. 2015; Tuon ct al. 2014). In both clinical settings (Canning et al. 2015; Liddle and Eagle 2014; Teixeira-Machado et al. 2015) and laboratory models, e.g. repeated administrations of 1-methyl-4-phenyl-1,2,3,6-tetrahydropyridine (MPTP), physical exercise ameliorated functional, neurochemical and other biomarkers of dopaminergic deficits (Archer and Fredriksson 2010, 2012, 2013a, b, c; Fredriksson et al. 2011) while the combination of exercise with Milmed® provided evidence of complete restoration of these deficits (Archer and Fredriksson 2013a; Archer et al. 2014). Furthermore, activity/exercise reduced the risk for future parkinsonism in different populations (Canning et al. 2014; Yang et al. 2015).

Physical exercise, sometimes referred to as general exercise, remains the most practicable, useful and non-invasive, non-pharmacological strategy for improvements in symptom-profiles, balance and posture and biomarkers in PD (Bello et al. 2014; Conradsson et al. 2014; Corcos et al. 2013; Dibble et al. 2009; Ganesan et al. 2014; Gobbi et al. 2009; Kara et al. 2012; Nocera et al. 2009; Wong-Yu Is and Mak 2013), whether robot-assisted (Paker et al. 2012; Picelli et al. 2012a, b, 2014) or of Tai Chi type training (Amano et al. 2013; Gao et al. 2014; Kim et al. 2014; Tsang 2013). Sumec et al. (2015) have reviewed the therapeutic effects of a range of physical exercise and neurofeedback-based technique interventions that improved postural and gait stability in PD patients, as well as improvements in quality-of-life, affective status, and other domains such as enjoyment and motivation. Resistance training, endurance training and other intensive training modalities are all applicable and safe for the purposes of PD patients, improving cardiovascular fitness, balance and walking performance (Uhrbrand et al. 2015). Six randomized and 2 nonrandomized controlled articles that originated from 3 trials of resistance exercise training interventions demonstrated that strength training was found to significantly improve muscle strength in PD patients (15–83.2 %) with significant improvements in mobility (11.4 %) and disease progression (Cruickshank et al. 2015). The Pisa syndrome (PS), an adverse effect of some medications, presents a postural deformity condition in which there is sustained involuntary flexion of the body and head to one side and slight rotation of the trunk so the person appears to lean to one side as if presenting the 'Leaning Tower of Pisa'. Postural control, balance, and gait disturbances in patients presenting combined PD and PS. Geroin et al. (2015) observed that patients with PD and PS experienced more difficulty achieving good postural alignment with gravity and greater velocity of body sway than the other (PD) groups, implying that rehabilitation programs for patients with PD and PS ought to include also exercises for spine alignment and dynamic postural training. Veritably, there are now a huge number of PD patients, presenting varying degrees of disorder severity, who have benefitted from the intervention wherein high 'dose' levels, or frequency, intensity and duration may be less important, of exercise was related to lower disease severity and better cognition than was observed in less-active or sedentary patients (Oguh et al. 2014). The gathering evidence in favour of exercise/activity as a major component of anti-parkinsonian interventions implies the focus upon multidisciplinary health care provision requires much further effort (Ebersbach 2015).

3 Exercise and DA-Integrity Relationship

In a study that assessed the relationship between exercise performance in a running wheel following neurotoxin treatment using the DA neurotoxin, MPTP, groups of mice were given interventions that consisted of either running-wheel exercise alone, or combined with the Milmed® yeast in suspension form or in the dried form. For the purposes of this account the efficacy of the interventions is not under investigation but rather the correlation between running capacity and DA-integrity. MPTP (3 × 30 mg/kg, s.c., 4 groups) or saline (vehicle 1 × 5 ml/kg, s.c., 1 group) were administered in a single dose regime over three consecutive weeks on Fridays. Three MPTP groups were given four 30-min periods/week (Mondays to Thursdays), of these two groups, MPTP + Exer + Yeast L and MPTP + Exer + Yeast D; the former were introduced to exercise and Milmed (oral injection) in the suspension form and the latter on the week following the 1st MPTP injection and the latter on the Monday prior to the 1st injection of MPTP onwards. One MPTP group, MPTP + Exer, was given access to exercise (running wheels) from the week following the 1st MPTP injection onwards. The fourth MPTP group, MPTP–NoEx, and the Vehicle group were only given access to exercise on a single day each week (Wednesdays, exercise test) from the week following the 1st MPTP injection onwards. The exercise/exercise + Milmed regime was maintained for a further 9 weeks. It was observed that exercise by itself ameliorated MPTP-induced deficits regarding motor function and dopamine loss only partially whereas in the groups combining exercise with twice weekly dosages of Milmed the MPTP-induced deficits were abolished by the 10th week of the intervention. It was observed that DA integrity was observed to be a direct function of ability to express running exercise in a treadmill wheel-running arrangement (Fig. 2).

The prodigiously strong correlation between distance run by each mouse in the tenth and final week of exercise + Milmed® (Yeast) treatment and striatal DA

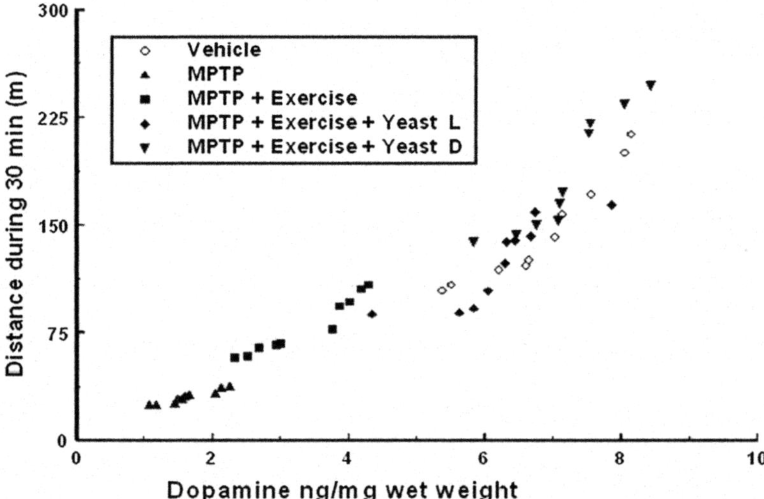

Fig. 2 Correlation, Pearson's product moment ($R = 0.965$, $df = 50$, $p < 0.00001$) between striatal DA concentrations (animals sacrificed week 12 after start of experiment) and distance run in running-wheels on the final Wednesday test (week 10). Yeast L = Milmed® in suspension; Yeast D = Milmed® in dried form

levels, following sacrifice one week later, indicate that DA-integrity is intimately involved in the capacity of individual for ruuning-wheel exercise performance. Nevertheless, in a somewhat speculative postulation, it has been claimed that exercise, through recovery of DA and glutamate transporters, thereby stabilizing the extraneuronal neurotransmitters, alleviated the motor deficits rather than sparing nigrostriatal neurons (Sconce et al. 2015). They employed a progressive MPTP dose regime (8, 16, 24, 36 mg/kg, over 4 weeks) with the voluntary wheel-running intervening half-way through the MPTP administration schedule. The biomarkers measured were vascicular monoamine transporter II (VMAT2), dopamine transporter (DAT), glutamate transporters (VGLUT1, GLT-1) and excitatory aminoacid carrier 1. Exercise-induced improvements were observed in all these markers; sadly, they did not measure striatal DA, instead it is suggested that locomotor (or running) behavior recovered through suppression of inflammation. Remarkably, locomotor behavior in the final week (week 10) of wheel-running is correlated also hugely to DA-levels ($r = 0.966$, $p < 0.00001$).

Physical exercise augments brain DA levels (Deslandes et al. 2009; Foley and Fleshner 2008; Sutoo and Akiyama 2003). It appears to preserve the integrity of brain structures through mobilization of 'pro-life' processes, including survival, maintenance, and functioning of progenitor cells and brain tissues. Just two weeks of exercise increased survival of DA neurons in the substantia nigra and upheld DA projections to striatum cortex (Yoon et al. 2007). Exercise stimulates DA mobilization and neurofactor (e.g. BDNF, GDNF, IGF-1 and FGF-2) synthesis (Monteiro-Junior et al. 2015). DA was elevated also by exercise schedules that improved cognition (Berse et al. 2015; Winter et al. 2007; see also Palliard et al. 2015). Finally, in spontaneous hypertensive rats, DA-receptor expression was decreased (Cho et al. 2014). DA and its metabolites, DOPAC and HVA, as well as tyrosine hydroxylase, were all elevated in rats during treadmill running in speed-dependent manner, i.e. greater speed, greater increase (Hattori et al. 1994). Further, in 6-hydroxydopamine-lesioned and intact rats, DA, DOPAC and HVA showed 130 and 150 % increase during treadmill running; running was found to be a satisfactory indicator of DA turnover (Hattori et al. 1993).

4 Epigenetic Influences

Exercise and nutrition are synergistic in mitigating disorder states with exercise releasing exosomes that contain miRNAs. Nutrition/vitamins B6 and B12 regulate the metabolism of homocysteine, an epigenetic byproduct of DNA/RNA/protein methylation (Tyagi and Joshua 2014). Despite issues linked to population selection and quantification of exercise, the overall pattern emerging appears to be a product of the utilization of global methylation as an outcome measure, not depicting changes in DNA methylation at the gene-specific level. Among the genes whose methylation levels were changed significantly by exercise were those involved in metabolism, muscle growth, hematopoeisis and inflammation, all exacerbating factors for PD, with intensity, duration and frequency of exercise important factors (Voisin et al. 2015). Not least of importance are the notions of "trainability" and "epigenetic silencing" that modulate the therapeutic value of exercise. Both survival and the delay of mortality are the benefits of physical activity programs with advantages to match smoking cessation in cardiovascular conditions. It has been observed that aberrant dopaminergic transmission, caused by prolonged administration of L-Dopa in PD, may activate PcG repressed genes in the brain thereby contributing to long-term maladaptive responses such as the motor complications, or dyskinesia; as noted above, physical exercise reduces the dose levels of L-Dopa required (Archer and Fredriksson 2010, 2012, 2013a, b) which ought to facilitate the removal of an adversive epigenetic influence. More directly, running exercise facilitated recovery from methamphetamine-induced and nigrostriatal dopaminergic damage to DA terminals (O'Dell et al. 2007, 2012).

References

Alcalay RN, Gu Y, Mejia-Santana H, Cote L, Marder KS, Scarmeas N (2012) The association between diet adherence and Parkinson's disease. Mov Disord 27:771–774. doi:10.1002/mds.24918

Antonini A, Schwarz J, Oertel WH, Beer HF, Madeja UD, Leenders KL (1994) [11C]raclopride and positron emission tomography in previously untreated patients with Parkinson's disease: Influence of L-dopa and lisuride therapy on striatal dopamine D2-receptors. Neurology 44(7):1325–1329

Amano S, Nocera JR, Vallabhajosula S, Juncos JL, Gregor RJ, Waddell DE, Wolf SL, Hass CJ (2013) The effect of Tai Chi exercise on gait initiation and gait performance in persons with Parkinson's disease. Parkinsonism Relat Disord 19(11):955–960. doi:10.1016/j.parkreldis.2013.06.007

Ammal Kaidey N, Tarannum S, Thomas B (2013) Epigenetic landscape of Parkinson's disease: emerging role in disease mechanisms and therapeutic modalities. Neurotherapeutics 10:698–708. doi:10.1007/s13311-013-0211-8

Anderson C, Checkoway H, Franklin GM, Beresford S, Smith-Weller T, Swanson PD (1999) Dietary factors in Parkinson's disease: the role of food groups and specific foods. Mov Disord 14(1):21–27

Andreadou E, Nikolaou C, Gournaras F, Rentzos M, Boufidou F, Tsoutsou A et al (2009) Serum uric acid in patients with Parkinson's disease: their relationship to treatment and disease duration. Clin Neurol Neurosurg 111:724–728. doi:10.1016/j.clineuro.2009.06.012

Archer T (2012) Influence of physical exercise on traumatic brain injury deficits: scaffolding effect. Neurotox Res 21(4):418–434. doi:10.1007/s12640-011-9297-0

Archer T (2014) Health benefits of physical exercise for children and adolescents. J Novel Physiother 4:203. doi:10.4172/2165-7025.1000203

Archer T, Fredriksson A (2010) Physical exercise attenuates MPTP-induced deficits in mice. Neurotox Res 18:313–327. doi:10.1007/s12640-010-9168-0

Archer T, Fredriksson A (2012) Delayed exercise-induced functional and neurochemical partial restoration following MPTP. Neurotox Res 21:210–221. doi:10.1007/s12640-011-9261-z

Archer T, Fredriksson A (2013a) The yeast product Milmed enhances the effect of physical exercise on motor performance and dopamine neurochemistry recovery in MPTP-lesioned mice. Neurotox Res 24, 393–406. doi:10.1007/s12640-013-9405-4

Archer T, Fredriksson A (2013b) Pharmacogenomics and personalized medicine in Parkinsonism. In: Barh D et al. (eds) Omics for personalized medicine, pp. 265–287, Springer India. doi:10. 1007/978-81-322-1184-6_14

Archer T, Fredriksson A (2013c) Physical exercise as intervention in Parkinsonism. In: Kostezewa RM (ed) Handbook of neurotoxicity, pp. 2255–2280. Springer + Business media, New York. doi:10.1007/978-1-4614-5836-4

Archer T, Garcia D (2014) Physical exercise influences academic performance and well-being in children and adolescents. Int J School Cogn Psychol 1:e102

Archer T, Garcia D (2015) Exercise and dietary restriction for promotion of neurohealth benefits. Health 7:136–152. doi:10.4236/health.2015.71016

Archer T, Kostrzewa RM (2015) Physical exercise alleviates health defects, symptoms, and biomarkers in schizophrenia spectrum disorder. Neurotox Res. PMID: 26174041. doi:10.1007/ s12640-015-9543-y

Archer T, Svensson K, Alricsson M (2012) Physical exercise ameliorates deficits induced by traumatic brain injury. Acta Neurol Scand 125:293–302. doi:10.1111/j.1600-0404.2011.01638.x

Archer T, Kostrzewa RM, Beninger RJ, Palomo T (2011) Staging neurodegenerative disorders: structural, regional, biomarker, and functional progressions. Neurotox Res. 19(2):211–234. doi:10.1007/s12640-010-9190-2

Archer T, Josefsson T, Lindwall M (2014) Effects of physical exercise on depressive symptoms and biomarkers in depression. CNS Neurol Disord Drug Targ. 13(10):1640–1653

Asikainen S, Rudgalvyte M, Heikkinen L, Louhiranta K, Lakso M, Wong G, Nass R (2010) Global microRNA expression profiling of Caenorhabditis elegans Parkinson's disease models. J Mol Neurosci 41(1):210–218. doi:10.1007/s12031-009-9325-1

Babenko O, Kovalchuk I, Metz GA (2012) Epigenetic programming of neurodegenerative diseases by an adverse environment. Brain Res 1444:96–111. doi:10.1016/j.brainres.2012.01.038

Barichella M, Cereda E, Pezzoli G (2009) Major nutritional issues in the management of Parkinson's disease. Mov Disord 24:1881–1892. doi:10.1002/mds.22705

Bega D, Gonzalez-Lapati P, Zadikoff C, Simuni T (2014) A review of the clinical evidence for complimentary and alternative therapies in Parkinson's disease. Curr Treat Options Neurol 16:314. doi:10.1007/s11940-014-0314-5

Bello O, Sanchez JA, Lopez-Alonso V, Márquez G, Morenilla L, Castro X, Giraldez M, Santos-García D, Fernandez-del-Olmo M (2014) The effects of treadmill or overground walking training program on gait in Parkinson's disease. Gait Posture 38(4):590–595. doi:10. 1016/j.gaitpost.2013.02.005

Berse T, Rolfes K, Barenberg J, Dutke S, Kuhlenbäumer G, Völker K, Winter B, Wittig M, Knecht S (2015) Acute physical exercise improves shifting in adolescents at school: evidence for a dopaminergic contribution. Front Behav Neurosci 9:196. doi:10.3389/fnbeh.2015.00196

Bousquet M, Saint-Pierre M, Julien C, Salem N Jr, Cicchetti F, Calon F (2008) Beneficial effects of dietary omega-3 polyunsaturated fatty acid on toxin-induced neuronal degeneration in an animal model of Parkinson's disease. FASEB J. 22(4):1213–1225

Bousquet M, Gibrat C, Saint-Pierre M, Julien C, Calon F, Cicchetti F (2009) Modulation of brain-derived neurotrophic factor as a potential neuroprotective mechanism of action of omega-3 fatty acids in a parkinsonian animal model. Prog Neuropsychopharmacol Biol Psychiatry 33(8):1401–1408. doi:10.1016/j.pnpbp.2009.07.018

Bousquet M, Calon F, Cicchetti F (2011a) Impact of ω-3 fatty acids in Parkinson's disease. Ageing Res Rev 10(4):453–463. doi:10.1016/j.arr.2011.03.001

Bousquet M, Gue K, Emond V, Julien P, Kang JX, Cicchetti F, Calon F (2011b) Transgenic conversion of omega-6 into omega-3 fatty acids in a mouse model of Parkinson's disease. J Lipid Res 52(2):263–271. doi:10.1194/jlr.M011692

Braak H, Sastre M, Del Tredici K (2007) Development of alpha-synuclein immunoreactive astrocytes in the forebrain parallels stages of intraneuronal pathology in sporadic Parkinson's disease. Acta Neuropathol 114(3):231–241

Braidy N, Subash S, Essa MM, Vaishav R, Al-Adawi S, Al-Asmi A, Al-Senawi H, Alobaidy AAR, Lakhtakia R, Guillemin GJ (2013) Neuroprotective effects of a variety of pomegranate juice extracts (PJE) against MPTP-induced cytotoxicity and oxidative stress in human primary neurons. Ocid Med Cell Longev. doi:10.1155/2014/576363

Brouwer M, Koeman T, van den Brandt PA, Kromhout H, Schouten LJ, Peters S, Huss A, Vermeulen R (2015) Occupational exposures and Parkinson's disease mortality in a prospective Dutch cohort. Occup Environ Med 72(6):448–455. doi:10.1136/oemed-2014-102209

Canning CG, Paul SS, Nieuwboer A (2014) Prevention of falls in Parkinson's disease: a review of fall risk factors and the role of physical interventions. Neurodegener Dis Manag 4(3):203–21. doi:10.2217/nmt.14.22

Canning CG, Sherrington C, Lord SR, Close JC, Heritier S, Heller GZ, Howard K, Allen NE, Latt MD, Murray SM, O'Rourke SD, Paul SS, Song J, Fung VS (2015) Exercise for falls prevention in Parkinson disease: a randomized controlled trial. Neurology 84(3):304–12. doi:10.1212/WNL.0000000000001155

Cansev M, Ulus IH, Wang L, Maher TJ, Wurtman RJ (2008) Restorative effects of uridine plus docosahexaenoic acid in a rat model of Parkinson's disease. Neurosci Res 62:206–209. doi:10.1016/j.neures.2008.07.005

Chen H, Zhang SM, Hernan MA, Willett WC, Ascherio A (2002) Diet and Parkinson's disease: a potential role of dairy products in men. Ann Neurol 52:793–801. doi:10.1002/ana.10381

Cersosimo MG, Benarroch EE (2012a) Pathological correlates of gastrointestinal dysfunction in Parkinson's disease. Neurobiol Dis 46(3):559–564. doi:10.1016/j.nbd.2011.10.014

Cersosimo MG, Benarroch EE (2012b) Autonomic involvement in Parkinson's disease: pathology, pathophysiology, clinical features and possible peripheral biomarkers. J Neurol Sci 313(1–2): 57–63. doi:10.1016/j.jns.2011.09.030

Cho HS, Baek DJ, Baek SS (2014) Effect of exercise on hyperactivity, impulsivity and dopamine D2 receptor expression in the substantia nigra and striatum of spontaneous hypertensiverats. J Exerc Nutrition Biochem 18(4):379–84. doi:10.5717/jenb.2014.18.4.379

Chen H, Zhang SM, Hernan MA, Willett WC, Ascherio A (2003) Dietary intakes of fat and risk of Parkinson's disease. Am J Epidemiol 157:1007–1014. doi:10.1093/aje/kwg073

Chen H, Zhang SM, Schwarrzschild MA, Hernan MA, Logroscino G, Willett WC et al (2004) Folate intake and risk of Parkinson's disease. Am J Epidemiol 160:368–375. doi:10.1093/aje/kwh213

Chen H, O'Reilly E, McCullough ML, Rodriguez C, Schwarzschild MA, Calle EE, Thun MJ, Ascherio A (2007) Consumption of dairy products and risk of Parkinson's disease. Am J Epidemiol 165:998–1006. doi:10.1093/ajefkwk089

Choi HK (2005) Dietary risk factors for rheumatic diseases. Curr Opin Rheumatol 17:141–146

Coimbra CG, Junqueira VB (2003) High doses of riboflavin and the elimination of dietary red meat promote the recovery of some motor functions in Parkinson's disease patients. Braz J Med Biol Res 36:1409–1417. doi:10.1590/S0100-879X2003001000019

Conradsson D, Löfgren N, Ståhle A, Franzén E (2014) Is highly challenging and progressive balance training feasible in older adults with Parkinson's disease. Arch Phys Med Rehab 95:1000–1003

Corcos DM, Robichaud JA, David FJ et al (2013) A two-year randomized controlled trial of progressive resistance exercise for Parkinson's disease. Mov Disord 28:1230–1240

Corrigan FM, Murray L, Wyatt CL, Shore RF (1998) Diorthosubstituted polychlorinated biphenyls in caudate nucleus in Parkinson's disease. Exp Neurol 150:339–342. doi:10.1006/exnr.1998.6776

Crews C, Hough P, Godward J, Brereton P, Lees M, Guiet S, Winkelmann W (2005) Study of the main constituents of some authentic walnut oils. J Agric Food Chem 53:4853–4860

Cruickshank TM, Reyes AR, Ziman MR (2015) A systematic review and meta-analysis of strength training in individuals with multiple sclerosis or Parkinson disease. Medicine (Baltimore) 94 (4):e411. doi:10.1097/MD.0000000000000411

Cugusi L, Solla P, Zedda F, Loi M, Serpe R, Cannas A, Marrosu F, Mercuro G (2014) Effects of an adapted physical activity program on motor and non-motor functions and quality of life in patients with Parkinson's disease. NeuroRehabilitation 35(4):789–94. doi:10.3233/NRE-141162

Davis JW, Grandinetti A, Waslien CI, Ross GW, White LR, Morens DM (1996) Observations on serum uric acid levels and the risk of idiopathic Parkinson's disease. Am J Epidemiol 144 (5):480–4

De Lau LM, Koudstaal PJ, Hofman A et al (2005) Serum uric acid levels and the risk of Parkinson's disease. Ann Neurol 58:797–800

Denny Joseph KM, Muralidhara M (2012) Fish oil prophylaxis attenuates rotenone-induced oxidative impairments and mitochondrial dysfunctions in rat brain. Food Chem Toxicol 50 (5):1529–37. doi:10.1016/j.fct.2012.01.020

Denny Joseph KM, Muralidhara M (2013) Enhanced neuroprotective effect of fish oil in combination with quercetin against 3-nitropropionic acid induced oxidative stress in rat brain. Prog Neuropsychopharmacol Biol Psychiatry 40:83–92. doi:10.1016/j.pnpbp.2012.08.018

Denny Joseph KM, Muralidhara M (2015) Combined oral supplementation of fish oil and quercetin enhances neuroprotection in a chronic rotenone rat model: relevance to Parkinson's disease. Neurochem Res 40(5):894–905. doi:10.1007/s11064-015-1542-0

Deslandes A, Moraes H, Ferreira C, Veiga H, Silveira H, Mouta R, Pompeu FA, Coutinho ES, Laks J (2009) Exercise and mental health: many reasons to move. Neuropsychobiology 59 (4):191–8. doi:10.1159/000223730

Desplats PA (2015) Perinatal programming of neurodevelopment: epigenetic mechanisms and the prenatal shaping of the brain. Adv Neurobiol 10:335–61. doi:10.1007/978-1-4939-1372-5_16

Desplats P, Spencer B, Coffee E, Patel P, Michael S, Patrick C, Adame A, Rockenstein E, Masliah E (2011) Alpha-synuclein sequesters Dnmt1 from the nucleus: a novel mechanism for epigenetic alterations in Lewy body diseases. J Biol Chem 286(11):9031–7. doi:10.1074/jbc.C110.212589

Desplats P, Patel P, Kosberg K, Mante M, Patrick C, Rockenstein E, Fujita M, Hashimoto M, Masliah E (2012a) Combined exposure to Maneb and Paraquat alters transcriptional regulation of neurogenesis-related genes in mice models of Parkinson's disease. Mol Neurodegener 7:49. doi:10.1186/1750-1326-7-49

Desplats P, Spencer B, Crews L, Pathel P, Morvinski-Friedmann D, Kosberg K, Roberts S, Patrick C, Winner B, Winkler J, Masliah E (2012b) α-Synuclein induces alterations in adult neurogenesis in Parkinson disease models via p53-mediated repression of Notch1. J Biol Chem 287(38):31691–702. doi:10.1074/jbc.M112.354522

Dibble LE, Hale TF, Marcus RL, Gerber JP, Lastayo PC (2009) High intensity eccentric resistance training decreases bradykinesia and improves quality of life in persons with Parkinson's disease: a preliminary study. Parkinsonism Rel Disorder 15:752–757

Ebersbach G (2015) Exercise matters in patients with PD—another piece of evidence. Nat. Rev. Neurol. 11:9–10. doi:10.1038/nrneurol.2014.231

Ellwanger JH, Molz P, Dallemole DR, Pereira dos Santos A, Müller TE, Cappelletti L, Gonçalves da Silva M, Franke SI, Prá D, Pêgas Henriques JA (2015) Selenium reduces bradykinesia and DNA damage in a rat model of Parkinson's disease. Nutrition 31(2):359–65. doi:10.1016/j.nut.2014.07.004

Felicio AC, Shih MC, Godeiro-Junior C, Andrade LA, Bressan RA, Ferraz HB (2009) Molecular imaging studies in Parkinson disease: reducing diagnostic uncertainty. Neurologist 15(1):6–16. doi:10.1097/NRL.0b013e318183fdd8

Essa MM, Subash S, Dhanalakshmi C, Manivasagam T, Al-Adawi S, Guillemin GJ, Thenmozhi AJ (2015) Dietary supplementation of walnut partially reverses 1-methyl-4-phenyl-1,2,3,6-tetrahydropyridine induced neurodegeneration in a mouse model of Parkinson's disease. Neurochem Res 40:1283–1293. doi:10.1007/s11064-015-1593-2

Fleming L, Mann JB, Bean J, Briggle T, Sanchez-Ramos JR (1994) Parkinson's disease and brain levels of organochlorine pesticides. Ann Neurol 36:100–103

Flores-Mancilla LE, Hernandez-Gonzalez M, Guevara MA, Benavides-Haro DE, Martinez-Arteaga P (2014) Long-term fish oil supplementation attenuates seizure activity in the amygdala induced by 3-mercaptoproprionic acid in adult male rats. Epilepsy Behav 33:126134. doi:10.1016/j.yebeh.2014.02.023

Foley T, Fleshner M (2008) Neuroplasticity of dopamine circuits after exercise: implications for centrak fatigue. Neuronol Med 10:67–80

Fredriksson A, Stigsdotter IM, Hurtig A, Ewalds-Kvist B, Archer T (2011) Running wheel activity restores MPTP-induced functional deficits. J Neural Tranms 118:407–420. doi:10.1007/s00702-010-0474-8:

Ganesan M, Sathyaprabha TN, Gupta A, Pal PK (2014) Effect of partial weight-supported treadmill gait training on balance in patients with Parkinson's disease. PM & R 6:22–33

Gaggelli E, Kozlowski H, Valensin D, Valensin G (2006) Copper homeostasis and neurodegenerative disorders (Alzheimer's, prion, and Parkinson's diseases and amyotrophic lateral sclerosis)

Gao Q, Leung A, Yang Y et al (2014) Effect of Tai Chi on balance and fall prevention in Parkinson's disease: a randomized controlled trial. Clin Rehabil 28:748–753

Garcia D, Archer T (2014) Positive affect and age as predictors of exercise compliance. PeerJ 2: e694. doi:10.7717/peerj.694

Garcia D, Jimmefors A, Mousavi F, Adrianson L, Rosenberg P, Archer T (2015) Self-regulatory mode (locomotion and assessment), well-being (subjective and psychological), and exercise behavior (frequency and intensity) in relation to high school pupils' academic achievement. PeerJ 3:e847. doi:10.7717/peerj.847

Geroin C, Smania N, Schena F, Dimitrova E, Verzini E, Bombieri F, Nardello F, Tinazzi M, Gandolfi M (2015) Does the Pisa syndrome affect postural control, balance, and gait in patients with Parkinson's disease? An observational cross-sectional study. Parkinsonism Relat Disord 21(7):736–41. doi:10.1016/j.parkreldis.2015.04.020

Gobbi LTB, Oliveira-Ferreira MDT, Caetano MJD et al (2009) Exercise programs improve mobility and balance in people with Parkinson's disease. Parkinsonism Rel Disorder 15, Suppl. 3, S49–S52

Gosslau A, En Jao DL, Huang MT, Ho CT, Evans D, Rawson NE et al (2011) Effects of the black tea polyphenol theaflavin-2 on apoptotic and inflammatory pathways in vitro and in vivo. Mol Nutr Food Res 55:198–208. doi:10.1002/mnfr.201000165

Hattori S, Li Q, Matsui N, Nishino H (1993) Treadmill running combined with microdialysis can evaluate motor deficit and improvement following dopaminergic grafts in 6-OHDA lesioned rats. Restor Neurol Neurosci 6(1):65–72. doi:10.3233/RNN-1993-6107

Hattori S, Naoi M, Nishino H (1994) Striatal dopamine turnover during treadmill running in the rat: relation to the speed of running. Brain Res Bull 35(1):41–9

Hawkes CH, Del Tredici K, Braak H (2007) Parkinson's disease: a dual-hit hypothesis. Neuropathol Appl Neurobiol 33(6):599–614

Hellenbrand W, Seidler A, Boeing H, Robra BP, Vieregge P, Nischan P et al (1996) Diet and Parkinson's disease. I: a possible role for the past intake of specific foods and food groups. Results from a self-administered food-frequency questionnaire in a case-control study. Neurology 47:636–643. doi:10.1212/WNL.47.3636

Hoang T, Choi DK, Nagai M, Wu DC, Nagata T, Prou D, Wilson GL, Vila M, Jackson-Lewis V, Dawson VL, Dawson TM, Chesselet MF, Przedborski S (2009) Neuronal NOS and cyclooxygenase-2 contribute to DNA damage in a mouse model of Parkinson disease. Free Radic Biol Med 47(7):1049–56. doi:10.1016/j.freeradbiomed.2009.07.013

International Parkinson's Disease Genomics Consortium (IPDGC); Wellcome Trust Case Control Consortium 2 (WTCCC2) (2011) A two-stage meta-analysis identifies several new loci for Parkinson's disease. PLoS Genet 7e:1002142. doi:10.1371/journal.pgen.1002142

Iwamoto M, Sato M, Kono M, Hirooka Y, Saka K, Takeshita A, Imaizumi K (2000) Walnuts lower serum cholesterol in Japanese men and women. J Nutr 130:171–176

James EL, Parkinson EK (2015) Serum metabolomics in animal models and human disease. Curr Opin Clin Nutr Metab Care PMID 26147529

Jankovic J (2008) Parkinson's disease: clinical features and diagnosis. J Neurol Neurosurg Psychiatry 79(4):368–376. doi:10.1136/jnnp.2007.131045

Jiang R, Manson JE, Stamfer MJ, Liu S, Willett WC, Hu FB (2002) Nut and pea nut butter consumption and risk of type 2 diabetes in women. JAMA 288:2554–2560

Kara B, Genc A, Colakoglu BD, Cakmur R (2012) The effect of supervised exercise on static and dynamic balance in Parkinson's disease patients. NeuroRehabilitation 30:351–357

Kaur H, Chauhan S, Sandhir R (2011) Protective effect of lycopene on oxidative stress and cognitive decline in rotenone induced model of Parkinson's disease. Neurochem Res 36:1435–1443. doi:10.1007/s11064-011-0469-3

Kim HD, Jae HD, Jeong JH (2014) Tai Chi exercise can improve the obstacle negotiating ability of people with Parkinson's disease: a preliminary study. J Phys Ther 26:1025–1030

Kyrozis A, Ghika A, Stathopoulos P, Vassilopoulos D, Trichopoulos D, Trichopoulou A (2013) Dietary and lifestyle variables in relation to incidence of Parkinson's disease in Greece. Eur J Epidemiol 28:67–77. doi:10.1007/s10654-012-9760-0

Lamotte G, Rafferty MR, Prodoehl J, Kohrt WM, Comella CL, Simuni T, Corcos DM (2015) Effects of endurance exercise training on the motor and non-motor features of Parkinson's disease: a review. J Parkinsons Dis 5(1):21–41. doi:10.3233/JPD-140425

Landgrave-Gomez J, Mercado-Gomez O, Guevara-Guzman R (2015) Epigenetic mechanisms in neurological and neurodegenerative diseases. Front Cell Neurosci. doi:10.3389/fncel.2015.00058

Lees AJ, Hardy J, Revesz T (2009) Parkinson's disease. Lancet 373(9680):2055–66. doi:10.1016/S0140-6736(09)60492-X

Liddle J, Eagles R (2014) Moderate evidence exists for occupational therapy-related interventions for people with Parkinson's disease in physical activity training, environmental cues and individualised programmes promoting personal control and quality of life. Aust Occup Ther J 61(4):287–8. doi:10.1111/1440-1630.12153

Low DA, Vichayanrat E, Iodice V, Mathias CJ (2014) Exercise hemodynamics in Parkinson's disease and autonomic dysfunction. Parkinsonism Relat Disord. 20(5):549–53. doi:10.1016/j.parkreldis.2014.02.006

Luchtman DW, Meng Q, Song C (2012) Ethyl-eicosapentaenoate (E-EPA) attenuates motor impairments and inflammation in the MPTP-probenecid mouse model of Parkinson's disease. Behav Brain Res 226(2):386–96. doi:10.1016/j.bbr.2011.09.033

Maraldi M, Vanzour D, Angeloni C (2014) Dietary polyphenols and their effects on cell biochemistry and pathophysiology. Oxid Med Cell Longev 2014:576363. doi:10.1155/2014/576363

Mattson MP (2014) Interventions that improve body and brain bioenergetics for Parkinson's disease risk reduction and therapy. J Parkinsons Dis 4(1):1–13. doi:10.3233/JPD-130335

Monteiro-Junior RS, Cevada T, Oliveira BRR, Lattari E, Portugal EMM, Carvalho A, Deslandes AC (2015) We need to move bra: Neurobiological hypotheses of physical exercise as a treatment for Parkinson's disease. Med Hypotheses. doi:10.1016/j.mehy.2015.07.011

Moradi S, Nima AA, Rapp Ricciardi M, Archer T, Garcia D (2014) Exercise, character strengths, well-being, and learning climate in the prediction of performance over a 6-month period at a call center. Front Psychol 5:497. doi:10.3389/fpsyg.2014.00497

Morelli M, Carta AR, Kachroo A, Schwarzschild MA (2010) Pathophysiological roles for purines: adenosine, caffeine and urate. Prog Brain Res 183:183–208. doi:10.1016/S0079-6123(10)83010-9

Morris M, Schoo A (2004) Optimizing exercise and physical activity in older adults. Butterworth Heinemann, Edinburgh

Morroni F, Tarrozi A, Sita G, Bolondi C, Zolezzi Moraga JM, Cantelli-Forti G et al (2013) Neuroprotective effect of sulforaphane in 6-hydroxydopamine-lesioned mouse model of Parkinson's disease. Neurotoxicology 36:63–71. doi:10.1016/j.neuro.2013.03.004

Muthaiyah B, Essa MM, Chauhan C, Chauhan V (2011) Protective effects of walnut extract against amyloid beta peptide-induced cell death and oxidative stress in PC12 cells. Neurochem Res 36:2096–2103

Muthaiyah B, Essa MM, Lee M, Chauhan V, Kaur K, Chauhan A (2014) Dietary supplementation of walnut improves memory deficits and learning skills in mouse model of Alzheimer's disease. J Alzheimer's Disease 42:1397–1405. doi:10.3233/JAD-140675

Nalls MA, Bras J, Hernandez DG, Keller MF, Majounie E, Renton AE, Saad M, Jansen I, Guerreiro R, Lubbe S, Plagnol V, Gibbs JR, Schulte C, Pankratz N, Sutherland M, Bertram L, Lill CM, DeStefano AL, Faroud T, Eriksson N, Tung JY, Edsall C, Nichols N, Brooks J, Arepalli S, Pliner H, Letson C, Heutink P, Martinez M, Gasser T, Traynor BJ, Wood N, Hardy J, Singleton AB; International Parkinson's Disease Genomics Consortium (IPDGC); Parkinson's Disease meta-analysis consortium. (2015) NeuroX, a fast and efficient genotyping platform for investigation of neurodegenerative diseases. Neurobiol Aging 36(3):1605.e7-12. doi: 10.1016/j.neurobiolaging.2014.07.028

Nalls MA, Keller MF, Hernandez DG, Chen L, Stone DJ, Singleton AB; Parkinson's Progression Marker Initiative (PPMI) investigators. (2015) Baseline genetic associations in the Parkinson's Progression Markers Initiative (PPMI). Mov Disord doi:10.1002/mds.26374

Nicolucci A, Balducci S, Cardelli P, Cavallo S, Fallucca S, Bazuro A, Simonelli P, Iacobini C, Zanuso S, Pugliese G; Italian Diabetes Exercise Study Investigators (2012) Relationship of exercise volume to improvements of quality of life with supervised exercise training in patients with type 2 diabetes in a randomised controlled trial: the Italian Diabetes and Exercise Study (IDES). Diabetologia 55(3):579–88. doi:10.1007/s00125-011-2425-9

Nielsen SB, Macchi F, Raccosta S, Langkilde AE, Giehm L, Kyrsting A, Svane AS, Manno M, Christiansen G, Nielsen NC, Oddershede L, Vestergaard B, Otzen DE (2013) Wildtype and A30P mutant alpha-synuclein form different fibril structures. PLoS One 8(7):e67713. doi:10.1371/journal.pone.0067713

Nocera J, Horvat M, Ray CT (2009) Effects of home-based exercise on postural control and sensory organization in individuals with Parkinson's disease. Parkinsonism Rel Disorder 15:742–745

O'Dell SJ, Gross NB, Fricks AN, Casiano BD, Nguyen TB, Marshall JF (2007) Running wheel exercise enhances recovery from nigrostriatal dopamine injury without inducing neuroprotection. Neuroscience 144(3):1141–51

O'Dell SJ, Galvez BA, Ball AJ, Marshall JF (2012) Running wheel exercise ameliorates methamphetamine-induced damage to dopamine and serotonin terminals. Synapse 66(1):71–80. doi:10.1002/syn.20989

Oguh O, Eisenstein A, Kwasny M, Simuni T (2014) Back to the basics: regular exercise matters in Parkinson's disease: results from the National Parkinson Foundation registry study. Parkinsonism Relat Disord 20:1221–1225

Ozsoy O, Seval-Celik Y, Hacioglu G, Yargicoglu P, Demir R, Agar A, Aslan M (2011) The influence and the mechanism of docosahexaenoic acid on a mouse model of Parkinson's disease. Neurochem Int 59(5):664–70. doi:10.1016/j.neuint.2011.06.012

Paker N, Bugdayci D, Goksenoglu G, Sen A, Kesiktas N (2012) Effect of robotic treadmill training on functional mobility, walking capacity, motor symptoms and quality of life in ambulatory patients with Parkinson's disease: a preliminary prospective longitudinal study. NeuroRehabilitation 33:323–328

Palliard T, Rolland Y, de Souto Barreto P (2015) Protective effects of physical exercise in Alzheimer's and Parkinson's disease: a narrative review. JClin Neurol 11:212–219. doi:10.3988/jcn.2015.11.3.212

Park LK, Ross GW, Petrovitch H, White LR, Masaki KH, Nelson JS et al (2005a) Consumption of milk and calcium in midlife and the future risk of Parkinson disease. Neurology 64:1047–1051. doi:10.1212/01.WLN0000154532.98495.BF

Park M, Ross GW, Petrovitch H, White LR, Masaki KH, Nelson JS, Tanner CM, Curb JD, Blanchette PL, Abbott RD (2005b) Consumption of milk and calcium in midlife and the future risk of Parkinson disease. Neurology 64(6):1047–51

Park LK, Friso S, Choi SW (2012) Nutritional influences on epigenetics and age-related disease. Proc Nutr Soc 71:75–83. doi:10.1017/S0029665111003302

Picelli A, Melotti C, Origano F, Waldner A, Fiaschi A, Santilli V, Smania N (2012a) Robot-assisted gait training in patients with Parkinson disease: a randomized controlled trial. Neurorehabil Neural Repair 26(4):353–361. doi:10.1177/1545968311424417

Picelli A, Melotti C, Origano F, Waldner A, Gimigliano R, Smania N (2012b) Does robotic gait training improve balance in Parkinson's disease? A randomized controlled trial. Parkinsonism Relat Disord 18(8):990–993. doi:10.1016/j.parkreldis.2012.05.010

Picelli A, Tamburin S, Passuello M, Waldner A, Smania N (2014) Robot-assisted arm training in patients with Parkinson's disease: a pilot study. J Neuroeng Rehabil 11:28. doi:10.1186/1743-0003-11-28

Pickrell AM, Pinto M, Hida A, Moraes CT (2011) Striatal dysfunctions associated with DNA damage in dopaminergic neurons in a mouse model of Parkinson's disease. J Neurosci 31:17640–17658

Priyadarshi A, Khuder SA, Schaub EA, Shrivastava S (2000) A meta-analysis of Parkinson's disease and exposure to pesticides. Neurotoxicology 21(4):435–440

Qian Y, Guan T, Tang X, Huang L, Huang M, Li Y, Sun H (2011) Maslinic acid, a natural triterpenoid compound from Olea europaea, protects cortical neurons against oxygen-glucose deprivation-induced injury. Eur J Pharmacol 670(1):148–153. doi:10.1016/j.ejphar.2011.07.037

Qian Y, Guan T, Huang M, Cao L, Li Y, Cheng H, Jin H, Yu D (2012) Neuroprotection by the soy isoflavone, genistein, via inhibition of mitochondria-dependent apoptosis pathways and reactive oxygen induced-NF-κB activation in a cerebral ischemia mouse model. Neurochem Int 60(8):759–67. doi:10.1016/j.neuint.2012.03.011

Qureshi IA, Mehler MF (2012) Emerging roles of non-coding RNAs in brain evolution, development, plasticity and disease. Nat Rev Neurosci 13(8):528–541. doi:10.1038/nrn3234

Qureshi IA, Mehler MF (2013a) Epigenetic mechanisms governing the process of neurodegeneration. Mol Aspects Med 34(4):875–882. doi:10.1016/j.mam.2012.06.011

Qureshi IA, Mehler MF (2013b) Understanding neurological disease mechanisms in the era of epigenetics. JAMA Neurol 70(6):703–710. doi:10.1001/jamaneurol.2013.1443

Qureshi IA, Mehler MF (2014) Epigenetic mechanisms underlying the pathogenesis of neurogenetic diseases. Neurotherapeutics 11(4):708–720. doi:10.1007/s13311-014-0302-1

Qureshi IA, Mehler MF (2015) Epigenetics and therapeutic targets mediating neuroprotection. Brain Res. pii: S0006-8993(15)00577-6. doi:10.1016/j.brainres.2015.07.034

Rasia RM, Bertoncini CW, Marsh D, Hoyer W, Cherny D, Zweckstetter M, Griesinger C, Jovin TM, Fernández CO (2005) Structural characterization of copper(II) binding to alpha-synuclein: Insights into the bioinorganic chemistry of Parkinson's disease. Proc Natl Acad Sci USA 102(12):4294–4299

Ros E, Nnez I, Perez-Heras A, Merce S, Gilabert R, Casals E, Deulofeu R (2004) Walnut diet improves endothelial functions in hypercholesterolemic subject. Circulation 109:1609–1614

Scalzo P, Kümmer A, Cardoso F, Teixeira AL (2010) Serum levels of interleukin-6 are elevated in patients with Parkinson's disease and correlate with physical performance. Neurosci Lett 468 (1):56–58. doi:10.1016/j.neulet.2009.10.062

Schlesinger I, Schlesinger N (2008) Uric acid in Parkinson's disease. Mov Disord 23:1653–1657. doi:10.1002/mds.22139

Sconce MD, Churchill MJ, Greene RE, Meshul CK (2015) Intervention with exercise restores motor deficits but not nigrostriatal loss in a progressive MPTP mouse model of Parkinson's disease. Neuroscience 299:156–174. doi:10.1016/j.neuroscience.2015.04.069

Seidl SE, Santiago JA, Bilyk H, Potashkin JA (2014) The emerging role of nutrition in Parkinson's disease. Front Aging Neurosci. doi:10.3389/fnagi.2014.00036

Shaltiel-Karyo R, Frenkel-Pinter M, Rockenstein E, Patrick C, Levy-Sakin M, Schiller A (2013) A blood-brain barrier (BBB) disrupter is also a potent alpha-synuclein (alpha-syn) aggregation inhibitor: a novel dual mechanism of mannitol for treatment of Parkinson disease (PD). J Biol Chem 288:17579–17588. doi:10.1074/jbc.M112.434787

Shen C, Guo Y, Lin C, Ding M (2013) Serum urate and the risk of Parkinson's disease: results from a meta-analysis. Can J Neurol Sci 40:73–79

Shumway-Cook A, Matsuda PN, Taylor C (2015) Investigating the validity of the environmental framework underlying the original and modified Dynamic Gait Index. Phys Ther 95(6):864–870. doi:10.2522/ptj.20140047

Sonsalla PK, Wong LY, Harris SL, Richardson JR, Khobahy I, Li W, Gadad BS, German DC (2012) Delayed caffeine treatment prevents nigral dopamine neuron loss in a progressive rat model of Parkinson's disease. Exp Neurol 234(2):482–487. doi:10.1016/j.expneurol.2012.01.022

Sonsalla PK, Coleman C, Wong LY, Harris SL, Richardson JR, Gadad BS, Li W, German DC (2013) The angiotensin converting enzyme inhibitor captopril protects nigrostriatal dopamine neurons in animal models of parkinsonism. Exp Neurol 250:376–383. doi:10.1016/j.expneurol.2013.10.014

Spaccarotella KJ, Kris-Etherton PM, Stone WL, Bagshaw DM, Fishell VK, West SG, Lawrence FR, Hartman TJ (2008) The effect of walnut intake on factors related to prostate and vascular health in older men. Nutr J 7:13. doi:10.1186/1475-2891-7-13

Sumec R, Filip P, Sheardova K, Bares M (2015) Psychological benefits of nonpharmacological methods aimed for improving balance in Parkinson's disease: a systematic review. Behav Neurol. doi:10.1155/2015/620674

Sutoo DE, Akiyama K (2003) Regulation of brain function by exercise. Neurbiol Dis 13:1–14

Tamilselvam K, Braidy N, Manivasagam T, Essa MM, Prasad NR, Karthikeyan S, Thenmozhi AJ, Selvaraju S, Guillemin GJ (2013) Neuroprotective effects of hesperidin, a plant flavanone, on rotenone-induced oxidative stress and apoptosis in a cellular model for Parkinson's disease. Oxid Med Cell Longev 2013:102741. doi:10.1155/2013/102741

Teixeira-Machado L, Araújo FM, Cunha FA, Menezes M, Menezes T, Melo DeSantana J (2015) Feldenkrais method-based exercise improves quality of life in individuals with Parkinson's disease: a controlled, randomized clinical trial. Altern Ther Health Med 21(1):8–14

Tian Y, Zhang Y, Zhang R, Qiao S, Fan J (2015) Resolvin D2 recovers neural injury by suppressing inflammatory mediators expression in lipopolysaccharide-induced Parkinson's disease rat model. Biochem Biophys Res Commun 460(3):799–805. doi:10.1016/j.bbrc.2015.03.109

Tsang WWN (2013) Tai Chi training is effective in reducing balance impairments and falls in patients with Parkinson's disease. J Physiother 59, art. 55

Tuon T, Valvassori SS, Dal Pont GC, Paganini CS, Pozzi BG, Luciano TF, Souza PS, Quevedo J, Souza CT, Pinho RA (2014) Physical training prevents depressive symptoms and a decrease in brain-derived neurotrophic factor in Parkinson's disease. Brain Res Bull 108:106–112. doi:10.1016/j.brainresbull.2014.09.006

Tyagi SC, Joshua IG (2014) Exercise and nutrition in myocardial matrix metabolism, remodeling, regeneration, epigenetics, microcirculation, and muscle. Can J Physiol Pharmacol 92(7):521–523. doi:10.1139/cjpp-2014-0197

Uhrbrand A, Stenager E, Pedersen MS, Dalgas U (2015) Parkinson's disease and intensive exercise therapy—a systematic review and meta-analysis of randomized controlled trials. J Neurol Sci 353(1–2):9–19. doi:10.1016/j.jns.2015.04.004

Vines A, Delattre AM, Lima MMS, Rodrigues LS, Suchecki D, Machado RB, Tufik S, Pereira SI, Zanata SM, Ferraz AC (2012) The role of 5-HT1A receptors in the fish oil-mediated incressed BDNF expression in the rat hippocampus and cortex: a possible antidepressant mechanism. Neuropharmacology 62:184–191. doi:10.1016/j.neuropharm.2011.06.017

Voisin S, Eynon N, Yan X, Bishop DJ (2015) Exercise training and DNA methylation in humans. Acta Physiol (Oxf) 213(1):39–59. doi:10.1111/apha.12414

Winter B, Breitenstein C, Mooren FC, Voelker K, Fobker M, Lechtermann A, Krueger K, Fromme A, Korsukewitz C, Floel A, Knecht S (2007) High impact running improves learning. Neurobiol Learn Mem 87(4):597–609

Wong-Yu Is IS, Mak M (2013) Effects of a context-specific physiotherapy exercise programme on enhancing balance performance and balance confidence in people with Parkinson's disease—a randomized controlled trial. Hong Kong Physiother J 31, 100.101

Wu T, Wang J, Wang C, Hallett M, Zang Y, Wu X, Chan P (2012) Basal ganglia circuits changes in Parkinson's disease patients. Neurosci Lett 524(1):55–59. doi:10.1016/j.neulet.2012.07.012

Xiao D, Cassin JJ, Healy B, Burdett TC, Chen JF, Fredholm BB, Schwarzschild MA (2011) Deletion of adenosine A_1 or $A(_2A)$ receptors reduces L-3,4-dihydroxyphenylalanine-induced dyskinesia in a model of Parkinson's disease. Brain Res 1367:310–318. doi:10.1016/j.brainres.2010.08.099

Yadav S, Gupta SP, Srivastava G, Srivastava PK, Singh MP (2012) Mucuna pruriens seed extract reduces oxidative stress in nigrostriatal tissue and improves neurobehavioral activity in paraquat-induced Parkinsonian mouse model. Neurochem Int 62(8):1039–1047. doi:10.1016/j.neuint.2013.03.015

Yakunin E, Loeb V, Kisos H, Biala Y, Yehuda S, Yaari Y, Selkoe DJ, Sharon R (2012) A-synuclein neuropathology is controlled by nuclear hormone receptors and enhanced by docosahexaenoic acid in a mouse model for Parkinson's disease. Brain Pathol 22(3):280–294. doi:10.1111/j.1750-3639.2011.00530.x

Yang F, Trolle Lagerros Y, Bellocco R, Adami HO, Fang F, Pedersen NL, Wirdefeldt K (2015) Physical activity and risk of Parkinson's disease in the Swedish National March Cohort. Brain. 138(Pt 2):269–275. doi:10.1093/brain/awu323

Yang ML, Hasadsri L, Woods WS, George JM (2010) Dynamic transport and localization of alpha-synuclein in primary hippocampal neurons. Mol Neurodegener 5(1):9. doi:10.1186/1750-1326-5-9

Ye Q, Ye L, Xu X, Huang B, Zhang X, Zhu Y, Chen X (2012) Epigallocatechin-3-gallate suppresses 1-methyl-4-phenyl-pyridine-induced oxidative stress in PC12 cells via the SIRT1/PGC-1α signaling pathway. BMC Complement Altern Med 12:82. doi:10.1186/1472-6882-12-82

Yoon MC, Shin MS, Kim TS, Kim BK, Ko IG, Sung YH, Kim SE, Lee HH, Kim YP, Kim CJ (2007) Treadmill exercise suppresses nigrostriatal dopaminergic neuronal loss in 6-hydroxydopamine-induced Parkinson's rats. Neurosci Lett 423(1):12–17

Lifelong Rodent Model of Tardive Dyskinesia—Persistence After Antipsychotic Drug Withdrawal

Richard M. Kostrzewa and Ryszard Brus

Abstract Tardive dyskinesia (TD), first appearing in humans after introduction of the phenothiazine class of antipsychotics in the 1950s, is now recognized as an abnormality resulting predominately by long-term block of dopamine (DA) D_2 receptors (R). TD is thus reproduced in primates and rodents by chronic administration of D_2-R antagonists. Through a series of studies predominately since the 1980s, it has been shown in rodent modeling of TD that when haloperidol or other D_2-R antagonist is added to drinking water, rats develop spontaneous oral dyskinesias, vacuous chewing movements (VCMs), after ~ 3 months, and this TD is associated with an increase in the number of striatal D_2-R. This TD persists for the duration of haloperidol administration and another ~ 2 months after haloperidol withdrawal. By neonatally lesioning dopaminergic nerves in brain in neonatal rats with 6-hydroxydopamine (6-OHDA), it has been found that TD develops sooner, at ~ 2 months, and also is accompanied by a much higher number of VCMs in these haloperidol-treated lesioned rats, and the TD persists lifelong after haloperidol withdrawal, but is not associated with an increased D_2-R number in the haloperidol-withdrawn phase. TD apparently is related in part to supersensitization of both D_1-R and serotoninergic 5-HT_2-R, which is also a typical outcome of neonatal 6-OHDA (n6-OHDA) lesioning. Testing during the haloperidol-withdrawn phase in n6-OHDA rats displaying TD reveals that receptor agonists and antagonists of a host of neuronal phenotypic classes have virtually no effect on spontaneous VCM number, except for 5-HT_2-R antagonists which acutely abate the incidence of VCMs in part. Extrapolating to human TD, it appears that (1) 5-HT_2-R supersensitization is the crucial alteration accounting for persistence of TD, (2) dopaminergic—perhaps age-related partial denervation—is a risk factor for

R.M. Kostrzewa (✉)
Department of Biomedical Sciences, Quillen College of Medicine,
East Tennessee State University, PO Box 70577, Johnson City, TN 37614, USA
e-mail: kostrzew@etsu.edu

R. Brus
Department of Nurse, High School of Strategic Planning, Koscielna 6,
41-303 Dabrowa Gornicza, Poland

© Springer International Publishing Switzerland 2015
Curr Topics Behav Neurosci (2016) 29: 353–362
DOI 10.1007/7854_2015_395
Published Online: 16 October 2015

the development of TD, and (3) 5-HT$_2$-R antagonists have the therapeutic potential to alleviate TD, particularly if/when an antipsychotic D$_2$-R blocker is withdrawn.

Keywords Tardive dyskinesia · Oral dyskinesia · Vacuous chewing movements · Antipsychotic · Haloperidol · Dopamine · Serotonin · 5-HT$_2$-Receptor · 6-hydroxydopamine · 6-OHDA · Receptor supersensitivity

Contents

1 Introduction 354
2 Dopamine and TD 355
3 DA D$_1$ Receptors and VCMs 355
4 5-HT-R and VCMs 356
5 Animal Modeling of TD 357
    5.1 Long-Term D$_2$-R Antagonist Administration, to Model TD 357
    5.2 Long-Term D$_2$-R Antagonist Administration Plus 6-OHDA Lesioning,
        to Model TD 357
6 Conclusions 359
References 359

1 Introduction

Tardive dyskinesia (TD) is movement disorder characterized by involuntary repetitive purposeless movements, most commonly of the lower face. These movements present as vacuous chewing (i.e., not directed onto any physical material), tongue-thrusting, grimacing, lip smacking/puckering, and excessive eye blinking. Less frequently, lower body dyskinesia may be seen involving fingers, limbs, and torso (Casey 1987; Jeste and Caligiuri 1993). The 'tardive' nature of TD relates to the nature of its onset, usually taking months or years to develop, and most often as a consequence with long-term administration of an antipsychotic drug (Baldessarini et al. 1988). In summary, TD most often occurs in patients with schizophrenia, after having been treated with an antipsychotic drug for a long period of time.

TD reportedly occurs at an incidence rate of 5 % per treatment year for patients treated with a classic antipsychotic, such as chlorpromazine, haloperidol, fluphenazine, or trifluoperazine (Morgenstern and Glazer 1993). This means that for every 100 patients treated chronically with such antipsychotics, 5 will develop TD after one year, 10 on average will display TD at two years, 15 will have TD after three years, etc. Newer atypical antipsychotics, such as aripiprazole, clozapine, olanzapine, quetiapine, risperidone, and ziprasidone, have a lesser incidence of TD (Rana et al. 2013).

If antipsychotic treatment is discontinued when TD symptoms first begin, TD may disappear. However, withdrawal of an antipsychotic drug in a patient with schizophrenia may not be a viable option. Overt treatment of TD is difficult and often unsuccessful. Sometimes, the initial presentation of TD can be suppressed by increasing the dose of antipsychotic, but this may have long-term adverse consequences. Except for symptom suppression in this manner, there is no effective treatment. Therefore, once TD occurs, it may be lifelong—continuing even if the antipsychotic drug is discontinued. This is severely troubling for the patient and his/her family and can affect prospects for securing work (van Harten and Tenback 2011).

2 Dopamine and TD

Antipsychotic drugs, the class most often associated with the development of TD, are antagonists of dopamine (DA) receptors (R) in brain, typically the DA D_2-R class, although DA D_1-R can be involved as well. Because antipsychotics have a long half-life and are administered daily for months or years at a time, DA-R are partially inactivated long term. The resultant deactivation of DA-R function and associated long-term imbalance of D_1-R and D_2-R actions are considered to be the major features underlying onset of TD (Waddington et al. 1983).

3 DA D_1 Receptors and VCMs

Experiments conducted over a period of decades have led to a reasonable understanding of the associations between DA-R and VCMs in general. Initially, it was shown that the acute administration of a D_2-R antagonist, but not a D_1-R antagonist, provoked an acute increase in the number of VCMs in rats (Rosengarten et al. 1983a, b, 1986; Rupniak et al. 1985; Arnt et al. 1987; Koshikawa et al. 1987; Molloy and Waddington 1988; Levin et al. 1989; Murray and Waddington 1989). Conversely, acute treatment with a D_2-R agonist did not acutely provoke VCMs, but very high dosage of a D_1-R agonist did evoke VCMs in rats (Rosengarten et al. 1983a, b, 1986). These studies in the 1980s demonstrate an association or balance between D_1-R and D_2-R and the incidence of VCMs.

In an unrelated series of studies, Breese and colleagues found that D_1-R could be supersensitized in rats that had been lesioned with the neurotoxin 6-hydroxydopamine (6-OHDA; desipramine pretreatment) as neonates (n6-OHDA) (Breese et al. 1985a, b). This treatment produced an approximate 99 % destruction of substantia nigra pars compacta (SNpc) and approximate 95–99 % dopaminergic denervation of striatum, while preserving noradrenergic innervation throughout the brain (Breese et al. 1985a, b, 1987; Doucet et al. 1986; Descarries et al. 1992, 1995). When studied as adults, the two initial doses of a D_1-R agonist, at one-week intervals, did not substantially alter behavioral actions, versus controls. However, the third weekly dose and all subsequent doses of a D_1-R agonist produced prominent stereotypic and locomotor effects. This delayed induction of D_1-R supersensitivity was designated as a '*priming*' phenomenon (Breese et al. 1985a, b, 1987; Criswell et al. 1989; Kostrzewa 1995; 1998).

Subsequently, we found that the first dose of a D_1-R agonist, administered to adult n6-OHDA rats, produced a prominent increase in the number of VCMs, and the dosage of the D_1-R agonist required for this effect was $\sim$ 1000-fold lower than the dose required in intact non-lesioned control rats. Moreover, the induction of VCMs was seen after the initial dose of the D_1-R agonist, thereby equating the VCM response as indicative of *overt* D_1-R supersensitization (Kostrzewa and Gong 1991). In vitro studies established that there was no change in the affinity of striatal D_1-R in the striatum of n6-OHDA rats (Breese et al. 1987; Duncan et al. 1987; Kostrzewa and Hamdi 1991; Radja et al. 1993b) and the number of striatal D_1-R was unchanged (Breese et al. 1987; Duncan et al. 1987; Kostrzewa and Hamdi 1991) or slightly decreased (Radja et al. 1993b). Thus, D_1-R supersensitivity is likely related to enhanced second messenger signaling in nerves with D_1-R and/or D_2-R, but not associated with DA receptor binding parameters.

In the n6-OHDA rat, D_2-R affinity is unaltered, while the number of D_2-R is reported to be unchanged (when spiperone is the ligand for in vitro study) (Breese et al. 1985a, b; Gong et al. 1994) or increased (when raclopride is the ligand) (Dewar et al. 1990; Radja et al. 1993a).

A D_2-R antagonist evoked a similarly enhanced VCM response in adult n6-OHDA rats versus control rats, reinforcing the recognized association between D_1-R and D_2-R balance in this response (Kostrzewa and Gong 1991).

4 5-HT-R and VCMs

Striatal dopaminergic denervation produced by n6-OHDA provokes reactive serotoninergic fiber proliferation (Breese et al. 1984; Stachowiak et al. 1984), so that the striatum in adult n6-OHDA-lesioned rats is serotoninergic hyperinnervated (Berger et al. 1985; Snyder et al. 1986; Luthman et al. 1987; Towle et al. 1989; Dewar et al. 1990) by approximately 20 % in caudal striatum and by 100 % in rostral striatum (Mrini et al. 1995). Also, $5\text{-HT}_{1B/1non-AB/2C}$-R number is increased throughout striatum by 30–60 %, while 5-HT_{2A}-R is increased in number in rostral striatum (Radja et al. 1993b).

On a functional level, and as it pertains to TD, there is an enhanced behavioral response to 5-HT-R agonists in adult n6-OHDA-lesioned rats which display dopaminergic denervated and serotoninergic hyperinnervated striatum. Notably, the largely 5-HT_{2C}-R agonist m-chlorophenylpiperazine (mCPP) evokes an enhanced induction in VCMs in n6-OHDA-lesioned rats versus intact controls (Kostrzewa and Hamdi 1991; Gong and Kostrzewa 1992; El Mansari et al. 1994; Plech et al. 1995), and the response is attenuated by the 5-HT_2-R antagonist mianserin but not by 5-HT_1-R or 5-HT_3-R antagonists (Gong and Kostrzewa 1992).

Perhaps of greater relevance is the fact that the 5-HT$_2$-R antagonist mianserin attenuated D$_1$-R agonist induction of VCMs in n6-OHDA-lesioned rats, while D$_1$-R antagonists failed to attenuate VCM induction by the 5-HT$_2$-R agonist mCPP (Gong et al. 1992). Also, when the serotoninergic neurotoxin 5,7-dihydroxytryptamine (5,7-DHT) was coadministered with 6-OHDA to neonatal rats—producing adulthood 90 % destruction of serotoninergic innervation to striatum—both D$_1$-R supersensitivity and 5-HT$_2$-R supersensitivity for VCM induction were eliminated (Brus et al. 1994). Additionally, when 5,7-DHT was administered to adult n6-OHDA-lesioned rats to destroy the serotoninergic fiber overgrowth of striatum, the effect of n5,7-DHT was replicated [unpublished]. This series of findings highlight the relevance of the serotoninergic system as a mediator or regulator of DA-R agonist effects on VCM induction.

5 Animal Modeling of TD

5.1 Long-Term D$_2$-R Antagonist Administration, to Model TD

Waddington et al. (1983) successfully produced TD in adult rats by adding an antipsychotic (i.e., largely D$_2$-R antagonist) to the drinking water for 6 months. After a period of several months, these rats developed classic VCMs, oral dyskinesias, which persisted for as long as the D$_2$-R antagonist was in the drinking water and for 2½ months after the antipsychotic drug was withdrawn. This model fulfilled the criteria of a TD: (1) The oral dyskinesia was spontaneous and purposeless and (2) the dyskinesia was truly 'tardive,' as it did not arise until the D$_2$-R antagonist had been administered for several months; also, the dyskinesia persisted for a protracted period after the D$_2$-R antagonist had been withdrawn for more than 2 months. The character, form, periodicity, and energy spectrum (1–2 Hz) of VCMs in long-term haloperidol or fluphenazine in rats, when assessed by a computerized video analysis system using fast Fourier analysis, were virtually identical in form to those observed for oral dyskinesias in human TD (Ellison and See 1989). The pathophysiology and phenomenology of rodent TD are very similar to those of human TD (Waddington 1990). The chronic haloperidol-treated rat has remained the gold standard for animal modeling of TD.

5.2 Long-Term D$_2$-R Antagonist Administration Plus 6-OHDA Lesioning, to Model TD

Spontaneous oral dyskinesias, VCMs, developed in adult rats in which the frontal cortex was ablated bilaterally. In these rats as well as in rats treated as adults with intracerebroventricular (icv) 6-OHDA, subsequent long-term (i.e., 10–12 weeks)

D_2-R antagonist administration induced a twofold increase in VCMs, as contrasted with intact rats treated solely with long-term D_2-R antagonist. In the 6-OHDA-lesioned adult rats that received chronic haloperidol, VCMs persisted for ten weeks after withdrawal of the D_2-R antagonist (Gunne et al. 1982), similar to the duration of persistent VCMs later observed in chronic haloperidol-treated rats (Waddington et al. 1983). These findings demonstrate that a brain injury or a more selective lesioning of dopaminergic neurons predisposed to D_2-R antagonist induction of oral dyskinesia.

On the basis of these findings, we hypothesized that long-term administration of a D_2-R antagonist would have an even more profound effect on oral dyskinesia in rats in which the dopaminergic neuropil was largely destroyed. This was tested by administering 6-OHDA (icv) to neonatal rats, thereby destroying >95 % of dopaminergic innervation to the striatum (Kostrzewa and Gong 1991; Gong et al. 1993; Huang et al. 1997). Starting 2 months after birth, haloperidol (1.5 mg/kg/day) was administered in drinking water for 10 months, in order to induce TD. At ~ 4 months after initiating haloperidol, there was a twofold increase in VCM number in non-lesioned control rats (i.e., from ~ 5 to ~ 15 VCMs per session), and this elevation in VCM number persisted for the duration of the study, during which time haloperidol was included in the drinking water. By comparison, in n6-OHDA rats receiving haloperidol in drinking water, there was an abrupt increase in VCM number, starting at ~ 3 months. Thus, onset of elevated VCM number occurred one month earlier in n6-OHDA rats with haloperidol versus intact controls. Moreover, in n6-OHDA rats, the number of 'spontaneous' VCMs rose $\sim$ threefold at 3 months (i.e., from ~ 10 to ~ 35 VCMs per session)—a level at least twofold greater than that of intact controls with haloperidol (Huang et al. 1997).

Haloperidol was discontinued after 11 months, and 8 or 9 days later, the majority of rats were taken for neurochemical analysis. As expected, there was an increase in the B_{max} for D_2-R ($[^3H]$raclopride binding) (i.e., reflecting an increase in D_2-R number) in the striatum of haloperidol-exposed rats, both intact and n6-OHDA-lesioned rats (Huang et al. 1997). A small number of the n6-OHDA rats were maintained for observation after haloperidol withdrawal from drinking water. While intact haloperidol-exposed rats revert to low levels of VCMs after 2.5 months after discontinuing haloperidol (Waddington et al. 1983), a high level of VCMs was maintained in the n6-OHDA rats for 8 months after having discontinued haloperidol. Significantly, these rats showed no elevation in the striatal B_{max} for D_2-R ($[^3H]$raclopride binding) (analysis was in conjunction/simultaneous with B_{max} determinations for all other groups), indicating that the maintained increase in VCM number during the haloperidol-withdrawn period is unrelated to D_2-R number (Huang et al. 1997).

When probed with a series of drugs at intervals during the haloperidol-withdrawn phase, adrenoceptor α_1 (phentolamine, phenoxybenzamine)- and α_2 (propranolol)-receptor antagonists were found to be ineffective in abating VCMs in rats. The same was so also for agonists or antagonists of the muscarinic-R (scopolamine), DA D_1-R [SCH 23390, (R)-(+)-8-chloro-2,3,4,5-tetrahydro-3-methyl-5-phenyl-1H-3-benzazepin-7-ol hemimaleate] and DA D_2-R (spiperone, metoclopramide), also histamine H_1-R (ranitidine, cyproheptadine), adenosine A_{2A}-R (theophylline), $GABA_A$-R (muscimol), NMDA-R (ketamine, MK-801), opioid μ-R (morphine, naloxone), as well as serotoninergic $5\text{-}HT_{1A}$-R (pindolol), and $5\text{-}HT_{2A}$-R (ketanserin) (Kostrzewa et al. 2007).

However, acute treatment by a variety of $5\text{-}HT_2$-R antagonists (mianserin, clozapine, mesulergine, ritanserin, mesulergine) effectively abated VCMs in the haloperidol-withdrawn rats (Kostrzewa et al. 2007). It is intriguing that chronic blockade of DA D_2-R is the impetus for TD, yet block of $5\text{-}HT_2$-R abates or at least partially attenuates the oral dyskinesias in TD. The effect may be related to an earlier finding in laboratory studies, demonstrating that DA-R supersensitization is mediated by serotoninergic systems, as indicated by the findings in n6-OHDA-lesioned rats that (1) a $5\text{-}HT_2$-R antagonist attenuates D_1-agonist-evoked oral dyskinesias, VCMs (Gong et al. 1992), and (2) a lesion of serotoninergic neurons in n6-OHDA rats eliminates D_1-agonist-evoked VCMs (Brus et al. 1994).

6 Conclusions

TD in humans is an outcome with an incidence of $\sim 5~\%$ per treatment year, usually with an antipsychotic drug, which classically is known to act at least in part by blocking DA D_2-R in brain. TD can be produced in non-human primates and in rodents by similar prolonged treatment with a D_2-R blocker (antagonist). Through a series of studies in rats, it is now known that brain injury or dopaminergic nerve injury predisposes toward TD, with oral dyskinesias becoming more numerous in rats so lesioned. This effect appears to be related to an increase in dopamine D_1-R and also $5\text{-}HT_2$-R supersensitivity. While chronic D_2-R block is associated with an increase in the relative number of D_2-R in rat brain, the resultant dyskinesia—at least during the period when the D_2-R blocker has been withdrawn—persists despite the fact that the relative number of D_2-R in brain (striatum) reverts to normal. And from a treatment perspective, it has now been demonstrated that blockers of $5\text{-}HT_2$-R, not dopamine D_1-R or D_2-R, have the ability to acutely abate the dyskinesia.

References

Arnt J, Hyttel J, Perregard J (1987) Dopamine D-1 receptor agonists combined with the selective D-2 agonist, quinpirole, facilitate the expression of oral stereotyped behaviour in rats. Eur J Pharmacol 133:137–145

Baldessarini RJ, Cohen BM, Teicher MH (1988) Significance of neuroleptic dose and plasma level in the pharmacological treatment of psychoses. Arch Gen Psychiatry, 45(1):79–91. Review. PMID: 2892478

Berger TW, Kaul S, Stricker EM, Zigmond MJ (1985) Hyperinnervation of the striatum by dorsal raphe afferents after dopamine-depleting brain lesions in neonatal rats. Brain Res 366:354–358

Breese GR, Baumeister AA, McCown TJ, Emerick SG, Frye GD, Crotty K, Mueller RA (1984) Behavioural differences between neonatal and adult 6-hydroxydopamine-treated rats to dopamine agonists: relevance to neurological symptoms in clinical syndromes with reduced brain dopamine. J Pharmacol Exp Ther 231:343–354

Breese GR, Baumeister A, Napier TC, Frye GD, Mueller RA (1985a) Evidence that D_1 dopamine receptors contribute to the supersensitive behavioral responses induced by L-dihydroxyphenylalanine in rats treated neonatally with 6-hydroxydopamine. J Pharmacol Exp Ther 235:287–295

Breese GR, Napier TC, Mueller RA (1985b) Dopamine agonist-induced locomotor activity in rats treated with 6-hydroxydopamine at differing ages: functional supersensitivity of D_1 dopamine receptors in neonatally lesioned rats. J Pharmacol Exp Ther 234:447–455

Breese GR, Duncan GE, Napier TC, Bondy SC, Iorio LC, Mueller RA (1987) 6-Hydroxydopamine treatments enhance behavioral responses to intracerebral microinjection of D_1- and D_2-dopamine agonists into the nucleus accumbens and striatum without changing dopamine antagonist binding. J Pharmacol Exp Ther 240:167–176

Brus R, Kostrzewa RM, Perry KW, Fuller RW (1994) Supersensitization of the oral response to SKF 38393 in neonatal 6-hydroxydopamine-lesioned rats is eliminated by neonatal 5,7-dihydroxytryptamine treatment. J Pharmacol Exp Ther 268:231–237

Casey DE (1987) Tardive dyskinesia, In: Meltzer HY (ed) Psychopharmacology: the third generation of progress. Raven Press, New York, pp 1411–1419

Criswell H, Mueller RA, Breese GR (1989) Priming of D_1-dopamine receptor responses: long-lasting behavioral supersensitivity to a D_1-dopamine agonist following repeated administration to neonatal 6-OHDA-lesioned rats. J Neurosci 9:125–133

Descarries L, Soghomonian J-J, Garcia S, Doucet G, Bruno JP (1992) Ultrastructural analysis of the serotonin hyperinnervation in adult rat neostriatum following neonatal dopamine denervation with 6-hydroxydopamine. Brain Res 569:1–13

Descarries L, Soucy J-P, Lafaille F, Mrini A, Tanguay R (1995) Evaluation of three transporter ligands as quantitative markers of serotonin innervation density in rat brain. Synapse 211:131–139

Dewar KM, Soghomonian J-J, Bruno JP, Descarries L, Reader TA (1990) Elevation of dopamine D_2 but not D_1 dopamine receptors in adult rat neostriatum after neonatal 6-hydroxydopamine denervation. Brain Res 536:287–296

Doucet G, Descarries L, Garcia S (1986) Quantification of the dopamine innervation in adult rat neostriatum. Neuroscience 19:427–445

Duncan GE, Criswell HE, McCown TJ, Paul IA, Mueller RA, Breese GR (1987) Behavioral and neurochemical responses to haloperidol and SCH-23390 in rats treated neonatally or as adults with 6-hydroxydopamine. J Pharmacol Exp Ther 243:1027–1034

El Mansari M, Radja F, Ferron A, Reader TA, Molina-Holgado E, Descarries L (1994) Hypersensitivity to serotonin and its agonists in serotonin-hyperinnervated neostriatum after neonatal dopamine denervation. Eur J Pharmacol 261:171–178

Ellison GD, See RE (1989) Rats administered chronic neuroleptics develop oral movements which are similar in form to those in humans with tardive dyskinesia. Psychopharmacology 98:564–566

Gong L, Kostrzewa RM (1992) Supersensitized oral responses to a serotonin agonist in neonatal 6-OHDA-treated rats. Pharmacol Biochem Behav 41:621–623

Gong L, Kostrzewa RM, Fuller RW, Perry KW (1992) Supersensitization of the oral response to SKF 38393 in neonatal 6-OHDA-lesioned rats is mediated through a serotonin system. J Pharmacol Exp Ther 261:1000–1007

Gong L, Kostrzewa RM, Brus R, Fuller RW, Perry KW (1993) Ontogenetic SKF 38393 treatments sensitize dopamine D_1 receptors in neonatal 6-OHDA-lesioned rats. Dev Brain Res 76:59–65

Gong L, Kostrzewa RM, Li C (1994) Neonatal 6-hydroxydopamine and adult SKF 38393 treatments alter dopamine D_1 receptor mRNA levels: absence of other neurochemical associations with the enhanced behavioral responses of lesioned rats. J Neurochem 63:1282–1290

Gunne LM, Growdon J, Glaeser B (1982) Oral dyskinesia in rats following brain lesions and neuroleptic drug administration. Psychopharmacology 77(2):134–139

Huang N-Y, Kostrzewa RM, Li C, Perry KW, Fuller RW (1997) Increased spontaneous oral dyskinesias persist in haloperidol-withdrawn rats neonatally lesioned with 6-hydroxydopamine: absence of an association with the B_{max} for [^{3}H]raclopride binding to neostriatal homogenates. J Pharmacol Exp Ther 280:268–276

Jeste DV, Caligiuri MP (1993) Tardive dyskinesia. Schizophr Bull 19:303–315

Koshikawa N, Aoki S, Tomiyama M, Maruyama Y, Kobayashi M (1987) Sulpiride injection into the dorsal striatum increases methamphetamine-induced gnawing in rats. Wur J Pharmacol 133:119–125

Kostrzewa RM (1995) Dopamine receptor supersensitivity. Neurosci Biobehav Rev 19:1–17

Kostrzewa RM, Gong L (1991) Supersensitized D_1 receptors mediate enhanced oral activity after neonatal 6-OHDA. Pharmacol Biochem Behav 39:677–682

Kostrzewa RM, Hamdi A (1991) Potentiation of spiperone-induced oral activity in rats after neonatal 6-hydroxydopamine. Pharmacol Biochem Behav 38:215–218

Kostrzewa RM, Reader TA, Descarries L (1998) Serotonin neural adaptations to ontogenetic loss of dopamine neurons in rat brain. J Neurochem 70:889–898

Kostrzewa RM, Huang NY, Kostrzewa JP, Nowak P, Brus R (2007) Modeling tardive dyskinesia: predictive 5-HT2C receptor antagonist treatment. Neurotox Res 11(1):41–50

Levin ED, See RE, South D (1989) Effects of dopamine D_1 and D_2 receptor antagonists on oral activity in rats. Pharmacol Biochem Behav 34:43–48

Luthman J, Botioli B, Tustsumi T, Verhofstad A, Jonsson G (1987) Sprouting of striatal serotonin nerve terminals following selective lesions of nigro-striatal dopamine neurons in neonatal rat. Brain Res Bull 19:269–274

Molloy AG, Waddington JL (1988) Behavioural responses to the selective D_1-dopamine receptor agonist R-SK&F 38393 and the selective D_2-agonist RU 24213 in young compared with aged rat. Br J Pharmacol 95:335–342

Morgenstern H, Glazer WM (1993) Identifying risk factors for tardive dyskinesia among long-term outpatients maintained with neuroleptic medications. Results of the Yale Tardive Dyskinesia Study. Arch Gen Psychiatry 50(9):723–733

Mrini A, Soucy J-P, Lafaille F, Lemoine P, Descarries L (1995) Quantification of the serotonin hyperinnervation in adult rat neostriatum after neonatal 6-hydroxydopamine lesion of nigral dopamine neurons. Brain Res 669:303–308

Murray AM, Waddington JL (1989) The induction of grooming and vacuous chewing by a series of selective D-1 dopamine receptor agonists: two directions of D-1:D-2 interaction. Eur J Pharmacol 160:377–387

Plech A, Brus R, Kalbfleisch JH, Kostrzewa RM (1995) Enhanced oral activity responses to intrastriatal SKF 38393 and *m*-CPP are attenuated by intrastriatal mianserin in neonatal 6-OHDA-lesioned rats. Psychopharmacology 119:466–473

Radja F, El Mansari M, Soghomonian J-J, Dewar KM, Ferron A, Reader TA, Descarries L (1993a) Changes in D_1 and D_2 receptors in adult rat neostriatum after neonatal dopamine denervation: quantitative data from ligand binding, in situ hybridization and iontophoresis. Neuroscience 57:635–648

Radja F, Descarries L, Dewar KM, Reader TA (1993b) Serotonin 5-HT$_1$ and 5-HT$_2$ receptors in adult rat brain after neonatal destruction of nigrostriatal dopamine neurons: a quantitative autoradiographic study. Brain Res 606:273–285

Rana AQ, Chaudry ZM, Blanchet PJ (2013) New and emerging treatments for symptomatic tardive dyskinesia. Drug Des Devel Ther 7:1329–1340. Review. PMID: 24235816

Rosengarten H, Schweitzer JW, Friedhoff AJ (1983a) Induction of oral dyskinesias in naïve rats by D_1 stimulation. Life Sci 33:2479–2482

Rosengarten H, Schweitzer JW, Egawa J, Friedhoff AJ (1983b) Diminished D_2 dopamine receptor function and the emergence of repetitive jaw movements. Adv Exp Med Biol 235:159–169

Rosengarten H, Schweitzer JW, Egawa J, Friedhoff AJ (1986) Diminished D_2 dopamine receptor function and the emergence of repetitive jaw movements. Adv Exp Med Biol 235:159–167

Rupniak NMJ, Jenner P, Marsden CD (1985) Pharmacological characterization of spontaneous or drug-associated purposeless chewing movements in rats. Psychopharmacology (Berlin) 85:71–79

Snyder AM, Zigmond MJ, Lund RD (1986) Sprouting of serotonergic afferents into striatum after dopamine depleting lesions in infant rats: a retrograde transport and immunocytochemical study. J Comp Neurol 245:274–281

Stachowiak MK, Bruno JP, Snyder AM, Stricker EM, Zigmond MJ (1984) Apparent sprouting of striatal serotonergic terminals after dopamine-depleting brain lesions in neonatal rats. Brain Res 291:164–167

Towle AC, Criswell HE, Maynard EH, Lauder JM, Joh RH, Mueller RA, Breese GR (1989) Serotonergic innervation of the rat caudate following a neonatal 6-hydroxydopamine lesion: an anatomical, biochemical and pharmacological study. Pharmacol Biochem Behav 34:367–374

van Harten PN, Tenback DE (2011) Tardive dyskinesia: clinical presentation and treatment. Int Rev Neurobiol 98:187–210. Review. PMID: 21907088

Waddington JL (1990) Spontaneous orofacial movements induced in rodents by very long-term neuroleptic drug administration: Phenomenology, pathophysiology and putative relationship to tardive dyskinesia. Psychopharmacology (Berlin) 101:431–447

Waddington JL, Cross AJ, Gamble SJ, Bourne RC (1983) Spontaneous orofacial dyskinesia and dopaminergic function in rats after 6 months of neuroleptic treatment. Science 220:530–532

Perinatal Influences of Valproate on Brain and Behaviour: An Animal Model for Autism

Peter Ranger and Bart A. Ellenbroek

Abstract Valproic acid or valproate (VPA) is an anti-convulsant and mood stabiliser effective in treating epilepsy and bipolar disorders. Although in adults VPA is well tolerated and safe, there is convincing evidence that it has teratogenic properties, ranging from mild neurodevelopmental changes to severe congenital malformations. In particular, studies involving humans and other animals have shown that prenatal exposure to VPA can induce developmental abnormalities reminiscent of autism spectrum disorder (ASD). In this chapter, we discuss the connection between VPA and ASD, evaluate the VPA animal model of ASD, and describe the possible molecular mechanisms underlying VPA's teratogenic properties.

Keywords Autism · ASD · Valproate · VPA · Valproic Acid · HDAC-I · ROS · Oxidative stress · Animal models · Behaviour · Teratogen

Contents

1 Introduction.. 364
2 Valproic Acid (VPA)... 364
3 ASD.. 365
4 Animal Models 367
5 Animal Models for ASD 367
6 VPA as an Animal Model of ASD.. 369
   6.1 Construct Validity 369
   6.2 Face Validity 369
   6.3 Predictive Validity 370
7 Mechanisms of Action of VPA.. 372
8 Histone Deacetylase Inhibition (HDAC-I).. 373

P. Ranger · B.A. Ellenbroek (✉)
School of Psychology, Victoria University of Wellington, PO Box 600,
Wellington, New Zealand
e-mail: bart.ellenbroek@vuw.ac.nz

© Springer International Publishing Switzerland 2015
Curr Topics Behav Neurosci (2016) 29: 363–386
DOI 10.1007/7854_2015_404
Published Online: 29 October 2015

9   Reactive Oxygen Species and Oxidative Stress 375
10 Experimental Models and Mitigating Factors... 377
11 Concluding Remarks 378
References.. 379

1   Introduction

The thalidomide disaster in the 1950s and 1960s signalled an unprecedented increase in the awareness that the effects of drugs which are relatively safe in adulthood can have severe consequences for the developing foetus. Originally marketed as a sedative, thalidomide was widely used to treat pregnancy-related nausea until it was discovered to lead to pronounced limb reductions, congenital heart disease and other developmental malformations (Kim and Scialli 2011). As exceptional as the thalidomide case was, it is now well recognised that many drugs can induce developmental alterations when given during pregnancy. These alterations can range from severe malformations to much more subtle changes in behaviour and personality. In this chapter, we focused on valproic acid (VPA), a drug that, like thalidomide, is considered relatively safe in adulthood.

2   Valproic Acid (VPA)

VPA is an anti-convulsant and mood stabiliser used predominantly to treat epilepsy, bipolar disorder, and migraine (Lloyd 2013; Mulleners et al. 2014; Trinka et al. 2014). However, its therapeutic potential in Alzheimer's disease, cancer, and HIV has also been explored (Lehrman et al. 2005; Qing et al. 2008; Hu et al. 2011; Avallone et al. 2014; Brodie and Brandes 2014; Grishina et al. 2015). In addition to its prophylactic properties, VPA is a known teratogen (Wyszynski et al. 2005; Koren et al. 2006; Morrow et al. 2006; Diav-Citrin et al. 2008; Meador et al. 2008). A systematic review of the literature concluded that taking VPA during pregnancy was associated with a 3.77-fold increased risk of major congenital malformations in offspring, relative to healthy women, a 2.59-fold increased risk relative to women treated with other anti-epileptic medication, and a 3.16-fold increased risk relative to those with untreated epilepsy (Koren et al. 2006). Together, the literature investigating VPA exposure and congenital malformations indicates an approximate threefold increase in major malformations in children exposed prenatally to VPA (Ornoy 2009).

The potential teratogenic effects of VPA provide a dilemma for pregnant women required to take VPA in their course of treatment (Hill et al. 2010; Tomson and Battino 2012). Although approximately 94 % of children born to mothers medicated with VPA are completely normal (Morrow et al. 2006), the decision to medicate with VPA must involve a calculation of the dangers imposed to a foetus by the drug against the dangers imposed to the mother and foetus by the disease (Meador et al. 2008; Hill et al. 2010). Conditions such as epilepsy provide real risks to pregnant mothers and their offspring due to the possibility of severe seizures (Tomson and Battino 2012), although the risk of congenital malformations resulting from maternal epilepsy alone is strongly disputed (Kaaja et al. 2003; Fried et al. 2004).

A teratogen is predominantly defined as an agent that causes malformation to, and/or disrupts the development of, the embryo or foetus. What constitutes a teratogenic *outcome* is somewhat less clear. Typically, discussion of outcomes surrounds congenital malformations such as neural tube defects present from birth. However, in addition to the increased likelihood of congenital malformations, children exposed prenatally to VPA are also significantly more likely to suffer from more subtle neurodevelopmental deficits (Williams et al. 2001; Wyszynski et al. 2005; Diav-Citrin et al. 2008; Ornoy 2009; Christensen et al. 2013). There is especially strong evidence suggesting that prenatal exposure to VPA significantly increases the likelihood of children developing autism spectrum disorders (ASD) (Christianson et al. 1994; Moore et al. 2000; Rasalam et al. 2005; Christensen et al. 2013; Roullet et al. 2013; Smith and Brown 2014). Most recently, data from a Danish population-based study concluded that prenatal exposure to VPA was associated with an absolute risk factor of 4.42 % for ASD (Christensen et al. 2013). Consequently, prenatal exposure to VPA has been thoroughly assessed as an animal model for this disorder and has been used in cell-cultures, tadpoles, zebrafish, and rodents (Miyazaki et al. 2005; Schneider and Przewlocki 2005; Bauman et al. 2010; Kim et al. 2011; Patterson 2011; Jacob et al. 2014; James et al. 2015). The focus of this chapter is primarily on the link between VPA and ASD. However, as we will discuss later on, there is convincing evidence to suggest that the congenital malformations and the ASD-like pathology depend, to a large degree, on similar mechanistic processes including alterations in epigenetic processes and oxidative stress (Dufour-Rainfray et al. 2011).

3 ASD

Before discussing the relationship between VPA and ASD, it is relevant to briefly describe the disorder. ASD is a heterogeneous neurodevelopmental disorder characterised by deficits in social interaction and communication, as well as an increase in restricted or repetitive behaviour (Fig. 1) (Levy et al. 2009b). Repetitiveness is seen not only in motor patterns, but also in cognition and thought. Although these three classes of symptoms form the core of ASD, it is important to realise that most ASD patients also show a plethora of other symptoms. These auxiliary symptoms can be psychiatric (including depression and anxiety), behavioural (including aggression), intellectual (lower IQ in many cases), and neurological (including epilepsy). These clinical symptoms manifest themselves at a very early age (often

Fig. 1 The three core symptom domains of ASD: social behaviour, communication, and stereotypy

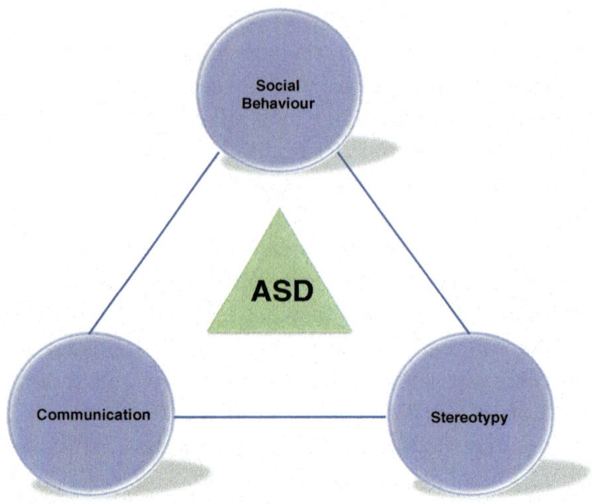

before the age of 3) and usually persist into adulthood (Kinney et al. 2008b; Levy et al. 2009a). One of the most interesting and troublesome aspects of ASD is that over the last fifty years, the incidence of ASD has increased significantly. Indeed, a study in the USA found that the incidence increased from 0.67 % in 2000 to 1.5 % in 2010, with males being 4–5 times more likely affected than females (Wingate et al. 2014). Although there are several different theories why the incidence may be rising (changes in diagnostic criteria, in awareness, or an actual increase in cases), one of the important consequences is that costs relating to the treatment and care of ASD patients are also dramatically rising. A recent study from Great Britain has estimated the total costs of children with ASD at £ 2.7 billion, and for adults at over £ 25 billion per year (Knapp et al. 2009).

In addition to the increased costs, treatment options for patients with ASD are fairly limited (Vorstman et al. 2014), especially in the area of pharmacological interventions. Currently, there are only two FDA-approved drugs for ASD: the second generation antipsychotics risperidone and aripiprazole, both specifically for the treatment of irritability and stereotyped behaviour. Furthermore, drugs such as antidepressants (especially selective serotonin reuptake inhibitors), anxiolytics, and even psychostimulants such as methylphenidate are being used to treat some of the auxiliary symptoms of ASD, such as aggression, anxiety, and irritability (Levy et al. 2009a; Hsia ct al. 2014; Murray et al. 2014; Vorstman et al. 2014). However, especially in relation to the social and communication domains, effective therapy is sadly lacking.

The precise causes of ASD are not well understood. However, what *is* known is that ASD has both genetic and environmental determinants, and that its aetiology is likely multifactorial (Tordjman et al. 2014). In fact, ASD is probably the most genetic of all major psychiatric disorders with heritability estimates of around 56-95 % and a substantial number of different risk genes have been identified in patients (Betancur 2011; Chaste et al. 2015; Colvert et al. 2015b). In addition, given that the concordance rate for monozygotic twins consistently falls short of 100 % (Ronald and Hoekstra 2011), non-genetic environmental factors must be involved in the aetiology of ASD as well. In line with this, several environmental factors have been found to significantly enhance the risk of ASD, including prenatal exposure to infectious agents (Atladottir et al. 2010), prenatal stress such as experiencing hurricanes (Kinney et al. 2008a), and, as previously mentioned, VPA.

4 Animal Models

The major reason for using animal models of ASD, and indeed for any disorder, is that they allow the testing of specific hypotheses and the identification of novel therapies (Bauman et al. 2010). An animal model's validity is usually judged on three distinct criteria:

First, *construct validity*: the extent to which the model replicates the theoretical underpinnings of the disorder, i.e. is the model constructed by the same factors that are known (or at least hypothesised) to cause the disorder in humans? Second, *face validity*: the extent to which the signs and symptoms of the model mimic those seen in humans with the disorder. Third, *predictive validity*: the extent to which the administration of pharmacological treatments that are effective in humans is also effective in reversing the signs and symptoms in the animal model (Ellenbroek and Cools 2000; van der Staay et al. 2009; Bauman et al. 2010; Nestler and Hyman 2010).

Thus, the ideal animal model for ASD would be generated by mimicking both genetic and environmental risk factors (construct validity), would produce the three major behavioural hallmarks of ASD, some of the co-morbid symptoms, and bio-chemical changes found in patients (face validity), and be responsive to medication that also works in patients (predictive validity). Creating the *perfect* animal model of any human disease, especially in the field of brain disorders, has been notoriously difficult, so a *"best-we-can-do"* approach is traditionally taken. The major barriers in creating animal models of ASD are the complex and unclear aetiology of the disorder, the heterogeneity of the symptomatology, and the lack of effective treatments for humans, making it difficult to achieve suitable construct, face, and predictive validity, respectively.

5 Animal Models for ASD

Before analysing the VPA animal model of ASD in more detail, we will briefly describe several other animal models for ASD, based on either genetic or environmental factors. As such models have been extensively described in recent reviews (Oddi et al. 2013; Ellegood and Crawley 2015; Servadio et al. 2015), we will only briefly mention the most relevant ones.

The maternal immune activation (MIA) model of ASD is one of the most prominent environmentally based models in the literature. There is good evidence indicating that maternal infection leads to an increase in ASD in offspring, and that this likely acts via the maternal immune response (Shi et al. 2003; Atladóttir et al. 2010; Bauman et al. 2014). Animal models using substances that elicit a maternal immune response, such as lipopolysaccharide (LPS, mimicking a bacterial infection) and polyriboinosinic-polyribocytidilic acid (Poly I:C, mimicking a viral infection), have demonstrated the ability to generate ASD-like symptomatology in animals, including all three of the major behavioural hallmarks of ASD and additional neuroanatomical and immunological changes observed in patients with ASD (Shi et al. 2009; Bauman et al. 2014; Ohkawara et al. 2015).

Dysregulation of serotonin (5-HT) is a consistent finding with ASD patients (Devlin et al. 2005). In fact, 5-HT has been suggested to be the neurochemical with the most consistently proven involvement in ASD (Lam et al. 2006). Tryptophan hydroxylase 2 knockout (TPH2-KO) mice, lacking any brain 5-HT, display all three of the major behavioural hallmarks of ASD; moreover, they exhibit key developmental delays related to central nervous system (CNS) functioning. These TPH2-KO mice have been suggested as a promising genetic model of ASD (Kane et al. 2012). Intriguingly, animals lacking the serotonin transporter, and therefore displaying significantly increased (extracellular) serotonin levels have also been proposed as animal models for ASD (Kinast et al. 2013).

The BTBR +tf/J inbred mouse strain exhibits behavioural symptom relevant to all three of the major hallmarks of ASD (Bolivar et al. 2007; McFarlane et al. 2008; Wöhr et al. 2011) and is used as a model of ASD. However, one key problem with the BTNR +tf/J mouse model of ASD is that it is unclear which other genetic mouse strains should be used as a control group for comparison (Patterson 2011).

Accumulating evidence suggests disruption of synaptic pathways plays a key role in patients with ASD (Betancur et al. 2009). Neuroligins (NLGN) are cell adhesion molecules that facilitate synapse formation (Chih et al. 2005). Interestingly, mutations in genes encoding NLGN3 and NLGN4 have been associated with ASD (Jamain et al. 2003). Modelling of these mutations in animals has produced multiple behavioural phenotypes relevant to ASD. Specifically, NLGN4 (the murine orthologue of NLGN4) knockout mice demonstrate selective deficits in two out of the three major behavioural hallmarks of ASD: social interaction and communication (Jamain et al. 2008). The data suggest that NLGN models of ASD show good construct validity, although other models have demonstrated superior face validity.

In addition to these models, many other genetic models have been developed (Ellegood et al. 2015), many of them based on single gene mutations with a link to ASD. Although these models therefore have some construct validity, it is important to realise that only very few cases ASD can be explained by a single genetic deficit (such as fragile X syndrome). Therefore, forward genetic models (i.e. those starting with the symptoms of ASD) may be more relevant. In addition, future research should be aimed at combining genetic and environmental factors in a single animal model, as ASD is caused by both genetic and environmental factors, and thus models which include both will have stronger construct validity. In fact, as was found in several humans studies as well, the true effects of both genetic and environmental factors may be uncovered when investigating their interactive properties that may never have been discovered when looking at the specific genetic or environmental factor in isolation. In other words, ignoring investigations of gene x environment interactions may lead to a failure to identify effects in both areas.

6   VPA as an Animal Model of ASD

Given the link between maternal treatment with VPA and ASD in humans (see above), prenatal exposure to VPA has been proposed as an animal model of ASD (Rodier et al. 1997; Schneider and Przewlocki 2005). The typical method of creating this model is to inject pregnant rats with a single dose of VPA around the time of the foetal neural tube closure, approximately gestational day 12 (Kim et al. 2011). However, the exact dose, method of injection, day of exposure, and whether the exposure is acute or chronic can vary from study to study; often leading to varying outcomes (Cohen et al. 2013; Štefánik et al. 2015). In fact, it has been proposed that the dose of VPA determines the mechanism of influence and thus the outcome (Johannessen and Johannessen 2003). The VPA animal model of ASD will now be discussed with regard to the previously mentioned validity criteria.

6.1   Construct Validity

Although the VPA model is created using a well-established environmental risk factor for ASD, it does not take into account the large genetic component of ASD (Colvert et al. 2015a). Thus, from a construct validity point of view, the model is clearly limited as it only incorporates an environmental factor involved in ASD and ignores the genetic contribution to the disorder. Moreover, VPA is likely only *one* of the environmental determinants that interact with genotype in the development of ASD (Bauman et al. 2010). Finally, as timing is a crucial element in the outcome of early environmental challenges (see also below), it is important to realise that in most studies only a single injection of VPA is administered, whereas in humans VPA is typically taken throughout pregnancy.

6.2   Face Validity

The VPA model achieves very good face validity. Numerous behavioural and biochemical outcomes associated with ASD in humans are produced by this model; importantly, these outcomes are observed in a variety of species, including "outbred" genetically heterogeneous rat and mouse strains with stronger translational validity to the genetically diverse human population. It is quite remarkable given VPA's broad spectrum of action and the complexity of ASD that the pattern of outcomes associated with this model overlaps the pattern of deficits observed in ASD so well.

Specifically, prenatal exposure to VPA can produce the following behavioural abnormalities that are associated with ASD in humans: lower sociability, deficits in communication, increased repetitive behaviour/stereotypies, pre-pulse-inhibition deficits, lowered sensitivity to pain, increased anxiety, and hyperlocomotor activity (Schneider and Przewlocki 2005; Schneider et al. 2008; Dufour-Rainfray et al. 2010; Gandal et al. 2010; Mehta et al. 2011; Choi et al. 2014; James et al. 2015). In addition, prenatal exposure to VPA produces the following biochemical, anatomical, or neuronal deficits, many of which are associated with ASD in humans: a reduction in Purkinje cells, cerebellar, and gastrointestinal abnormalities (Rodier et al. 1997; Ingram et al. 2000; Kim et al. 2013a), deficits in the Akt/mTOR pathway (Nicolini et al. 2015), increased cortical thickness and number of neurons in the neocortex (Sabers et al. 2015), an increase in the basolateral nucleus of the amygdala (Loohuis et al. 2015), a reduction in spine density in the hippocampus (Takuma et al. 2014), decreased cortical brain-derived neurotrophic factor (BDNF) mRNA (Roullet et al. 2010) abnormal serotonergic differentiation, migration, and maturation (Miyazaki et al. 2005), hyperserotonemia (Narita et al. 2002), both increased and decreased hippocampal serotonin levels (Narita et al. 2002; Dufour-Rainfray et al. 2010), and the failure of serotonin expression (Jacob et al. 2014) (see Table 1).

The VPA model has even been able to replicate the gender imbalance found in ASD. VPA exposure in animals has a more detrimental impact on behaviour, morphology, and the immune system in males than it does in females (Schneider et al. 2008; Kataoka et al. 2013a; Kim et al. 2013b; Mowery et al. 2015). The reasons for this differential impact of VPA are uncertain (Schneider et al. 2008; Mowery et al. 2015). However, the preponderance of evidence suggests the likely answer is that natural differences between the sexes exacerbate or protect against the teratogenic impact of VPA. In other words, female-specific biochemical patterns during critical developmental periods may protect against VPA (Schneider et al. 2008). Indeed, female oestrogen has been described as protective against harmful toxins implicated in the onset of ASD (Geier et al. 2010).

In short, the VPA animal model's strength rests upon its very high degree of face validity. The model is able to produce the varied symptomatology of ASD, and its unique gender expression profile.

6.3 Predictive Validity

The ability to achieve pharmacological predictive validity in ASD animal models is currently impossible, due to the lack of effective treatments of the core symptoms in

Table 1 ASD-like behavioural and neurochemical alterations induced by prenatal VPA administration in animals

Reference	Effect	Species
Choi et al. (2014)	Hyperlocomotor activity	Rat (Sprague-Dawley)
Gandal et al. (2010)	↓ Social interaction ↓ USV ↑ Rep. behav.	Mice
Mehta et al. (2011)	↑ Anxiety ↑ Rep. behav.	Mice
Schneider and Przewlocki (2005)	↓ Sensitivity to pain ↓ p.p.i ↓ Social behav. ↑ Stereotypies	Rat (Wistar)
Schneider et al. (2008)	↓ Sensitivity to pain ↑ Rep. behav. ↑ Anxiety ↓ Social behav.	Rat (Wistar)
James et al. (2015)	Abnormal social behaviour	Xenopus laevis Tadpoles
Ingram et al. (2000)	↓ Purkinje cells	Rat (Long Evan)
Kim et al. (2013a)	Gastrointestinal abnormalities	Rat (Sprague-Dawley)
Rodier et al. (1997)	Cerebellar abnormalities	Rat
Nicolini et al. (2015)	Deficits in the Akt/mTOR pathway	Rat (Wistar Han)
Sabers et al. (2015)	↑ Cortical thickness ↑ Neurons neocortex	Rat (Wistar)
Loohuis et al. (2015)	↑ In the basolateral nucleus of the amygdala	Rat (Wistar)
Takuma et al. (2014)	↓ Spine density in hippocampus	Mice
Roullet et al. (2010)	↓ BDNF mRNA	Mice
Miyazaki et al. (2005)	Abnormal 5HT differentiation, migration and maturation	Rat (Wistar)
Narita et al. (2002)	Hyperserotonemia ↑ 5HT in hippocampus	Rat (Sprague-Dawley)
Dufour-Rainfray et al. (2010)	↓ 5HT in hippocampus	Rat (Wistar)
Jacob et al. (2014)	Failure of 5HT expression	Zebrafish

humans. However, VPA animal models have offered predictions in the other direction, i.e. predictions as to what might benefit humans based upon their impact in the VPA model. Several research groups have been able to show attenuation of the deficits produced by the VPA model, all via differing methods.

Specifically, the acetylcholinesterase inhibitor (AChEI) donepezil ameliorated social deficits, repetitive behaviour, and hyperactivity in mice prenatally exposed to VPA (Kim et al. 2014). Ciproxifan (CPX), an H3R antagonist, improved social behavioural deficits and repetitive behaviours in mice (Baronio et al. 2015). Moreover, treatment with atomoxetine (ATX), a norepinephrine reuptake inhibitor, reversed the hyperactivity induced by prenatal VPA exposure (Choi et al. 2014). In addition, post-natal environmental enrichment has also been shown to reverse a wide array of the expected outcomes of the VPA rat model, including social behavioural deficits, repetitive behaviour, and anxiety (Schneider et al. 2006).

Finally, antioxidants have been investigated for their attenuating properties. Astaxanthin was able to ameliorate VPA-induced deficits in social behaviour, and lowered sensitivity to pain, as well as significantly reducing oxidative stress in the liver and brain (Al-Amin et al. 2015). In addition, green tea extract attenuated some of the effects in a rodent VPA model, including cognitive and motor deficits, hyperlocomotion, and anxiety (Banji et al. 2011).

Taken together, these findings suggest that the cholinergic system, the histaminergic system, and the norepinephrine transporter may all play important roles in attenuating the deficits observed in ASD and, in turn, offer promising pharmacological targets in the drug discovery process. Further, environmental enrichment and specific antioxidants may be effective in helping patients with ASD. However, with regard to *environmental enrichment*, the translational value for humans is currently unclear. The identification of new targets for drug treatment is one of the key aims of animal models of disease (van der Staay et al. 2009), and in this respect the VPA model of ASD seems to offer great potential. However, clearly clinical trials with these compounds will need to be performed to test the validity of these predictions.

In summary, the prenatal exposure to VPA model of ASD is as good a model as there currently is for ASD. The strength of the model rests on its high degree of face validity across a range of different species and its ability to work in heterogeneous "outbred" rat strains that exhibit a genetic heterogeneity more representative of the human population. Although the VPA model has limited construct validity, given the complexity and heterogeneity of the disorder this is not surprising. In addition, despite the lack of treatments for ASD currently preventing the examination of predictive validity in animal models of this disorder, the VPA model has identified several distinct new targets/pathways for potential therapeutic intervention worthy of future research.

7 Mechanisms of Action of VPA

We have seen that prenatal exposure to VPA is a well-established animal model for ASD; however, the molecular mechanism(s) underlying these effects are far from clear (Jeong et al. 2003; Fujiki et al. 2013; Fathe et al. 2014; Bollino et al. 2015). The literature on the mechanisms of VPA is filled with entirely different explanations, conflicting results, failed replications, and lingering unproven hypotheses. However, from the complexity, at least three points can be gleaned: VPA has a broad spectrum of action, it works via multiple different mechanisms, and it

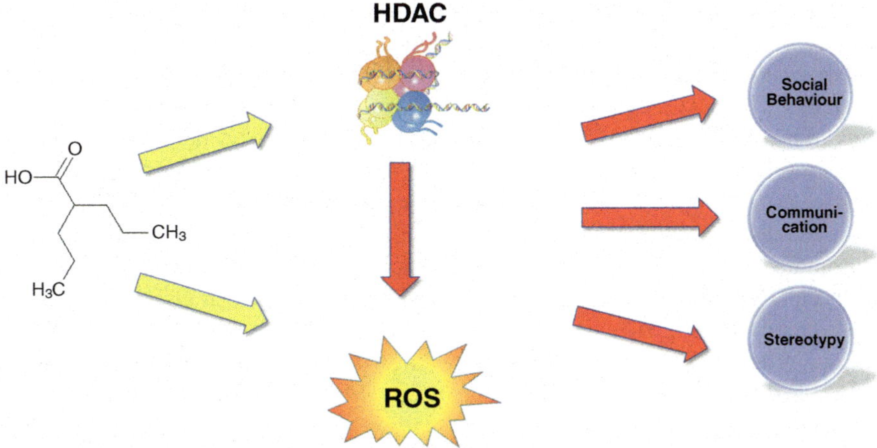

Fig. 2 Although the precise mechanism(s) by which VPA affects (neuro) development are as yet unknown, inhibition of histone deacylation (*HDAC*) and increase in reactive oxygen species (*ROS*) are two processes likely to contribute. As discussed in the text, there is evidence supporting an interaction between these two processes, with alterations in HDAC leading to changes in ROS

involves multiple different neurotransmitters, proteins, and enzymes (Johannessen and Johannessen 2003).

Although a variety of mechanisms for VPA-induced teratogenesis have been proposed, including VPA interference of folate metabolism (Wegner and Nau 1992) and inhibition of folate receptors (Fathe et al. 2014), here we focus on the two major mechanisms for which there is compelling evidence (see Fig. 2). A discussion of the mechanisms underlying VPA's therapeutic properties is beyond the scope of this chapter.

8   Histone Deacetylase Inhibition (HDAC-I)

One of the most convincing mechanisms through which VPA could exert its teratogenic activity is through inhibition of histone deacetylase (Phiel et al. 2001; Menegola et al. 2005; Eikel et al. 2006; Tung and Winn 2010; Fujiki et al. 2013; Lloyd 2013). Deoxyribonucleic acid (DNA) molecules are surprisingly long and in order to fit within the small confines of the cell nucleus, it has to be dramatically compressed. This compression is accomplished by tightly wrapping the DNA molecule around proteins known as histones, to form nucleosomes: the repeating units of chromatin (Kornberg 1977; Li and Reinberg 2011). However, this high degree of compression makes it more difficult for gene transcription to occur, as this involves large proteins (such as transcription factors) binding to DNA before RNA polymerase can bind and initiate gene transcription. Thus, gene expression is regulated, among other factors, by how tightly DNA is wrapped around the core histones. Several different modifications of the core histones are known to influence this wrapping, including histone methylation, histone phosphorylation, and histone acetylation. This latter process is regulated by two classes of enzymes: Histone acetyltransferases (HATs), which attach acetyl groups to lysine residues on histones, and histone deacetylases (HDACs) which subsequently remove these acetyl groups. As acetylation reduces the positive charge of lysine residues, the chromatin becomes less tight (as DNA is negatively charged) and therefore gene transcription is promoted. Conversely, by removing the acetyl groups, HDACs lead to a more compact DNA-histone package and, in turn, reduced gene transcription (Lloyd 2013; Ivanov et al. 2014).

Four different classes of HDACs are known in mammals: the Zn-dependent HDAC-I, II, and IV, and the NAD^+-dependent HADC-III (also known as sirtuins). VPA has been shown to non-selectively inhibit most HDAC, (Phiel et al. 2001; Menegola et al. 2005; Fujiki et al. 2013), thus preventing HDACs from removing acetyl groups and consequently resulting in hyperacetylation and increased *gene activation*. By disturbing the natural HAT and HDAC dynamic processes, VPA has the ability to impact many different genes at the same time (Lloyd 2013). These HDAC-inhibitory properties are thought to underlie the teratogenic influence of VPA (but also its potential therapeutic effects in cancer and cognitive deficits). Indeed, like VPA, other HDAC-Is such as TSA (trichostatin A) and sodium butyrate have teratogenic effects. Interestingly, whereas VPA analogues that retain HDAC-inhibitory activity also induce teratogenicity, analogues that lack this effect do not (Gurvich et al. 2005).

Histone modifications that can alter gene expression come under the umbrella term of epigenetics. Epigenetics is defined as the study of changes of function to the genome that modifies the expression of genes but does not change the nucleotide sequence (Ivanov et al. 2014; Tordjman et al. 2014). It is now believed that specific environmental factors can induce changes in gene expression via epigenetic mechanisms. These changes in gene expression are functionally expressed by the affected organism and can be responsible for a variety of phenotypes, both positive and negative. Importantly, histone modifications can be long-lasting, thus affecting gene transcription for prolonged periods of time. Epigenetics plays an important role in the broader explanatory model of gene x environment interactions, and therefore, VPA can be viewed as an environmental agent that has the capability to interact with specific genes that lead to an ASD phenotype.

The main question that follows from VPA's influence on HDAC is: why does HDAC-I-induced *gene activation* lead to teratogenic effects? At present, it is very difficult to answer this question for a number of reasons. First, depending on the type of histone that is acetylated (there are four basic histones (H2A, H2B, H3, and H4) that make up the nucleosome, plus H1 that connects nucleosomes together), histone acetylation can induce different effects. Second, the histone acetylation can occur in many different genes, thus leading to an increased transcription of many genes. Third, epigenetic changes, such as histone acetylation, may be very local, i.e. differ between different brain regions or within a single brain region even between different types of cells. Finally, HDACs are known to interact with other epigenetic alterations (such as histone methylation and DNA methylation). Altogether, the puzzle of which genes are activated by VPA and when and where has not been solved. However, the types of genes that were activated can be inferred from the behavioural or biochemical outcomes. In other words, if we observe a teratogenic outcome, we can infer that the types of genes that lead to this outcome were the ones activated in this instance. The field of toxicogenomics has the potential to make significant strides in our understanding of VPA-responsive genes.

9 Reactive Oxygen Species and Oxidative Stress

Reactive oxygen species (ROS) may also play an important role in the mechanism of VPA-induced teratogenesis (Tung and Winn 2011). ROS are highly reactive molecules that, in excess, have the capacity to damage many elements of a cell (Andersen 2004; Wells et al. 2009). There are a variety of endogenous mechanisms by which ROS are generated, including mitochondrial respiration and the immune response system (Wells et al. 2009; Lloyd 2013), but ROS can also be enhanced exogenously by xenobiotics, including VPA (Na et al. 2003; Defoort et al. 2006; Kawai and Arinze 2006; Wells et al. 2009; Tung and Winn 2011). Although generation of ROS is both normal and beneficial, excessive generation of ROS can have major detrimental effects through either disruption of signal transduction and/or damage to lipids, DNA, RNA, proteins, and carbohydrates (Wells et al. 2009). A variety of defence mechanisms exist that help fight against the excess generation of ROS and regulate this environment, including antioxidant enzymes and compounds, and direct and indirect repair systems (Sies 1997; Davies 2000). When the generation of ROS overwhelms the multitiered defence mechanisms, a state of oxidative stress ensues and deleterious effects to the host can occur. The balancing act of ROS generation versus the host's defence mechanisms to keep a regulated and healthy ROS environment reflects what is referred to as the *oxygen paradox*—the concept that although aerobic life requires oxygen to survive, oxygen is also intrinsically dangerous to its existence (Davies 1995).

Importantly, the embryo and foetus have lower antioxidant enzyme levels, and in turn, a lowered defence system against excess generation of ROS (Winn and Wells 1999; Wells et al. 2009). This lowered defence system theoretically leaves the embryo and foetus with a higher susceptibility to the effects of ROS-generating mechanisms or xenobiotics, such as VPA (Zaken et al. 2000). Numerous reports have demonstrated that exposure to VPA increases the production of ROS and has negative consequences for cell survival and development (Na et al. 2003; Defoort et al. 2006; Tung and Winn 2011). One mechanism whereby this may be achieved is through enzymatic bioactivation (Winn and Wells 1997). Xenobiotics can be bioactivated by certain enzymes that *are* highly prevalent in the embryo, such as prostaglandin H synthase (PHS) and lipoxygenase (LPO) and converted to free radical reactive intermediates that commence ROS generation (Wells et al. 1997).

If the excess ROS generation overwhelms the host's defence mechanisms and oxidative stress results, then adverse developmental effects may be produced (Wells et al. 2010). Put simply, the teratogenic effect of VPA could result from a combination of an undeveloped defence mechanism and VPA's ability to initiate ROS production.

Furthermore, VPA has the ability to interfere with the defence mechanisms themselves. Superoxide dismutase (SOD) and glutathione (GSH) are two important antioxidants involved in the defence against ROS, and a downregulation of both SOD and GSH have been observed following VPA exposure (Zhang et al. 2010; Hsieh et al. 2012). GSSG (glutathione in its oxidised form) and its ratio with GSH can be used as a measure of oxidative stress, with increases in GSSG: GSH ratio indicative of increased oxidative stress. Dose-dependent increases in embryonic GSSG: GSH ratio have been observed following VPA exposure at doses ≥ 100 µg/ml (Zhang et al. 2010). Together, these data suggest VPA's ability to alter antioxidant homoeostasis in the embryo may play an important role in VPA's teratogenic influence.

ROS can directly induce DNA double-strand breaks (Winn 2003). Homologous recombination (HR) is a DNA repair mechanism that can repair DNA double-strand breaks (Haber 1999). However, HR is not an entirely error-free procedure and has the potential to contribute to detrimental genetic changes. Thus, increased levels of HR theoretically would increase the odds of important genes in the developmental process being disrupted at critical time points, possibly resulting in teratogenesis (Defoort et al. 2006). Interestingly, VPA has been demonstrated to cause oxidative stress and, in turn, increase HR levels in vitro. Furthermore, the antioxidative enzyme *catalase*, one of the cellular defence mechanisms against oxidative stress, completely blocked the increased HR following VPA treatment (Defoort et al. 2006). These data suggest HR could be an underlying mechanism of VPA-induced teratogenesis and that oxidative stress plays an important role (Defoort et al. 2006).

Finally, the role of oxidative stress and ROS in the mechanism of VPA-induced teratogenesis is further supported by data demonstrating the attenuating effects of certain antioxidants in prenatal VPA animal models. For instance, green tea extract was found to exhibit neuroprotective effects, possibly due to its antioxidant properties (Banji et al. 2011). Likewise, the antioxidant Vitamin E attenuated the VPA-induced teratogenic effects in mice (Al Deeb et al. 2000). Embryonic models have suggested the main mechanisms of Vitamin E in attenuating VPA-induced teratogenicity are through the inhibition of ROS and the restoration of GSH (Hsieh et al. 2014). In addition, the neuroprotective antioxidant Astaxanthin (Liu and Osawa 2009) was seen to improve ASD-related behavioural outcomes in mice; an effect also attributed to its antioxidant properties (Al-Amin et al. 2015).

Overall these data provide convincing evidence for a role of HDAC inhibition and ROS production in VPA's effects on the unborn foetus. However, it should be realised that these processes are not necessarily independent (Fig. 2). As HDAC inhibition is likely to affect the expression of a multitude of genes, it is at least conceivable that the alterations in ROS production and the subsequent oxidative stress may be secondary to the inhibition of HDAC. In line with this, studies in the cancer field have clearly shown that HDAC inhibitors can increase ROS production and programmed cell death (Carew et al. 2008).

10 Experimental Models and Mitigating Factors

The broad spectrum of action and complex nature of VPA make investigation of the underlying mechanisms and linking them to the behavioural and biochemical changes very difficult. The heterogeneity in the data on the outcomes and mechanisms of VPA is substantial. It is therefore vitally important to study the factors contributing to the variability of the effects of VPA (Bielecka and Obuchowicz 2008; Roullet et al. 2013).

Cell, animal, and human research on VPA have demonstrated several key points:

First, response to VPA differs as a function of the developmental age, brain region, and gender investigated (Bittigau et al. 2002; Kataoka et al. 2013a). For instance, the apoptotic effects of VPA in 14 different brain regions were studied in rat pups exposed to VPA at various developmental stages. Results revealed the response to VPA differed as a function of both developmental age and brain region (Bittigau et al. 2002). Regional specific neuronal cell loss has also been observed in mouse models of VPA (Kataoka et al. 2013a). In addition, mice treated with VPA at gestational day (GD)12.5 led to social interaction deficits in male, but not female mice, highlighting the importance of gender in VPA exposure (Kataoka et al. 2013a).

The second key point is that response to VPA is both differentiation stage and cell-type dependent (Wang et al. 2011; Fujiki et al. 2013). For instance, VPA was found to have a proapoptotic effect on embryonic stem cell-derived neural progenitor cells of glutamatergic neurons, but this effect was not observed in their neuronal progeny (Fujiki et al. 2013). Moreover, a neuron–astrocyte culture mix treated with VPA induced apoptotic effects that were not observed in a simple neuron-enriched culture, implicating the importance of cell-type in VPA-induced neurodegeneration (Wang et al. 2011).

The third key point is that even seemingly small experimental changes can lead to not just different but opposing findings. A clear example of this comes from two different research groups investigating VPA exposure at GD9 and measuring hippocampal serotonin at post-natal day (PND)50 (Narita et al. 2002; Dufour-Rainfray et al. 2010). Whereas one paper found an *increase* in serotonin in Sprague-Dawley rats following 800 mg/kg VPA (Narita et al. 2002), the other found a 46 % *decrease* in serotonin in Wistar rats following 600 mg/kg VPA (Dufour-Rainfray et al. 2010). The discrepant findings were hypothesised to be a result of the differences in the experimental procedure (Dufour-Rainfray et al. 2010).

The final key point is that VPA-induced outcomes are highly dependent on dosage and timing, or the window of exposure, to the drug (Jeong et al. 2003; Johannessen and Johannessen 2003; Takuma et al. 2014). The *amount* of VPA administered has been repeatedly shown to affect the outcome of the drug in humans, with higher doses associated with higher rates of teratogenicity (Koren et al. 2006; Diav-Citrin et al. 2008; Meador et al. 2008). The timing of VPA administration also has significant implications for the response to the drug. For instance, mice treated with VPA at GD12.5, but not GD9 and GD14.5, exhibited ASD-like symptomatology, including deficits in social interaction (Kataoka et al. 2013b). Another, particularly striking, example of the role of timing in VPA-induced-outcome was seen in a paper exposing mice prenatally to VPA and then treating these same mice with VPA post-natally. VPA-exposed mice had deficits in novel object recognition, and decreased spine density in the hippocampus. Remarkably, post-natal chronic treatment of VPA attenuated both deficits (Takuma et al. 2014). In other words, the very drug that created the deficits prenatally attenuated the deficits when given post-natally. On a related note, studies with ASD patients have found that although prenatal exposure to VPA can increase the likelihood of ASD, treatment with Divalproex (a derivative of VPA) was actually seen to benefit patients—helping with repetitive behaviours and irritability (Hollander et al. 2006, 2010). One reason for the strong effect of timing may relate to VPA's HDAC-I and ROS-inducing properties, and the fact that different genes may be activated at different developmental periods. The effect of timing may be exactly why VPA is used prophylactically in adulthood for several conditions, but still causes damage when exposure occurs prenatally.

Taken together, the data suggests strongly that VPA has a particularly high sensitivity for experimental variables and therefore even slightly different experimental models can produce very different results. It is clear that the timing, dosage, cell-type, differentiation stage, strain-type, gender, and brain region studied can all have a meaningful impact on the outcome of research using VPA. These mitigating factors likely explain some of the diversities in the VPA literature.

11 Concluding Remarks

Prenatal exposure to VPA is as good an animal model as there is for ASD. The VPA model has limitations at several levels in the traditional concept of animal model validity, most prominently with regard to construct and predictive validity. However, these limitations say more about the state of knowledge on ASD, the heterogeneity of ASD itself, and the traditional criteria of validity, than the VPA model in question.

The mechanisms under which VPA exerts its teratogenic influence are not well understood. At this point, the preponderance of evidence suggests that VPA's HDAC-inhibitory properties are likely the major source of its influence. However, given VPA's enormously broad spectrum of action, it is probable that multiple mechanisms contribute, but to varying degrees. In addition, the genetic make-up of mother and foetus combined with other environmental factors will be critical in determining the susceptibility to VPA-induced teratogenesis.

Future research would benefit from combining this model with various genetic animal models, in order to investigate gene x environment interactions and improve its construct validity and explanatory power. In addition, more research investigating chronic prenatal VPA exposure, as opposed to a single exposure, would be of great value. The major reason for this would be because it would more accurately represent the human condition of VPA administration; this is especially important considering the enormous role that timing plays in response to VPA. More toxicogenomic research aimed at identifying VPA-responsive genes, when and for how long these genes are responsive, and then mapping these genes onto biochemical and behavioural outcomes would be enormously beneficial. Research of this nature would allow a clearer understanding of the molecular mechanisms of VPA-induced teratogenesis and may also reveal biomarkers that indicate genetic susceptibility to such teratogenic effects. The identification of biomarkers should be a high priority for research, as they can be used as targets and exploited for therapeutic action.

The ability of VPA to act as a model for ASD demonstrates how even a single exposure to a neuroteratogen at developmentally critical time points can lead to permanent biochemical and behavioural outcomes in offspring. The high sensitivity to experimental variables is indicative of the complexity of VPA, the developmental process, and the interaction between them.

It is perhaps fitting that one of the more complex and poorly understood disorders in neuroscience, ASD, is linked to one of the most complex and poorly understood environmental agents: VPA. That their respective complexities overlap in such a manner is nothing short of remarkable.

References

Al Deeb S, Al Moutaery K, Arshaduddin M, Tariq M (2000) Vitamin E decreases valproic acid induced neural tube defects in mice. Neurosci Lett 292:179–182

Al-Amin MM, Rahman MM, Khan FR, Zaman F, Reza HM (2015) Astaxanthin improves behavioral disorder and oxidative stress in prenatal valproic acid-induced mice model of autism. Behav Brain Res 286:112–121

Andersen JK (2004) Oxidative stress in neurodegeneration: cause or consequence?

Atladottir HO, Thorsen P, Ostergaard L, Schendel DE, Lemcke S, Abdallah M, Parner ET (2010) Maternal infection requiring hospitalization during pregnancy and autism spectrum disorders. J Autism Dev Disord 40:1423–1430

Avallone A, Piccirillo MC, Delrio P, Pecori B, Di Gennaro E, Aloj L, Tatangelo F, Valentina D, Granata C, Cavalcanti E (2014) Phase 1/2 study of valproic acid and short-course radiotherapy plus capecitabine as preoperative treatment in low-moderate risk rectal cancer-V-shoRT-R3 (Valproic acid-short RadioTherapy-rectum 3rd trial). BMC Cancer 14:875

Banji D, Banji OJ, Abbagoni S, Hayath MS, Kambam S, Chiluka VL (2011) Amelioration of behavioral aberrations and oxidative markers by green tea extract in valproate induced autism in animals. Brain Res 1410:141–151

Baronio D, Castro K, Gonchoroski T, de Melo GM, Nunes GDF, Bambini-Junior V, Gottfried C, Riesgo R (2015) Effects of an H3R antagonist on the animal model of autism induced by prenatal exposure to valproic acid. PLoS ONE 10:e0116363

Bauman MD, Crawley JN, Berman RF (2010) Autism: animal models. eLS

Bauman MD, Iosif A-M, Smith SE, Bregere C, Amaral DG, Patterson PH (2014) Activation of the maternal immune system during pregnancy alters behavioral development of rhesus monkey offspring. Biol Psychiatry 75:332–341

Betancur C (2011) Etiological heterogeneity in autism spectrum disorders: more than 100 genetic and genomic disorders and still counting. Brain Res 1380:42–77

Betancur C, Sakurai T, Buxbaum JD (2009) The emerging role of synaptic cell-adhesion pathways in the pathogenesis of autism spectrum disorders. Trends Neurosci 32:402–412

Bielecka AM, Obuchowicz E (2008) Antiapoptotic action of lithium and valproate. Pharmacol Rep 60:771–782

Bittigau P, Sifringer M, Genz K, Reith E, Pospischil D, Govindarajalu S, Dzietko M, Pesditschek S, Mai I, Dikranian K (2002) Antiepileptic drugs and apoptotic neurodegeneration in the developing brain. Proc Natl Acad Sci 99:15089–15094

Bolivar VJ, Walters SR, Phoenix JL (2007) Assessing autism-like behavior in mice: variations in social interactions among inbred strains. Behav Brain Res 176:21–26

Bollino D, Balan I, Aurelian L (2015) Valproic acid induces neuronal cell death through a novel calpain-dependent necroptosis pathway. J Neurochem 133:174–186

Brodie SA, Brandes JC (2014) Could valproic acid be an effective anticancer agent? The evidence so far. Expert Rev Anticancer Ther 14:1097–1100

Carew JS, Giles FJ, Nawrocki ST (2008) Histone deacetylase inhibitors: mechanisms of cell death and promise in combination cancer therapy. Cancer Lett 269:7–17

Chaste P, Klei L, Sanders SJ, Hus V, Murtha MT, Lowe JK, Willsey AJ, Moreno-De-Luca D, Yu TW, Fombonne E, Geschwind D, Grice DE, Ledbetter DH, Mane SM, Martin DM, Morrow EM, Walsh CA, Sutcliffe JS, Martin CL, Beaudet AL, Lord C, State MW, Cook EH, Devlin B (2015) A genome-wide association study of autism using the simons simplex collection: does reducing phenotypic heterogeneity in autism increase genetic homogeneity? Biol Psychiatry 77:775–784

Chih B, Engelman H, Scheiffele P (2005) Control of excitatory and inhibitory synapse formation by neuroligins. Science 307:1324–1328

Choi CS, Hong M, Kim KC, Kim J-W, Yang SM, Seung H, Ko MJ, Choi D-H, You JS, Shin CY (2014) Effects of atomoxetine on hyper-locomotive activity of the prenatally valproate-exposed rat offspring. Biomol Ther 22:406

Christensen J, Grønborg TK, Sørensen MJ, Schendel D, Parner ET, Pedersen LH, Vestergaard M (2013) Prenatal Valproate Exposure and Risk of Autism Spectrum Disorders and Childhood Autism. JAMA 309:1696–1703

Christianson AL, Chester N, Kromberg JG (1994) Fetal valproate syndrome: clinical and neuro-developmental features in two sibling pairs. Dev Med Child Neurol 36:361–369

Cohen OS, Varlinskaya EI, Wilson CA, Glatt SJ, Mooney SM (2013) Acute prenatal exposure to a moderate dose of valproic acid increases social behavior and alters gene expression in rats. Int J Dev Neurosci 31:740–750

Colvert E, Tick B, McEwen F, Stewart C, Curran SR, Woodhouse E, Gillan N, Hallett V, Lietz S, Garnett T (2015a) Heritability of autism spectrum disorder in a UK population-based twin sample. JAMA Psychiatry 72:415–423

Colvert E, Tick B, McEwen F, Stewart C, Curran SR, Woodhouse E, Gillan N, Hallett V, Lietz S, Garnett T (2015b) Heritability of autism spectrum disorder in a UK population-based twin sample. JAMA Psychiatry

Davies KJ (1995) Oxidative stress: the paradox of aerobic life. In: Biochemical Society Symposia. Portland Press-London, vol 61, pp 1–32

Davies KJ (2000) Oxidative stress, antioxidant defenses, and damage removal, repair, and replacement systems. IUBMB Life 50:279–289

Defoort EN, Kim PM, Winn LM (2006) Valproic acid increases conservative homologous recombination frequency and reactive oxygen species formation: a potential mechanism for valproic acid-induced neural tube defects. Mol Pharmacol 69:1304–1310

Devlin B, Cook E, Coon H, Dawson G, Grigorenko E, McMahon W, Minshew N, Pauls D, Smith M, Spence M (2005) Autism and the serotonin transporter: the long and short of it. Mol Psychiatry 10:1110–1116

Diav-Citrin O, Shechtman S, Bar-Oz B, Cantrell D, Arnon J, Ornoy A (2008) Pregnancy Outcome after In Utero Exposure to Valproate. CNS Drugs 22:325–334

Dufour-Rainfray D, Vourc'h P, Le Guisquet A-M, Garreau L, Ternant D, Bodard S, Jaumain E, Gulhan Z, Belzung C, Andres CR (2010) Behavior and serotonergic disorders in rats exposed prenatally to valproate: a model for autism. Neurosci Lett 470:55–59

Dufour-Rainfray D, Vourc'h P, Tourlet S, Guilloteau D, Chalon S, Andres CR (2011) Fetal exposure to teratogens: evidence of genes involved in autism. Neurosci Biobeh Rev 35:1254–1265

Eikel D, Lampen A, Nau H (2006) Teratogenic effects mediated by inhibition of histone deacetylases: evidence from quantitative structure activity relationships of 20 valproic acid derivatives. Chem Res Toxicol 19:272–278

Ellegood J, Crawley JN (2015) Behavioral and neuroanatomical phenotypes in mouse models of autism. Neurotherapeutics 12:521–533

Ellegood J, Anagnostou E, Babineau BA, Crawley JN, Lin L, Genestine M, DiCicco-Bloom E, Lai JKY, Foster JA, Penagarikano O, Geschwind DH, Pacey LK, Hampson DR, Laliberte CL, Mills AA, Tam E, Osborne LR, Kouser M, Espinosa-Becerra F, Xuan Z, Powell CM, Raznahan A, Robins DM, Nakai N, Nakatani J, Takumi T, van Eede MC, Kerr TM, Muller C, Blakely RD, Veenstra-VanderWeele J, Henkelman RM, Lerch JP (2015) Clustering autism: using neuroanatomical differences in 26 mouse models to gain insight into the heterogeneity. Mol Psychiatry 20:118–125

Ellenbroek B, Cools A (2000) Animal models for the negative symptoms of schizophrenia. Behav Pharmacol 11:223–234

Fathe K, Palacios A, Finnell RH (2014) Brief report novel mechanism for valproate-induced teratogenicity. Birth Defects Res A 100:592–597

Fried S, Kozer E, Nulman I, Einarson TR, Koren G (2004) Malformation rates in children of women with untreated epilepsy. Drug Saf 27:197–202

Fujiki R, Sato A, Fujitani M, Yamashita T (2013) A proapoptotic effect of valproic acid on progenitors of embryonic stem cell-derived glutamatergic neurons. Cell Death Dis 4:e677

Gandal MJ, Edgar JC, Ehrlichman RS, Mehta M, Roberts TP, Siegel SJ (2010) Validating γ oscillations and delayed auditory responses as translational biomarkers of autism. Biol Psychiatry 68:1100–1106

Geier DA, Kern JK, Geier MR (2010) The biological basis of autism spectrum disorders: Understanding causation and treatment by clinical geneticists. Acta Neurobiol Exp (Wars) 70:209–226

Grishina O, Schmoor C, Döhner K, Hackanson B, Lubrich B, May AM, Cieslik C, Müller MJ, Lübbert M (2015) DECIDER: prospective randomized multicenter phase II trial of low-dose decitabine (DAC) administered alone or in combination with the histone deacetylase inhibitor valproic acid (VPA) and all-trans retinoic acid (ATRA) in patients >60 years with acute myeloid leukemia who are ineligible for induction chemotherapy. BMC Cancer 15:430

Gurvich N, Berman MG, Wittner BS, Gentleman RC, Klein PS, Green JB (2005) Association of valproate-induced teratogenesis with histone deacetylase inhibition in vivo. FASEB J 19:1166–1168

Haber JE (1999) DNA recombination: the replication connection. Trends Biochem Sci 24:271–275

Hill DS, Wlodarczyk BJ, Palacios AM, Finnell RH (2010) Teratogenic effects of antiepileptic drugs

Hollander E, Soorya L, Wasserman S, Esposito K, Chaplin W, Anagnostou E (2006) Divalproex sodium vs. placebo in the treatment of repetitive behaviours in autism spectrum disorder. Int J Neuropsychopharmacol 9:209–213

Hollander E, Chaplin W, Soorya L, Wasserman S, Novotny S, Rusoff J, Feirsen N, Pepa L, Anagnostou E (2010) Divalproex sodium vs placebo for the treatment of irritability in children and adolescents with autism spectrum disorders. Neuropsychopharmacology 35:990–998

Hsia Y, Wong AY, Murphy DG, Simonoff E, Buitelaar JK, Wong IC (2014) Psychopharmacological prescriptions for people with autism spectrum disorder (ASD): a multinational study. Psychopharmacology 231:999–1009

Hsieh C-L, Wang H-E, Tsai W-J, Peng C-C, Peng RY (2012) Multiple point action mechanism of valproic acid-teratogenicity alleviated by folic acid, vitamin C, And N-acetylcysteine in chicken embryo model. Toxicology 291:32–42

Hsieh CL, Chen KC, Lin PX, Peng CC, Peng RY (2014) Resveratrol and vitamin E rescue valproic acid-induced teratogenicity: the mechanism of action. Clin Exp Pharmacol Physiol 41:210–219

Hu J-P, Xie J-W, Wang C-Y, Wang T, Wang X, Wang S-L, Teng W-P, Wang Z-Y (2011) Valproate reduces tau phosphorylation via cyclin-dependent kinase 5 and glycogen synthase kinase 3 signaling pathways. Brain Res Bull 85:194–200

Ingram JL, Peckham SM, Tisdale B, Rodier PM (2000) Prenatal exposure of rats to valproic acid reproduces the cerebellar anomalies associated with autism. Neurotoxicol Teratol 22:319–324

Ivanov M, Barragan I, Ingelman-Sundberg M (2014) Epigenetic mechanisms of importance for drug treatment. Trends Pharmacol Sci 35:384–396

Jacob J, Ribes V, Moore S, Constable SC, Sasai N, Gerety SS, Martin DJ, Sergeant CP, Wilkinson DG, Briscoe J (2014) Valproic acid silencing of ascl1b/Ascl1 results in the failure of serotonergic differentiation in a zebrafish model of fetal valproate syndrome. Dis Models Mech 7:107–117

Jamain S, Quach H, Betancur C, Råstam M, Colineaux C, Gillberg IC, Soderstrom H, Giros B, Leboyer M, Gillberg C (2003) Mutations of the X-linked genes encoding neuroligins NLGN3 and NLGN4 are associated with autism. Nat Genet 34:27–29

Jamain S, Radyushkin K, Hammerschmidt K, Granon S, Boretius S, Varoqueaux F, Ramanantsoa N, Gallego J, Ronnenberg A, Winter D (2008) Reduced social interaction and ultrasonic communication in a mouse model of monogenic heritable autism. Proc Natl Acad Sci 105:1710–1715

James EJ, Gu J, Ramirez-Vizcarrondo CM, Hasan M, Truszkowski TL, Tan Y, Oupravanh PM, Khakhalin AS, Aizenman CD (2015) Valproate-Induced Neurodevelopmental Deficits in Xenopus laevis Tadpoles. J Neurosci 35:3218–3229

Jeong MR, Hashimoto R, Senatorov VV, Fujimaki K, Ren M, Lee MS, Chuang D-M (2003) Valproic acid, a mood stabilizer and anticonvulsant, protects rat cerebral cortical neurons from spontaneous cell death: a role of histone deacetylase inhibition. FEBS Lett 542:74–78

Johannessen CU, Johannessen SI (2003) Valproate: past, present, and future. CNS Drug Rev 9:199–216

Kaaja E, Kaaja R, Hiilesmaa V (2003) Major malformations in offspring of women with epilepsy. Neurology 60:575–579

Kane MJ, Angoa-Peréz M, Briggs DI, Sykes CE, Francescutti DM, Rosenberg DR, Kuhn DM (2012) Mice genetically depleted of brain serotonin display social impairments, communication deficits and repetitive behaviors: possible relevance to autism. PLoS ONE 7:e48975

Kataoka S, Takuma K, Hara Y, Maeda Y, Ago Y, Matsuda T (2013) Autism-like behaviours with transient histone hyperacetylation in mice treated prenatally with valproic acid. Int J Neuropsychopharmacol 16:91–103

Kawai Y, Arinze IJ (2006) Valproic acid-induced gene expression through production of reactive oxygen species. Cancer Res 66:6563–6569

Kim JH, Scialli AR (2011) Thalidomide: the tragedy of birth defects and the effective treatment of disease. Toxicol Sci 122:1–6

Kim KC, Kim P, Go HS, Choi CS, Yang S-I, Cheong JH, Shin CY, Ko KH (2011) The critical period of valproate exposure to induce autistic symptoms in Sprague-Dawley rats. Toxicol Lett 201:137–142

Kim J-W, Choi CS, Kim KC, Park JH, Seung H, Joo SH, Yang SM, Shin CY, Park SH (2013a) Gastrointestinal tract abnormalities induced by prenatal valproic acid exposure in rat offspring. Toxicol Res 29:173

Kim KC, Kim P, Go HS, Choi CS, Park JH, Kim HJ, Jeon SJ, Han SH, Cheong JH, Ryu JH (2013b) Male-specific alteration in excitatory post-synaptic development and social interaction in pre-natal valproic acid exposure model of autism spectrum disorder. J Neurochem 124:832–843

Kim J-W, Seung H, Kwon KJ, Ko MJ, Lee EJ, Oh HA, Choi CS, Kim KC, Gonzales EL, You JS (2014) Subchronic treatment of donepezil rescues impaired social, hyperactive, and stereotypic behavior in valproic acid-induced animal model of autism. PLoS ONE 9:e104927

Kinast K, Peeters D, Kolk SM, Schubert D, Homberg JR (2013) Genetic and pharmacological manipulations of the serotonergic system in early life: neurodevelopmental underpinnings of autism-related behavior. Front Cell Neurosci 7:72

Kinney DK, Miller AM, Crowley DJ, Huang E, Gerber E (2008a) Autism prevalence following prenatal exposure to hurricanes and tropical storms in Louisiana. J Autism Dev Disord 38:481–488

Kinney DK, Munir KM, Crowley DJ, Miller AM (2008b) Prenatal stress and risk for autism. Neurosci Biobehav Rev 32:1519–1532

Knapp M, Romeo R, Beecham J (2009) Economic cost of autism in the UK. Autism 13:317–336

Koren G, Nava-Ocampo AA, Moretti ME, Sussman R, Nulman I (2006) Major malformations with valproic acid. Can Fam Physician 52:441–442

Kornberg RD (1977) Structure of chromatin. Annu Rev Biochem 46:931–954

Lam KS, Aman MG, Arnold LE (2006) Neurochemical correlates of autistic disorder: a review of the literature. Res Dev Disabil 27:254–289

Lehrman G, Hogue IB, Palmer S, Jennings C, Spina CA, Wiegand A, Landay AL, Coombs RW, Richman DD, Mellors JW (2005) Depletion of latent HIV-1 infection in vivo: a proof-of-concept study. Lancet 366:549–555

Levy SE, Mandell DS, Schultz RT (2009) Autism. Lancet 374:1627–1638

Li G, Reinberg D (2011) Chromatin higher-order structures and gene regulation. Curr Opin Genet Dev 21:175–186

Liu X, Osawa T (2009) Astaxanthin protects neuronal cells against oxidative damage and is a potent candidate for brain food

Lloyd KA (2013) A scientific review: mechanisms of valproate-mediated teratogenesis. Bioscience Horizons 6:hzt003

Loohuis NO, Kole K, Glennon J, Karel P, Van der Borg G, Van Gemert Y, Van den Bosch D, Meinhardt J, Kos A, Shahabipour F (2015) Elevated microRNA-181c and microRNA-30d levels in the enlarged amygdala of the valproic acid rat model of autism. Neurobiol Dis 80:42–53

McFarlane H, Kusek G, Yang M, Phoenix J, Bolivar V, Crawley J (2008) Autism-like behavioral phenotypes in BTBR T + tf/J mice. Genes Brain Behav 7:152–163

Meador K, Reynolds MW, Crean S, Fahrbach K, Probst C (2008) Pregnancy outcomes in women with epilepsy: a systematic review and meta-analysis of published pregnancy registries and cohorts. Epilepsy Res 81:1–13

Mehta MV, Gandal MJ, Siegel SJ (2011) mGluR5-antagonist mediated reversal of elevated stereotyped, repetitive behaviors in the VPA model of autism. PLoS ONE 6:e26077

Menegola E, Di Renzo F, Broccia ML, Prudenziati M, Minucci S, Massa V, Giavini E (2005) Inhibition of histone deacetylase activity on specific embryonic tissues as a new mechanism for teratogenicity. Birth Defects Res B 74:392–398

Miyazaki K, Narita N, Narita M (2005) Maternal administration of thalidomide or valproic acid causes abnormal serotonergic neurons in the offspring: implication for pathogenesis of autism. Int J Dev Neurosci 23:287–297

Moore S, Turnpenny P, Quinn A, Glover S, Lloyd D, Montgomery T, Dean J (2000) A clinical study of 57 children with fetal anticonvulsant syndromes. J Med Genet 37:489–497

Morrow J, Russell A, Guthrie E, Parsons L, Robertson I, Waddell R, Irwin B, McGivern R, Morrison P, Craig J (2006) Malformation risks of antiepileptic drugs in pregnancy: a prospective study from the UK Epilepsy and Pregnancy Register. J Neurol Neurosurg Psychiatry 77:193–198

Mowery TM, Wilson SM, Kostylev PV, Dina B, Buchholz JB, Prieto AL, Garraghty PE (2015) Embryological exposure to valproic acid disrupts morphology of the deep cerebellar nuclei in a sexually dimorphic way. Int J Dev Neurosci 40:15–23

Mulleners WM, McCrory DC, Linde M (2014) Antiepileptics in migraine prophylaxis: an updated Cochrane review. Cephalalgia 0333102414534325

Murray ML, Hsia Y, Glaser K, Simonoff E, Murphy DG, Asherson PJ, Eklund H, Wong IC (2014) Pharmacological treatments prescribed to people with autism spectrum disorder (ASD) in primary health care. Psychopharmacology 231:1011–1021

Na L, Wartenberg M, Nau H, Hescheler J, Sauer H (2003) Anticonvulsant valproic acid inhibits cardiomyocyte differentiation of embryonic stem cells by increasing intracellular levels of reactive oxygen species. Birth Defects Res A 67:174–180

Narita N, Kato M, Tazoe M, Miyazaki K, Narita M, Okado N (2002) Increased monoamine concentration in the brain and blood of fetal thalidomide-and valproic acid–exposed rat: putative animal models for autism. Pediatr Res 52:576–579

Nestler EJ, Hyman SE (2010) Animal models of neuropsychiatric disorders. Nat Neurosci 13:1161–1169

Nicolini C, Ahn Y, Michalski B, Rho JM, Fahnestock M (2015) Decreased mTOR signaling pathway in human idiopathic autism and in rats exposed to valproic acid. Acta neuropathologica Commun 3:3

Oddi D, Crusio WE, D'Amato FR, Pietropaolo S (2013) Monogenic mouse models of social dysfunction: implications for autism. Behav Brain Res 251:75–84

Ohkawara T, Katsuyama T, Ida-Eto M, Narita N, Narita M (2015) Maternal viral infection during pregnancy impairs development of fetal serotonergic neurons. Brain Dev 37:88–93

Ornoy A (2009) Valproic acid in pregnancy: how much are we endangering the embryo and fetus? Reprod Toxicol 28:1–10

Patterson PH (2011) Modeling autistic features in animals. Pediatr Res 69:34R–40R

Phiel CJ, Zhang F, Huang EY, Guenther MG, Lazar MA, Klein PS (2001) Histone deacetylase is a direct target of valproic acid, a potent anticonvulsant, mood stabilizer, and teratogen. J Biol Chem 276:36734–36741

Qing H, He G, Ly PT, Fox CJ, Staufenbiel M, Cai F, Zhang Z, Wei S, Sun X, Chen C-H (2008) Valproic acid inhibits Aβ production, neuritic plaque formation, and behavioral deficits in Alzheimer's disease mouse models. J Exp Med 205:2781–2789

Rasalam A, Hailey H, Williams J, Moore S, Turnpenny P, Lloyd D, Dean J (2005) Characteristics of fetal anticonvulsant syndrome associated autistic disorder. Dev Med Child Neurol 47:551–555

Rodier PM, Ingram JL, Tisdale B, Croog VJ (1997) Linking etiologies in humans and animal models: studies of autism. Reprod Toxicol 11:417–422

Ronald A, Hoekstra RA (2011) Autism spectrum disorders and autistic traits: a decade of new twin studies. Am J Med Genet B Neuropsychiatr Genet 156B:255–274

Roullet FI, Wollaston L, deCatanzaro D, Foster JA (2010) Behavioral and molecular changes in the mouse in response to prenatal exposure to the anti-epileptic drug valproic acid. Neuroscience 170:514–522

Roullet FI, Lai JK, Foster JA (2013) In utero exposure to valproic acid and autism–a current review of clinical and animal studies. Neurotoxicol Teratol

Sabers A, Bertelsen FC, Scheel-Krüger J, Nyengaard JR, Møller A (2015) Corrigendum to "Long-term valproic acid exposure increases the number of neocortical neurons in the developing rat brain" [Neurosci Lett 580(2014):12–16]: a possible new animal model of autism. Neurosci Lett 588:203–207

Schneider T, Przewlocki R (2005) Behavioral alterations in rats prenatally exposed to valproic acid: animal model of autism. Neuropsychopharmacology 30:80–89

Schneider T, Turczak J, Przewłocki R (2006) Environmental enrichment reverses behavioral alterations in rats prenatally exposed to valproic acid: issues for a therapeutic approach in autism. Neuropsychopharmacology 31:36–46

Schneider T, Roman A, Basta-Kaim A, Kubera M, Budziszewska B, Schneider K, Przewłocki R (2008) Gender-specific behavioral and immunological alterations in an animal model of autism induced by prenatal exposure to valproic acid. Psychoneuroendocrinology 33:728–740

Servadio M, Vanderschuren L, Trezza V (2015) Modeling autism-relevant behavioral phenotypes in rats and mice: Do 'autistic' rodents exist? Behav Pharmacol 26:522–540

Shi L, Fatemi SH, Sidwell RW, Patterson PH (2003) Maternal influenza infection causes marked behavioral and pharmacological changes in the offspring. J Neurosci 23:297–302

Shi L, Smith SE, Malkova N, Tse D, Su Y, Patterson PH (2009) Activation of the maternal immune system alters cerebellar development in the offspring. Brain Behav Immun 23:116–123

Sies H (1997) Oxidative stress: oxidants and antioxidants. Exp Physiol 82:291–295

Smith V, Brown N (2014) Prenatal valproate exposure and risk of autism spectrum disorders and childhood autism. Archives of disease in childhood-education & practice edition edpract-2013–305636

Štefánik P, Olexová L, Kršková L (2015) Increased sociability and gene expression of oxytocin and its receptor in the brains of rats affected prenatally by valproic acid. Pharmacol Biochem Behav 131:42–50

Takuma K, Hara Y, Kataoka S, Kawanai T, Maeda Y, Watanabe R, Takano E, Hayata-Takano A, Hashimoto H, Ago Y (2014) Chronic treatment with valproic acid or sodium butyrate attenuates novel object recognition deficits and hippocampal dendritic spine loss in a mouse model of autism. Pharmacol Biochem Behav 126:43–49

Tomson T, Battino D (2012) Teratogenic effects of antiepileptic drugs. Lancet Neurol 11:803–813

Tordjman S, Somogyi E, Coulon N, Kermarrec S, Cohen D, Bronsard G, Bonnot O, Weismann-Arcache C, Botbol M, Lauth B (2014) Gene × environment interactions in autism spectrum disorders: role of epigenetic mechanisms. Front Psychiatry 5

Trinka E, Höfler J, Zerbs A, Brigo F (2014) Efficacy and safety of intravenous valproate for status epilepticus: a systematic review. CNS Drugs 28:623–639

Tung EW, Winn LM (2010) Epigenetic modifications in valproic acid-induced teratogenesis. Toxicol Appl Pharmacol 248:201–209

Tung EW, Winn LM (2011) Valproic acid increases formation of reactive oxygen species and induces apoptosis in postimplantation embryos: a role for oxidative stress in valproic acid-induced neural tube defects. Mol Pharmacol 80:979–987

van der Staay FJ, Arndt SS, Nordquist RE (2009) Evaluation of animal models of neurobehavioral disorders. Behav Brain Funct 5:11

Vorstman JA, Spooren W, Persico AM, Collier DA, Aigner S, Jagasia R, Glennon JC, Buitelaar JK (2014) Using genetic findings in autism for the development of new pharmaceutical compounds. Psychopharmacology 231:1063–1078

Wang C, Luan Z, Yang Y, Wang Z, Cui Y, Gu G (2011) Valproic acid induces apoptosis in differentiating hippocampal neurons by the release of tumor necrosis factor-α from activated astrocytes. Neurosci Lett 497:122–127

Wegner C, Nau H (1992) Alteration of embryonic folate metabolism by valproic acid during organogenesis: implications for mechanism of teratogenesis. Neurology 42:17–24

Wells PG, Kim PM, Laposa RR, Nicol CJ, Parmana T, Winn LM (1997) Oxidative damage in chemical teratogenesis. Mutation Res/Fundam Mol Mech Mutagenesis 396:65–78

Wells PG, McCallum GP, Chen CS, Henderson JT, Lee CJ, Perstin J, Preston TJ, Wiley MJ, Wong AW (2009) Oxidative stress in developmental origins of disease: teratogenesis, neurodevelopmental deficits, and cancer. Toxicol Sci 108:4–18

Wells PG, Lee CJ, McCallum GP, Perstin J, Harper PA (2010) Receptor-and reactive intermediate-mediated mechanisms of teratogenesis. In: Adverse drug reactions, Springer, Berlin, pp 131–162

Williams G, King J, Cunningham M, Stephan M, Kerr B, Hersh JH (2001) Fetal valproate syndrome and autism: additional evidence of an association. Dev Med Child Neurol 43:202–206

Wingate M, Kirby RS, Pettygrove S, Cunniff C, Schulz E, Ghosh T, Robinson C, Lee L-C, Landa R, Constantino J (2014) Prevalence of autism spectrum disorder among children aged 8 years-autism and developmental disabilities monitoring network, 11 sites, United States, 2010. MMWR Surveillance Summaries 63

Winn LM (2003) Homologous recombination initiated by benzene metabolites: a potential role of oxidative stress. Toxicol Sci 72:143–149

Winn LM, Wells PG (1997) Evidence for embryonic prostaglandin H synthase-catalyzed bioactivation and reactive oxygen species-mediated oxidation of cellular macromolecules in phenytoin and benzo [a] pyrene teratogenesis. Free Radic Biol Med 22:607–621

Winn LM, Wells PG (1999) Maternal administration of superoxide dismutase and catalase in phenytoin teratogenicity 1. Free Radic Biol Med 26:266–274

Wöhr M, Roullet FI, Crawley JN (2011) Reduced scent marking and ultrasonic vocalizations in the BTBR T + tf/J mouse model of autism. Genes Brain Behavior 10:35–43

Wyszynski D, Nambisan M, Surve T, Alsdorf R, Smith C, Holmes L (2005) Increased rate of major malformations in offspring exposed to valproate during pregnancy. Neurology 64:961–965

Zaken V, Kohen R, Ornoy A (2000) The development of antioxidant defense mechanism in young rat embryos in vivo and in vitro. Early Pregnancy 4:110–123

Zhang B, Wang X, Nazarali A (2010) Ascorbic acid reverses valproic acid-induced inhibition of hoxa2 and maintains glutathione homeostasis in mouse embryos in culture. Cell Mol Neurobiol 30:137–148

Applications of the Neonatal Quinpirole Model to Psychosis and Convergence upon the Dopamine D_2 Receptor

Russell W. Brown and Daniel J. Peterson

Abstract This mini review focuses on the importance of the dopamine D_2-like receptor family and its importance in psychosis. Past findings from this laboratory along with collaborators have been that neonatal quinpirole (a dopamine D_2-like receptor agonist) results in increases in dopamine D_2 receptor sensitivity that persists throughout the animal's lifetime. Findings from this model have been shown to have particular application and validity to schizophrenia, but may have broader implications toward other psychoses, which is reviewed in the present manuscript. In the present review, we also highlight other models of psychoses that have been centered on the subchronic administration of quinpirole to rats in order to model certain psychoses, which has uncovered some interesting and valid behavioral findings. This review highlights the importance of the combination of behavioral findings and neurobiological mechanisms focusing on neural plasticity in discovering underlying pathologies in these disorders that may lead to treatment discoveries, as well as the value of animal models across all psychoses.

Keywords Dopamine D_2 receptor · Neonatal · Quinpirole · D_2 receptor supersensitivity · Psychosis

Contents

1 Introduction 388
2 Schizophrenia 389
3 Plasticity Mechanisms 391
4 The Dopamine D_2 Receptor as a Focal Point of Psychopathology 391

R.W. Brown (✉)
Department of Biomedical Science, James H. Quillen College of Medicine,
East Tennessee State University, Johnson City, TN 37614-1702, USA
e-mail: brown1@mail.etsu.edu

D.J. Peterson
Department of Psychology, East Tennessee State University, Johnson City,
TN 37614-1702, USA

© Springer International Publishing Switzerland 2015
Curr Topics Behav Neurosci (2016) 29: 387–402
DOI 10.1007/7854_2015_394
Published Online: 16 October 2015

5   Bipolar Disorder 393
6   Obsessive–Compulsive Disorder (OCD).. 393
7   Attention-Deficit Hyperactive Disorder (ADHD)... 395
8   Substance Abuse.. 396
9   Conclusions.. 397
References.. 397

1   Introduction

Approximately 25 years ago, Dr. Richard Kostrzewa and colleagues demonstrated that neonatal treatment of rats with quinpirole, a dopamine D_2/D_3 agonist, produces an increase in dopamine D_2-like receptor sensitivity (Kostrzewa and Brus 1991), and later demonstrated this change persisted throughout the animal's lifetime (Kostrzewa 1995). This increase in dopamine D_2 receptor sensitivity is consistent with increases in D_2 receptor sensitivity in schizophrenia (Seeman et al. 2006), and over the past decade or so, our laboratory has utilized this model in studies aimed at elucidating mechanisms of substance abuse comorbidity in schizophrenia. First and foremost, however, the animal model we use must be valid. As stated by Monteiro and Feng (2015) in a recent review in *Biological Psychiatry*, animal models of neuropsychiatric disorders should exhibit at least one of the following characteristics: atypical behaviors that resemble human symptoms (face validity); shared biological grounds with human conditions, such as mutation of a specific gene (construct validity); or successful response to the same therapeutic agents prescribed to patients, allowing outcome predictability (predictive validity). Findings from the neonatal quinpirole model have attained all three types of validity, in cognitive impairment and prepulse inhibition deficits (Brown et al. 2002; Maple et al. 2015), significant decreases of neurotrophic factors (Thacker et al. 2006; Brown et al. 2008) as well as decreases in expression of the regulator of G-protein signaling, RGS9 (Maple et al. 2007), and finally, olanzapine treatment alleviated cognitive impairment and decreases of neurotrophic factor protein (Thacker et al. 2006)—consistent with findings in schizophrenics (Cuesta et al. 2009; Rizos et al. 2010). This mini review will be directed toward the changes in the dopamine system that exist across several different psychoses including schizophrenia, and how these changes may have relevance to not only schizophrenia, but also other psychopathological disorders related to dysfunction in the dopamine D_2 receptor family. Finally, we will discuss animal models of each disorder and the possible applications from these findings toward the discovery of underlying neurobiological mechanisms that may lead to future treatments.

2 Schizophrenia

Patients suffering from schizophrenia exhibit an exceedingly broad behavioral syndrome that includes abnormal ideation, thought disorders, altered perception, sensory dysfunction, and often flattened affect (Picchioni and Murray 2007). A neural hallmark of the syndrome is generally proposed to be dopamine hyperactivity, specifically increases in dopamine D_2 receptor function although it is well established that serotonergic dysfunction (Selvaraj et al. 2014) and hypofunction at the glutamatergic N-methyl D-aspartate (NMDA) receptor (Cohen et al. 2015) are also present. The problem with modeling this disorder in rodents is that there are so many alterations in neural communication present as well as neuropathology, and there is no way to model all of these changes in one system. There have been several rodent models of schizophrenia that use acute administrations of pharmacological agents that produce robust increases in dopamine release and/or NMDA receptor hypofunction, such as amphetamine or phencyclidine (PCP; Robinson and Becker 1986; Janhunen et al. 2015). The primary issue with such a model is that there are not long-term changes in neurotransmission or plasticity that are known to be present in schizophrenia. The most prevalent rodent model of schizophrenia involves inducing neonatal hippocampal lesions to rats at postnatal day (P)7, which was developed by Barbara Lipska and Daniel Weinberger at NIMH (2002). As with many other models, this has yielded accurate and important behavioral data relative to both social and cognitive sequelae of the disorder (Sams-Dodd et al. 1997; Lyon et al. 2012). The primary weakness with the hippocampal lesion model is that there is no evidence of cell death in the hippocampus of human schizophrenics (Harrison 1999).

Regardless, increased activation of the dopamine D_2 receptor has been shown to play a major role in abnormal behaviors observed in schizophrenia (Nikam and Awasthi 2008). Additionally, all effective antipsychotic drugs are D_2-like receptor antagonists (Julien et al. 2015). Our laboratory and a collaborating laboratory have demonstrated that neonatal treatment with quinpirole, a dopamine D_2/D_3 agonist, administered to rats from postnatal days (P)1–11, 1–21, or 21–35 produces an increase in sensitivity of the D_2 receptor that is persistent through adulthood, and this is independent of a change in D_2 receptor number. Findings from our laboratory have shown that neonatal quinpirole alters the cholinergic system (ChAT), nerve growth factor (NGF), and brain-derived neurotrophic factor (BDNF) as well as genetic expression of these three proteins along with changes in expression of RGS9-2 in dopaminergic terminal areas of the brain (Maple et al. 2007). This suggests that hyperactivity of the dopamine system ultimately modulates other neurobiological systems that may also play a role in the disorder. All of these above effects are consistent with the known neuropathology of schizophrenia, which were reviewed recently in a special issue of *Developmental Neuroscience* (Brown et al. 2012, see Table 1).

Interestingly, one possible inconsistency is reflected in data from collaborators that demonstrated significant upregulation of α7 nicotinic receptors (nAChRs) in the striatum and hippocampus in early adolescence (Tizabi et al. 1999). In this study, Tizabi and colleagues administered quinpirole HCl (1 mg/kg) or saline from (P)1–21 in male and female rats, identical to the treatment regimen used in our previous work. Animals were raised to (P)30, and brain tissue was harvested for analyses of nAChR binding. Frontal cortex, striatum, hippocampus, and cerebellum were obtained and analyzed. Regarding $\alpha 4\beta 2$ nAChRs, neonatal quinpirole treatment elevated [^{3}H]cytisine binding only in cerebellum. No significant effects were detected in any other area examined. Regarding $\alpha 7$ nAChR binding, neonatal quinpirole treatment resulted in elevated [^{125}I]α-bungarotoxin binding in hippocampus (65 % compared to neonatal saline-treated animals) and striatum (94 % compared to neonatal saline-treated animals). No significant effect was detected in any other brain area. Critically, this finding demonstrates that neonatal quinpirole produces a significant 94 % increase in striatal $\alpha 7$ nAChRs without producing a change in $\alpha 4\beta 2$ nAChRs in this same brain area. The striatum is a dopaminergic terminal area and is important in nicotine sensitization and the rewarding effects of nicotine (Mao and McGehee 2010).

One issue with the above findings is that the increase in hippocampal $\alpha 7$ nAChR is inconsistent with postmortem findings in hippocampus in schizophrenics (Freedman et al. 2000). However, it is not known whether nAChRs are changed at earlier time points of the life span in schizophrenia. What we hypothesize as particularly important is that an increase in striatal $\alpha 7$ nAChR availability in adolescence is especially critical to the susceptibility of smoking in schizophrenia. The striatum is known to play a role in the reinforcing and rewarding effects of addictive drugs, along with the nucleus accumbens. Based on the fact that the striatum and NAcc are both heavily innervated by the dopamine system and the inputs to both regions are primarily from dopaminergic cell bodies of the midbrain, we suspect that the $\alpha 7$ nAChR upregulation that is occurring in the striatum in adolescent rats treated neonatally with quinpirole is also occurring in the NAcc in these animals. However, this has not yet been analyzed. We suspect that the increase in $\alpha 7$ nAChR availability may play an important role in smoking behavior in schizophrenia, but may also affect dopaminergic plasticity in these brain regions, especially based on the fact the dopamine system is going through rapid development during the adolescent period (see Andersen and Teicher 2000).

Related to this issue, $\alpha 7$ nAChRs have been primarily identified as calcium (Ca^{+2}) channels. Calcium is known to play an important role in neurotransmitter release through its entry into presynaptic terminals due to the arrival of the action potential, its binding to calmodulin and the synaptic vesicle containing neurotransmitter, and transport of the vesicle to the terminal membrane for attachment and release. Important to brain plasticity and synaptic strength, $\alpha 7$ nAChRs have been shown to be localized on glutamatergic terminals in both the ventral tegmental area (VTA) and striatum. The mechanism that has been identified through which $\alpha 7$ nAChRs affects dopamine release in both the VTA and striatum is through its effects on glutamate. Kaiser and Wonnacott (2000) have shown that nicotine's agonist action at the $\alpha 7$ receptor increases glutamate release, which in turn excites dopamine terminals, stimulating dopamine release. Further support for this model has shown glutamate stimulates the release of [^{3}H]dopamine from rat striatal slices (Jones et al. 1993) and this effect appears to be mediated by both AMPA/kainate and N-methyl-D-aspartate (NMDA) receptors present on dopaminergic nerve terminals (Kendrick et al. 1996; Smolders et al. 1996). Therefore, drugs acting on α7 nAChRs (e.g., nicotine) or changing glutamatergic function may influence plasticity in this region, which may have an important impact during development.

3 Plasticity Mechanisms

Developmental models always have the advantage of modeling a neurological disorder because of their long-term and persistent changes on the rewiring of the central nervous system. Important players in brain plasticity are neurotrophic factors, and one neurotrophic factor of particular interest in our work has been BDNF. BDNF is present in appreciable amounts throughout the CNS and is known to play an important role in the survival, differentiation, and maintenance of developing and mature neurons, and in the formation of synaptic circuitry in the brain (Huang and Reichardt 2001). BDNF is especially important in the modification of synaptic transmission (Binder and Scharfman 2004). Among postmortem tissue studies, BDNF levels are decreased in adult hippocampus of schizophrenics (Durany et al. 2001). We have also shown a decrease of hippocampal BDNF due to neonatal quinpirole treatment in the neonatal quinpirole model (Thacker et al. 2006). Possibly, even more important are the "downstream" changes produced by changes in BDNF. BDNF binds to the membranous tyrosine kinase B receptor and is released from both neurons and glial cells (Cowansage et al. 2010). Of interest is mammalian (or mechanistic) target of rapamycin (mTOR) and its corresponding complexes mTORC1 and mTORC2. It has been established that increases in mTOR can lead to the transcription of glutamatergic AMPA receptors (Wang et al. 2006) as well as influences on downstream targets such as ribosomal protein S6 which is directly involved in cell growth (Magnuson et al. 2012). Therefore, mTOR can have direct influence on both synaptic strength and synaptic growth. There has been interest in targeting mTOR in psychosis, and although it has yet to be established that there are changes in mTOR in different psychoses, all data appear to point to involvement of this important complex in neural plasticity.

4 The Dopamine D$_2$ Receptor as a Focal Point of Psychopathology

There are many disorders in which changes in dopamine D$_2$ receptor function have been shown, including bipolar (Salvadore et al. 2010), obsessive–compulsive disorder (OCD) (Nikolaus et al. 2010), and attention-deficit hyperactivity disorder (Ford 2014). The changes in the D$_2$ receptor and its function will be reviewed for each of these disorders below. This is not to say that increases in dopamine D_2 receptor sensitivity may be relevant to all of these disorders, but it does appear that a change in dopamine D_2-like receptor functioning is a point at which psychopathology converges. Like the dopamine D_1 receptor family, dopamine D_2 receptors are metabotropic G-protein-coupled receptors. The dopamine D_2-like receptor family was first identified on the basis of its high affinity for antipsychotic drugs. Unlike the dopamine D_1 receptor family, dopamine D_2 receptors are known for their ability to inhibit adenylyl cyclase (Kebabian and Calne 1979). The highest density of D_2 receptors in the brain are in the striatum, nucleus accumbens, substantia nigra, and olfactory tubercle as well as the olfactory bulbs. In addition, unlike dopamine D_1 receptors, dopamine D_2 receptors are located both pre- and postsynaptically, with the presynaptic subtype essentially serving as an autoreceptor inhibiting dopamine release. Several drugs of abuse, including amphetamine and nicotine, have been shown to desensitize the D_2 autoreceptor in addition to increasing dopamine release (Seutin et al. 1991; Schmitz et al. 2001). This action has been hypothesized to be the basis of the reinforcing properties of these drugs.

Dr. Phillip Seeman has published several reviews over the past decade on the importance of the dopamine D_2 receptor in psychosis. Interestingly, although early work revealed that there was a significant increase of the dopamine D_2 receptor in postmortem analyses of striatum in schizophrenics, more recent work has indicated that dopamine D_2 receptors are only slightly elevated in schizophrenia. However, there have been reports that demonstrate increased dopamine responding to drugs of abuse, such as amphetamine (Laruelle et al. 1999) or enhanced psychotic responding to amphetamine (Thompson et al. 2013). Regardless, what is known is that all antipsychotic drugs block the dopamine D_2 receptor with some affinity and are especially effective at alleviating positive symptoms of the disorder (Tollefson 1996).

In addition, there have been several reports of changes in RGS in schizophrenia. RGS proteins activate the breakdown of guanosine triphosphate (GTP) that is transiently attached to the Gi and Gq subunits of G-protein and effectively act as GTPase activators to shorten or terminate the action of an agonist (Neubig and Siderovski 2002; Neubig 2002). Regulation of GTPase activity, for example, by regulation of RGS protein expression or localization, is an effective cellular strategy formulating the sensitivity of GPCRs to agonist stimulation (Hollinger and Hepler 2002). Specifically, RGS9-2 has been identified as being colocalized with dopamine D_2 class dopamine receptors located on medium spiny neurons of the striatum and accelerates the termination of D_2-triggered events (Kovoor et al. 2005). Further, this same group has shown that RGS9-2 regulates D_2 cellular functions and inhibits dopamine-mediated internalization of dopamine D_2 receptors (Celver et al. 2010). One study has shown decreases of RGS9-2 in postmortem analyses of schizophrenics (Seeman et al. 2006), although others have failed to report this change (Greenbaum et al. 2010; Okahisa et al. 2011). Importantly, the gene for RGS9 is located in the identical chromosomal region that contains at least one other marker linked to schizophrenia (Cardno et al. 2001). We have also reported that neonatal quinpirole treatment results in significant decreases of RGS9-2 expression in the frontal cortex, striatum, and nucleus accumbens of adult male rats (Maple et al. 2007). The decreases of RGS9 were robust in all three regions, consistent with persistent dopamine D_2 receptor supersensitivity.

5   Bipolar Disorder

As mentioned, currently available antipsychotic drugs all antagonize the dopamine D_2 receptor with some affinity. Although antipsychotic drugs have been used in the treatment of schizophrenia, many typical and more recently atypical antipsychotic drugs have been utilized to treat bipolar disorder, suggesting that a neurobiological mechanism of this disorder is related to changes in the dopamine D_2 receptor. Several genetic polymorphisms in bipolar disorder are related to the dopamine system, and these include alterations in the genes that code for the dopamine D_2 receptor (Beaulieu and Gainetdinov 2011) as well as the dopamine transporter (DAT; Pinsonneault et al. 2011). However, without a defined neurobiological mechanism, bipolar disorder has traditionally been difficult to model in a rodent. Interestingly, there is a quinpirole model of bipolar disorder that has been published, in which quinpirole is administered to adult rats and the biphasic locomotor response to acute quinpirole treatment was alleviated by common treatments for bipolar disorder, including valproate and carbamazepine (Shaldubina et al. 2002). The problem with the analysis of only locomotor activity is limited, because assessment of purely locomotor activity provides only limited information regarding the etiology of a complex disorder such as bipolar disorder. Therefore, a number of manipulations have been studied in relation to mania, including sleep deprivation (Szabo et al. 2009) and social defeat stress (Einat 2007b). Such manipulations, while causing hyperactivity, also lead to other symptoms such as increased aggression and changes in sexual activity, as well as molecular alterations in systems that are affected by mood stabilizers such as the glycogen synthase kinase-3 gene (for reviews see Beaulieu et al. 2011). As with manic depressive disorder, genetic studies have been performed to assess mania (Chen et al. 2010; Einat 2007a; Malkesman et al. 2009; Saul et al. 2012), many of which take the form of assessing strain differences or disrupting circadian rhythms (Roybal et al. 2007). Ultimately, utilization of a variety of approaches will be required to gain better insight into the underlying pathophysiology of bipolar disorder.

6   Obsessive–Compulsive Disorder (OCD)

OCD is an anxiety disorder that affects approximately 2–3 % of the population and causes an impairment in social and occupational functioning (Ruscio et al. 2010). The disorder is characterized by distress- and anxiety-provoking obsessions (repetitive intrusive thoughts) and compulsions (repetitive ritualistic behavior), which are performed to diminish anxiety (American Psychiatric Association 2013). One of the most prevalent and replicated findings in OCD is the disruption of cortico-striato-thalamocortical circuitry. Essentially, what has been discovered is that there are hyperactive circuitry communications within subregions of these anatomical areas that are manifested in the behavioral disruptions observed in OCD (for review, see Monteiro and Feng 2015). Of course, both serotonin (5-HT) and dopamine are known to play major roles in this communication (along with glutamate), and much of the research focus and treatment have centered on these neurotransmitter systems.

Treatment for OCD often involves the class of drugs known as selective serotonin reuptake inhibitors (SSRIs), which are also a common pharmacotherapy for depression (Denys et al. 2004a, b; Aouizerate et al. 2005). This treatment suggests the involvement of the serotoninergic system, and neuroimaging studies have strengthened the notion of serotonergic dysfunction in OCD by providing evidence for reduced availability of 5-HT transporters (SERTs) in the midbrain, thalamus, and brainstem; and reduced availability of 5-HT$_{2A}$ receptors in prefrontal, parietal, and temporal brain regions (Hesse et al. 2005; Perani et al. 2008). However, an estimated 50–60 % of patients do not respond to this treatment and require additional treatment with atypical antipsychotics, most of which work on both the dopamine and serotoninergic systems (Denys et al. 2004a; Denys 2006; Fineberg et al. 2005). Abnormalities in the dopamine system have also been observed in OCD patients, such as increased DAT levels in the striatum and reduced availability of the D$_1$ and D$_2$ receptors in striatum (Kim et al. 2003; Denys et al. 2004b; van der Wee et al. 2004; Olver et al. 2009).

There are relatively few animal models that exist for OCD, but there is support for the current hypothesis that manipulations in 5-HT (Andersen et al. 2010; Schilman et al. 2010) or dopamine neurotransmission (Einat and Szechtman 1995; Hoffman and Rueda Morales 2012) are effective models with some behavioral validity. There are a number of genetic models in the mouse, including the *Sapap3* knockout. *Sapap3* is a scaffolding protein normally enriched at corticostriatal glutamatergic synapses, and deletion of this gene results in selective synaptic dysfunction in the striatum (Wan et al. 2014). This model has shown enhanced anxiety on a number of behavioral assays as well as compulsive, self-injurious grooming (Welch et al. 2007). However, in the more behaviorally adept rat, one of the more common OCD models in existence is the quinpirole sensitization model developed by Eilam, Szechtman, and colleagues (Eilam et al. 2006). This model involves behaviorally testing rats after subchronic administration of quinpirole (0.125–0.25 mg/kg) to adult rats and has shown compulsive checking behavior typically measured in an open field. This behavioral effect in rats is alleviated by 5-HT agonists (Tucci et al. 2014). The weakness with this model is primarily that increases in D$_2$ receptor function are not observed in OCD, but there is reduced availability in D$_2$ receptors. Regardless, this model appears to have some behavioral validity and suggests that increases in D$_2$ receptor function may be relevant to analyzing behavioral effects in an OCD rodent model.

7 Attention-Deficit Hyperactive Disorder (ADHD)

Similar to OCD at the neurobiological level, there is evidence suggesting that ADHD is the result of abnormal dopamine and 5-HT functioning (Sagvolden et al. 2005; Zeeb et al. 2009). Though the effects of dopamine on prefrontal functioning are quite complicated, the general consensus is that dopamine exerts a strong regulatory effect on prefrontal neuronal activity (Sagvolden et al. 2005). The effect of dopamine depends on the state of the prefrontal cells (e.g., hyperpolarized non-firing or depolarized firing state). Dopamine is postulated to mediate the value of delayed rewards (Wade et al. 2000). When comparing patients who have had injuries or diseases of the prefrontal cortex with individuals clinically diagnosed with ADHD, there are similarities in attention deficiencies and distractibility (Winstanley et al. 2006). Additionally, there are data from functional magnetic resonance imaging (fMRI) that show atypical fronto-striatal activity during go/no-go tasks with ADHD (Tripp and Wickens 2009). Whereas several studies have shown genetic abnormalities across the dopamine system, the role of the dopamine D_2 receptor consistently presents mixed results (Volkow et al. 2009). However, D_2 polymorphisms have been implicated in some of the behavioral deficits associated with ADHD, such as reward and motivation (Haenlein and Caul 1987: Johansen et al. 2009).

According to the American Psychiatric Association (2002), treatment for ADHD often involves the administration of stimulants including dextroamphetamine (d-AMP) or methylphenidate (MPH). These stimulants inhibit the dopamine transporter (DAT) and the norepinephrine transporter, ultimately inhibiting dopamine and norepinephrine reuptake and producing an increase of the neurotransmitters' presence at postsynaptic receptors (Russell et al. 2005; Sharma and Couture 2014). Interestingly enough, imaging studies have shown a similar neurobiology between ADHD and substance abusers, along with stimulant treatment reducing both ADHD and substance abuse symptomatology (Frodl 2010). However, it should be noted that long-term use of stimulants remains a topic of controversy. Both MPH and d-AMP are known to cause physical harm and dependence, though this does not seem to occur when taken as prescribed and may ultimately reduce the risk of substance abuse when administered at a clinically relevant dose (Nutt et al. 2007; Sharma and Couture 2014).

Currently, there are several animal models of ADHD. These rodent models range from various strains, social isolation rearing, pollutant exposure, hippocampal X-irradiation during infancy, and neurotoxic brain lesions (Sagvolden 2000). While these models are able to effectively model some of the behavioral deficits associated with ADHD, they often fail to model the genetic abnormalities or are inconsistent with the etiology of ADHD in the clinical population. Genetic models include a DAT knockout mouse, Naples High/Low Excitability rats, and the most prevalently used model, the spontaneously hypertensive rat (SHR) (Sagvolden 2000). Numerous studies have demonstrated that SHRs display behavioral characteristics analogous to the behaviors of individuals diagnosed with ADHD (Russell et al.

2005). Research assessing the validity of the SHR across measures of sustained attention (Aase and Sagvolden 2006), increased motor activity and behavioral variability (Wultz and Sagvolden 1992; Mook et al. 1993; Saldana and Neuringer 1998), as well as impulsiveness (Hand et al. 2009) has shown that SHRs demonstrate behavior analogous to that of individuals with ADHD. The dopamine system has also been shown to be disturbed in the SHR. More specifically, dopamine turnover is lower in the substantia nigra, VTA, and frontal cortex of adult SHRs which suggests that dopamine release may be decreased and impaired (Russell et al. 2005). Binding studies have also shown that D_2 receptor density increased the striatum of the SHR (Chiu et al. 1982). An analysis of the role of the D_2 receptor in psychostimulant treatment in a developmental model, as ADHD has been reclassified as a developmental disorder, will likely provide insight toward the role of dopamine in ADHD.

8 Substance Abuse

In a previous review, we discussed substance abuse comorbidity in schizophrenia and the consequences of substance abuse in this population along with the underlying neurobiological mechanisms. We focused much of this review toward nicotine, because the majority (70–90 %) of the schizophrenic population smokes cigarettes and smokes them heavily (McCreadie and Kelly 2000). However, when it comes to psychosis, increased substance abuse is not unique to the schizophrenic population, because there is a substantial increase in alcohol and psychostimulant use across many behavioral disorders (Mueser and Gingerich 2013). Not surprisingly across all disorders, symptoms are worsened when drugs are abused, with the costs of care being dramatically increased (Dixon 1999).

A more recent focus at the National Institute on Drug Abuse (NIDA) has been on the increased use of cannabis in psychotic populations. With the exception of alcohol, cannabis is the most commonly abused substance worldwide, with estimates at approximately 5 million daily users. In the USA, it was also the most commonly used illicit drug by children 12–17 years (7.9 %) in 2011(Schneider and Koch 2003). Cannabis contains more than 70 different cannabinoids, and the main psychoactive ingredient of cannabis is delta-9-tetrahydrocannabinol (THC). Substantial epidemiological evidence suggests a link between cannabis use and the risk of schizophrenia (Andréasson et al. 1987; Johnston et al. 2012), and individuals diagnosed with schizophrenia are 10 times more likely to use cannabis than the general population (Henquet et al. 2005). There is also a strong relationship between cannabis exposure and symptoms of psychosis (Iritani 2007). Risk of psychosis has been positively associated with the frequency of cannabis use (50–200 % in the most frequent users; Harrison 1999), and earlier age of onset of psychotic symptoms has been associated with earlier initiation of cannabis use.

Surprisingly, there are relatively few studies on substance abuse comorbidity in animal models of psychosis. However, it has been shown that there is enhanced sensitivity to the behavioral response to other substances. For example, the neonatal ventral hippocampal lesion (NVHL) model of schizophrenia has been shown to demonstrate enhanced nicotine and cocaine sensitization as well as enhanced sensitivity to alcohol (Conroy et al. 2007; Berg and Chambers 2008; Chambers et al. 2013). Likewise, have several reports in our model also demonstrate enhanced dopamine release to both amphetamine (Cope et al. 2010) and nicotine (Brown et al. 2012). The consequences of these changes are yet to be determined, which may lead to unique treatments of substance abuse comorbidity in psychosis.

9 Conclusions

In summary, it is clear that the dopamine D_2 receptor system plays an important role in several psychoses. Moreover, the D_2 receptor has been the target for antipsychotic drugs since the mid-1950s. As with any neurotransmitter system, changes in dopamine D_2 receptor function lead to a number of other changes in brain involving other neurotransmitter systems, brain plasticity, and genetic expression. It is in these neurobiological plasticity changes that mechanisms will likely be discovered and in the future possibly lead to pharmacological targets for the development of treatments for these devastating disorders.

References

Aase H, Sagvolden T (2006) Infrequent, but not frequent, reinforcers produce more variable responding and deficient sustained attention in young children with attention deficit/hyperactivity disorder (ADHD). J Child Psychol Psychiatry 47:457–471

American Psychiatric Association (2002) Practice guideline for the treatment of patients with bipolar disorder. American Psychiatric Pub

American Psychiatric Association (2013) Diagnostic and Statistical Manual of Mental Disorders, 5th edn. American Psychiatric Publishing, Arlington

Andersen SL, Teicher MH (2000) Sex differences in dopamine receptors and their relevance to ADHD. Neurosci Biobehav Rev 24:137–141

Andersen SL, Greene-Colozzi EA, Sonntag KC (2010) A novel, multiple symptom model of obsessive-compulsive-like behaviors in animals. Biol Psychiatry 68:741–747

Andréasson S, Engström A, Allebeck P, Rydberg U (1987) Cannabis and schizophrenia longitudinal study of Swedish conscripts. The Lancet 330:1483–1486

Aouizerate B, Guehl D, Cuny E, Rougier A, Burbaud P, Tignol J, Bioulac B (2005) Updated overview of the putative role of the serotoninergic system in obsessive-compulsive disorder. Neuropsychiatr Dis Treat 1:231–243

Beaulieu JM, Gainetdinov RR (2011) The physiology, signaling, and pharmacology of dopamine receptors. Pharmacol Rev 63:182–217

Beaulieu JM, Del'guidice T, Sotnikova TD, Lemasson M, Gainetdinov RR (2011) Beyond camp: the regulation of Akt and GSK3 by dopamine receptors. Front Mol Neurosci. 4:38

Berg SA, Chambers RA (2008) Accentuated behavioral sensitization to nicotine in the neonatal ventral hippocampal lesion model of schizophrenia. Neuropharmacology 8:1201–1207

Binder DK, Scharfman HE (2004) Brain-derived neurotrophic factor. Growth Factors 22:123–131

Brown RW, Gass JT, Kostrzewa RM (2002) Ontogenetic quinpirole treatments produce spatial memory deficits and enhance skilled reaching in adult rats. Pharmacol Biochem Behav 72:591–600

Brown RW, Perna MK, Maple AM, Wilson TD, Miller BE (2008) Adulthood olanzapine treatment fails to alleviate decreases of ChAT and BDNF RNA expression in rats quinpirole-primed as neonates. Brain Res 20:66–77

Brown RW, Maple AM, Perna MK, Sheppard AB, Cope ZA, Kostrzewa RM (2012) Schizophrenia and substance abuse comorbidity: nicotine addiction and the neonatal quinpirole model. Dev Neurosci 34(2–3):140–151

Cardno AG, Holmans PA, Rees MI, Jones LA, McCarthy GM, Hamshere ML, Williams NM, Norton N, Williams HJ, Fenton I, Murphy KC, Sanders RD, Gray MY, O'Donovan MC, McGuffin P, Owen MJ (2001) A genomewide linkage study of age at onset in schizophrenia. Am J Med Genet 105:439–45

Celver J, Sharma M, Kovoor A (2010) RGS9-2 mediates specific inhibition of agonist-induced internalization of D2-dopamine receptors. J Neurochem 114:739–749

Chambers RA, McClintick JN, Sentir AM, Berg SA, Runyan M, Choi KH, Edenberg HJ (2013) Cortical-striatal gene expression in neonatal hippocampal lesion (NVHL)-amplified cocaine sensitization. Genes Brain Behav 12:564–575

Chen G, Henter ID, Manji HK (2010) Translational research in bipolar disorder: emerging insights from genetically based models. Mol Psychiatry 15:883–895

Chiu P, Rajakumar G, Chiu S, Kwan CY, Mishra RK (1982) Enhanced [3 H] spiroperidol binding in striatum of spontaneously hypertensive rat (SHR). Eur J Pharmacol 82:243–244

Cohen SM, Tsien RW, Goff DC, Halassa MM (2015) The impact of NMDA receptor hypofunction on GABAergic neurons in the pathophysiology of schizophrenia. Schizophr Res 167:98–107

Conroy SK, Rodd Z, Chambers RA (2007) Ethanol sensitization in a neurodevelopmental lesion model of schizophrenia in rats. Pharmacol Biochem Behav 86:386–394

Cope ZA, Huggins KN, Sheppard AB, Noel DM, Roane DS, Brown RW (2010) Neonatal quinpirole treatment enhances locomotor activation and dopamine release in the nucleus accumbens core in response to amphetamine treatment in adulthood. Synapse 64:289–300

Cowansage KK, LeDoux JE, Monfils MH (2010) Brain-derived neurotrophic factor: a dynamic gatekeeper of neural plasticity. Curr Mol Pharmacol 3:12–29

Cuesta MJ, Jalón EG, Campos MS, Peralta V (2009) Cognitive effectiveness of olanzapine and risperidone in first-episode psychosis. Br J Psychiatry 194:439–445

Denys D (2006) Pharmacotherapy of obsessive-compulsive disorder and obsessive-compulsive spectrum disorders. Psychiatr Clin North Am 29:553–584

Denys D, de Geus F, van Megen HJ, Westenberg HG (2004a) A double-blind, randomized, placebo-controlled trial of quetiapine addition in patients with obsessive-compulsive disorder refractory to serotonin reuptake inhibitors. J Clin Psychiatry 65:1040–1048

Denys D, van der Wee N, Janssen J, De Geus F, Westenberg HG (2004b) Low level of dopaminergic D2 receptor binding in obsessive-compulsive disorder. Biol Psychiatry 55:1041–1045

Dixon L (1999) Dual diagnosis of substance abuse in schizophrenia: prevalence and impact on outcomes. Schizophr Res 35(Suppl):S93–S100

Durany N, Michel T, Zöchling R, Boissl KW, Cruz-Sánchez FF, Riederer P, Thome J (2001) Brain-derived neurotrophic factor and neurotrophin 3 in schizophrenic psychoses. Schizophr Res 52:79–86

Eilam D, Zor R, Szechtman H, Hermesh H (2006) Rituals, stereotypy and compulsive behavior in animals and humans. Neurosci Biobehav Rev 30:456–471

Einat H (2007a) Establishment of a battery of simple models for facets of bipolar disorder: a practical approach to achieve increased validity, better screening and possible insights into endophenotypes of disease. Behav Genet 37:244–255

Einat H (2007b) Different behaviors and different strains: potential new ways to model bipolar disorder. Neurosci Biobehav Rev 31:850–857

Einat H, Szechtman H (1995) Perseveration without hyperlocomotion in a spontaneous alternation task in rats sensitized to the dopamine agonist quinpirole. Physiol Behav 57:55–59

Fineberg NA, Sivakumaran T, Roberts A, Gale T (2005) Adding quetiapine to SRI in treatment-resistant obsessive–compulsive disorder: a randomized controlled treatment study. Int Clin Psychopharmacol 20:223–226

Ford CP (2014) The role of D2-autoreceptors in regulating dopamine neuron activity and transmission. Neuroscience 282C:13–22

Freedman R, Adams CE, Leonard S (2000) The alpha7-nicotinic acetylcholine receptor and the pathology of hippocampal interneurons in schizophrenia. Chem Neuroanat 2000(20):299–306

Frodl T (2010) Comorbidity of ADHD and substance use disorder (SUD): a neuroimaging perspective. J Attention Disord 14(2):109–120

Greenbaum L, Pelov I, Teltsh O, Lerer B, Kohn Y (2010) No association between regulator of G-protein signaling 9 (RGS9) and schizophrenia in a Jewish population. Psychiatr Genet 20:47–48

Haenlein M, Caul WF (1987) Attention deficit disorder with hyperactivity: a specific hypothesis of reward dysfunction. J Am Acad Child Adolesc Psychiatry 26:356–362

Hand DJ, Fox AT, Reilly MP (2009) Differential effects of d-amphetamine onimpulsive choice in spontaneously hypertensive and wistar-kyoto rats. Behav Pharmacol 20:549–553

Harrison PJ (1999) The neuropathology of schizophrenia. A critical review of the data and their interpretation. Brain 122:593–624

Henquet C, Murray R, Linszen D, van Os J (2005) The environment and schizophrenia: the role of cannabis use. Schizophr Bull 31:608–612

Hesse S, Muller U, Lincke T, Barthel H, Villmann T, Angermeyer MC et al (2005) Serotonin and dopamine transporter imaging in patients with obsessive-compulsive disorder. Psychiatry Res 140:63–72

Hoffman KL, Rueda Morales RI (2012) D1 and D2 dopamine receptor antagonists decrease behavioral bout duration, without altering the bout's repeated behavioral components, in a naturalistic model of repetitive and compulsive behavior. Behav Brain Res 230:1–10

Hollinger S, Hepler JR (2002) Cellular regulation of RGS proteins: modulators and integrators of G protein signaling. Pharmacol Rev 54:527–559

Iritani S (2007) Neuropathology of schizophrenia: A mini review. Neuropathology 27:604–608

Huang EJ, Reichardt LF (2001) Neurotrophins: roles in neuronal development and function. Annu Rev Neurosci 24:677–736

Janhunen SK, Svärd H, Talpos J, Kumar G, Steckler T, Plath N, Lerdrup L, Ruby T, Haman M, Wyler R, Ballard TM (2015) The subchronic phencyclidine rat model: relevance for the assessment of novel therapeutics for cognitive impairment associated with schizophrenia. Psychopharmacology

Johansen E, Killeen PR, Russell VA, Tripp G, Wickens JR, Tannock R, Sagvolden T (2009) Origins of altered reinforcement effects in ADHD. Behav Brain Functions 5:7–11

Johnston L, Bachman J, O'Malley P, Schulenberg J (2012) Monitoring the future: a continuing study of American youth (12th-Grade Survey), ICPSR34409-v2. Ann Arbor MI Inter-Univ Consort Polit Soc Res Distrib 11–20

Jones CA, Zempléni E, Davis B, Reynolds GP (1993) Glutamate stimulates dopamine release from cortical and limbic rat brain in vitro. Eur J Pharmacol 242:183–187

Julien RM, Advokat C, Comaty J (2015) A primer of drug action, 12th edn. Worth Publishers, New York

Kaiser S, Wonnacott S (2000) Alpha-bungarotoxin-sensitive nicotinic receptors indirectly modulate [(3)H]dopamine release in rat striatal slices via glutamate release. Mol Pharmacol 58:312–318

Kebabian JW, Calne DB (1979) Multiple receptors for dopamine. Nature 277:93–96

Kim CH, Koo MS, Cheon KA, Ryu YH, Lee JD, Lee HS (2003) Dopamine transporter density of basal ganglia assessed with [123I] IPT SPET in obsessive-compulsive disorder. Eur J Nucl Med Mol Imaging 30(12):1637–1643

Kostrzewa RM (1995) Dopamine receptor supersensitivity. Neurosci Biobehav Rev 19:1–17

Kostrzewa RM, Brus R (1991) Ontogenic homologous supersensitization of quinpirole-induced yawning in rats. Pharmacol Biochem Behav 39:517–519

Kovoor A, Seyffarth P, Ebert J, Barghshoon S, Chen CK, Schwarz S, Axelrod JD, Cheyette BN, Simon MI, Lester HA, Schwarz J (2005) D2 dopamine receptors colocalize regulator of G-protein signaling 9-2 (RGS9-2) via the RGS9 DEP domain, and RGS9 knock-out mice develop dyskinesias associated with dopamine pathways. J Neurosci 25:2157–2165

Laruelle M, Abi-Dargham A, Gil R, Kegeles L, Innis R (1999) Increased dopamine transmission in schizophrenia: relationship to illness phases. Biol Psychiatry 46:56–72

Lipska BK, Weinberger DR (2002) A neurodevelopmental model of schizophrenia: neonatal disconnection of the hippocampus. Neurotox Res 4:469–475

Lyon L, Saksida LM, Bussey TJ (2012) Spontaneous object recognition and its relevance to schizophrenia: a review of findings from pharmacological, genetic, lesion and developmental rodent models. Psychopharmacology 220:647–672

Magnuson B, Ekim B, Fingar DC (2012) Regulation and function of ribosomal protein S6 kinase (S6K) within mTOR signalling networks. Biochem J 441:1–21

Malkesman O, Austin DR, Chen G, Manji HK (2009) Reverse translational strategies for developing animal models of bipolar disorder. Disease Models Mechanisms. 2:238–245

Mao D, McGehee DS (2010) Nicotine and behavioral sensitization. J Mol Neurosci 40:154–163

Maple AM, Perna MK, Parlaman JP, Stanwood GD, Brown RW (2007) Ontogenetic quinpirole treatment produces long-lasting decreases in the expression of Rgs9, but increases Rgs17 in the striatum, nucleus accumbens and frontal cortex. Eur J Neurosci 26(9):2532–2538

Maple AM, Smith KJ, Perna MK, Brown RW (2015) Neonatal quinpirole treatment produces prepulse inhibition deficits in adult male and female rats. Pharmacol Biochem Behav 137:93–100

McCreadie RG, Kelly C (2000) Patients with schizophrenia who smoke. Private disaster, public resource. Br J Psychiatry 176:109–116

Monteiro P, Feng G (2015) Learning from animal models of obsessive-compulsive disorder. Biol Psychiatry

Mook DM, Jeffrey J, Neuringer A (1993) Spontaneously hypertensive rats (SHR) readily learn to vary but not repeat instrumental responses. Behav Neural Biol 59:126–135

Mueser KT, Gingerich S (2013) Treatment of co-occurring psychotic and substance use disorders. Soc Work Public Health 28:424–439

Neubig RR (2002) Regulators of G protein signaling (RGS proteins): novel central nervous system drug targets. J Pept Res 60:312–316

Neubig RR, Siderovski DP (2002) Regulators of G-protein signalling as new central nervous system drug targets. Nat Rev Drug Discov 1:187–197

Nikam SS, Awasthi AK (2008) Evolution of schizophrenia drugs: a focus on dopaminergic systems. Curr Opin Investig Drugs 9:37–46

Nikolaus S, Antke C, Beu M, Müller HW (2010) Cortical GABA, striatal dopamine and midbrain serotonin as the key players in compulsive and anxiety disorders-results from in vivo imaging studies. Rev Neurosci 21:119–139

Nutt D, King LA, Saulsbury W, Blakemore C (2007) Development of a rational scale to assess the harm of drugs of potential misuse. The Lancet 369:1047–1053

Okahisa Y, Kodama M, Takaki M, Inada T, Uchimura N, Yamada M, Iwata N, Iyo M, Sora I, Ozaki N, Ujike H (2011) Association between the regulator of G-protein signaling 9 gene and patients with methamphetamine use disorder and schizophrenia. Curr Neuropharmacol 9:190–194

Olver JS, O'Keefe G, Jones GR, Burrows GD, Tochon-Danguy HJ, Ackermann U, Norman TR (2009) Dopamine D1 receptor binding in the striatum of patients with obsessive–compulsive disorder. J Affect Disord 114:321–326

Perani D, Garibotto V, Gorini A, Moresco RM, Henin M, Panzacchi A, Fazio F (2008) In vivo PET study of 5HT 2A serotonin and D2 dopamine dysfunction in drug naive obsessive-compulsive disorder. Neuroimage 42:306–314

Picchioni MM, Murray RM (2007) Schizophrenia. BMJ 335:91–95

Pinsonneault JK, Han DD, Burdick KE, Kataki M, Bertolino A, Malhotra AK, Gu HH, Sadee W (2011) Dopamine transporter gene variant affecting expression in human brain is associated with bipolar disorder. Neuropsychopharmacology. 36:1644–1655

Rizos EN, Papadopoulou A, Laskos E, Michalopoulou PG, Kastania A, Vasilopoulos D, Katsafouros K, Lykouras L (2010) Reduced serum BDNF levels in patients with chronic schizophrenic disorder in relapse, who were treated with typical or atypical antipsychotics. World J Biol Psychiatry 11:251–255

Robinson TE, Becker JB (1986) Enduring changes in brain and behavior produced by chronic amphetamine administration: a review and evaluation of animal models of amphetamine psychosis. Brain Res 396:157–198

Roybal K, Theobold D, Graham A, DiNieri JA, Russo SJ, Krishnan V, McClung CA (2007) Mania-like behavior induced by disruption of CLOCK. Proc Natl Acad Sci 104:6406–6411

Ruscio AM, Stein DJ, Chiu WT, Kessler RC (2010) The epidemiology of obsessive-compulsive disorder in the national comorbidity survey replication. Mol Psychiatry 15:53–63

Russell V, de Villiers A, Sagvolden T, Lamm M, Taljaard J (1995) Altered dopaminergic function in the prefrontal cortex, nucleus accumbens and caudate-putamen of an animal model of attention-deficit hyperactivity disorder—the spontaneously hypertensive rat. Brain Res 676:343–351

Russell VA, Sagvolden T, Johansen EB (2005) Animal models of attention-deficit hyperactivity disorder. Behav Brain Functions 1(1):9

Sagvolden T (2000) Behavioral validation of the spontaneously hypertensive rat (SHR) as an animal model of attention-deficit/hyperactivity disorder (AD/HD). Neurosci Biobehav Rev 24:31–39

Sagvolden T, Asse H, Johansen EB, Russell VA (2005) A dynamic developmental theory of attention-deficit/hyperactivity disorder (ADHD) predominately hyperactive/impulsive and combined subtypes. Behav Brain Sci 28(397):486

Saldana L, Neuringer A (1998) Is instrumental variability abnormally high in children exhibiting ADHD and aggressive behavior? Behav Brain Res 94:51–59

Salvadore G, Quiroz JA, Machado-Vieira R, Henter ID, Manji HK, Zarate CA (2010) The neurobiology of the switch process in bipolar disorder: a review. J Clin Psychiatry 71:1488–1501

Sams-Dodd F, Lipska BK, Weinberger DR (1997) Neonatal lesions of the rat ventral hippocampus result in hyperlocomotion and deficits in social behaviour in adulthood. Psychopharmacology 132:303–310

Saul MC, Gessay GM, Gammie SC (2012) A new mouse model for mania shares genetic correlates with human bipolar disorder. PLoS ONE 7:e38128

Schilman EA, Klavir O, Winter C, Sohr R, Joel D (2010) The role of the striatum in compulsive behavior in intact and orbitofrontal-cortex-lesioned rats: possible involvement of the serotonergic system. Neuropsychopharmacology 35:1026–1039

Schmitz Y, Lee CJ, Schmauss C, Gonon F, Sulzer D (2001) Amphetamine distorts stimulation-dependent dopamine overflow: effects on D2 autoreceptors, transporters, and synaptic vesicle stores. J Neurosci 21:5916–5924

Schneider M, Koch M (2003) Chronic pubertal, but not adult chronic cannabinoid treatment impairs sensorimotor gating, recognition memory, and the performance in a progressive ratio task in adult rats. Neuropsychopharmacology 28:1760–1769

Seeman P, Schwarz J, Chen JF, Szechtman H, Perreault M, McKnight GS, Roder JC, Quirion R, Boksa P, Srivastava LK, Yanai K, Weinshenker D, Sumiyoshi T (2006) Psychosis pathways converge via D2 high dopamine receptors. Synapse 60:319–346

Selvaraj S, Arnone D, Cappai A, Howes O (2014) Alterations in the serotonin system in schizophrenia: a systematic review and meta-analysis of postmortem and molecular imaging studies. Neurosci Biobehav Rev 45:233–245

Seutin V, Verbanck P, Massotte L, Dresse A (1991) Acute amphetamine-induced subsensitivity of A10 dopamine autoreceptors in vitro. Brain Res 558:141–144

Shaldubina A, Einat H, Szechtman H, Shimon H, Belmaker RH (2002) Preliminary evaluation of oral anticonvulsant treatment in the quinpirole model of bipolar disorder. J Neural Transm 109 (3):433–440

Sharma A, Couture J (2014) A review of the pathophysiology, etiology, and treatment of attention-deficit hyperactivity disorder (ADHD). Ann Pharmacother 48(2):209–225

Smolders I, Sarre S, Vanhaesendronck C, Ebinger C, Michotte C (1996) Extracellular striatal dopamine and glutamate after decortication and kainate receptor stimulation, as measured by microdialysis. J Neurochem 66:2373–2380

Szabo ST, Machado-Vieira R, Yuan P, Wang Y, Wei Y, Falke C, Du J (2009) Glutamate receptors as targets of protein kinase C in the pathophysiology and treatment of animal models of mania. Neuropharmacology 56:47–55

Thacker SK, Perna MK, Ward JJ, Schaefer TL, Williams MT, Kostrzewa RM, Brown RW (2006) The effects of adulthood olanzapine treatment on cognitive performance and neurotrophic factor content in male and female rats neonatally treated with quinpirole. Eur J Neurosci 24:2075–2083

Thompson JL, Urban N, Slifstein M, Xu X, Kegeles LS, Girgis RR, Beckerman Y, Harkavy-Friedman JM, Gil R, Abi-Dargham A (2013) Striatal dopamine release in schizophrenia comorbid with substance dependence. Mol Psychiatry. 18:909–915

Tizabi Y, Copeland RL Jr, Brus R, Kostrzewa RM (1999) Nicotine blocks quinpirole-induced behavior in rats: psychiatric implications. Psychopharmacology 145:433–441

Tollefson GD (1996) Update on new atypical antipsychotics. J Clin Psychiatry 57:318–319

Tripp G, Wickens JR (2009) Neurobiology of ADHD. Neuropharmacology 57:579–589

Tucci MC, Dvorkin-Gheva A, Sharma R, Taji L, Cheon P, Peel J, Kirk A, Szechtman H (2014) Separate mechanisms for development and performance of compulsive checking in the quinpirole sensitization rat model of obsessive-compulsive disorder (OCD). Psychopharmacology 231:3707–3718

van der Wee NJ, Stevens H, Hardeman JA, Mandl RC, Denys DA, van Megen HJ, Westenberg HM (2004) Enhanced dopamine transporter density in psychotropic naive patients with obsessive-compulsive disorder shown by [123I] β-CIT SPECT. Am J Psychiatry 161:2201–2206

Volkow ND, Wang G-J, Kollins SH, Wigal TL, Newcorn JH, Telang F (2009) Evaluating dopamine reward pathway in ADHD: clinical implications. JAMA 302:1084–1091

Wade TR, de Wit H, Richards JB (2000) Effects of dopaminergic drugs on delayed reward as a measure of impulsive behavior in rats. Psychopharmacology 150:90–101

Wan Y, Ade KK, Caffall Z, Ilcim Ozlu M, Eroglu C, Feng G, Calakos N (2014) Circuit-selective striatal synaptic dysfunction in the Sapap3 knockout mouse model of obsessive-compulsive disorder. Biol Psychiatry 75:623–630

Wang Y, Barbaro MF, Baraban SC (2006) A role for the mTOR pathway in surface expression of AMPA receptors. Neurosci Lett 401:35–39

Welch JM, Lu J, Rodriguiz RM, Trotta NC, Peca J, Ding JD, Feliciano C, Chen M, Adams JP, Luo J, Dudek SM, Weinberg RJ, Calakos N, Wetsel WC, Feng G (2007) Cortico-striatal synaptic defects and OCD-like behaviours in Sapap3-mutant mice. Nature 448:894–900

Winstanley CA, Eagle DM, Robbins TW (2006) Behavioral models of impulsivity in relation to ADHD: translation between clinical and preclinical studies. Clin Psychol Rev 26:379–395

Wultz B, Sagvolden T (1992) The hyperactive spontaneously hypertensive rat learns to sit still, but not to stop bursts of responses with short interresponse times. Behav Genet 22:415–433

Zeeb FD, Robbins TW, Winstanley CA (2009) Serotonergic and dopaminergicmodulation of gambling behavior as assessed using a novel rat gambling task. Neuropsychopharmacology 34:1–15

Postnatal Phencyclidine (PCP) as a Neurodevelopmental Animal Model of Schizophrenia Pathophysiology and Symptomatology: A Review

B. Grayson, S.A. Barnes, A. Markou, C. Piercy, G. Podda and J.C. Neill

Abstract Cognitive dysfunction and negative symptoms of schizophrenia remain an unmet clinical need. Therefore, it is essential that new treatments and approaches are developed to recover the cognitive and social impairments that are seen in patients with schizophrenia. These may only be discovered through the use of carefully validated, aetiologically relevant and translational animal models. With recent renewed interest in the neurodevelopmental hypothesis of schizophrenia, postnatal administration of *N*-methyl-*D*-aspartate receptor (NMDAR) antagonists such as phencyclidine (PCP) has been proposed as a model that can mimic aspects of schizophrenia pathophysiology. The purpose of the current review is to examine the validity of this model and compare it with the adult subchronic PCP model. We review the ability of postnatal PCP administration to produce behaviours (specifically cognitive deficits) and neuropathology of relevance to schizophrenia and their subsequent reversal by pharmacological treatments. We review studies investigating effects of postnatal PCP on cognitive domains in schizophrenia in rats. Morris water maze and delayed spontaneous alternation tasks have been used for working memory, attentional set-shifting for executive function, social novelty discrimination for selective attention and prepulse inhibition of acoustic startle for sensorimotor gating. In addition, we review studies on locomotor activity and neuropathology. We also include two studies using dual hit models incorporating postnatal PCP and two studies on social behaviour deficits following postnatal PCP. Overall, the evidence we provide supports the use of postnatal PCP to model cognitive and neuropathological disturbances of relevance to schizophrenia.

B. Grayson (✉) · C. Piercy · G. Podda · J.C. Neill
Manchester Pharmacy School, University of Manchester, Oxford Road,
Manchester M13 9PT, UK
e-mail: ben.grayson@manchester.ac.uk

J.C. Neill
e-mail: joanna.neill@manchester.ac.uk

S.A. Barnes · A. Markou
Department of Psychiatry, University of California San Diego, 9500 Gilman Drive,
La Jolla, San Diego, CA 92093-0603, USA

© Springer International Publishing Switzerland 2015
Curr Topics Behav Neurosci (2016) 29: 403–428
DOI 10.1007/7854_2015_403
Published Online: 29 October 2015

To date, there is a lack of evidence to support a significant advantage of postnatal PCP over the adult subchronic PCP model and full advantage has not been taken of its neurodevelopmental component. When thoroughly characterised, it is likely that it will provide a useful neurodevelopmental model to complement other models such as maternal immune activation, particularly when combined with other manipulations to produce dual or triple hit models. However, the developmental trajectory of behavioural and neuropathological changes induced by postnatal PCP and their relevance to schizophrenia must be carefully mapped out. Overall, we support further development of dual (or triple) hit models incorporating genetic, neurodevelopmental and appropriate environmental elements in the search for more aetiologically valid animal models of schizophrenia and neurodevelopmental disorders (NDDs).

Keywords Postnatal PCP · Cognition · Schizophrenia · Social behaviour · Neurodevelopment · Rat

Contents

1 Introduction 404
2 Effects of Neonatal PCP on Behaviour of Relevance to Schizophrenia 409
   2.1 Sensorimotor Gating 409
   2.2 Locomotor Activity 410
3 Effect of Neonatal PCP on Cognitive Domains Affected in Schizophrenia 410
   3.1 Working Memory 410
   3.2 Executive Function 412
   3.3 Effects of Neonatal PCP on Attention in a Social Novelty Discrimination
       Paradigm 414
4 Effects of Neonatal PCP on Social Behaviour 415
5 Neonatal PCP Combined with Other Environmental Manipulations 416
   5.1 Neonatal PCP and Isolation Rearing 416
   5.2 Neonatal PCP and Poly I:C 417
6 Effects of Neonatal PCP on Neuropathology 417
7 Discussion 420
   7.1 Comparison with the Adult PCP Model 420
8 Conclusion 422
References 423

1 Introduction

Our aim in this review is to evaluate the use of postnatal phencyclidine (PCP, angel dust) to induce behavioural and neuropathological disturbances of relevance to schizophrenia. Here, we consider neuropathological, cognitive, social and other behavioural changes induced by administration of the N-methyl-D-aspartate receptor

(NMDAR) antagonist, PCP, to rodents neonatally and make a brief comparison with the adult PCP model. A recent and comprehensive genetic study of schizophrenia identified several associations between genes involved in glutamate neurotransmission and schizophrenia (Schizophrenia Working Group of the Psychiatric Genomics 2014). Post-mortem studies in patients reveal reductions in NMDAR subunits NR1 and NR2C, essential to the functionality of NMDARs (Weickert et al. 2013). Given the type of evidence described above, endogenous disruptions in glutamate neurotransmission in schizophrenia patients may therefore lead to persistent alterations in excitatory/inhibitory (E/I) balance and contribute to the complex symptomatology observed in this disorder. These changes may be mimicked in animals by the administration of NMDAR antagonists. We and others have recently reviewed the relevance of the adult subchronic PCP model for mimicking symptomatology and pathophysiology of relevance to schizophrenia (Neill et al. 2010, 2014). Other NMDAR antagonists have been employed, but PCP is most commonly used, and a subchronic dosing regimen of PCP (most usually twice daily for 7 days followed by 7-days drug free) given to adult rodents is arguably the best validated and most widely employed model currently used in schizophrenia research. This model, however, lacks a neurodevelopmental (and genetic) component, which is clearly a very important part of the aetiology of schizophrenia (Rapoport et al. 2012; Fatemi and Folsom 2009; Harrison and Weinberger 2005; Lewis and Levitt 2002; Harrison 1997) and other neurodevelopmental disorders—NDDs (see (Knuesel et al. 2014) for an extensive and excellent review of maternal immune activation in rodents for modelling schizophrenia and other NDDs). Neonatal administration of PCP clearly provides an element of neurodevelopment in the context of the pathology induced by NMDAR antagonism. One advantage is enabling the study of developmental trajectory of dysfunction in brain and behaviour with intervention at specific time points. However, it remains to be determined whether the neonatal PCP model provides any other advantages over the adult PCP model. Here, we examine the effects of neonatal PCP on behaviour (focusing on cognition) and social interaction, representing negative and cognitive symptoms and current unmet clinical needs, in addition to neuropathology.

A PubMed® search of the literature covering the last ten years (09.06.2005–09.06.2015) using **postnatal or neonatal phencyclidine** and **rat** as keywords identified 79 publications. In comparison, using **subchronic** or **subchronic phencyclidine** and **rat** identified 108 publications. Using **postnatal or neonatal phencyclidine** and **mouse** as keywords, PubMed identified 21 publications compared to 32 publications using the keywords **subchronic or subchronic phencyclidine** and **mouse.** Surprisingly, PubMed revealed no reviews published on postnatal or neonatal phencyclidine in rats and/or mice over the last 10 years compared with 3 review publications when using the keywords **subchronic or subchronic phencyclidine** and **rat or mouse.**

With renewed interest in the neurodevelopmental hypothesis of schizophrenia (reviewed in (Knuesel et al. 2014)), the lack of reviews on postnatal PCP is rather surprising; therefore, this review is somewhat overdue.

The neurodevelopmental hypothesis of schizophrenia suggests that the late second trimester of pregnancy is when critical development of the central nervous system (CNS) occurs in the human foetus that can be adversely affected by environmental events, resulting in the development of schizophrenia and other NDDs in later life. This period of development in the human brain corresponds to the first two weeks of postnatal life in the rat (Clancy et al. 2007). The critical first two weeks is a stage at which the rodent is particularly sensitive to early insults such as viral infection or environmental disturbances that can lead to alterations in normal brain development, enhancing the risk of developing schizophrenia-like alterations in adulthood (Rapoport et al. 2012). One way to alter neurodevelopment in the laboratory is to administer pharmacological agents, such as NMDAR antagonists, during this critical developmental stage which results in profound effects on CNS development and function (Haberny et al. 2002). This finding concurs with the observation in humans that exposure to NMDAR antagonists, such as PCP, during the late second trimester is a risk factor for developing schizophrenia in later life (Mouri et al. 2007). Based on these findings, early postnatal administration of PCP on postnatal days (PND) 7, 9 and 11 (see Table 1) or repeated daily administration on PND 5–15 have been suggested as neurodevelopmental insults to model schizophrenia-like deficits in rodents (Wang et al. 2001; Sircar 2003).

Cognitive and social impairments observed in schizophrenia are believed to be the core features of the disease that correlate highly with functional outcome (Fett et al. 2011) and at present are poorly treated with available antipsychotic medication (Keefe 2007; Dodell-Feder et al. 2015). As a result, it is essential that new treatments and approaches are developed to recover the behavioural, cognitive and social impairments that are seen in patients with schizophrenia, which is therefore the focus of our review. These may only be developed through the use of carefully validated, aetiologically relevant and translational animal models partly validated by their neuropathological similarities to the disease, also covered in this review. Specifically, we describe studies on prepulse inhibition (PPI) of the startle response, locomotor activity, working memory, executive function, social novelty discrimination as a test of selective attention, social behaviour (surprisingly, only two studies were found on this) and neuropathology. We also include a short section on dual hit models where another environmental insult has been combined with postnatal PCP in an attempt to produce a more robust, translational dual hit animal model.

Table 1 Summary of the effects of NMDAR blockade during postnatal days (PND) 7, 9, 11 on behaviour, cognition and neurobiology

Species and and strain	Treatment regimen	Sensorimotor gating	Locomotor activity	Memory	Executive function	Social	Neuropathology
Sprague-Dawley rat	PCP 10 mg/kg on PND 7, 9, 11	PPI deficits Wang et al. (2001; Anastasio and Johnson 2008) No PPI deficits Rasmussen et al. (2007; Boctor and Ferguson 2009)	Baseline activity unaffected. Increased sensitivity to acute PCP challenge Boctor and Ferguson (2010)	Spatial alternation deficits Wang et al. (2001; Wiley et al. 2003)			Cortical apoptosis; impaired striatal synaptogenesis Wang et al. (2004)
	PCP 8.7 mg/kg on PND 7, 9, 11			Spatial working memory deficits in males but not females Andersen and Pouzet (2004)	Spatial reversal learning deficits in males but not females Andersen and Pouzet (2004)		
Lister Hooded rat	PCP 20 mg/kg on PND 7, 9, 11	PPI deficit Kjaerby et al. (2013)					Reduced frequency of mIPSCs in cortical layer II/III pyramidal neurons (Kjaerby et al. 2014) Reduced PV cell expression in mPFC Kaalund et al. (2013)
	PCP 10–20 mg/kg on PND 7, 9, 11				ASST deficits Broberg et al. (2008, 2009)		
Wistar rat	PCP 10 mg/kg on PND 7, 9, 11					Deficits in social novelty discrimination Terranova et al. (2005; Harich et al. 2007; Griebel et al. 2012; Clifton et al. 2013)	

(continued)

Table 1 (continued)

Species and and strain	Treatment regimen	Sensorimotor gating	Locomotor activity	Memory	Executive function	Social	Neuropathology
G42 × C57BL/6 mice	Ketamine 30 mg/kg on PND 7, 9, 11						Reduced cortical PV expression Powell et al. (2012)
Crlj:CD1 mice	PCP 10 mg/kg on PND 7, 9, 11		Baseline activity unaffected. Increased sensitivity to acute PCP challenge Nakatani-Pawlak et al. (2009)	Spatial working memory deficits; no deficit in novel object recognition Nakatani-Pawlak et al. (2009)		Reduced social interaction Nakatani-Pawlak et al. (2009)	Reduced PV cell expression in frontal cortex and hippocampus CA1; Reduced dendritic spine density in frontal cortex, hippocampal CA1 and nucleus accumbens Nakatani-Pawlak et al. (2009)

ASST attentional set-shifting task; *CA1* cornus ammonis subregion 1 of the hippocampus; *mIPSCs* miniature inhibitory postsynaptic currents; *PFC* prefrontal cortex; *NMDAR N*-methyl-*D*-asparate receptor; *PCP* phencyclidine; *PND* postnatal day; *PPI* prepulse inhibition of the startle response; *PV* parvalbumin

2 Effects of Neonatal PCP on Behaviour of Relevance to Schizophrenia

2.1 Sensorimotor Gating

PPI of the startle response is a measure used in the assessment of sensorimotor gating deficits in schizophrenia. When an animal is exposed to a weak prestimulus, it is less reactive to a subsequent startling stimulus as described by Geyer et al. (2001). In a study by Anastasio and Johnson (2008) when female Sprague-Dawley rats were injected with PCP at 10 mg/kg s.c. on PND 7, 9 and 11, a significant deficit in PPI was produced on PND 24–26. This PCP-induced deficit was prevented by acute administration of the atypical antipsychotic olanzapine (1 mg/kg, s. c) and risperidone (0.25 mg/kg, s.c) 30 min prior to postnatal PCP dosing (Anastasio and Johnson 2008). In agreement with this, Kjaerby and colleagues (Kjaerby et al. 2013) also produced a PPI deficit by the administration of PCP (20 mg/kg, s.c) to both male and female Lister Hooded rats on PND 7, 9 and 11. This PCP-induced deficit was subsequently reversed by the metabotropic glutamate receptor 5, mGluR5, positive allosteric modulator ADX47273 (5 mg/kg, s.c) and a selective partial agonist of the α7 nicotinic acetylcholine receptor, SSR180711 (3 mg/kg, s.c.), given twice daily for 7 days in adolescence, at 4–5 weeks of age. It has been suggested that a disrupted GABA neurotransmitter system is involved in the pathophysiology of schizophrenia (Benes and Berretta 2001), and these drugs are known to enhance γ-aminobutyric acid (GABA)ergic transmission. These results support this theory in that ADX47273 reversed the PCP-induced PPI deficit (Kjaerby et al. 2014). In a study by Wang et al. (2001) when female Sprague-Dawley rats were injected with PCP (10 mg/kg, s.c) on PND 7, 9 and 11, a significant deficit in PPI was observed on PND 24–28. This PCP-induced deficit was prevented by olanzapine (2 mg/kg, s.c.) given prior to PCP administered on PND 7, 9 and 11 and also reversed by 12-day (PND 12–23) treatment with olanzapine (2 mg/kg, s.c.). Together, these data suggest that postnatal PCP results in a PPI deficit that is both prevented and reversed by atypical antipsychotic treatment and by novel drug targets.

Conversely, a study conducted by Rasmussen and colleagues (2007) demonstrated that PCP administration (10 mg/kg, s.c.) to male and female Sprague-Dawley rats on PND 7, 9 and 11 produced no significant impairments in PPI when tested on PND 32. However, there was a PPI deficit in both male and female rats when they were given a second acute single dose of PCP (10 mg/kg, s.c) on PND 45 compared with animals not given another acute dose of PCP. These animals were initially tested after a longer washout period compared to other groups which may explain the weaker effect of postnatal PCP treatment on PPI (Rasmussen et al. 2007). A study by Boctor and Ferguson (2009) provides some support for the work of Rasmussen and colleagues as they demonstrated that PCP administration (10 mg/kg, s.c.) to male and female Sprague-Dawley rats on PND 7, 9 and 11 induced no change in PPI of the acoustic startle response when tested on PND 25.

Rats in this study underwent extensive handling by the experimenter from PND 0–4, and the authors suggest that this could have induced stress in the animals at such a young age. In support of this, neonatal handling has been shown to attenuate adult PCP-induced deficits in the open field and forced swim test on PND 75 (5 mg/kg for 5 days) (Tejedor-Real et al. 2007).

2.2 Locomotor Activity

Locomotion is measured using two key paradigms, real-time recording of activity by the experimenter and using photo-beams in a specifically constructed cage (Forrest et al. 2014). In a study by Zhang et al. (2012), male and female Sprague-Dawley rats were treated with PCP (10 mg/kg, s.c.) on PND 2–16. Rats were handled daily for 10 min on PND 24–30, and locomotor activity was measured on PND 30 using an automated set-up. The animals' locomotor activity was measured at 10, 30 and 60 min and significantly increased in PCP-treated animals compared to controls at the 60-min time interval.

In contrast, Sircar and Soliman (2003) found no significant effect on locomotor activity following postnatal PCP alone. Male and female Sprague-Dawley rats were administered PCP (1, 3 or 5 mg/kg, i.p) on PND 5–15. Rats were tested at adulthood (PND 70–120), whereby both control and PCP-treated animals were administered an acute dose of PCP (10 mg/kg, i.p) on the day of testing and locomotor activity was monitored for a total of 95 min. In both sexes of rats, acute PCP challenge significantly increased locomotor activity in both postnatal PCP and control groups, an effect that was greater in female rats compared to males (Sircar and Soliman 2003). It has been previously shown that female rats are more sensitive to the effects of PCP which could account for these results (Nabeshima et al. 1984). In addition, Boctor and Ferguson (2010) demonstrated that postnatal PCP (10 mg/kg s.c., PND 7, 9, 11) in male and female Sprague-Dawley rats induced locomotor sensitisation to an acute dose of PCP without affecting baseline activity.

3 Effect of Neonatal PCP on Cognitive Domains Affected in Schizophrenia

3.1 Working Memory

Working memory has been defined as the ability to transiently hold and manipulate information to guide goal-directed behaviour and is partly dependent on functionality of the dorsolateral prefrontal cortex (Baddeley 1992; Lett et al. 2014). Disturbances in working memory are a core feature of the cognitive dysfunction seen in schizophrenia patients and their unaffected relatives (Lett et al. 2014; Lee and Park 2005).

3.1.1 Morris Water Maze

The Morris water maze can be used to assess spatial working memory in rodents. An early study in 2003 by Sircar (2003) examined the effects of repeated postnatal PCP, whereby both male and female Sprague-Dawley rats were administered PCP (1 or 5 mg/kg, i.p.) daily between PND 5 and 15. Rats were subsequently weaned on PND 21 and tested for their performance in the Morris water maze on PND 35 (adolescents) or PND 60 (adults). Repeated postnatal PCP impaired the acquisition of spatial learning in adolescents and adults of both sexes. This impairment in performance was accompanied by increased [^{3}H]MK-801-labelled NMDARs measured by saturation binding in the hippocampus and frontal cortex. These data support the hypothesis that PCP given during the postnatal period produces deficits in working memory by disrupting the developing glutamatergic system. Another study examined the effects of chronic 12-day treatment with D-serine (640 or 1280 mg/kg, s.c.) after postnatal administration of PCP (8.7 mg/kg, s.c.) on PND 7, 9, and 11 in both male and female Sprague-Dawley rats on spatial working memory in the Morris water maze (Andersen and Pouzet 2004). Rats were weaned on PND 21 and behavioural experiments commenced on PND 56, that is when rats had just reached adulthood. Male PCP-treated rats were slightly impaired during the spatial reference memory task but strongly impaired during the reversal and spatial working memory tasks. Chronic treatment with D-serine, a selective glycine/*N*-methyl-*D*-aspartate receptor agonist, at both doses reversed the PCP-induced cognitive deficit. Conversely, female rats were not significantly affected by postnatal PCP treatment in this task. A study by Secher et al. (2009) examined the effect of a neural cell adhesion molecule-derived fibroblast growth factor receptor agonist (FGL) on spatial learning and memory deficits induced by postnatal PCP treatment. Male Sprague-Dawley rats were administered a high dose of PCP (30 mg/kg, s.c.) on PND 7, 9 and 11. These rats were also treated chronically with FGL at 10 mg/kg, s.c. on PND 4, 6, 8 and 10. FGL treatment was continued on PND 12 and then administered twice per week until PND 49, 4-days prior to the start of behavioural testing. Rats were tested as adults in the reference memory, reversal learning, working memory and visible-platform tasks of the Morris water maze. The PCP-treated rats demonstrated a robust impairment in working memory and reversal learning. However, there was no effect of postnatal PCP treatment on the long-term memory component of the reference memory test. Chronic FGL treatment had no effect on the reversal learning deficit but significantly restored the working memory deficits to almost control levels. These results suggest that postnatal PCP treatment produced deficits in cognition relevant to schizophrenia and that the working memory deficits were selectively protected. The mechanism of action of FGL using this dosing regimen may be due to its growth factor-like and neuroprotective properties which could protect and provide growth support to the neurons affected during the neonatal PCP treatment. Furthermore, administration of FGL after the PCP treatment provides trophic neuronal support and thus may enhance recovery of the affected neurons and networks. Finally, FGL administration prior to behavioural testing can positively affect cognition as this molecule is known to display memory-enhancing properties (Cambon et al. 2004).

3.1.2 Delayed Spontaneous Alternation

Wang et al. (2001) used the delayed spatial alternation task to assess the effects of treatment with PCP (10 mg/kg, s.c.) on PND 7, 9 and 11 in female Sprague-Dawley rats. Rats were tested between PND 33 and 70 for acquisition of a delayed spatial learning task using an inter-trial interval of 10 s. Results showed that initial accuracy did not differ between PCP and control rats; however, accuracy improved faster in the controls, showing that PCP reduced the rate of learning. Between PND 38 and 49, the PCP-treated rats showed reduced accuracy; however, by PND 70, both groups were performing at the same level. In a later study by Wiley et al. (2003), they performed a similar study again in female Sprague-Dawley rats treated with PCP at 10 mg/kg, given i.p. on PND 7, 9 and 11 and then trained over 10 days (starting on PND 34) using the same 10-s inter-trial interval. Rats that had been treated with postnatal PCP reached acquisition criteria for the delayed spatial alternation task later compared to controls. Subsequent pharmacological challenges with NMDAR antagonists, PCP, ketamine- and the dopamine-releasing agent amphetamine, decreased accuracy in both PCP-treated and control rats. Nicotine treatment failed to affect performance in either group. MK-801, the high-affinity NMDA open-channel blocker, produced a more robust reduction in accuracy in the PCP-treated rats compared to controls. When delays were extended to 30 s, acute treatment with PCP (1-3 mg/kg, i.p.) also significantly decreased accuracy in PCP-treated rats more than in controls. These results show that postnatal PCP treatment induces long-term deficits in cognition that are not always evident under baseline conditions, but that the deficit may be revealed by pharmacological challenge and manipulation of task difficulty.

3.2 Executive Function

In our literature search, we have found that executive function in neonatal PCP-treated rats is most frequently assessed using the attentional set-shifting task (ASST). This task is a measure of cognitive flexibility and executive function and is prefrontal cortex dependent. It is a highly translational task, which has been back-translated from the clinic into the preclinical setting (Goetghebeur and Dias 2014). The first study describing the effects of early postnatal PCP administration in the rodent ASST was carried out by Broberg and colleagues in 2008 (Broberg et al. 2008). In this study, both male and female Lister Hooded rats were administered PCP (10 and 20 mg/kg, s.c.) on PND 7, 9 and 11 and were tested at adulthood for their ability to perform the ASST. Both male and female rats were tested between PND 53–67 and PND 82–93, respectively. Results showed that postnatal PCP treatment impaired the ability to shift attentional set when compared to controls in both sexes. This impairment was demonstrated by a selective increase in trials to reach criterion on the extra dimensional shift (EDS) phase of the ASST. These results support the validity of the early postnatal PCP regimen to produce cognitive deficits of relevance to schizophrenia and other disorders. In a later study by the same laboratory, Broberg et al. (2009) examined the ability of acute treatment with an ampakine (CX516) at doses of 5, 10, 20 and 40 mg/kg, s.c. and the second-generation antipsychotic sertindole (1.25 mg/kg, p.o.) to reverse ASST performance deficits in male Lister Hooded rats induced by early postnatal PCP (20 mg/kg, s.c. PND 7, 9 and 11). Results again demonstrated that postnatal PCP induced a significant EDS phase deficit that was dose dependently reversed by CX516 and by sertindole. Recently, studies in male Lister Hooded rats carried out by Redrobe and colleagues (Redrobe et al. 2012) reported on the effects of PND 7, 9 and 11 administration of PCP (20 mg/kg, s.c.) in the ASST. Their results again demonstrate a significant deficit in the EDS phase induced by postnatal PCP. Treatment with a negative modulator of the $GABA_A$ α5 receptor RO4938581 (1 mg/kg, p.o.), 60 min before behavioural testing, resulted in a significant reversal of the postnatal PCP-induced deficit. These data support a potential clinical role of ampakines and $GABA_A$ α5-negative modulators in the treatment of cognitive deficits, especially in patients with executive function deficits. A later study from the same laboratory (Eskildsen et al. 2014) examined the novel selective brain penetrant phosphodiesterase2A inhibitor LuAF64280 (1–20 mg/kg, s.c.) on the EDS phase deficit induced by postnatal PCP administration (20 mg/kg, s.c.) on PND 7, 9 and 11 in male Lister Hooded rats. The results confirm a significant deficit in the EDS phase of the ASST induced by postnatal PCP administration. Treatment with LuAF64280 (1 and 10 mg/kg, s.c.), 30 min before behavioural testing resulted in a significant reversal of the postnatal PCP-induced cognitive deficit in the ASST. These data are in support of the role of phosphodiesterase2A in cognitive processes associated with schizophrenia.

Although not a great deal of studies have been performed on effects of postnatal PCP on locomotor activity, PPI and cognitive deficits of relevance to schizophrenia, most of the studies described above demonstrate robust deficits in working memory and executive function amenable to pharmacological reversal. In particular, attentional set-shifting deficits appear to be a robust cognitive deficit induced by postnatal PCP, as seen in the adult model (Neill et al. 2010). Locomotor activity and PPI changes appear less robust and require further treatment with the NMDAR antagonist to produce the deficit. However, when a PPI deficit is induced, it is amenable to pharmacological reversal. Surprisingly, few studies have been conducted on behaviour of relevance to positive symptoms, as with the adult model. This is a clear gap in the field in general, due to the lack of validated translational tests in this area.

3.3   Effects of Neonatal PCP on Attention in a Social Novelty Discrimination Paradigm

Terranova et al. (2005) compared the activity of the putative atypical antipsychotic SSR181507 (a mixed dopamine D_2 receptor antagonist and 5-HT_{1A} receptor agonist) against reference compounds, on disturbances of novelty discrimination in a social context in adult (PND 51–119) male Wistar rats that had previously received postnatal treatment with PCP (10 mg/kg, s.c.) on PND 7, 9 and 11. In this test, adult rats were presented with a familiar juvenile rat for a period of 30 min (P1), and at the end of P1, a second novel juvenile rat was then introduced for a period of 5 min (P2). The ability of the adult rat to discriminate between the two juveniles presented at the same time in P2 was measured. Results from this study show that postnatal PCP-treated rats over 5 successive experiments have robust and persistent impairments in novelty discrimination. Acute treatment with the atypical antipsychotics clozapine (0.1 mg/kg, i.p.), amisulpiride (1 and 3 mg/kg, i.p.) and the putative atypical antipsychotic SSR181507 (3 mg/kg, i.p.) significantly normalised the impairments in novelty discrimination. The classical antipsychotic agent haloperidol (0.1-0.3 mg/kg, i.p.), anticholinesterase tacrine (1 mg/kg, i.p.) and tricyclic antidepressant imipramine (16 mg/kg) were ineffective in this paradigm. The authors suggest that the discriminative capacity of adult rats in this test could be useful for exploring selective attention deficits for rats. Therefore, reversal of the postnatal PCP-induced attentional deficits may be predictive of clinical activity to restore information-processing deficits observed in schizophrenia. A study by Harich et al. (2007) evaluated the effects of LY-354740 (mGluR2/3 agonist) and LY-487379 (mGluR2 potentiator) on deficits in social novelty discrimination in male Wistar rats treated with PCP (10 mg/kg, s.c.) on PND 7, 9 and 11. Rats were tested at adulthood between PND 70–100 using a 0 s inter-trial interval, i.e. novel juvenile rats (4 weeks old) were immediately placed into the test arena. The results showed that control rats spent significantly more time exploring the novel than the familiar juvenile. This social novelty discrimination was impaired in rats that received postnatal PCP treatment. Treatment with clozapine (0.3–3 mg/kg) and the glycine transporter, GlyT1, inhibitor SSR-504734 (1–10 mg/kg), reversed the PCP-induced social novelty discrimination impairment. Furthermore, LY-354740 (1–10 mg/kg) and LY-487379 (3–30 mg/kg) also restored social novelty discrimination in postnatal PCP-treated rats. These data suggest that drugs targeting the glutamatergic system may reverse the long-term developmental attentional deficits produced by postnatal PCP. Griebel et al. (2012) assessed the cognitive effects of SAR110894, a potent histamine H_3-receptor antagonist, in adult male Wistar rats that had received PCP (10 mg/kg, s.c.) on PND 7, 9 and 11. Rats were tested at adulthood for their cognitive ability in a social novelty discrimination task using a 30-min inter-trial interval. Results demonstrate that rats treated postnatally with PCP spent significantly less time exploring a novel juvenile rat compared to controls. The authors interpret this effect as an impairment of selective attention. Treatment with SAR110894 dose dependently improved this impairment suggesting that H_3

receptor antagonism could be useful in treating certain disorders with attentional deficits such as schizophrenia and attention deficit hyperactivity disorder (ADHD). A more recent study by Clifton et al. (2013) evaluated the effects of the positive allosteric modulators of the mGlu5 receptor, CDPPB (0.16–40 mg/kg, i.p.) and ADX47273 (0.16–10 mg/kg, i.p.), in adult male Wistar rats that had previously received treatment with PCP (10 mg/kg, s.c.) on PND 7, 9 and 11. At adulthood (PND 58), the rats were assessed for their cognitive performance in a social novelty discrimination task using a 30-min inter-trial interval. Rats treated postnatally with PCP spent an equal time investigating the novel and familiar juvenile in the retention phase of the task, indicating a failure in the ability to discriminate between the two. Acute treatment with CDPPB (2.5 and 10 mg/kg, i.p.) and ADX47273 (2.5 mg/kg, i. p.) significantly increased the time spent exploring the novel animal compared to PCP-treated controls indicating enhanced social novelty discrimination. These observations support a role for mGluR5 receptors in the control of attentional processing in a social context. All the studies above show that postnatal PCP on PND 7, 9 and 11 induces robust deficits in selective attention as measured by social novelty discrimination at adulthood and that this deficit is reversed by various pharmacological treatments. This provides a useful means of assessing novel targets for improving attentional deficits in a social context of relevance to schizophrenia and other disorders. However, no studies using this paradigm have yet examined the deficit at various stages of development or the neurobiology of this effect.

4　Effects of Neonatal PCP on Social Behaviour

We found only two studies investigating effects of postnatal PCP on social withdrawal per se. White et al. (2009) postulated that, since rapid changes occur in dopaminergic and glutamatergic systems during PND 40–60 (Andersen et al. 2000; Farber et al. 1995, 2002), treatment with PCP during later development in adolescence would induce similar behavioural deficits when rats were tested at adulthood. Male Wistar rats were treated on PND 50–51 with PCP (9 mg/kg, i.p.) or saline, twice per day at 12-h intervals for two consecutive days (a total of 4 injections). The rats were tested for social interaction 3, 7, 14 and 28 days after the last injection of PCP. On the test day, two unfamiliar rats (one test rat and one conspecific) were placed into the test arena and behaviour was video recorded. Analysis of the first 8 min of the test revealed that PCP-treated rats had a significantly reduced social interaction compared to controls on days 7, 14 and 28. These data provide evidence that treatment with PCP during later development leads to long-lasting deficits in social behaviour at adulthood. Results from the social novelty discrimination studies above suggest that postnatal PCP on days 7, 9 and 11 leads to reduced social interaction in rats as with the adult PCP model. These studies have been conducted in mice, but not rats. Nakatani-Pawlak et al. (2009) investigated the involvement of glutamic acid in neural development by administration of PCP (10 mg/kg, s.c.) to male Cr1j:CD1 (ICR) mice on postnatal days 7, 9 and 11. The mice were tested for social interaction at adulthood, postnatal week 16. In the evaluation of social behaviour, two unfamiliar mice from the same treatment group were placed into a clean cage and assessed for cumulative time of contact (investigation of genitalia, sniffing and social grooming) for a period of 20 min. Results show that PCP-treated mice had a significantly reduced frequency of social interaction compared to controls. This effect was attenuated by acute treatment with the atypical antipsychotic clozapine (0.3–3.0 mg/kg, p.o.), the partial NMDAR agonist D-cycloserine (0.3–10 mg/kg, p.o.), the $GABA_A$ receptor antagonist flumazenil (3 mg/kg, s.c.) and by a $GABA_B$ receptor antagonist SHC50911 (1 mg/kg, i.p). These data demonstrate that postnatal PCP treatment in mice results in impaired social behaviour at adulthood that can be reversed by pharmacological treatments.

5 Neonatal PCP Combined with Other Environmental Manipulations

5.1 Neonatal PCP and Isolation Rearing

Rats reared in isolation from weaning until adulthood show several behavioural changes including increased locomotor activity, anxiogenesis and enhanced sensitivity to psychoactive drugs such as amphetamine and cocaine (Jones et al. 1990; Smith et al. 1997) in addition to sensorimotor gating deficits as measured by reduced PPI of acoustic startle response (Geyer et al. 1993; Cilia et al. 2001, 2005). The effect of isolation rearing on cognitive performance has also been investigated (Schrijver and Wurbel 2001; Dalley et al. 2002; Weiss et al. 2004; Li et al. 2007; McLean et al. 2010). In an attempt to create a more translational, robust animal model of schizophrenia, two interventions relating to important stages of human development, i.e. late stage second trimester of pregnancy and from approximately 3 years of age until adulthood (postnatal PCP on days 7, 9 and 11 and isolation rearing from day 21), have been combined (Gaskin et al. 2014). PCP (10 mg/kg, s. c.) was administered on PND 7, 9 and 11 to male Lister Hooded rats that were subsequently weaned on PND 23 into group housing or isolation. Six weeks after weaning, i.e. at adulthood, rats were tested for novelty and PCP-induced (3.2 mg/kg, i.p.) locomotor activity, novel object discrimination using a 2-h inter-trial interval, PPI of acoustic startle and contextual memory in a conditioned emotional response paradigm. When rats were assessed for their locomotor response to a novel environment, it was found that only the isolation-reared rats demonstrated a significant increase in locomotor activity compared to group-housed controls. However, postnatal PCP treatment, isolation rearing or combined isolation rearing with postnatal PCP treatment did not induce locomotor sensitisation to a subsequent acute PCP injection. In the novel object recognition studies, postnatal PCP treatment alone failed to induce novel object recognition deficits, whereas the isolation-reared rats alone and combined isolation rearing with postnatal PCP-treated rats were unable to significantly differentiate between the novel and familiar object. However, rats in the combined isolation rearing with postnatal PCP-treated group were not significantly more impaired compared to the isolation-reared controls, potentially due to the nature of the paradigm. In the PPI study, only the combined isolation rearing with postnatal PCP-treated rats demonstrated an impaired PPI response. The lack of object recognition deficits and locomotor sensitisation to acute PCP could be explained by the increased length of time between PCP administration and behavioural testing. Contextual memory was not significantly affected by postnatal PCP administration; however, freezing behaviour was significantly reduced by isolation rearing, but an even greater effect occurred when combined with postnatal PCP treatment. In this study, postnatal PCP and isolation rearing alone induced behavioural deficits in adult rats and combined treatment induced cognitive impairments but not significantly more so than isolation rearing or post-natal PCP administration alone, with the exception of PPI where only the combination had an effect.

5.2 *Neonatal PCP and Poly I:C*

In a recent study, Hida et al. (2014) combined polyriboinosinic-polyribocytidylic acid (Poly I:C) at 5 mg/kg s.c. for 5 days, on PND 2–6, in male and female C57BL/6 J mice with PCP at 10 mg/kg s.c. for 7 days, on PND 35–41, i.e. at adolescence. The combination of Poly I:C and adolescent PCP produced more pronounced increases in psychostimulant-induced locomotor activity and deficits in novel object recognition than either treatment alone and was the only intervention to reduce social interaction and increase the glutamate/aspartate transporter (GLAST) protein in the prefrontal cortex. The effect on GLAST is suggested by the authors to be indicative of NMDAR antagonist-induced disruption of Ca^{2+} transport into astrocytes. Lending further support to the validity of this model for gluta-matergic dysfunction associated with schizophrenia, the deficit in NOR induced by this two hit model was rescued by a glutamate transporter inhibitor, dl-Threo-beta-benzyloxyaspartate (DL-TBOA), injected directly into the prefrontal cortex. This study provides robust evidence for enhanced translational value of this dual hit model than either intervention alone. These authors and Knuesel and colleagues (2014) provide an excellent overview of MIA and dual hit models for schizophrenia and NDDs.

6 Effects of Neonatal PCP on Neuropathology

A complex interplay between excitatory and inhibitory neurotransmission maintains optimal (E/I) balance. Disrupting normal glutamatergic neurotransmission during critical stages in brain development leads to persistent alterations in E/I balance that are similar to those observed in schizophrenia patients. The inhibitory GABAergic system modulates excitatory glutamatergic pyramidal cell firing generating cortical rhythms necessary for normal network function (Rudy et al. 2011; Bartos et al. 2007). Schizophrenia patients display alterations in both excitatory (Glantz and Lewis 2000; Sweet et al. 2009) and inhibitory (Lewis et al. 2012) neuronal populations and in oscillatory activity (Uhlhaas and Singer 2010), suggesting that an E/I imbalance may be central to the disorder. Functional maturation of the GABAergic system emerges during early neurodevelopment (Huang 2009; Lee et al. 1998; Emson et al. 1979; Doischer et al. 2008; Okaty et al. 2009), and NMDAR-mediated glutamatergic transmission is essential for this process (Ben-Ari et al. 1997; Le Magueresse and Monyer 2013). Disruption of normal glutamatergic transmission during the critical period of postnatal neurodevelopment, when the GABAergic network is functionally maturing (Hoftman and Lewis 2011), may lead to persistent alterations in E/I balance, disruption in cortical network function and the emergence of symptoms relevant to schizophrenia.

As previously discussed, administration of NMDAR antagonists during critical stages in postnatal neurodevelopment can be used to disrupt normal glutamate neurotransmission. Interestingly, this treatment regimen induces a number of pathophysiological alterations that are also observed in schizophrenia patients. Parvalbumin-positive (PV) interneurons are a subclass of fast-spiking GABAergic interneurons, expressing the calcium-binding protein, PV, reduced in schizophrenia patients (Reynolds et al. 2004; Zhang and Reynolds 2002). Postnatal PCP (10–20 mg/kg, s.c.) or ketamine (30 mg/kg s.c.) treatment in rats (Lister Hooded, Sprague-Dawley, Wistar) or mice (C57BL/6, Crlj:CD1) administered between PND 2–12 has been shown to induce long-term reductions in PV cell expression in multiple brain regions, including the medial prefrontal cortex, hippocampus and nucleus accumbens (Nakatani-Pawlak et al. 2009; Kaalund et al. 2013; Powell et al. 2012; Radonjić et al. 2013; Wang et al. 2008). In mice, a potential mechanism underlying the diminished expression of PV in GABAergic interneurons after ketamine treatment may be oxidative stress (Powell et al. 2012; Behrens et al. 2007; Behrens and Sejnowski 2009). Identification of oxidative stress in schizophrenia patients (Emiliani et al. 2014) suggests that this mechanism could contribute to the PV deficiency also observed in this clinical population which in turn contributes to the impairment in inhibitory regulation of pyramidal neurons observed after neonatal PCP treatment. Indeed, computational modelling has shown that reduced PV expression can result in diminished gamma-band oscillatory activity (Volman et al. 2011), suggesting that PV deficits impair excitatory regulation. While research has focused on assessing the impact of postnatal NMDAR blockade on PV cells, inhibitory interneurons are a heterogeneous population of cells (Rudy et al. 2011) and other types of inhibitory interneurons are also altered in schizophrenia (Volk et al. 2012). Interestingly, alterations in somatostatin-positive interneurons have also been reported after postnatal PCP treatment (Radonjić et al. 2013). Further, postnatal PCP treatment (20 mg/kg, s.c., PND 7, 9, 11) in Lister Hooded rats induced a reduction in the frequency of miniature inhibitory postsynaptic currents in layer II/III pyramidal neurons. These findings confirm that an impairment in the ability to regulate excitatory neurotransmission is evident after disrupting normal glutamatergic transmission during neurodevelopment (Kjaerby et al. 2014).

There is evidence to suggest that postnatal NMDAR antagonism induces neurotoxicity and apoptosis (Mouri et al. 2007). Indeed, it has been suggested that the loss of PV cells observed after postnatal PCP, given on PND 7 only (10 mg/kg, s.c., in Sprague-Dawley rats) determined by peroxidase staining, may be an effect on postmitotic neurons and the result of cell death (Wang et al. 2008). Interestingly, it has been shown that postnatal ketamine treatment reduced PV cell expression in the absence of cell death (Powell et al. 2012). G42 X C57BL/6 mice that express green fluorescent protein (GFP) in PV cells were treated with ketamine (30 mg/kg, s.c.) on PND 7, 9 and 11. These mice were shown to exhibit PV deficits with no alteration in GFP expression, indicating that the cells were still present and that the reduction in PV was not due to cell death. While apoptosis may account for the reduction of PV cells in previous studies utilising PCP, it is possible that the interneurons are still present but were developmentally immature and that PV was below the threshold for detection. NMDAR antagonists are suggested to block NMDA signalling on receptors located on GABAergic interneurons (Homayoun and Moghaddam 2007). As postnatal glutamate transmission is required for the functional development of these neurons (Ben-Ari et al. 1997; Le Magueresse and Monyer 2013), it is possible that postnatal NMDAR antagonism perturbs the maturation of these cells. Using the Cre-LoxP system, conditional ablation of the functional subunit, NR1, from NMDARs located on GABAergic interneurons in mice resulted in reduced expression of PV and the emergence of schizophrenia-like phenotypes (Belforte et al. 2010). Moreover, this collection of alterations was only evident when the deletion occurred in early neurodevelopment and not when the animals were older, supporting the developmental component of the phenotype (Belforte et al. 2010). In addition, recent research has identified that, along with ionotropic NMDARs, metabotropic glutamate receptor activity plays an important role in normal development of the PV network. Conditional ablation of the mGluR5 from PV cells, beginning in the second postnatal week, induced neurochemical, neurophysiological and behavioural alterations associated with neurodevelopmental disorders, including schizophrenia (Barnes et al. 2015). Collectively, these findings demonstrate that disruption of normal postnatal glutamatergic transmission, either by ionotropic- or by metabotropic-mediated glutamate transmission, during critical stages in neurodevelopment, induces persistent deficits in inhibitory neurons, which in turn leads to behavioural alterations observed in schizophrenia.

In addition to the persistent impairment in inhibitory neurotransmission, administration of NMDAR antagonists during neurodevelopment results in disrupted glutamate neurotransmission as evidenced by the pharmacological studies described above and by neuropathological studies described in this section. Postnatal PCP (10 mg/kg, s.c., PND 7, 9 and 11) treatment reduced the dendritic spine density of pyramidal neurons located in the frontal cortex and hippocampus of

Crlk:CD1 mice (Nakatani-Pawlak et al. 2009). Reductions in dendritic spine density of spiny neurons within the nucleus accumbens were also reported (Nakatani-Pawlak et al. 2009). Dendritic spine development involves synaptogenesis (Cline 2001) and is dependent on glutamatergic neurotransmission (Le Magueresse and Monyer 2013; Cline 2005). Synaptogenesis and dendritic spine development are also altered in schizophrenia (Glantz and Lewis 2000; Sweet et al. 2009; Arnold et al. 2005). Evidence suggests that postnatal PCP (10 mg/kg, s.c., PND 7, 9 and 11, Sprague-Dawley rats) or ketamine (up to 25 mg/kg, s.c., PND 10, NMRI mice) disrupts neuronal development and processes involved in synaptogenesis (Viberg et al. 2008; Wang et al. 2004). Disrupting normal glutamatergic neurotransmission during postnatal development may, therefore, lead to alterations in synaptogenesis and impairments in dendritic spine development.

These findings demonstrate that administration of NMDAR antagonists during postnatal neurodevelopment leads to neuropathological alterations in adulthood that replicate those often reported in schizophrenia patients. These findings suggest that postnatal NMDAR treatment represents a useful inducing condition for studying the cognitive, behavioural and neuropathological disruptions associated with schizophrenia.

7 Discussion

7.1 Comparison with the Adult PCP Model

Repeated administration of NMDAR antagonists in adulthood produces many of the neuropathological and behavioural deficits described above. However, much more research emphasis has been placed on NMDAR antagonist treatment during adulthood, and consequently, a more comprehensive data set has been acquired with this manipulation compared to the neurodevelopmental NMDAR antagonist manipulation. Indeed, reductions in PV cell expression and/or glutamic acid decarboxylase isoform 67 (GAD67) (Behrens et al. 2007; Amitai et al. 2012; Abdul-Monim et al. 2007; Jenkins et al. 2010), grey matter deficits (Barnes et al. 2014) and alterations in functional connectivity (Dawson et al. 2014) have been reported after repeated NMDAR antagonism in adulthood. Persistent alterations in E/I balance are hypothesised to contribute to structural and functional alterations observed in psychiatric disorders (Plitman et al. 2014). These findings confirm the fundamental role of normal glutamate signalling and the functionality of GABAergic networks. However, while the administration of NMDAR antagonists during adulthood replicates many of the neuropathological alterations observed in schizophrenia (Neill et al. 2010), an important limitation to consider is that schizophrenia is widely considered to be a disorder of neurodevelopmental origin

(Fatemi and Folsom 2009; Lewis and Levitt 2002; Harrison 1997). Postnatal NMDAR administration may, therefore, represent a more aetiologically valid inducing condition relevant to the pathogenesis of schizophrenia than the adult model. Indeed, some deficits (e.g. PPI deficits) are not evident after protracted NMDAR blockade in adulthood (Barnes et al. 2014; Egerton et al. 2008; Martinez et al. 1999) but have been observed in some studies that used postnatal PCP treatment (Wang et al. 2001; Anastasio and Johnson 2008; Kjaerby et al. 2013). While not all studies show PPI deficits (see Sect. 2.1), the presence of PPI deficits after postnatal, but not adult, PCP treatment supports the notion that deficits in sensorimotor gating may rely on a developmental component. How robust PPI deficits are in either model, however, remains to be clearly ascertained, and reports are emerging of several unpublished negative studies which highlight the need to publish our negative findings and enhanced mechanisms to do so.

In terms of other behavioural alterations, again the adult model has been studied more widely, and from the information provided in this review, it appears that no advantage is conferred by the postnatal PCP model in this respect. This may well be due to the overall lack of studies to date. It is surprising that such limited work has been conducted on visual memory, a domain of cognition impaired in schizophrenia and assessed in animals by novel object recognition. Indeed, we have been unable to include a section on this as we found so few papers investigating effects of postnatal PCP in this test, only in association with a dual hit model (Hida et al. 2014). The majority of behavioural work appears to have been conducted in rats, of course this species is better suited to behavioural work, particularly when social behaviour is being studied, due to high levels of conspecific-directed aggression in mice not usually seen in rats. One behavioural test used here and not in the adult model is the social novelty discrimination test for selective attention in a social context. The deficits induced by postnatal PCP in this test appear particularly robust and amenable to reversal by pharmacological treatments. The relevance of this test for schizophrenia symptomatology, its relative simplicity, lack of confounding features and addition of pharmacological intervention makes it a particularly valuable addition to a cognitive test battery. Interestingly, the studies we found using this test with neonatal PCP have all been conducted in an industry setting supporting the robust nature of the deficit and relatively high throughput.

The full advantage of a neurodevelopmental model has not been fully realised to date with postnatal PCP. Few studies have investigated neuropathology and behaviour at the various stages of development or included specific interventions at critical time points of relevance to the illness, e.g. in adolescence, representing the prodromal stage of the illness in humans (see Gill et al. (2014) for a recent study of this type using the methylazoxymethanol (MAM) neurodevelopmental model). A detailed analysis of the effects of postnatal PCP on behaviour and pathophysiology of relevance to schizophrenia at the various stages of development is warranted. The inclusion of a second intervention to produce a dual hit model has been attempted by only two laboratories to date with postnatal PCP, this clearly has significant advantages over the single hit model, and it is our view that further work in this area will provide more robust modelling of schizophrenia and NDDs in animals. Full developmental analysis and dual/triple hit models have clear relevance for the aetiology and pathology of schizophrenia, and it is surprising that more work has not taken full advantage of the benefits of using a neurodevelopmental model.

One striking difference between these studies and the adult PCP model is the impressive inclusion of both sexes in these studies. In general, most studies in the adult model use male animals with a few laboratories employing females only (Neill et al. 2010, 2014; Meltzer et al. 2013; Grayson et al. 2015). This is an issue of tremendous importance as sex differences "occur at all levels of biological organisation" (Prendergast et al. 2014). Both genders suffer from schizophrenia, there are clear gender differences in vulnerability to neuropsychiatric disorders, coping with and management of illness, and men and women respond differently to pharmacotherapy and have different pharmacokinetic profiles (see an earlier volume in this series for a comprehensive review (Neill and Kulkarni 2011)). The misconception in animal work is that data sets from females are more variable due to the presence of hormonal fluctuations occurring during the 4-day oestrous cycle in rodents. However, in contrast, a recent large analysis of over 293 articles, monitoring morphological, physiological, molecular and behavioural traits without controlling for stage of the oestrous cycle, found that data sets from *male* mice are more variable when compared with females (Prendergast et al. 2014). Fortunately, this issue is becoming more widely recognised and measures are in place to rectify the unfortunate reliance on males in animal research with increasing recognition that data produced from males cannot be generalised to females of the species. Indeed, Clayton and Collins recently reported on NIH plans to rectify this situation for preclinical research as they have successfully achieved for clinical research (Clayton and Collins 2014).

As also applies to work with the adult PCP model, it will be important to add antipsychotic treatment to testing where efficacy of add-on pharmacotherapy for cognition and negative symptoms in schizophrenia is to be evaluated, and mechanisms of antipsychotic-induced brain dysfunction are of interest. See (Cotel et al. 2015) for a recent in vivo investigation of clinically relevant doses of antipsychotics on activated microglia, a marker of neuroinflammation. Grace and colleagues have begun to explore the impact of antipsychotic treatment on efficacy of pharmacological agents to restore function using the MAM neurodevelopmental model (Gill et al. 2014).

8 Conclusion

When comparing methodological differences throughout the studies, it is important to note that experimental conditions of PCP at a dose of 10 mg/kg given s.c. on PND 7, 9, 11 in Sprague-Dawley rats have been most frequently used in the experiments described in this review (see Table 1). Most studies dose animals on PND 7, 9 and 11, some studies use Lister Hooded rats, and 20 mg/kg sc and

10 mg/kg i.p. have also been used. A dose of 10 mg/kg s.c. in male Wistar rats has been used exclusively for social novelty studies.

Taken together, the research findings summarised above demonstrate that, despite the methodological and test differences observed between studies (e.g. type of test, type of NMDAR antagonist, timing of the treatment, age at testing, strain and sex of rodent), deficits in working memory, prefrontal cortex-dependent executive function, selective attention and neuropathological changes such as reduced PV cell density have been reported in the majority of these studies, while deficits in PPI and effects on locomotor activity appear less reliable. Both sexes are generally included in the studies described in our review, which is a very important step forward in preclinical studies, hitherto lacking. Again, we encourage researchers to follow this lead in their development of other animal models of all disorders, not just for neuropsychiatry. Finally, as the postnatal PCP model is less well characterised than the adult PCP model, and so far its neurodevelopmental aspects have not been fully realised, it is difficult to assess whether it really confers any advantages over the adult model. Further, more neurodevelopmentally focused studies using this model will be required to fully assess its value in this field. It is our prediction that, in combination with other timed interventions producing two and three hit models, it could provide a very useful addition to existing models.

Disclosures Jo Neill has received expenses to attend conferences and fees for lecturing and consultancy work (including attending advisory boards) from the manufacturers of various neuropsychiatric drugs. B Grayson, C Piercy, G Podda and Jo Neill are employees of the University of Manchester. During the last 3 years, AM has received research contract support from Astra-Zeneca, Bristol-Myers-Squibb and Forest Laboratories and honoraria from AbbVie, Germany. SAB and AM are employees of the University of California San Diego.

References

Abdul-Monim Z, Neill JC, Reynolds GP (2007) Sub-chronic psychotomimetic phencyclidine induces deficits in reversal learning and alterations in parvalbumin-immunoreactive expression in the rat. J Psychopharmacol 21(2):198–205

Amitai N et al (2012) Repeated phencyclidine administration alters glutamate release and decreases GABA markers in the prefrontal cortex of rats. Neuropharmacology 62(3): 1422–1431

Anastasio NC, Johnson KM (2008) Atypical anti-schizophrenic drugs prevent changes in cortical *N*-methyl-*D*-aspartate receptors and behavior following sub-chronic phencyclidine administration in developing rat pups. Pharmacol Biochem Behav 90(4):569–577

Andersen JD, Pouzet B (2004) Spatial memory deficits induced by perinatal treatment of rats with PCP and reversal effect of D-serine. Neuropsychopharmacology 29(6):1080–1090

Andersen SL et al (2000) Dopamine receptor pruning in prefrontal cortex during the periadolescent period in rats. Synapse 37(2):167–169

Arnold SE, Talbot K, Hahn CG (2005) Neurodevelopment, neuroplasticity, and new genes for schizophrenia. Prog Brain Res 147:319–345

Baddeley A (1992) Working memory. Science 255(5044):556–559

Barnes SA et al. (2014) Impaired limbic cortico-striatal structure and sustained visual attention in a rodent model of schizophrenia

Barnes SA et al. (2015) Disruption of mGluR5 in parvalbumin-positive interneurons induces core features of neurodevelopmental disorders. Mol Psychiatry doi:10.1038/mp.2015.113

Bartos M, Vida I, Jonas P (2007) Synaptic mechanisms of synchronized gamma oscillations in inhibitory interneuron networks. Nat Rev Neurosci 8(1):45–56

Behrens MM, Sejnowski TJ (2009) Does schizophrenia arise from oxidative dysregulation of parvalbumin-interneurons in the developing cortex? Neuropharmacology 57(3):193–200

Behrens MM et al (2007) Ketamine-induced loss of phenotype of fast-spiking interneurons is mediated by NADPH-oxidase. Sci Signal 318(5856):1645

Belforte JE et al (2010) Postnatal NMDA receptor ablation in corticolimbic interneurons confers schizophrenia-like phenotypes. Nat Neurosci 13(1):76–83

Ben-Ari Y et al (1997) GABAA, NMDA and AMPA receptors: a developmentally regulated `ménage à trois'. Trends Neurosci 20(11):523–529

Benes FM, Berretta S (2001) GABAergic interneurons: implications for understanding schizophrenia and bipolar disorder. Neuropsychopharmacology 25(1):1–27

Boctor SY, Ferguson SA (2009) Neonatal NMDA receptor antagonist treatments have no effects on prepulse inhibition of postnatal day 25 Sprague-Dawley rats. Neurotoxicology 30(1):151–154

Boctor SY, Ferguson SA (2010) Altered adult locomotor activity in rats from phencyclidine treatment on postnatal days 7, 9 and 11, but not repeated ketamine treatment on postnatal day 7. NeuroToxicology 31(1):42–54

Broberg BV et al (2008) Evaluation of a neurodevelopmental model of schizophrenia–early postnatal PCP treatment in attentional set-shifting. Behav Brain Res 190(1):160–163

Broberg BV et al (2009) Reversal of cognitive deficits by an ampakine (CX516) and sertindole in two animal models of schizophrenia–sub-chronic and early postnatal PCP treatment in attentional set-shifting. Psychopharmacology 206(4):631–640

Cambon K et al (2004) A synthetic neural cell adhesion molecule mimetic peptide promotes synaptogenesis, enhances presynaptic function, and facilitates memory consolidation. J Neurosci 24(17):4197–4204

Cilia J et al (2001) Long-term evaluation of isolation-rearing induced prepulse inhibition deficits in rats. Psychopharmacology 156(2–3):327–337

Cilia J et al (2005) Reversal of isolation-rearing-induced PPI deficits by an alpha7 nicotinic receptor agonist. Psychopharmacology 182(2):214–229

Clancy B et al (2007) Extrapolating brain development from experimental species to humans. Neurotoxicology 28(5):931–937

Clayton JA, Collins FS (2014) Policy: NIH to balance sex in cell and animal studies. Nature 509 (7500):282–283

Clifton NE et al (2013) Enhancement of social novelty discrimination by positive allosteric modulators at metabotropic glutamate 5 receptors: adolescent administration prevents adult-onset deficits induced by neonatal treatment with phencyclidine. Psychopharmacology 225(3):579–594

Cline HT (2001) Dendritic arbor development and synaptogenesis. Curr Opin Neurobiol 11 (1):118–126

Cline H (2005) Synaptogenesis: a balancing act between excitation and inhibition. Curr Biol 15: R203–R205

Cotel MC et al. (2015) Microglial activation in the rat brain following chronic antipsychotic treatment at clinically relevant doses. Eur Neuropsychopharmacol

Dalley JW et al (2002) Specific abnormalities in serotonin release in the prefrontal cortex of isolation-reared rats measured during behavioural performance of a task assessing visuospatial attention and impulsivity. Psychopharmacology 164(3):329–340

Dawson N et al (2014) Sustained NMDA receptor hypofunction induces compromised neural systems integration and schizophrenia-like alterations in functional brain networks. Cereb Cortex 24(2):452–464

Dodell-Feder D, Tully LM, Hooker CI (2015) Social impairment in schizophrenia: new approaches for treating a persistent problem. Curr Opin Psychiatry 28(3):236–242

Doischer D et al (2008) Postnatal differentiation of basket cells from slow to fast signaling devices. J Neurosci 28(48):12956–12968

Egerton A et al (2008) Subchronic and chronic PCP treatment produces temporally distinct deficits in attentional set shifting and prepulse inhibition in rats. Psychopharmacology 198(1):37–49

Emiliani FE, Sedlak TW, Sawa A (2014) Oxidative stress and schizophrenia: recent breakthroughs from an old story. Curr Opin Psychiatry 27(3):185–190

Emson PC et al (1979) Development of vasoactive intestinal polypeptide (VIP) containing neurones in the rat brain. Brain Res 177(3):437–444

Eskildsen J et al (2014) Discovery and optimization of Lu AF58801, a novel, selective and brain penetrant positive allosteric modulator of alpha-7 nicotinic acetylcholine receptors: attenuation of subchronic phencyclidine (PCP)-induced cognitive deficits in rats following oral administration. Bioorg Med Chem Lett 24(1):288–293

Farber NB et al (1995) Age-specific neurotoxicity in the rat associated with NMDA receptor blockade: potential relevance to schizophrenia? Biol Psychiatry 38(12):788–796

Farber NB et al (2002) Antiepileptic drugs and agents that inhibit voltage-gated sodium channels prevent NMDA antagonist neurotoxicity. Mol Psychiatry 7(7):726–733

Fatemi SH, Folsom TD (2009) The neurodevelopmental hypothesis of schizophrenia, revisited. Schizophr Bull 35(3):528–548

Fett AK et al (2011) The relationship between neurocognition and social cognition with functional outcomes in schizophrenia: a meta-analysis. Neurosci Biobehav Rev 35(3):573–588

Forrest AD, Coto CA, Siegel SJ (2014) Animal models of psychosis: current state and future directions. Curr Behav Neurosci Rep 1(2):100–116

Gaskin PL, Alexander SP, Fone KC (2014) Neonatal phencyclidine administration and post-weaning social isolation as a dual-hit model of 'schizophrenia-like' behaviour in the rat. Psychopharmacology 231(12):2533–2545

Geyer MA et al (1993) Isolation rearing of rats produces a deficit in prepulse inhibition of acoustic startle similar to that in schizophrenia. Biol Psychiatry 34(6):361–372

Geyer MA et al (2001) Pharmacological studies of prepulse inhibition models of sensorimotor gating deficits in schizophrenia: a decade in review. Psychopharmacology 156(2–3):117–154

Gill KM et al. (2014) Prior antipsychotic drug treatment prevents response to novel antipsychotic agent in the methylazoxymethanol acetate model of schizophrenia. Schizophr Bull 40(2):341–350

Glantz LA, Lewis DA (2000) Decreased dendritic spine density on prefrontal cortical pyramidal neurons in schizophrenia. Arch Gen Psychiatry 57:65–73

Goetghebeur PJD, Dias R (2014) The attentional set-shifting test paradigm in rats for the screening of novel pro-cognitive compounds with relevance for cognitive deficits in schizophrenia. Curr Pharm Des 20(31):5060–5068

Grayson B et al (2015) Assessment of disease-related cognitive impairments using the novel object recognition (NOR) task in rodents. Behav Brain Res 285:176–193

Griebel G et al (2012) SAR110894, a potent histamine H(3)-receptor antagonist, displays procognitive effects in rodents. Pharmacol Biochem Behav 102(2):203–214

Haberny KA et al (2002) Ontogeny of the N-methyl-D-aspartate (NMDA) receptor system and susceptibility to neurotoxicity. Toxicol Sci 68(1):9–17

Harich S, Gross G, Bespalov A (2007) Stimulation of the metabotropic glutamate 2/3 receptor attenuates social novelty discrimination deficits induced by neonatal phencyclidine treatment. Psychopharmacology 192(4):511–519

Harrison PJ (1997) Schizophrenia: a disorder of neurodevelopment? Curr Opin Neurobiol 7(2):285–289

Harrison PJ, Weinberger DR (2005) Schizophrenia genes, gene expression, and neuropathology: on the matter of their convergence. Mol Psychiatry 10(1):40–68; image 5

Hida H et al (2014) Combination of neonatal PolyI: C and adolescent phencyclidine treatments is required to induce behavioral abnormalities with overexpression of GLAST in adult mice. Behav Brain Res 258:34–42

Hoftman GD, Lewis DA (2011) Postnatal developmental trajectories of neural circuits in the primate prefrontal cortex: identifying sensitive periods for vulnerability to schizophrenia. Schizophr Bull 37(3):493–503

Homayoun H, Moghaddam B (2007) NMDA receptor hypofunction produces opposite effects on prefrontal cortex interneurons and pyramidal neurons. J Neurosci 27(43):11496–11500

Huang ZJ (2009) Activity-dependent development of inhibitory synapses and innervation pattern: role of GABA signalling and beyond. J Physiol 587(Pt 9):1881–1888

Jenkins TA, Harte MK, Reynolds GP (2010) Effect of subchronic phencyclidine administration on sucrose preference and hippocampal parvalbumin immunoreactivity in the rat. Neurosci Lett 471(3):144–147

Jones GH, Marsden CA, Robbins TW (1990) Increased sensitivity to amphetamine and reward-related stimuli following social isolation in rats: possible disruption of dopamine-dependent mechanisms of the nucleus accumbens. Psychopharmacology 102(3):364–372

Kaalund SS et al (2013) Differential expression of parvalbumin in neonatal phencyclidine-treated rats and socially isolated rats. J Neurochem 124(4):548–557

Keefe RS (2007) Cognitive deficits in patients with schizophrenia: effects and treatment. J Clin Psychiatry 68(Suppl 14):8–13

Kjaerby C et al (2013) Repeated potentiation of the metabotropic glutamate receptor 5 and the alpha 7 nicotinic acetylcholine receptor modulates behavioural and GABAergic deficits induced by early postnatal phencyclidine (PCP) treatment. Neuropharmacology 72:157–168

Kjaerby C et al (2014) Impaired GABAergic inhibition in the prefrontal cortex of early postnatal phencyclidine (PCP)-treated rats. Cereb Cortex 24(9):2522–2532

Knuesel I et al (2014) Maternal immune activation and abnormal brain development across CNS disorders. Nat Rev Neurol 10(11):643–660

Le Magueresse C, Monyer H (2013) GABAergic interneurons shape the functional maturation of the cortex. Neuron 77(3):388–405

Lee J, Park S (2005) Working memory impairments in schizophrenia: a meta-analysis. J Abnorm Psychol 114(4):599–611

Lee EY et al (1998) Postnatal development of somatostatin- andneuropeptide y-immunoreactive neurons in rat cerebralcortex: a double-labeling immunohistochemical study. Int J Dev Neurosci 16(1):63–72

Lett TA et al (2014) Treating working memory deficits in schizophrenia: a review of the neurobiology. Biol Psychiatry 75(5):361–370

Lewis DA, Levitt P (2002) Schizophrenia as a disorder of neurodevelopment. Annu Rev Neurosci 25:409–432

Lewis DA et al (2012) Cortical parvalbumin interneurons and cognitive dysfunction in schizophrenia. Trends Neurosci 35(1):57–67

Li N, Wu X, Li L (2007) Chronic administration of clozapine alleviates reversal-learning impairment in isolation-reared rats. Behav Pharmacol 18(2):135–145

McLean S et al (2010) Isolation rearing impairs novel object recognition and attentional set shifting performance in female rats. J Psychopharmacol 24(1):57–63

Mouri A et al (2007) Phencyclidine animal models of schizophrenia: approaches from abnormality of glutamatergic neurotransmission and neurodevelopment. Neurochem Int 51(2–4):173–184

Martinez ZA et al (1999) Effects of sustained phencyclidine exposure on sensorimotor gating of startle in rats. Neuropsychopharmacology 21(1):28–39

Meltzer HY et al (2013) Translating the N-methyl-D-aspartate receptor antagonist model of schizophrenia to treatments for cognitive impairment in schizophrenia. Int J Neuropsychopharmacol 16(10):2181–2194

Nabeshima T et al (1984) Role of sex hormones in sex-dependent differences in phencyclidine-induced stereotyped behaviors in rats. Eur J Pharmacol 105(3–4):197–206

Nakatani-Pawlak A et al (2009) Neonatal phencyclidine treatment in mice induces behavioral, histological and neurochemical abnormalities in adulthood. Biol Pharm Bull 32(9):1576–1583

Neill JC et al (2010) Animal models of cognitive dysfunction and negative symptoms of schizophrenia: focus on NMDA receptor antagonism. Pharmacol Ther 128(3):419–432

Neill JC, Kulkarni J (2011) Biological basis of sex differences in psychopharmacology. Preface Curr Top Behav Neurosci 8:5–7

Neill JC et al (2014) Acute and chronic effects of NMDA receptor antagonists in rodents, relevance to negative symptoms of schizophrenia: a translational link to humans. Eur Neuropsychopharmacol 24(5):822–835

Okaty BW et al (2009) Transcriptional and electrophysiological maturation of neocortical fast-spiking GABAergic interneurons. J Neurosci 29(21):7040–7052

Plitman E et al (2014) Glutamate-mediated excitotoxicity in schizophrenia: a review. Eur Neuropsychopharmacol 24(10):1591–1605

Prendergast BJ, Onishi KG, Zucker I (2014) Female mice liberated for inclusion in neuroscience and biomedical research. Neurosci Biobehav Rev 40:1–5

Powell SB, Sejnowski TJ, Behrens MM (2012) Behavioral and neurochemical consequences of cortical oxidative stress on parvalbumin-interneuron maturation in rodent models of schizophrenia. Neuropharmacology 62(3):1322–1331

Radonjić N et al (2013) Perinatal phencyclidine administration decreases the density of cortical interneurons and increases the expression of neuregulin-1. Psychopharmacology 227(4): 673–683

Rapoport JL, Giedd JN, Gogtay N (2012) Neurodevelopmental model of schizophrenia: update 2012. Mol Psychiatry 17(12):1228–1238

Rasmussen BA et al (2007) Long-term effects of developmental PCP administration on sensorimotor gating in male and female rats. Psychopharmacology 190(1):43–49

Redrobe JP et al (2012) Negative modulation of GABAA alpha5 receptors by RO4938581 attenuates discrete sub-chronic and early postnatal phencyclidine (PCP)-induced cognitive deficits in rats. Psychopharmacology 221(3):451–468

Reynolds GP et al (2004) Calcium binding protein markers of GABA deficits in schizophrenia–postmortem studies and animal models. Neurotox Res 6(1):57–61

Rudy B et al (2011) Three groups of interneurons account for nearly 100 % of neocortical GABAergic neurons. Dev Neurobiol 71(1):45–61

Schizophrenia Working Group of the Psychiatric Genomics Consortium (2014) Biological insights from 108 schizophrenia-associated genetic loci. Nature 511(7510):421–427

Schrijver NC, Wurbel H (2001) Early social deprivation disrupts attentional, but not affective, shifts in rats. Behav Neurosci 115(2):437–442

Secher T et al (2009) Effect of an NCAM mimetic peptide FGL on impairment in spatial learning and memory after neonatal phencyclidine treatment in rats. Behav Brain Res 199(2):288–297

Sircar R (2003) Postnatal phencyclidine-induced deficit in adult water maze performance is associated with N-methyl-D-aspartate receptor upregulation. Int J Dev Neurosci 21(3):159–167

Sircar R, Soliman KF (2003) Effects of postnatal PCP treatment on locomotor behavior and striatal D2 receptor. Pharmacol Biochem Behav 74(4):943–952

Smith JK, Neill JC, Costall B (1997) Post-weaning housing conditions influence the behavioural effects of cocaine and d-amphetamine. Psychopharmacology 131(1):23–33

Sweet RA et al (2009) Reduced dendritic spine density in auditory cortex of subjects with schizophrenia. Neuropsychopharmacology 34(2):374–389

Tejedor-Real P et al (2007) Neonatal handling prevents the effects of phencyclidine in an animal model of negative symptoms of schizophrenia. Biol Psychiatry 61(7):865–872

Terranova JP et al (2005) SSR181507, a dopamine D(2) receptor antagonist and 5-HT(1A) receptor agonist, alleviates disturbances of novelty discrimination in a social context in rats, a putative model of selective attention deficit. Psychopharmacology 181(1):134–144

Uhlhaas PJ, Singer W (2010) Abnormal neural oscillations and synchrony in schizophrenia. Nat Rev Neurosci 11:100–113

Volk DW et al (2012) Deficits in transcriptional regulators of cortical parvalbumin neurons in schizophrenia. Am J Psychiatry 169(10):1082–1091

Volman V, Behrens MM, Sejnowski TJ (2011) Downregulation of parvalbumin at cortical GABA synapses reduces network gamma oscillatory activity. J Neurosci 31(49):18137–18148

Viberg H et al (2008) Neonatal ketamine exposure results in changes in biochemical substrates of neuronal growth and synaptogenesis, and alters adult behavior irreversibly. Toxicology 249(2–3):153–159

Wang C et al (2004) Blockade of N-methyl-D-aspartate receptors by phencyclidine causes the loss of corticostriatal neurons. Neuroscience 125(2):473–483

Wang C et al (2001) Long-term behavioral and neurodegenerative effects of perinatal phencyclidine administration: implications for schizophrenia. Neuroscience 107(4):535–550

Wang CZ et al (2008) Postnatal phencyclidine administration selectively reduces adult cortical parvalbumin-containing interneurons. Neuropsychopharmacology 33(10):2442–2455

Weickert CS et al (2013) Molecular evidence of N-methyl-D-aspartate receptor hypofunction in schizophrenia. Mol Psychiatry 18(11):1185–1192

Weiss IC et al (2004) Effect of social isolation on stress-related behavioural and neuroendocrine state in the rat. Behav Brain Res 152(2):279–295

White IM et al (2009) Brief exposure to methamphetamine (METH) and phencyclidine (PCP) during late development leads to long-term learning deficits in rats. Brain Res 1266:72–86

Wiley JL et al (2003) Pharmacological challenge reveals long-term effects of perinatal phencyclidine on delayed spatial alternation in rats. Prog Neuropsychopharmacol Biol Psychiatry 27(5):867–873

Zhang ZJ, Reynolds GP (2002) A selective decrease in the relative density of parvalbumin-immunoreactive neurons in the hippocampus in schizophrenia. Schizophr Res 55(1–2):1–10

Zhang R et al (2012) Myelination deficit in a phencyclidine-induced neurodevelopmental model of schizophrenia. Brain Res 1469:136–143

Pathological Implications of Oxidative Stress in Patients and Animal Models with Schizophrenia: The Role of Epidermal Growth Factor Receptor Signaling

Tadasato Nagano, Makoto Mizuno, Keisuke Morita
and Hiroyuki Nawa

Abstract Proinflammatory cytokines perturb brain development and neurotransmission and are implicated in various psychiatric diseases, such as schizophrenia and depression. These cytokines often induce the production of reactive oxygen species (ROS) and regulate not only cell survival and proliferation but also inflammatory process and neurotransmission. Under physiological conditions, ROS are moderately produced in mitochondria but are rapidly scavenged by reducing agents in cells. However, brain injury, ischemia, infection, or seizure-like neural activities induce inflammatory cytokines and trigger the production of excessive amounts of ROS, leading to abnormal brain functions and psychiatric symptoms. Protein phosphatases, which are involved in the basal silencing of cytokine receptor activation, are the major targets of ROS. Consistent with this, several ROS scavengers, such as polyphenols and unsaturated fatty acids, attenuate both cytokine signaling and psychiatric abnormalities. In this review, we list the inducers, producers, targets, and scavengers of ROS in the brain and discuss the interaction between ROS and cytokine signaling implicated in schizophrenia and its animal models. In particular, we present an animal model of schizophrenia established by perinatal exposure to epidermal growth factor and illustrate the pathological role of ROS and antipsychotic actions of ROS scavengers, such as emodin and edaravone.

T. Nagano
Faculty of Human Life Studies, University of Niigata Prefecture,
471 Ebigase, Higashi-ku, Niigata 950-8680, Japan

M. Mizuno
Aichi Human Service Center, Institute for Developmental Research,
Kasugai, Aichi 480-0392, Japan

K. Morita · H. Nawa (✉)
Department of Molecular Biology, Brain Research Institute, Niigata University,
Asahimachi-Dori 1-757, Niigata 951-8585, Japan
e-mail: hnawa@bri.niigata-u.ac.jp

© Springer International Publishing Switzerland 2015
Curr Topics Behav Neurosci (2016) 29: 429–446
DOI 10.1007/7854_2015_399
Published Online: 17 October 2015

Keywords Psychosis · Epidermal growth factor · Reactive oxygen species · Radical scavenger · Edaravone · Emodin · Trolox

Contents

1 Oxidative Stress Is Implicated in the Pathogenesis of Schizophrenia 430
2 Pharmacologic Actions of Antioxidants in Patients with and Animal Models
   for Schizophrenia... 432
3 Strong Interactions Between ROS and Cytokine Signaling... 433
4 Establishment of Animal Models of Schizophrenia by Exposure to EGF
   and NRG-1... 436
5 Pharmacologic Actions of ROS Scavengers in the EGF-Exposure-Generated Model.... 439
6 Conclusion 441
References... 441

1 Oxidative Stress Is Implicated in the Pathogenesis of Schizophrenia

Oxygen radicals are molecules essential for energy production, metabolic processes, and germicide in our body; however, they are also toxic to cellular components, such as membrane lipids and DNA (Brewer et al. 2015). Therefore, our tissues and cells contain a large variety of reducing agents (i.e., scavengers) as well as reactive oxygen species (ROS)-degrading enzymes (Chan 2001; Table 1). This is especially true in the brain, which produces high energy accompanying ROS generation and is enriched with these antioxidants to protect the tissue from oxidative stress. Following brain injury, ischemia, infection, or seizures, NADPH oxidase (NOX) is activated in microglia and GABAergic neurons and produces large amounts of ROS in the brain (Kaur et al. 2015). When ROS is not scavenged immediately, they act not only on cell components (membrane lipids and DNA) but also on the molecules that carry ROS (redox) sensors, such as the protein tyrosine phosphatase PTP1B, nuclear factor kappa B (NF-κB), potassium channels, hypoxia-inducible factors, or N-methyl-D-aspartic acid (NMDA) receptor. For example, ROS acts on free cysteine residues of the NMDA receptor, altering its channel conformation to decrease NMDA-mediated cation movements, presumably attenuating neuronal cell death (Sanchez et al. 2000; Dukoff et al. 2014). This anti-ROS process functions as a self-defense system in the brain; however, when ROS generation is prolonged, the reduction in NMDA receptor channel activity can perturb normal neurotransmission and brain function. Chronic suppression of NMDA receptor activity is known to result in abnormal cognition and behaviors relevant to schizophrenia (Mohn et al. 1999; Takahashi et al. 2006).

Several reports on ROS generation show that contents of reducing agents are decreased and oxidizing agents are increased in patients and animal models with schizophrenia (Yao and Keshavan 2011; Bokkon and Antal 2011). For example,

Table 1 Oxidative and antioxidative agents and enzymes

Reactive oxygen species generators (enzymes)
Cytochromes
NAD(P)H oxidase
Cyclooxygenase
Lipooxygenase
Monoamine oxidase, etc.
Oxidative agents
O_2^-: Superoxide anion
$\cdot OH$: Hydroxyl radical
H_2O_2: Hydrogen peroxide
$ONOO^-$: Peroxynitrite, etc.
Antioxidative enzymes
Superoxide dismutase
Catalase, etc.
Antioxidants
Thioredoxin
Glutathione
Vitamins C, A, E
N-acetyl cysteine
Cysteine, etc.

the products of reactions between ROS and membrane lipids (i.e., peroxidized lipids) are increased in patients in the acute phase of schizophrenia (Mahadik and Scheffer 1996; Ben Othmen et al. 2008). ROS also reacts with the guanosine residue of DNA in nuclei, converting it to 8-hydroxy-2-deoxyguanosine and leading to DNA fragmentation. The brain contents of 8-hydroxy-2-deoxyguanosine and fragmented DNA are elevated in patients with schizophrenia (Nishioka and Arnold 2004; Buttner et al. 2007). Blood levels of pentosidine, an advanced glycation end product, are also elevated as a result of sugar oxidation and are higher in patients with schizophrenia (Arai et al. 2010, 2014).

Conversely, the total amount of ROS scavengers (TAR, the total antioxidative response) is reduced in patients with schizophrenia (Ustundag et al. 2006). Blood concentrations of vitamins C and E are decreased in patients with schizophrenia (Suboticanec et al. 1990; McCreadie et al. 1995). In addition to these small molecules, the levels of enzymes that produce or degrade ROS are also altered. The activities of superoxide dismutase (SOD) and catalase are downregulated in the erythrocytes of patients with schizophrenia (Ben Othmen et al. 2008). Glutamate cysteine ligase is also decreased in patients with schizophrenia (Gysin et al. 2007). These findings suggest that the magnitude of ROS generation appears to be upregulated in patients with schizophrenia.

The pathologic interactions between ROS and psychosis have been investigated in animal models for drug-induced psychosis to monitor ROS generation following exposure to psychostimulants such as amphetamine, phencyclidine, ketamine, or the NMDA receptor blocker dizocilpine (MK-801). Behrens et al. (2007) showed that a ketamine-induced abnormality in GABAergic neurotransmission and behavior involves ROS generation by NOX. Zuo et al. (2007) also observed increases in hydroxyl radicals in the brain following MK-801 and ketamine challenges. It is established that psychostimulants upregulate ROS production in the brain and contribute to psychostimulant-driven behavioral impairments.

However, caution is necessary when relating these findings to schizophrenia. One problem is the disease specificity of ROS pathology (Cobb and Cole 2015). The pathologic implications of ROS are not limited to schizophrenia. ROS is also implicated in Alzheimer's disease and Parkinson's disease (Lovell and Markesbery 2007; Nakabeppu et al. 2007). Neural activity itself involves higher energy metabolism and results in higher ROS production, as evident in patients with epilepsy (Puttachary et al. 2015). Medication for schizophrenia also enhances the production of ROS; the metabolism of antipsychotic drugs necessitates cytochrome P450 recruitment and also promotes dopamine release and monoamine oxidase activation, all of which drive ROS generation (Martins et al. 2008; Reinke et al. 2004; Table 1). In this context, the pathologic and pharmacologic roles of ROS generation in patients with schizophrenia are controversial and remain to be characterized further (Naviaux 2012).

2 Pharmacologic Actions of Antioxidants in Patients with and Animal Models for Schizophrenia

In contrast to arguments supporting the pathologic role of ROS in patients with schizophrenia, many reports more consistently describe the antipsychotic actions of reducing agents. A variety of reducing agents and radical scavengers, including ω-3 fatty acids, *N*-acetyl cysteine, minocycline, and vitamins C and E, have been administered to patients, and their therapeutic effects evaluated. Based on this theory, Sivrioglu et al. (2007) supplemented the food of patients with schizophrenia with ω-3 fatty acids and reported a reduction in antipsychotic side effects and an improvement in psychotic symptoms. Farokhnia et al. (2013) administered a more potent antioxidant, *N*-acetyl cysteine, to patients with schizophrenia and observed its effectiveness on negative symptoms. In combination with atypical antipsychotics, minocycline was administered to patients in the acute phase, resulting in an improvement in cognition and negative symptoms (Miyaoka et al. 2007; Levkovitz et al. 2010).

The antipsychotic effects of reducing agents have also been tested in several animal models. Both MK-801 and methamphetamine induce hyperlocomotion and social interaction deficits and cause neurodegeneration in various parts of the brain (Ozyurt et al. 2007). The administration of vitamin E into MK-801-challenged rats decreased cell death in the cingulated cortex (Zhang et al. 2006; Willis and Ray 2007), whereas the administration of the radical scavenger edaravone inhibited cell death in striatal neurons (Kawasaki et al. 2006). In addition to cell death, the behavioral deficits induced by phencyclidine are also ameliorated by the scavenger of a SOD mimetic (Wang et al. 2003). Neonatal hippocampal lesions result in behavioral impairments at the postpubertal stage and serve as an animal model of schizophrenia (Lipska et al. 1993). The antipsychotic effects of *N*-acetyl cysteine have been verified in this animal model (Cabungcal et al. 2014). Despite discrepancies in the arguments supporting an etiologic or pathologic role of ROS in schizophrenia, almost all antioxidative agents exhibit beneficial effects on cognitive abnormalities relevant to schizophrenia, although the cellular and molecular mechanisms underlying these pharmacologic phenomena remain unclear.

3 Strong Interactions Between ROS and Cytokine Signaling

ROS signaling is known to link to cytokine production and cytokine receptor signaling (Landskron et al. 2014). Proinflammatory cytokines and growth factors activate NOX and produce ROS. Brain neurons and microglia release ROS in response to cytokines. Conversely, generated ROS stimulate cytokine production by activating the redox sensor NF-κB (Lian and Zheng 2009) or accelerate cytokine receptor signaling by inhibiting the redox sensor phosphatase. Notably, ROS alone induces the phosphorylation and activation of cytokine receptors in the absence of cytokine ligands such as epidermal growth factor (EGF). In this review, we discuss the interaction between ROS and EGF signaling most extensively demonstrated in the field of carcinogenesis (Chiarugi et al. 2003; Goldkorn et al. 2005).

EGF signaling is highly implicated and serves as a drug target in tumor biology, because this cytokine potently produces ROS, which are known to be DNA mutagens. An EGF challenge to cancer cells and cultured neurons elevates intracellular ROS concentrations (Bae et al. 1997; Cha et al. 2000). Table 2 shows the potency of the cytokines that induce ROS production in cultured neocortical

Table 2 Comparison of cytokine-induced reactive oxygen species production

DHE oxidation levels (% control)	
Control	100 ± 6.3
EGF	132 ± 5.4
Interleukin-1	125 ± 5.8
Interleukin-6	130 ± 6.1
Neuregulin-1	130 ± 8
BDNF	142 ± 8
Ionomycin (positive control)	153 ± 8.7

Cultured cortical neurons were exposed to various cytokines in the presence of dihydroethidium. Fluorescence intensity (560 nm) was measured in randomly selected 10 cells (n = 8)

DHE dihydroethidium; *EGF* epidermal growth factor

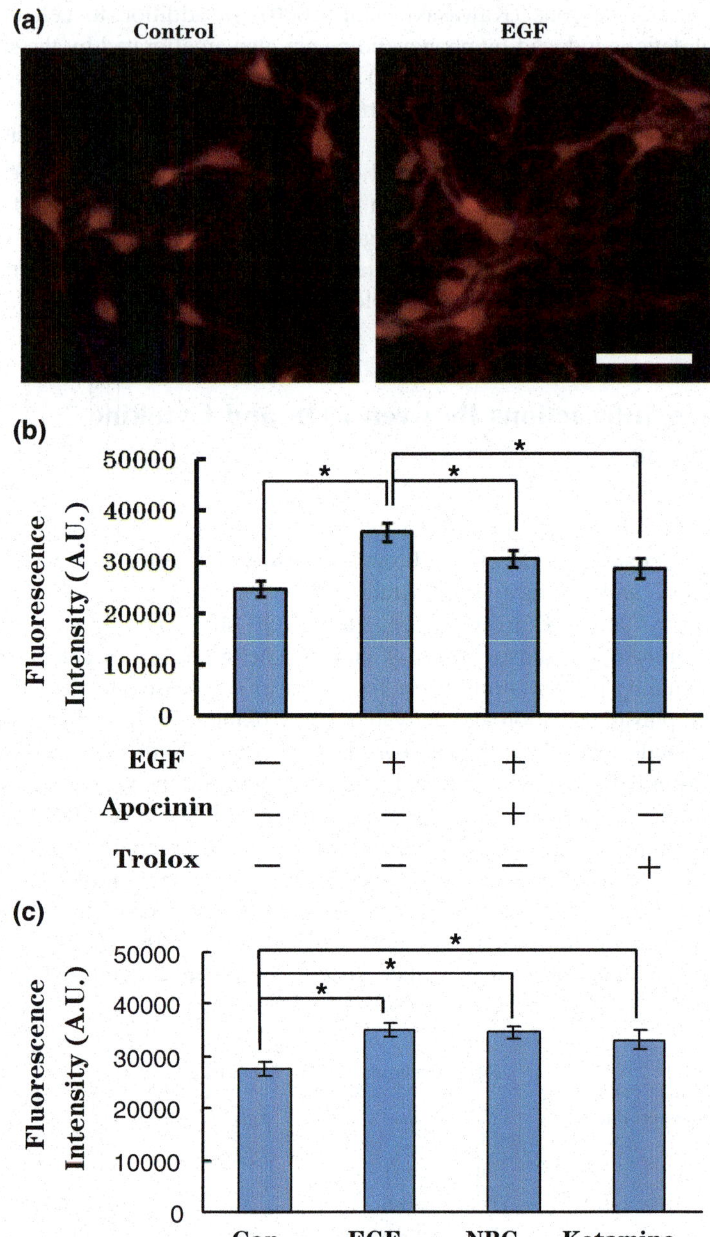

◄ Fig. 1 The effects of EGF and antioxidants on the levels of ROS in neocortical cultures. **a** Cultured cortical neurons (DIV7) were stimulated with or without 20 ng/ml EGF. ROS production was assessed by mitotracker, which sensed mitochondrial membrane potential. *Scale bar* 50 μm. **b** Cultured cortical neurons were treated with antioxidants (0.5 mM apocynin, 100 μM Trolox) in the absence or presence of 20 ng/ml EGF. **c** Cultured cortical neurons were stimulated with 20 ng/ml EGF, 20 ng/ml neuregulin-1, or 1 μM ketamine. The MitoTracker fluorescence intensity was measured in 10 cells per well using Image J (National Institutes of Health, Bethesda, MD, USA; arbitrary units, A.U.). $n = 8$, $*P < 0.05$, *EGF* epidermal growth factor

neurons. EGF oxidizes the ROS indicator dihydroethidium and changes the indicator to a fluorescent form. This EGF-triggered increase in ROS levels is blocked by the co-application of antioxidants (apocynin, Trolox; Fig. 1). These results indicate that EGF indeed induces ROS production in the brain.

In such inflammatory conditions, the main source of ROS is NOX, which is highly potentiated by EGF signals (Fan et al. 2005a, b; Chen et al. 2008). De Yulia et al. (2005) proposed that hydrogen peroxide is produced by the EGF receptor (EGFR, also known as ErbB1) itself. Conversely, ROS converts the thiol group of a cysteine residue to sulfuric acid in the core region of the tyrosine phosphatase PTP1B, inactivating this enzyme (Yip et al. 2010). Compared with many receptor-type protein tyrosine kinases, the EGFR undergoes a significant amount of basal autophosphorylation in the absence of its ligand and requires phosphatases for its silencing in a basal condition. Thus, the ROS-driven inactivation of PTP1B alone increases phosphorylation of the EGFR, resulting in the activation of EGF signaling without ligands (Lee et al. 1998; De Wit et al. 2001). Thus, hydrogen peroxide-oxidized low-density lipoprotein can also phosphorylate EGFR in the absence of EGF and evoke EGFR signaling (Suc et al. 1998). If acting on peripheral cells, hydrogen peroxide may promote cell proliferation and induce carcinogenesis. In this context, the EGFR signaling cascade appears to be one of the most crucial ROS targets.

Figure 2 summarizes the production of and interaction between EGF and ROS. The ligand EGF binds to EGFR and activates calcium signaling, in turn activating the ROS generators NOX, cyclooxygenase, and cytochromes in mitochondria. The ROS produced acts on phosphatases such as PTP1B and enhances and prolongs the EGF signal transduction. The enhanced phosphorylation of EGFR also hinders its internalization (Ravid et al. 2002). Additionally, ROS acts on the redox site of the transcription factor NF-κB and accelerates NF-κB-mediated gene expression, leading to the synthesis of mRNA for EGF and other cytokines. When ROS reacts with ion channels, neurotransmission in the brain is perturbed unless ROS are trapped by various scavenger molecules (glutathione, thioredoxin, SOD, catalase, etc.).

As shown in Table 2, the EGF homologue neuregulin-1 (NRG-1) triggers ROS production in the brain. In contrast to EGF, NRG-1 binds to the ErbB3 or ErbB4 receptor tyrosine kinases, but NRG-1 signal transduction also involves ROS-mediated autoregulation. However, compared with EGF, NRG-1-driven ROS production is more persistent (Goldsmit et al. 2001).

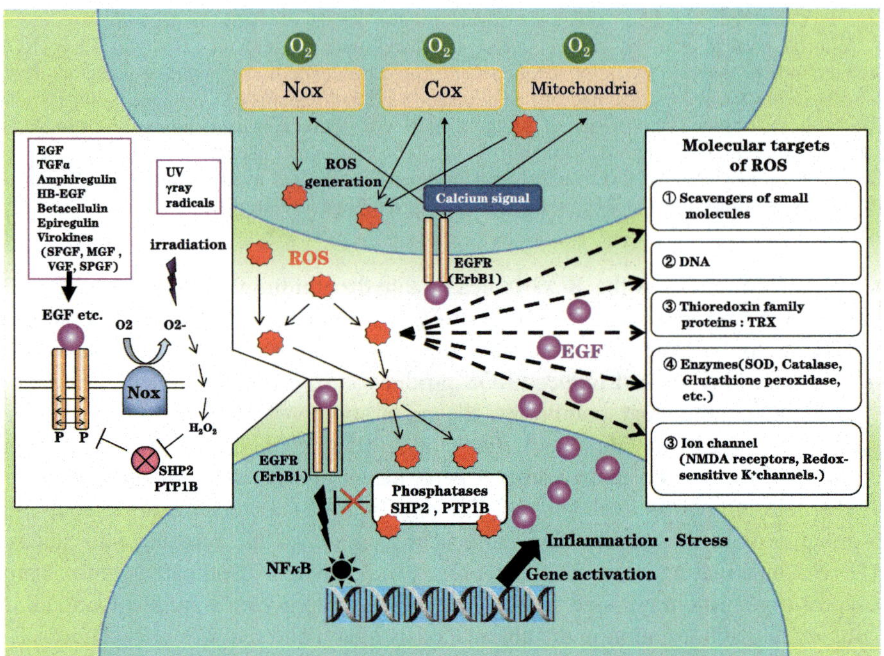

Fig. 2 Schematic diagram of the interaction between EGF and ROS. EGF and its homologues (transforming growth factor alpha, amphiregulin, heparin-binding EGF-like growth factor, betacellulin, epiregulin, neuregulins 1–6) as well as poxvirus virokines (SMDF, VGF, etc.) directly and indirectly interact with the EGFR (also known as ErbB1) to evoke calcium signaling. This activates NOX, cyclooxgenase (*COX*), and mitochondrial cytochrome c, which produce ROS. In turn, ROS act on and inhibit protein tyrosine phosphatases (SHP2, the protein tyrosine phosphatase PTP1B, etc.) and promote the phosphorylation and activation of EGFR, leading to the production of other cytokines and the further activation of NOX, COX, and mitochondrial cytochromes. *EGF* epidermal growth factor; *EGFR* EGF receptor; *NOX* NADPH oxidase; *ROS* reactive oxygen species

4 Establishment of Animal Models of Schizophrenia by Exposure to EGF and NRG-1

Maternal viral infection and obstetric complication are implicated as environmental risk factors for schizophrenia (Nawa and Takei 2006; Iwakura and Nawa 2013). EGF is known to be highly enriched in human amniotic fluids and contributes to obstetric complication (Varner et al. 1996). EGF derivatives are known to be encoded in viral genome of poxviruses and might be supplied to human fetus and neonates via their infection (Tzahar et al. 1998). Genetic linkage studies and neuropathologic investigations also suggest an association between schizophrenia and EGF and its derivative NRG-1 (Futamura et al. 2002; Anttila et al. 2004; Stefansson et al. 2004; Groenestege et al. 2007). On the premise of the neurodevelopmental

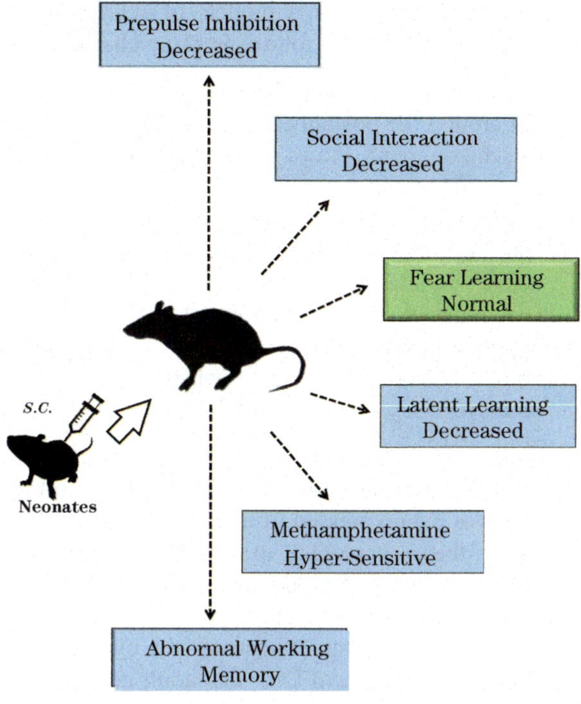

Fig. 3 Behavioral impairments in the animal model of schizophrenia established by perinatal exposure to EGF. *EGF* epidermal growth factor. Various behavioral abnormalities emerge at the post-pubertal stage

hypothesis for this illness, therefore, we analyzed animal models of schizophrenia established by perinatal exposure to EGF and its derivatives (Futamura et al. 2003; Tohmi et al. 2004; Watanabe et al. 2004; Tsuda et al. 2008; Kato et al. 2011). Treatment of neonatal rats and mice with EGF, NRG-1, and their homologues and paralogues (i.e., transforming growth factor alpha, epiregulin) all showed the behavioral deficits relevant to schizophrenia at the postpubertal stage (Fig. 3). Injected factors penetrate the immature blood–brain barrier and bind to brain neurons carrying EGFR or ErbB receptors, such as GABAergic and dopaminergic neurons (Abe et al. 2009; Namba et al. 2009). In addition to dopaminergic neurons, GABAergic neurons and glial cells also respond to these cytokines, but their responses gradually diminish after the cessation of cytokine treatment (Nagano et al. 2007; Abe et al. 2011). The most prominent and persistent influences of cytokine treatment are those on dopaminergic neurons, which cause the animals to display cognitive and behavioral impairments (Sotoyama et al. 2011, 2013). These behavioral impairments include abnormalities in sensorimotor gating (prepulse inhibition [PPI]), social interaction, exploratory movement, latent inhibition of learning, and sensitivity to psychostimulants (methamphetamine and/or MK-801), although the directions and magnitudes of these deficits are significantly altered by the genetic background of the animal species used (Mizuno et al. 2004; Tohmi et al. 2005). This observation agrees with the current theory that the onset of schizophrenia involves both genetic and environmental factors.

Although many animal models of schizophrenia have been established, our animal models have the following three characteristic features (Nawa et al. 2014):

- The emergence of behavioral abnormalities occurs not before or during puberty, but around the postpubertal stage. The EGF-exposure-generated monkey model of schizophrenia shows striking behavioral deficits from 5 years of age (Sakai et al. 2014).
- EGF- and NRG-1-exposure-generated models have no apparent deficits in learning. In the eight-arm radial maze test, context fear-learning paradigm, and active avoidance test, these models are indistinguishable from controls (Futamura et al. 2003; Kato et al. 2011).
- Atypical antipsychotics, such as risperidone and clozapine, are effective for the above behavioral deficits, whereas the pharmacologic action of typical antipsychotics (i.e., haloperidol) is limited (Futamura et al. 2003, Sotoyama et al. 2013).

We found similar behavioral abnormalities in transgenic mice in which EGF or NRG-1 was overexpressed from their transgenes (Kato et al. 2010; Eda et al. 2013). However, these approaches to cytokine administration obscure their cellular target (s). The behavioral abnormalities may therefore be ascribed not only to the dysfunction of midbrain dopaminergic neurons but also to that of other peripheral organs. To limit the target organ of the cytokines, we conducted intracerebroventricular administration of EGF into adult rats and verified the reproducibility of the behavioral impairments achieved by perinatal pre-exposure to EGF (Mizuno et al. 2007). This result suggests that peripherally administered EGF targets brain neurons in the neonatal EGF model for schizophrenia.

Subsequent analyses of these models revealed that NRG-1 mainly acts on dopaminergic neurons in the ventral tegmental area of the midbrain, whereas EGF interacts with those in the substantia nigra to promote dopamine synthesis and terminal arborization (Abe et al. 2009; Iwakura et al. 2011a, b). Neonatal mice treated with NRG-1 exhibit a hyperdopaminergic state in the prelimbic cortex (i.e., the medial prefrontal cortex) as adults (Kato et al. 2011). Similarly, neonatal rats treated with EGF display a hyperdopaminergic state in the globus pallidus (Sotoyama et al. 2011). Dopaminergic innervation and release in the globus pallidus are significantly elevated in this model. However, their PPI deficits and abnormal dopamine release are simultaneously normalized by the local administration of an antipsychotic drug to the globus pallidus (Sotoyama et al. 2011, 2013). The globus pallidus is enriched with dopamine D2 receptors and is the major target of antipsychotic drugs in the indirect pathway of the basal ganglia.

5 Pharmacologic Actions of ROS Scavengers in the EGF-Exposure-Generated Model

As mentioned earlier, EGF is a potent ROS producer, and knockdown of protein phosphatases by ROS in turn enhances phosphorylation-based signaling cascades emanating from EGFR. In agreement, Finch et al. (2006) found that catalase degrades hydrogen peroxide and reduces EGF signaling. To assess the direct interaction between ROS production and EGF signaling, we administered the three ROS scavengers (Trolox, edaravone, and 2,2,6,6-tetramethyl-4-piperidinol-N-oxyl [TEMPOL]) during EGF treatment in neonatal rats. The effects of these scavengers on the PPI deficits of EGF-treated rats were evaluated at the adult stage (Fig. 4). Co-administration of TEMPOL and edaravone with EGF resulted in attenuation of the PPI deficits of the EGF-exposure-generated model. This result suggests that ROS production during neonatal EGF treatment is required to generate the model.

We also examined the medication effects of ROS scavengers (emodin, Trolox, edaravone) on the behavioral impairments of the EGF- pretreatment model at the adult stage (Fig. 5). Emodin is a polyphenol ROS scavenger known as 3-methyl-1,6,8-trihydroxyanthraquinone and attenuates EGFR signaling (Mizuno et al. 2008). Edaravone is otherwise known as 3-methyl-1-phenyl-2-pyrazolin-5-one and is prescribed to patients who have suffered a stroke to mitigate the expansion of cerebral ischemic injuries. This drug is verified to reduce EGFR signaling in cancer cells (Suzuki et al. 2005). Subchronic treatment with emodin at

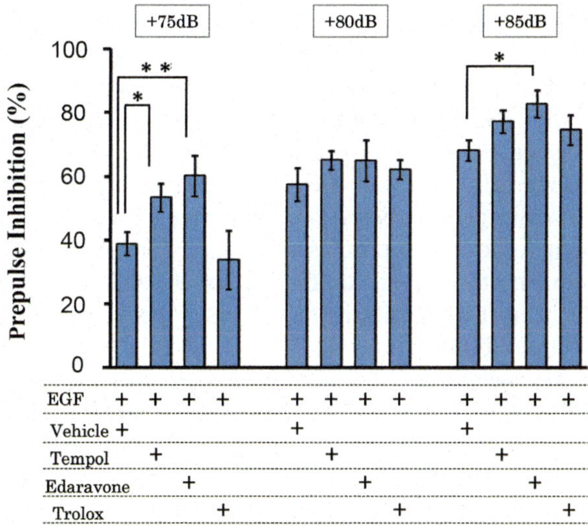

Fig. 4 Competition by antioxidants with neonatal EGF treatment. Epidermal growth factor (EGF; 0.875 μg/g, administered subcutaneously) or saline was administered daily on postnatal days 2–11 together with antioxidants (3 mg/kg TEMPOL, 10 mg/kg edaravone, and 10 mg/kg Trolox). Interference with the effects of EGF on prepulse inhibition scores was monitored at the adult stage. $n = 5$ in each group, *$P < 0.05$ and **$P < 0.01$, *EGF* epidermal growth factor

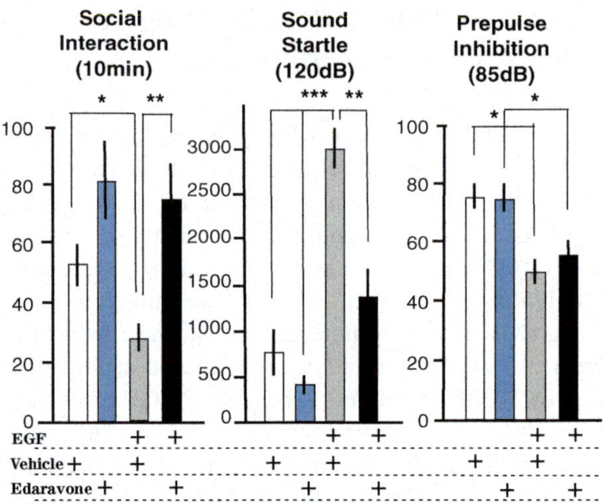

Fig. 5 Antipsychotic effects of edaravone on the EGF model relevant to schizophrenia. Epidermal growth factor (0.875 μg/g) or saline was administered subcutaneously daily on postnatal days 2–11. Adult rats were treated with or without the radical scavenger edaravone (5 mg/kg/day for 7 days, administered intraperitoneally). $n = 6$–11 in each group, $*P < 0.05$, $**P < 0.01$, and $***P < 0.001$. *EGF* epidermal growth factor

the adult stage suppressed the acoustic startle response and abolished the PPI deficits of the EGF-pretreatment model (Mizuno et al. 2008). The medication effects of emodin on social interaction impairments were undetectable. However, we observed an antipsychotic-like activity of emodin in the methamphetamine model (Mizuno et al. 2010). In contrast to emodin, intraperitoneal injection of edaravone improved social interaction scores and decreased an abnormally high acoustic startle response of the EGF-pretreatment model, although it had no effect on their PPI deficits (Fig. 5). Both the results verified the pharmaceutical effectiveness of ROS scavengers on behavioral impairments relevant to schizophrenia. However, the antipsychotic-like profiles of the individual scavengers differed significantly. Presumably, the preference for the target ROS species differs between those ROS scavengers and results in the diversity between their pharmacologic profiles (Suzuki et al. 2005).

Although the therapeutic mechanism of ROS scavengers remains undetermined, we believe the EGFR signaling cascade to be one of their molecular targets, as suggested by Suzuki et al. (2005). We postulate that an elevation in EGFR signaling (e.g., ErbB1, ErbB2) is sustained until the adult stage even when EGF treatment is completed during the neonatal stage (Futamura et al. 2002). The hypothesis stems from the fact that the administration of specific EGFR blockers also ameliorated the behavioral deficits of the EGF-exposure-generated model (Mizuno et al. 2008, 2013). In light of the strong interaction between EGF signaling and ROS production, further study is warranted into which plays the crucial role in the pathogenesis of both the EGF-exposure-generated model and patients with schizophrenia.

6 Conclusion

There are tight interactions between ROS and cytokine signaling. In particular, such interactions are evident for the EGF signal cascade in which basal EGFR autophosphorylation is not negligible. ROS itself inactivates its phosphatases and provokes EGF receptor signaling without any ligands. Conversely, EGFR phosphorylation efficiently activates various ROS-generating enzymes such as NOX and COX to produce excess amounts of ROS. The cytokine–ROS system appears to be recruited in the pathologic conditions of brain injury, infection, and seizures to protect the organ with their given neurotrophic actions as well as to promote the regeneration processes from the insults. When the activation of the cytokine–ROS system is prolonged, the unfavorable side effects emerge in dopaminergic and glutamatergic development/neurotransmission, leading to abnormal brain functions. Our latest results from the EGF model indicate that oxygen radical scavengers are beneficial for the medication of their cognitive deficits. However, the pharmacological profiles differ significantly among scavengers. We hope that safe and novel antipsychotic drugs will be developed from any of oxygen radical scavengers.

Acknowledgments We are grateful to Tomoyuki Sugano and Akira Yarimizu for their technical assistances. The authors' original work was partly supported by Grants-in-Aid for Scientific Research on Innovative Areas and for Challenging Exploratory Research (No. 24116010, No. 21659273) and a grant for Promotion of Niigata University Research Projects. Human recombinant EGF was kindly provided by Higeta Shoyu Co. Ltd. Except this gift, all the authors have no conflicts of interest to declare.

References

Abe Y, Namba H, Zheng Y, Nawa H (2009) In situ hybridization reveals developmental regulation of ErbB1-4 mRNA expression in mouse midbrain: implication of ErbB receptors for dopaminergic neurons. Neuroscience 161(1):95–110

Abe Y, Namba H, Kato T, Iwakura Y, Nawa H (2011) Neuregulin-1 signals from the periphery regulate AMPA receptor sensitivity and expression in GABAergic interneurons in developing neocortex. J Neurosci 31(15):5699–5709

Anttila S, Illi A, Kampman O, Mattila KM, Lehtimäki T, Leinonen E (2004) Association of EGF polymorphism with schizophrenia in Finnish men. NeuroReport 15(7):1215–1218

Arai M, Yuzawa H, Nohara I, Ohnishi T, Obata N, Iwayama Y, Haga S, Toyota T, Ujike H, Arai M, Ichikawa T, Nishida A, Tanaka Y, Furukawa A, Aikawa Y, Kuroda O, Niizato K, Izawa R, Nakamura K, Mori N, Matsuzawa D, Hashimoto K, Iyo M, Sora I, Matsushita M, Okazaki Y, Yoshikawa T, Miyata T, Itokawa M (2010) Enhanced carbonyl stress in a subpopulation of schizophrenia. Arch Gen Psychiatry 67(6):589–597

Arai M, Miyashita M, Kobori A, Toriumi K, Horiuchi Y, Itokawa M (2014) Carbonyl stress and schizophrenia. Psychiatry Clin Neurosci 68(9):655–665

Bae YS, Kang SW, Seo MS, Baines IC, Tekle E, Chock PB, Rhee SG (1997) Epidermal growth factor (EGF)-induced generation of hydrogen peroxide. Role in EGF receptor-mediated tyrosine phosphorylation. J Biol Chem 272(1):217–221

Behrens MM, Ali SS, Dao DN, Lucero J, Shekhtman G, Quick KL, Dugan LL (2007) Ketamine-induced loss of phenotype of fast-spiking interneurons is mediated by NADPH-oxidase. Science 318(5856):1645–1647

Ben Othmen L, Mechri A, Fendri C, Bost M, Chazot G, Gaha L, Kerkeni A (2008) Altered antioxidant defense system in clinically stable patients with schizophrenia and their unaffected siblings. Prog Neuropsychopharmacol Biol Psychiatry 32(1):155–159

Bokkon I, Antal I (2011) Schizophrenia: redox regulation and volume neurotransmission. Curr Neuropharmacol 9:289–300

Brewer TF, Garcia FJ, Onak CS, Carroll KS, Chang CJ (2015) Chemical approaches to discovery and study of sources and targets of hydrogen peroxide redox signaling through NADPH oxidase proteins. Annu Rev Biochem 84:765–790

Buttner N, Bhattacharyya S, Walsh J, Benes FM (2007) DNA fragmentation is increased in non-GABAergic neurons in bipolar disorder but not in schizophrenia. Schizophr Res 93(1–3): 33–41

Cabungcal JH, Counotte DS, Lewis EM, Tejeda HA, Piantadosi P, Pollock C, Calhoon GG, Sullivan EM, Presgraves E, Kil J, Hong LE, Cuenod M, Do KQ, O'Donnell P (2014) Juvenile antioxidant treatment prevents adult deficits in a developmental model of schizophrenia. Neuron 83(5):1073–1084

Cha YK, Kim YH, Ahn YH, Koh JY (2000) Epidermal growth factor induces oxidative neuronal injury in cortical culture. J Neurochem 75(1):298–303

Chan PH (2001) Reactive oxygen radicals in signaling and damage in the ischemic brain. J Cereb Blood Flow Metab 21(1):2–14

Chen W, Shang WH, Adachi Y, Hirose K, Ferrari DM, Kamata T (2008) A possible biochemical link between NADPH oxidase (Nox) 1 redox-signalling and ERp72. Biochem J 416(1):55–63

Chiarugi P, Pani G, Giannoni E, Taddei L, Colavitti R, Raugei G, Symons M, Borrello S, Galeotti T, Ramponi G (2003) Reactive oxygen species as essential mediators of cell adhesion: the oxidative inhibition of a FAK tyrosine phosphatase is required for cell adhesion. J Cell Biol 161(5):933–944

Cobb CA, Cole MP (2015) Oxidative and nitrative stress in neurodegeneration. Neurobiol Dis. doi:10.1016/j.nbd.2015.04.020

De Wit R, Makkinje M, Boonstra J, Verkleij AJ, Post JA (2001) Hydrogen peroxide reversibly inhibits epidermal growth factor (EGF) receptor internalization and coincident ubiquitination of the EGF receptor and Eps15. FASEB J 15(2):306–308

De Yulia GJ, Jr Cárcamo JM, Bórquez-Ojeda O, Shelton CC, Golde DW (2005) Hydrogen peroxide generated extracellularly by receptor-ligand interaction facilitates cell signaling. Proc Natl Acad Sci USA 102(14):5044–5049

Dukoff DJ, Hogg DW, Hawrysh PJ, Buck LT (2014) Scavenging ROS dramatically increase NMDA receptor whole-cell currents in painted turtle cortical neurons. J Exp Biol 217(Pt 18): 3346–3355

Eda T, Mizuno M, Araki K, Iwakura Y, Namba H, Sotoyama H, Kakita A, Takahashi H, Satoh H, Chan SY, Nawa H (2013) Neurobehavioral deficits of epidermal growth factor-overexpressing transgenic mice: impact on dopamine metabolism. Neurosci Lett 547:21–25

Fan CY, Katsuyama M, Yabe-Nishimura C (2005a) PKCdelta mediates up-regulation of NOX1, a catalytic subunit of NADPH oxidase, via transactivation of the EGF receptor: possible involvement of PKCdelta in vascular hypertrophy. Biochem J 390(Pt 3):761–767

Fan C, Katsuyama M, Nishinaka T, Yabe-Nishimura C (2005b) Transactivation of the EGF receptor and a PI3 kinase-ATF-1 pathway is involved in the upregulation of NOX1, a catalytic subunit of NADPH oxidase. FEBS Lett 579(5):1301–1305

Farokhnia M, Azarkolah A, Adinehfar F, Khodaie-Ardakani MR, Hosseini SM, Yekehtaz H, Tabrizi M, Rezaei F, Salehi B, Sadeghi SM, Moghadam M, Gharibi F, Mirshafiee O, Akhondzadeh S (2013) N-acetylcysteine as an adjunct to risperidone for treatment of negative symptoms in patients with chronic schizophrenia: a randomized, double-blind, placebo-controlled study. Clin Neuropharmacol 36(6):185–192

Finch JS, Tome ME, Kwei KA, Bowden GT (2006) Catalase reverses tumorigenicity in a malignant cell line by an epidermal growth factor receptor pathway. Free Radic Biol Med 40 (5):863–875

Futamura T, Toyooka K, Iritani S, Niizato K, Nakamura R, Tsuchiya K, Someya T, Kakita A, Takahashi H, Nawa H (2002) Abnormal expression of epidermal growth factor and its receptor in the forebrain and serum of schizophrenic patients. Mol Psychiatry 7(7):673–682

Futamura T, Kakita A, Tohmi M, Sotoyama H, Takahashi H, Nawa H (2003) Neonatal perturbation of neurotrophic signaling results in abnormal sensorimotor gating and social interaction in adults: implication for epidermal growth factor in cognitive development. Mol Psychiatry 8(1):19–29

Goldkorn T, Ravid T, Khan EM (2005) Life and death decisions: ceramide generation and EGF receptor trafficking are modulated by oxidative stress. Antioxid Redox Signal 7(1–2):119–128

Goldsmit Y, Erlich S, Pinkas-Kramarski R (2001) Neuregulin induces sustained reactive oxygen species generation to mediate neuronal differentiation. Cell Mol Neurobiol 21(6):753–769

Groenestege WM, Thébault S, van der Wijst J, van den Berg D, Janssen R, Tejpar S, van den Heuvel LP, van Cutsem E, Hoenderop JG, Knoers NV, Bindels RJ (2007) Impaired basolateral sorting of pro-EGF causes isolated recessive renal hypomagnesemia. J Clin Invest 117 (8):2260–2267

Gysin R, Kraftsik R, Sandell J, Bovet P, Chappuis C, Conus P, Deppen P, Preisig M, Ruiz V, Steullet P, Tosic M, Werge T, Cuénod M, Do KQ (2007) Impaired glutathione synthesis in schizophrenia: convergent genetic and functional evidence. Proc Natl Acad Sci USA 104 (42):16621–16626

Iwakura Y, Zheng Y, Sibilia M, Abe Y, Piao YS, Yokomaku D, Wang R, Ishizuka Y, Takei N, Nawa H (2011a) Qualitative and quantitative re-evaluation of epidermal growth factor-ErbB1 action on developing midbrain dopaminergic neurons in vivo and in vitro: target-derived neurotrophic signaling (Part 1). J Neurochem 118(1):45–56

Iwakura Y, Wang R, Abe Y, Piao YS, Shishido Y, Higashiyama S, Takei N, Nawa H (2011b) Dopamine-dependent ectodomain shedding and release of epidermal growth factor in developing striatum: target-derived neurotrophic signaling (Part 2). J Neurochem 118(1):57–68

Iwakura Y, Nawa H (2013) ErbB1-4-dependent EGF/neuregulin signals and their cross talk in the central nervous system: pathological implications in schizophrenia and Parkinson's diseasse. Front Cell Neurosci 7:4

Kato T, Abe Y, Sotoyama H, Kakita A, Kominami R, Hirokawa S, Ozaki M, Takahashi H, Nawa H (2011) Transient exposure of neonatal mice to neuregulin-1 results in hyperdopaminergic states in adulthood: implication in neurodevelopmental hypothesis for schizophrenia. Mol Psychiatry 16(3):307–320

Kato T, Kasai A, Mizuno M, Fengyi L, Shintani N, Maeda S, Yokoyama M, Ozaki M, Nawa H (2010) Phenotypic characterization of transgenic mice overexpressing neuregulin-1. PLoS ONE 5(12):e14185

Kaur U, Banerjee P, Bir A, Sinha M, Biswas A, Chakrabarti S (2015) Reactive oxygen species, redox signaling and neuroinflammation in Alzheimer's disease: the NF-κB connection. Curr Top Med Chem 15(5):446–457

Kawasaki T, Ishihara K, Ago Y, Nakamura S, Itoh S, Baba A, Matsuda T (2006) Protective effect of the radical scavenger edaravone against methamphetamine-induced dopaminergic neurotoxicity in mouse striatum. Eur J Pharmacol 542(1–3):92–99

Landskron G, De la Fuente M, Thuwajit P, Thuwajit C, Hermoso MA (2014) Chronic inflammation and cytokines in the tumor microenvironment. J Immunol Res 2014:149185

Lee SR, Kwon KS, Kim SR, Rhee SG (1998) Reversible inactivation of protein-tyrosine phosphatase 1B in A431 cells stimulated with epidermal growth factor. J Biol Chem 273 (25):15366–15372

Levkovitz Y, Mendlovich S, Riwkes S, Braw Y, Levkovitch-Verbin H, Gal G, Fennig S, Treves I, Kron S (2010) A double-blind, randomized study of minocycline for the treatment of negative and cognitive symptoms in early-phase schizophrenia. J Clin Psychiatry 71(2):138–149

Lian M, Zheng X (2009) HSCARG regulates NF-kappaB activation by promoting the ubiquitination of RelA or COMMD1. J Biol Chem 284(27):17998–18006

Lipska BK, Jaskiw GE, Weinberger DR (1993) Postpubertal emergence of hyperresponsiveness to stress and to amphetamine after neonatal excitotoxic hippocampal damage: a potential animal model of schizophrenia. Neuropsychopharmacology 9(1):67–75

Lovell MA, Markesbery WR (2007) Oxidative DNA damage in mild cognitive impairment and late-stage Alzheimer's disease. Nucleic Acids Res 35(22):7497–7504

Mahadik SP, Scheffer RE (1996) Oxidative injury and potential use of antioxidants in schizophrenia. Prostaglandins Leukot Essent Fatty Acids 55(1–2):45–54

Martins MR, Petronilho FC, Gomes KM, Dal-Pizzol F, Streck EL, Quevedo J (2008) Antipsychotic-induced oxidative stress in rat brain. Neurotox Res 13:63–69

McCreadie RG, MacDonald E, Wiles D, Campbell G, Paterson JR (1995) The Nithsdale Schizophrenia Surveys. XIV: plasma lipid peroxide and serum vitamin E levels in patients with and without tardive dyskinesia, and in normal subjects. Br J Psychiatry 167:610–617

Miyaoka T, Yasukawa R, Yasuda H, Hayashida M, Inagaki T, Horiguchi J (2007) Possible antipsychotic effects of minocycline in patients with schizophrenia. Prog Neuropsychopharmacol Biol Psychiatry 31(1):304–307

Mizuno M, Malta RS Jr, Nagano T, Nawa H (2004) Conditioned place preference and locomotor sensitization after repeated administration of cocaine or methamphetamine in rats treated with epidermal growth factor during the neonatal period. Ann NY Acad Sci 1025:612–618

Mizuno M, Sotoyama H, Narita E, Kawamura H, Namba H, Zheng Y, Eda T, Nawa H (2007) A cyclooxygenase-2 inhibitor ameliorates behavioral impairments induced by striatal administration of epidermal growth factor. J Neurosci 27(38):10116–10127

Mizuno M, Kawamura H, Takei N, Nawa H (2008) The anthraquinone derivative Emodin ameliorates neurobehavioral deficits of a rodent model for schizophrenia. J Neural Transm 115(3):521–530

Mizuno M, Kawamura H, Ishizuka Y, Sotoyama H, Nawa H (2010) The anthraquinone derivative emodin attenuates methamphetamine-induced hyperlocomotion and startle response in rats. Pharmacol Biochem Behav 97(2):392–398

Mizuno M, Sotoyama H, Namba H, Shibuya M, Eda T, Wang R, Okubo T, Nagata K, Iwakura Y, Nawa H (2013) ErbB inhibitors ameliorate behavioral impairments of an animal model for schizophrenia: implication of their dopamine-modulatory actions. Transl Psychiatry 30(3):e252

Mohn AR, Gainetdinov RR, Caron MG, Koller BH (1999) Mice with reduced NMDA receptor expression display behaviors related to schizophrenia. Cell 98(4):427–436

Nagano T, Namba H, Abe Y, Aoki H, Takei N, Nawa H (2007) In vivo administration of epidermal growth factor and its homologue attenuates developmental maturation of functional excitatory synapses in cortical GABAergic neurons. Eur J Neurosci 25(2):380–390

Namba H, Zheng Y, Abe Y, Nawa H (2009) Epidermal growth factor administered in the periphery influences excitatory synaptic inputs onto midbrain dopaminergic neurons in postnatal mice. Neuroscience 158(4):1731–1741

Naviaux RK (2012) Oxidative shielding or oxidative stress? J Pharmacol Exp Ther 342(3):608–618

Nawa H, Takei N (2006) Recent progress in animal modeling of immune inflammatory processes in schizophrenia: implication of specific cytokines. Neurosci Res 56(1):2–13

Nawa H, Sotoyama H, Iwakura Y, Takei N, Namba H (2014) Neuropathologic implication of peripheral neuregulin-1 and EGF signals in dopaminergic dysfunction and behavioral deficits relevant to schizophrenia: their target cells and time window. Biomed Res Int. 2014:697935

Nakabeppu Y, Tsuchimoto D, Yamaguchi H, Sakumi K (2007) Oxidative damage in nucleic acids and Parkinson's disease. J Neurosci Res 85(5):919–934

Nishioka N, Arnold SE (2004) Evidence for oxidative DNA damage in the hippocampus of elderly patients with chronic schizophrenia. Am J Geriatr Psychiatry 12(2):167–175

Ozyurt B, Ozyurt H, Akpolat N, Erdogan H, Sarsilmaz M (2007) Oxidative stress in prefrontal cortex of rat exposed to MK-801 and protective effects of CAPE. Prog Neuropsychopharmacol Biol Psychiatry 31(4):832–838

Puttachary S, Sharma S, Stark S, Thippeswamy T (2015) Seizure-induced oxidative stress in temporal lobe epilepsy. Biomed Res Int 2015:745613

Ravid T, Sweeney C, Gee P, Carraway KL 3rd, Goldkorn T (2002) Epidermal growth factor receptor activation under oxidative stress fails to promote c-Cbl mediated down-regulation. J Biol Chem 277(34):31214–31219

Reinke A, Martins MR, Lima MS, Moreira JC, Dal-Pizzol F, Quevedo J (2004) Haloperidol and clozapine, but not olanzapine, induces oxidative stress in rat brain. Neurosci Lett 372(1–2): 157–160

Sanchez RM, Wang C, Gardner G, Orlando L, Tauck DL, Rosenberg PA, Aizenman E, Jensen FE (2000) Novel role for the NMDA receptor redox modulatory site in the pathophysiology of seizures. J Neurosci 20(6):2409–2417

Sakai M, Kashiwahara M, Kakita A, Nawa H (2014) An attempt of non-human primate modeling of schizophrenia with neonatal challenges of epidermal growth factor. J Addict Res Ther 5:170

Sivrioglu EY, Kirli S, Sipahioglu D, Gursoy B, Sarandöl E (2007) The impact of omega-3 fatty acids, vitamins E and C supplementation on treatment outcome and side effects in schizophrenia patients treated with haloperidol: an open-label pilot study. Prog Neuropsychopharmacol Biol Psychiatry 31(7):1493–1499

Sotoyama H, Namba H, Chiken S, Nambu A, Nawa H (2013) Exposure to the cytokine EGF leads to abnormal hyperactivity of pallidal GABA neurons: implications for schizophrenia and its modeling. J Neurochem 126(4):518–528

Sotoyama H, Zheng Y, Iwakura Y, Mizuno M, Aizawa M, Shcherbakova K, Wang R, Namba H, Nawa H (2011) Pallidal hyperdopaminergic innervation underlying D2 receptor-dependent behavioral deficits in the schizophrenia animal model established by EGF. PLoS ONE 6(10): e25831

Stefansson H, Steinthorsdottir V, Thorgeirsson TE, Gulcher JR (2004) Stefansson K (2004) Neuregulin 1 and schizophrenia. Ann Med 36(1):62–71

Suboticanec K, olnegovic-Smalc V, Korbar M, Mestrovic B, Buzina R (1990) Vitamin C status in chronic schizophrenia. Biol Psychiatry 28:959–966

Suc I, Meilhac O, Lajoie-Mazenc I, Vandaele J, Jürgens G, Salvayre R, Nègre-Salvayre A (1998) Activation of EGF receptor by oxidized LDL. FASEB J 12(9):665–671

Suzuki R, Gopalrao RK, Maeda H, Rao P, Yamamoto M, Xing Y, Mizobuchi S, Sasaguri S (2005) MCI-186 inhibits tumor growth through suppression of EGFR phosphorylation and cell cycle arrest. Anticancer Res 25(2A):1131–1138

Takahashi M, Kakita A, Futamura T, Watanabe Y, Mizuno M, Sakimura K, Castren E, Nabeshima T, Someya T, Nawa H (2006) Sustained brain-derived neurotrophic factor up-regulation and sensorimotor gating abnormality induced by postnatal exposure to phencyclidine: comparison with adult treatment. J Neurochem 99(3):770–780

Tohmi M, Tsuda N, Mizuno M, Takei N, Frankland PW, Nawa H (2005) Distinct influences of neonatal epidermal growth factor challenge on adult neurobehavioral traits in four mouse strains. Behav Genet 35(5):615–629

Tohmi M, Tsuda N, Watanabe Y, Kakita A, Nawa H (2004) Perinatal inflammatory cytokine challenge results in distinct neurobehavioral alterations in rats: implication in psychiatric disorders of developmental origin. Neurosci Res 50(1):67–75

Tsuda N, Mizuno M, Yamanaka T, Komurasaki T, Yoshimoto M, Nawa H (2008) Common behavioral influences of the ErbB1 ligands transforming growth factor alpha and epiregulin administered to mouse neonates. Brain Dev 30(8):533–543

Tzahar E, Moyer JD, Waterman H, Barbacci EG, Bao J, Levkowitz G, Shelly M, Strano S, Pinkas-Kramarski R, Pierce JH, Andrews GC, Yarden Y (1998) Pathogenic poxviruses reveal viral strategies to exploit the ErbB signaling network. EMBO J 17(20):5948–5963

Ustundag B, Atmaca M, Kirtas O, Selek S, Metin K, Tezcan E (2006) Total antioxidant response in patients with schizophrenia. Psychiatry Clin Neurosci 60(4):458–464

Varner MW, Dildy GA, Hunter C, Dudley DJ, Clark SL, Mitchell MD (1996) Amniotic fluid epidermal growth factor levels in normal and abnormal pregnancies. J Soc Gynecol Investig. 3:17–19

Wang C, McInnis J, West JB, Bao J, Anastasio N, Guidry JA, Ye Y, Salvemini D, Johnson KM (2003) Blockade of phencyclidine-induced cortical apoptosis and deficits in prepulse inhibition by M40403, a superoxide dismutase mimetic. J Pharmacol Exp Ther 304(1):266–271

Watanabe Y, Hashimoto S, Kakita A, Takahashi H, Ko J, Mizuno M, Someya T, Patterson PH, Nawa H (2004) Neonatal impact of leukemia inhibitory factor on neurobehavioral development in rats. Neurosci Res 48(3):345–353

Willis CL, Ray DE (2007) Antioxidants attenuate MK-801-induced cortical neurotoxicity in the rat. Neurotoxicology 28(1):161–167

Yao JK, Keshavan MS (2011) Antioxidants, redox signaling, and pathophysiology in schizophrenia: an integrative view. Antioxid Redox Signal 15(7):2011–2035

Yip SC, Saha S, Chernoff J (2010) PTP1B: a double agent in metabolism and oncogenesis. Trends Biochem Sci 35(8):442–449

Zhang L, Kitaichi K, Fujimoto Y, Nakayama H, Shimizu E, Iyo M, Hashimoto K (2006) Protective effects of minocycline on behavioral changes and neurotoxicity in mice after administration of methamphetamine. Prog Neuropsychopharmacol Biol Psychiatry 30(8): 1381–1393

Zuo DY, Wu YL, Yao WX, Cao Y, Wu CF, Tanaka M (2007) Effect of MK-801 and ketamine on hydroxyl radical generation in the posterior cingulate and retrosplenial cortex of free-moving mice, as determined by in vivo microdialysis. Pharmacol Biochem Behav 86(1):1–7

CPSIA information can be obtained
at www.ICGtesting.com
Printed in the USA
LVOW01*1926230217
525239LV00002B/12/P

9 783319 341347